T0336268

SPECIAL FUNCTIONS AND ORTHOGONAL POLYNOMIALS

The subject of special functions is often presented as a collection of disparate results, rarely organized in a coherent way. This book emphasizes general principles that unify and demarcate the subjects of study. The authors' main goals are to provide clear motivation, efficient proofs, and original references for all of the principal results.

The book covers standard material, but also much more. It shows how much of the subject can be traced back to two equations – the hypergeometric equation and the confluent hypergeometric equation – and it details the ways in which these equations are canonical and special. There is extended coverage of orthogonal polynomials, including connections to approximation theory, continued fractions, and the moment problem, as well as an introduction to new asymptotic methods. The book includes chapters on Meijer G-functions and elliptic functions. The final chapter introduces Painlevé transcendents, which have been termed the "special functions of the twenty-first century."

Richard Beals was Professor of Mathematics at the University of Chicago and at Yale University. He is the author or co-author of books on mathematical analysis, linear operators, and inverse scattering theory, and has authored more than a hundred research papers in areas including partial differential equations, mathematical economics, and mathematical psychology.

Roderick Wong is Chair Professor of Mathematics at City University of Hong Kong. He is the author of books on asymptotic approximations of integrals and applied analysis. He has published over 140 research papers in areas such as asymptotic analysis, singular perturbation theory, and special functions.

CAMBRIDGE STUDIES IN ADVANCED MATHEMATICS

All the titles listed below can be obtained from good booksellers or from Cambridge University Press.
For a complete series listing visit: www.cambridge.org/mathematics.

Special Functions and Orthogonal Polynomials

RICHARD BEALS
Yale University

RODERICK WONG
City University of Hong Kong

CAMBRIDGE UNIVERSITY PRESS

CAMBRIDGE
UNIVERSITY PRESS

University Printing House, Cambridge CB2 8BS, United Kingdom

One Liberty Plaza, 20th Floor, New York, NY 10006, USA

477 Williamstown Road, Port Melbourne, VIC 3207, Australia

4843/24, 2nd Floor, Ansari Road, Daryaganj, Delhi - 110002, India

79 Anson Road, #06-04/06, Singapore 079906

Cambridge University Press is part of the University of Cambridge.

It furthers the University's mission by disseminating knowledge in the pursuit of
education, learning and research at the highest international levels of excellence.

www.cambridge.org
Information on this title: www.cambridge.org/9781107106987

© Cambridge University Press 2016

First published 2016

A catalogue record for this publication is available from the British Library

Library of Congress Cataloging in Publication data
Names: Beals, Richard, 1938– | Wong, Roderick, 1944–
Title: Special functions and orthogonal polynomials / Richard Beals, Yale
University, and Roderick S.C. Wong, City University of Hong Kong
Description: Cambridge: Cambridge University Press, [2016] |
Series: Cambridge studies in advanced mathematics; 153 |
Includes bibliographical references and index.
Identifiers: LCCN 2015050715 | ISBN 9781107106987 (hardback)
Subjects: LCSH: Orthogonal polynomials. | Functions, Special. |
Mathematical analysis.
Classification: LCC QA404.5.B3227 | DDC 515/.55–dc23
LC record available at http://lccn.loc.gov/2015050715

ISBN 978-1-107-10698-7 Hardback

Contents

Preface

This book originated as *Special Functions: A Graduate Text*. The current version is considerably enlarged: the number of chapters devoted to orthogonal polynomials has increased from two to four; Meijer G-functions and Painlevé transcendents are now treated.

As we noted in the earlier book, the subject of special functions lacks a precise delineation, but it has a long and distinguished history. The remarks at the end of each chapter discuss the history, with numerous references and suggestions for further reading.

This book covers most of the standard topics and some that are less standard. We have tried to provide context for the discussion by emphasizing unifying ideas. The text and the problems provide proofs or proof outlines for nearly all the results and formulas.

We have also tried to keep the prerequisites to a minimum: a reasonable familiarity with power series and integrals, convergence, and the like. Some proofs rely on the basics of complex function theory, which are reviewed in the first appendix. Some familiarity with Hilbert space ideas, in the L^2 framework, is useful. The chapters on elliptic functions and on Painlevé transcendents rely more heavily than the rest of the book on concepts from complex analysis. The second appendix contains a quick development of basic results from Fourier analysis, including the Mellin transform.

The first chapter provides a general context for the discussion of the linear theory, especially in connection with special properties of the hypergeometric and confluent hypergeometric equations. Chapter 2 treats the gamma and beta functions at some length, with an introduction to the Riemann zeta function. Chapter 3 covers the relevant material from the theory of ordinary differential equations, including a characterization of the classical polynomials as eigenfunctions, and a discussion of separation of variables for equations involving the Laplacian.

The next four chapters are concerned with orthogonal polynomials on a real interval. Chapter 4 introduces the general theory, including three-term recurrence relations, Padé approximants, continued fractions, and Favard's theorem. The classical polynomials (Hermite, Laguerre, Jacobi) are treated in detail in Chapter 5, including asymptotic distribution of zeros. Chapter 6 introduces finite difference analogues of the classification theorem, yielding the classical discrete polynomials as well as neoclassical versions and the Askey scheme. Two methods of obtaining asymptotic results are presented in Chapter 7. In particular, the Riemann–Hilbert method is carried through for Hermite polynomials.

Chapters 8 through 11 contain a detailed treatment of the confluent hypergeometric equation, the hypergeometric equation, and special cases. These include Weber functions, Whittaker functions, Airy functions, cylinder functions (Bessel, Hankel, . . .), spherical harmonics, and Legendre functions. Among the topics are linear relations, various transformations, integral representations, and asymptotics. Chapter 13 contains proofs of asymptotic results for these functions and for the classical polynomials.

In Chapter 12 we extend an earlier discussion of the special "recursive" property of the hypergeometric and confluent hypergeometric equations to equations of arbitrary order. This property characterizes the generalized hypergeometric equation. The corresponding solutions, the generalized hyper-geometric functions, are covered in more detail than in the first version. Elliptic integrals, elliptic functions of Jacobi and Weierstrass, and theta functions are treated in Chapter 14.

The principal new topics, Meijer G-functions and Painlevé transcendents, have current theoretical and practical interest.

Meijer G-functions, which are special solutions of generalized hypergeo-metric equations, are introduced in Chapter 12. They generalize the classic Mellin–Barnes integral representations. The G-functions occur in probability and physics, and play a large role in compiling tables of integrals.

Chapter 15 has an extensive introduction to the classical and modern theory of Painlevé equations and their solutions, with emphasis on the second Painlevé equation, PII. Painlevé's method is introduced and PII is derived in detail. The isomonodromy method and Bäcklund transformations are introduced, and used to obtain rational solutions and information about general solutions. The Riemann–Hilbert method is used to derive a connection formula for solutions of PII(0). Applications include differential geometry, random matrix theory, integrable systems, and statistical physics.

The earlier book contained a concise summary of each chapter. These have been omitted here, partly to save space, and partly because the summaries often proved to be more annoying than helpful in use of the book for reference.

The first-named author acknowledges the efforts of some of his research collaborators, especially Peter Greiner, Bernard Gaveau, Yakar Kannai, David

Sattinger, and Jacek Szmigielski, who managed over a period of years to convince him that special functions are not only useful but beautiful.

The authors thank Jacek Szmigielski, Mourad Ismail, Richard Askey, and an anonymous reviewer for helpful comments on the earlier manuscript. The first-named author is grateful to the Department of Mathematics and to the Liu Bie Ju Centre for Mathematical Sciences at City University of Hong Kong for help and hospitality during the preparation of both versions of this book. The second-named author is happy to acknowledge all his former students, collaborators, and assistants who helped with this project.

1

Orientation

The concept of a "special function" has no precise definition. From a practical point of view, a special function is a function of one variable that is (a) not one of the "elementary functions" – algebraic functions, trigonometric functions, the exponential, the logarithm, and functions constructed algebraically from these functions – and is (b) a function about which one can find information in many of the books about special functions. A large amount of such information has been accumulated over a period of three centuries. Like such elementary functions as the exponential and trigonometric functions, special functions come up in numerous contexts. These contexts include both pure mathematics and applications, ranging from number theory and combinatorics to probability and physical science.

The majority of special functions that are treated in many of the general books on the subject are solutions of certain second-order linear differential equations. Indeed, these functions were discovered through the study of physical problems: vibrations, heat flow, equilibrium, and so on. The associated equations are partial differential equations of second-order. In some coordinate systems these equations can be solved by separation of variables, leading to the second-order ordinary differential equations in question. (Solutions of the analogous *first-order* linear differential equations are elementary functions.)

Despite the long list of adjectives and proper names attached to this class of special functions (hypergeometric, confluent hypergeometric, cylinder, parabolic cylinder, spherical, Airy, Bessel, Hankel, Hermite, Kelvin, Kummer, Laguerre, Legendre, Macdonald, Neumann, Weber, Whittaker, . . .), each of them is closely related to one of two families of equations: the confluent hypergeometric equation(s)

$$x u''(x) + (c - x) u'(x) - a u(x) = 0 \tag{1.0.1}$$

and the hypergeometric equation(s)

$$x(1 - x) u''(x) + [c - (a + b + 1)x] u'(x) - ab u(x) = 0. \tag{1.0.2}$$

1

The parameters a, b, c are real or complex constants.

Some solutions of these equations are polynomials: up to a linear change of variables, they are the "classical orthogonal polynomials." Again there are many names attached: Chebyshev, Gegenbauer, Hermite, Jacobi, Laguerre, Legendre, ultraspherical. In this introductory chapter we discuss one context in which these equations, and (up to normalization) no others, arise. We also shall see how two equations can, in principle, give rise to such a menagerie of functions.

Some special functions are *not* closely connected to linear differential equations. These exceptions include the gamma function, the beta function, elliptic functions, and the Painlevé transcendents.

The gamma and beta functions evaluate certain integrals. They are indispensable in many calculations, especially in connection with the class of functions mentioned earlier, as we illustrate below.

Elliptic functions arise as solutions of a simple *nonlinear* second-order differential equation, and also in connection with integrating certain algebraic functions. They have a wide range of applications, from number theory to integrable systems.

The Painlevé transcendents are solutions of a class of nonlinear second-order equations that share a crucial property with the equations that characterize elliptic functions, in that the solutions are single-valued in certain fixed domains, independent of the initial conditions.

1.1 Power series solutions

The general homogeneous linear second-order equation is

$$p(x)u''(x) + q(x)u'(x) + r(x)u(x) = 0, \qquad (1.1.1)$$

with p not identically zero. We assume here that the coefficient functions p, q, and r are holomorphic (analytic) in a neighborhood of the origin.

If a function u is holomorphic in a neighborhood of the origin, then the function on the left side of (1.1.1) is also holomorphic in a neighborhood of the origin. The coefficients of the power series expansion of this function can be computed from the coefficients of the expansions of the functions p, q, r, and u. Under these assumptions, (1.1.1) is equivalent to the sequence of equations obtained by setting the coefficients of the expansion of the left side equal to zero. Specifically, suppose that the coefficient functions p, q, r have series expansions

$$p(x) = \sum_{k=0}^{\infty} p_k x^k, \quad q(x) = \sum_{k=0}^{\infty} q_k x^k, \quad r(x) = \sum_{k=0}^{\infty} r_k x^k,$$

and u has the expansion

$$u(x) = \sum_{k=0}^{\infty} u_k x^k.$$

Then the constant term and the coefficients of x and x^2 on the left side of (1.1.1) are

$$2p_0 u_2 + q_0 u_1 + r_0 u_0, \tag{1.1.2}$$

$$6p_0 u_3 + 2p_1 u_2 + 2q_0 u_2 + q_1 u_1 + r_1 u_0 + r_0 u_1,$$

$$12p_0 u_4 + 6p_1 u_3 + 2p_2 u_2 + 3q_0 u_3 + 2q_1 u_2 + q_2 u_1 + r_0 u_2 + r_1 u_1 + r_2 u_0,$$

respectively. The sequence of equations equivalent to (1.1.1) is the sequence

$$\sum_{j+k=n, k\geq 0} (k+2)(k+1)p_j u_{k+2} + \sum_{j+k=n, k\geq 0} (k+1)q_j u_{k+1}$$

$$+ \sum_{j+k=n, k\geq 0} r_j u_k = 0, \qquad n = 0,1,2,\ldots \tag{1.1.3}$$

We say that (1.1.1) is *recursive* if it has a nonzero solution u that is holomorphic in a neighborhood of the origin, and the equations (1.1.3) determine the coefficients $\{u_n\}$ by a simple recursion: the nth equation determines u_n in terms of u_{n-1} alone. Suppose that (1.1.1) is recursive. Then the first of the equations (1.1.2) should involve u_1 but not u_2, so $p_0 = 0$, $q_0 \neq 0$. The second equation should not involve u_3 or u_0, so $r_1 = 0$. Similarly, the third equation shows that $q_2 = r_2 = 0$. Continuing, we obtain

$$p_0 = 0, \quad p_j = 0, \ j \geq 3; \quad q_j = 0, \ j \geq 2; \quad r_j = 0, \ j \geq 1.$$

Collecting terms, we see that the nth equation is

$$[(n+1)np_1 + (n+1)q_0]\, u_{n+1} + [n(n-1)p_2 + nq_1 + r_0]\, u_n = 0.$$

For special values of the parameters p_1, p_2, q_0, q_1, r_0, one of these coefficients may vanish for some value of n. In such a case, either the recursion breaks down, or the solution u is a polynomial. We assume that this does not happen. Thus

$$u_{n+1} = -\frac{n(n-1)p_2 + nq_1 + r_0}{(n+1)np_1 + (n+1)q_0}\, u_n. \tag{1.1.4}$$

Assume $u_0 \neq 0$. If $p_1 = 0$ but $p_2 \neq 0$, the series $\sum_{n=0}^{\infty} u_n x^n$ diverges for all $x \neq 0$ (ratio test). Therefore, up to normalization – a linear change of coordinates and a multiplicative constant – we may assume that $p(x)$ has one of the two forms $p(x) = x(1-x)$ or $p(x) = x$.

If $p(x) = x(1-x)$, then (1.1.1) has the form

$$x(1-x)u''(x) + (q_0 + q_1 x)u'(x) + r_0 u(x) = 0.$$

Constants a and b can be chosen so that this becomes (1.0.2).

If $p(x) = x$ and $q_1 \neq 0$ we may replace x by a multiple of x and take $q_1 = -1$. Then (1.1.1) has the form (1.0.1).

Finally, suppose $p(x) = x$ and $q_1 = 0$. If also $r_0 = 0$, then (1.1.1) is a first-order equation for u'. Otherwise, we may replace x by a multiple of x and take $r_0 = 1$. Then (1.1.1) has the form

$$xu''(x) + cu'(x) + u(x) = 0. \tag{1.1.5}$$

This equation is not obviously related to either (1.0.1) or (1.0.2). However, it can be shown that it becomes a special case of (1.0.1) after a change of variable and a "gauge transformation" (see Exercise 5).

In summary: up to certain normalizations, an equation of the form (1.1.1) is recursive if and only if it has one of the three forms (1.0.1), (1.0.2), or (1.1.5). Moreover, (1.1.5) can be transformed to a case of (1.0.1).

Let us note briefly the answer to the analogous question for a homogeneous linear *first-order* equation

$$q(x)u'(x) + r(x)u(x) = 0, \tag{1.1.6}$$

with q not identically zero. This amounts to taking $p = 0$ in the argument above. The conclusion is again that q is a polynomial of degree at most one, with $q_0 \neq 0$, while $r = r_0$ is constant. Up to normalization, $q(x)$ has one of the two forms $q(x) = 1$ or $q(x) = x - 1$. Thus the equation has one of the two forms

$$u'(x) - au(x) = 0; \qquad (x-1)u'(x) - au(x) = 0,$$

with solutions
$$u(x) = ce^{ax}, \quad u(x) = c(x-1)^a,$$

respectively.

The analogous question for homogeneous linear equations of *arbitrary* order is taken up in Chapter 12, Section 12.2.

Let us return to the confluent hypergeometric equation (1.0.1). The power series solution with $u_0 = 1$ is sometimes denoted $M(a,c;x)$. It can be calculated easily from the recursion (1.1.4). The result is

$$M(a,c;x) = \sum_{n=0}^{\infty} \frac{(a)_n}{(c)_n n!} x^n, \quad c \neq 0, -1, -2, \ldots \tag{1.1.7}$$

Here the "shifted factorial" or "Pochhammer symbol" $(a)_n$ is defined by

$$(a)_0 = 1, \qquad (a)_n = a(a+1)(a+2)\cdots(a+n-1), \tag{1.1.8}$$

so that $(1)_n = n!$. The series (1.1.7) converges for all complex x (ratio test), so M is an entire function of x.

The special nature of (1.0.1) is reflected in the special nature of the coefficients of M. It leads to a number of relationships among these functions when the parameters (a,b) are varied. For example, a comparison of coefficients shows that the three "contiguous" functions $M(a,c;x)$, $M(a+1, c;x)$, and $M(a,c-1;x)$ are related by

$$(a-c+1)M(a,c;x) - aM(a+1,c;x) + (c-1)M(a,c-1;x) = 0. \quad (1.1.9)$$

Similar relations hold whenever the respective parameters differ by integers.

Equations (1.0.2) and (1.1.1) have solutions with expansions similar to (1.1.7), as do the generalizations considered in Chapter 12.

1.2 The gamma and beta functions

The gamma function

$$\Gamma(a) = \int_0^\infty e^{-t} t^{a-1} \, dt, \quad \operatorname{Re} a > 0,$$

satisfies the functional equation $a\,\Gamma(a) = \Gamma(a+1)$. More generally, the shifted factorial (1.1.8) can be written as

$$(a)_n = \frac{\Gamma(a+n)}{\Gamma(a)}.$$

It is sometimes convenient to use this form in series like (1.1.7).

A related function is the beta function, or beta integral,

$$B(a,b) = \int_0^1 s^{a-1}(1-s)^{b-1} \, ds, \quad \operatorname{Re} a > 0, \ \operatorname{Re} b > 0,$$

which can be evaluated in terms of the gamma function:

$$B(a,b) = \frac{\Gamma(a)\Gamma(b)}{\Gamma(a+b)};$$

see the next chapter. These identities can be used to obtain a representation of the function M in (1.1.7) as an integral, when $\operatorname{Re} c > \operatorname{Re} a > 0$. In fact,

$$\frac{(a)_n}{(c)_n} = \frac{\Gamma(a+n)}{\Gamma(a)} \cdot \frac{\Gamma(c)}{\Gamma(c+n)}$$

$$= \frac{\Gamma(c)}{\Gamma(a)\Gamma(c-a)} B(a+n, c-a)$$

$$= \frac{\Gamma(c)}{\Gamma(a)\Gamma(c-a)} \int_0^1 s^{n+a-1}(1-s)^{c-a-1} \, ds.$$

Therefore

$$M(a,c;x) = \frac{\Gamma(c)}{\Gamma(a)\Gamma(c-a)} \int_0^1 \left\{ s^{a-1}(1-s)^{c-a-1} \sum_{n=0}^{\infty} \frac{(sx)^n}{n!} \right\} ds$$

$$= \frac{\Gamma(c)}{\Gamma(a)\Gamma(c-a)} \int_0^1 s^{a-1}(1-s)^{c-a-1} e^{sx} ds. \tag{1.2.1}$$

This integral representation is useful in obtaining information that is not evident from the power series expansion (1.1.7). Other integral representations occur naturally in the context of the Meijer G-functions in Chapter 12.

1.3 Three questions

First question: *How can it be that so many of the functions mentioned in the introduction to this chapter can be associated with just two equations, (1.0.1) and (1.0.2)?*

Part of the answer is that different solutions of the same equation may have different names. An elementary example is the equation

$$u''(x) - u(x) = 0. \tag{1.3.1}$$

One might wish to normalize a solution by imposing a condition at the origin like

$$u(0) = 0 \quad \text{or} \quad u'(0) = 0,$$

leading to $u(x) = \sinh x$ or $u(x) = \cosh x$, respectively, or a condition at infinity like

$$\lim_{x \to -\infty} u(x) = 0 \quad \text{or} \quad \lim_{x \to +\infty} u(x) = 0,$$

leading to $u(x) = e^x$ or $u(x) = e^{-x}$, respectively. Similarly, Bessel functions, Neumann functions, and both kinds of Hankel functions are four solutions of a single equation, distinguished by conditions at the origin or at infinity.

The rest of the answer to the question is that one can transform solutions of one second-order linear differential equation into solutions of another, in two simple ways. One such transformation is a change of variables. For example, starting with the equation

$$u''(x) - 2xu'(x) + \lambda u(x) = 0, \tag{1.3.2}$$

suppose $u(x) = v(x^2)$. It is not difficult to show that (1.3.2) is equivalent to the equation

$$yv''(y) + \left(\frac{1}{2} - y\right)v'(y) + \frac{1}{4}\lambda v(y) = 0,$$

which is the case $a = -\frac{1}{4}\lambda$, $c = \frac{1}{2}$ of (1.0.1). Therefore even solutions of (1.3.2) can be identified with certain solutions of (1.0.1). The same is true of odd solutions: see Exercise 12. An even simpler example is the change $u(x) = v(ix)$ in (1.3.1), leading to $v'' + v = 0$, and the trigonometric and complex exponential solutions $\sin x$, $\cos x$, e^{ix}, e^{-ix}.

The second type of transformation is a "gauge transformation." For example, if the function u in (1.3.2) is written in the form

$$u(x) = e^{x^2/2} v(x),$$

then (1.3.2) is equivalent to an equation with no first-order term:

$$v''(x) + (1 + \lambda - x^2) v(x) = 0. \tag{1.3.3}$$

Each of the functions mentioned in the third paragraph of the introduction to this chapter is a solution of an equation that can be obtained from (1.0.1) or (1.0.2) by one or both of a change of variable and a gauge transformation.

Second question: *What does one want to know about these functions?*

As we noted above, solutions of an equation of the form (1.1.1) can be chosen uniquely through various normalizations, such as behavior as $x \to 0$ or as $x \to \infty$. The solution (1.1.7) of (1.0.1) is normalized by the condition $u(0) = 1$. Having explicit formulas, like (1.1.7) for the function M, can be very useful. On the other hand, understanding the behavior as $x \to +\infty$ is not always straightforward. The integral representation (1.2.1) allows one to compute this behavior for M (see Exercise 18). This example illustrates why it can be useful to have an integral representation (with an integrand that is well understood).

Any three solutions of a second-order linear equation (1.1.1) satisfy a linear relationship, and one wants to compute the coefficients of such a relationship. An important tool in this and in other aspects of the theory is the computation of the Wronskian of two solutions u_1, u_2:

$$W(u_1, u_2)(x) \equiv u_1(x) u_2'(x) - u_2(x) u_1'(x).$$

In particular, these two solutions are linearly independent if and only if the Wronskian does not vanish.

Because of the special nature of (1.0.1) and (1.0.2) and the equations derived from them, solutions satisfy various linear relationships like (1.1.9). One wants to determine a set of relationships that generate all such relationships.

Finally, the coefficient of the zero-order term in equations like (1.0.1), (1.0.2), and (1.3.3) is an important parameter, and one often wants to know how a given normalized solution like $M(a, c; x)$ varies as the parameter approaches $\pm\infty$. In (1.3.3), denote by v_λ the even solution normalized by $v_\lambda(0) = 1$. As

$1 + \lambda = \mu^2 \to +\infty$, over any bounded interval the equation looks like a small perturbation of the equation $v'' + \mu^2 v = 0$. Therefore it is plausible that

$$v_\lambda(x) \sim A_\lambda(x) \cos(\mu x + B_\lambda) \quad \text{as } \lambda \to +\infty,$$

with $A_\lambda(x) > 0$. We want to compute the "amplitude function" $A_\lambda(x)$ and the "phase constant" B_λ. A word about notation like this in the preceding equation: the meaning of the statement

$$f(x) \sim A g(x) \quad \text{as } x \to \infty$$

is

$$\lim_{x \to \infty} \frac{f(x)}{g(x)} = A, \quad A \neq 0.$$

This is in slight conflict with the notation for an *asymptotic series expansion*:

$$f(x) \sim g(x) \sum_{n=0}^{\infty} a_n x^{-n} \quad \text{as } x \to \infty.$$

This means that for every positive integer N, truncating the series at $n = N$ gives an approximation to order x^{-N-1}:

$$\frac{f(x)}{g(x)} - \sum_{n=0}^{N} a_n x^{-n} = O(x^{-N-1}) \quad \text{as } x \to \infty.$$

As usual, the "big O" notation

$$h(x) = O(k(x)) \quad \text{as } x \to \infty$$

means that there are constants A, B such that

$$\left| \frac{h(x)}{k(x)} \right| \leq A \quad \text{if } x \geq B.$$

The similar "small o" notation

$$h(x) = o(k(x))$$

means that

$$\lim_{x \to \infty} \frac{h(x)}{k(x)} = 0.$$

Third question: *Is this list of functions or related equations exhaustive, in any sense?*

A partial answer has been given: the requirement that the equation be "recursive" leads to just three cases, (1.0.1), (1.0.2), and (1.1.5), and the third of these three equations reduces to a case of the first equation. Two other answers are given in Chapter 3.

The first of the two answers in Chapter 3 starts with a question of mathematics: given that a differential operator of the form that occurs in (1.1.1),

$$p(x)\frac{d^2}{dx^2} + q(x)\frac{d}{dx} + r(x),$$

is self-adjoint with respect to a weight function on a (bounded or infinite) interval, under what circumstances will the eigenfunctions be polynomials? An example is the operator in (1.3.2), which is self-adjoint with respect to the weight function $w(x) = e^{-x^2}$ on the line:

$$\int_{-\infty}^{\infty} \left[u''(x) - 2xu'(x)\right] v(x)e^{-x^2}\, dx = \int_{-\infty}^{\infty} u(x)\left[v''(x) - 2xv'(x)\right] e^{-x^2}\, dx.$$

The eigenvalues are $\lambda = 2,4,6,\ldots$, in (1.3.2) and the Hermite polynomials are eigenfunctions. Up to normalization, the equation associated with such an operator is one of the three equations (1.0.1), (1.0.2) (after a simple change of variables), or (1.3.2). Moreover, as suggested above, (1.3.2) can be converted to two cases of (1.0.1).

(One can ask the same question in connection with *difference equations*, with the derivative d/dx replaced by difference operators:

$$\Delta_+ u(m) = u(m+1) - u(m), \qquad \Delta_- u(m) = u(m) - u(m-1).$$

The polynomials that arise in this way are the classical discrete orthogonal polynomials of Charlier, Krawtchouk, Meixner, and Chebyshev–Hahn. Taking differences in the complex direction instead leads to the remaining "semi-classical" orthogonal polynomials.)

The second of the two answers in Chapter 3 starts with a question of mathematical physics: given the Laplace equation

$$\Delta u(\mathbf{x}) = 0$$

or the Helmholtz equation

$$\Delta u(\mathbf{x}) + \lambda u(\mathbf{x}) = 0,$$

say in three variables, $\mathbf{x} = (x_1, x_2, x_3)$, what equations arise by separating variables in various coordinate systems (Cartesian, cylindrical, spherical, parabolic cylindrical)? Each of the equations so obtained can be related to either (1.0.1) or (1.0.2) by a gauge transformation and/or a change of variables.

1.4 Other special functions

The remaining special functions to be discussed in this book are also associated with differential equations. Generalized hypergeometric functions and Meijer G-functions are solutions of linear equations of any order that have the recursive property discussed in Section 1.1. Elliptic functions and Painlevé transcendents are solutions of special second-order equations that are not linear.

One of the simplest nonlinear second-order differential equations of mathematical physics is the equation that describes the motion of an ideal pendulum, which can be normalized to

$$2\theta''(t) = -\sin\theta(t). \tag{1.4.1}$$

Multiplying equation (1.4.1) by $\theta'(t)$ and integrating gives

$$[\theta'(t)]^2 = a + \cos\theta(t) \tag{1.4.2}$$

for some constant a. Let $u = \sin\frac{1}{2}\theta$. Then (1.4.2) takes the form

$$[u'(t)]^2 = A\left[1 - u(t)^2\right]\left[1 - k^2u(t)^2\right]. \tag{1.4.3}$$

By rescaling time t, we may take the constant A to be 1. Solving for t as a function of u leads to the integral form

$$t = \int_{u_0}^{u} \frac{dx}{\sqrt{(1 - x^2)(1 - k^2x^2)}}. \tag{1.4.4}$$

This is an instance of an elliptic integral – an integral of the form

$$\int_{u_0}^{u} R\left(x, \sqrt{P(x)}\right) dx, \tag{1.4.5}$$

where P is a polynomial of degree 3 or 4 with no repeated roots, and R is a rational function (quotient of polynomials) in two variables. If P had degree 2, then (1.4.5) could be integrated by a trigonometric substitution. For example,

$$\int_{0}^{u} \frac{dx}{\sqrt{1 - x^2}} = \sin^{-1} u;$$

equivalently,

$$t = \int_{0}^{\sin t} \frac{dx}{\sqrt{1 - x^2}}.$$

Elliptic functions are the analogues, for the case where P has degree 3 or 4, of the trigonometric functions in the case of degree 2.

All second-order equations discussed up to this point have the form

$$u'' = \frac{P(x, u, u')}{Q(x, u, u')},$$

where P and Q are polynomial functions of their arguments. The behavior, in the complex x-plane, of the general solution of such an equation will typically depend in a complicated way on initial conditions $u(x_0)$, $u'(x_0)$. For example, the location of branch points may change. A certain number of such equations have the property that this does not happen, but that the general solutions are not expressible in terms of other known functions, such as elliptic functions. The solutions of these equations are known as the Painlevé transcendents. They occur naturally in a number of contexts in physics (e.g. integrable systems) and mathematics (e.g. random matrix theory).

1.5 Exercises

1. Suppose that u is a solution of (1.0.1) with parameters (a, c). Show that the derivative u' is a solution of (1.0.1) with parameters $(a+1, c+1)$.
2. Suppose that u is a solution of (1.0.2) with parameters (a, b, c). Show that the derivative u' is a solution of (1.0.1) with parameters $(a+1, b+1, c+1)$.
3. Show that the power series solution to (1.1.5) with $u_0 = 1$ is

$$u(x) = \sum_{n=0}^{\infty} \frac{1}{(c)_n \, n!} (-x)^n, \qquad c \neq 0, -1, -2, \ldots$$

4. In Exercise 3, suppose $c = \frac{1}{2}$. Show that

$$u(x) = \cosh\left[2(-x)^{\frac{1}{2}}\right].$$

5. Compare the series solution in Exercise 3 with the series expansion (9.1.1). This suggests that if u is a solution of (1.1.5), then

$$u(x) = x^{-\frac{1}{2}v} v\left(2\sqrt{x}\right), \qquad v = c - 1,$$

where v is a solution of Bessel's equation (3.6.10). Verify this fact directly. Together with the results in Section 3.7, this confirms that (1.1.5) can be transformed to (1.0.1).
6. Show that the power series solution to (1.0.1) with $u_0 = 1$ is given by (1.1.7).
7. Show that the power series solution to (1.0.2) with $u_0 = 1$ is the *hypergeometric function*

$$u(x) = F(a, b, c; x) = \sum_{n=0}^{\infty} \frac{(a)_n (b)_n}{(c)_n \, n!} x^n.$$

8. In the preceding exercise, suppose that $a = c$. Find a closed form expression for the sum of the series. Relate this to the differential equation (1.0.2) when $a = c$.

9. Consider the first-order equation (1.1.6) under the assumption that q and r are polynomials and neither is identically zero. Show that all solutions have the form

$$u(x) = P(x)\exp R(x),$$

where P is a polynomial and R is a rational function (quotient of polynomials).

10. Suppose that an equation of the form (1.1.1) with holomorphic coefficients has the property that the nth equation (1.1.3) determines u_{n+2} from u_n alone. Show that up to normalization, the equation can be put into one of the following two forms:

$$u''(x) - 2xu'(x) + 2\lambda u(x) = 0;$$

$$(1 - x^2)u''(x) + axu'(x) + bu(x) = 0.$$

11. Determine the power series expansion of the even solution ($u(-x) = u(x)$) of the first equation in Exercise 10, with $u(0) = 1$. Write $u(x) = v(x^2)$ and show that $v(y) = M(a, c; y)$ for suitable choices of the parameters a and c.

12. Determine the power series expansion of the odd solution ($u(-x) = -u(x)$) of the first equation in Exercise 10, with $u'(0) = 1$. Write $u(x) = xv(x^2)$ and show that $v(y) = M(a, c; y)$ for suitable choices of the parameters a and c.

13. Let $x = 1 - 2y$ in the second equation of Exercise 10 and show that, in terms of y, the equation takes the form (1.0.2) for some choice of the parameters a, b, c.

14. When does (1.0.1) have a (nonzero) polynomial solution? What about (1.0.2)?

15. Show that $\Gamma(n) = (n-1)!$, $n = 1, 2, 3, \ldots$

16. Show that one can use the functional equation for the gamma function to extend it as a meromorphic function on the half-plane $\{\operatorname{Re} a > -1\}$ with a simple pole at $a = 0$.

17. Show that the gamma function can be extended to a meromorphic function on the whole complex plane with simple poles at the nonpositive integers.

18. Use the change of variables $t = (1 - s)x$ in the integral representation (1.2.1) to prove the asymptotic result

$$M(a, c; x) \sim \frac{\Gamma(c)}{\Gamma(a)} x^{a-c} e^x \quad \text{as } x \to +\infty.$$

19. Derive an integral representation for the series solution to (1.0.2) with $u(0) = 1$ (Exercise 7), assuming $\operatorname{Re} c > \operatorname{Re} a > 0$.

20. Show that the change of variables $u(x) = v(2\sqrt{x})$ converts (1.0.1) to an equation with leading coefficient 1 and zero-order coefficient of the form $r(y) - a$.

21. Suppose the coefficient p in (1.1.1) is positive. Suppose that $y = y(x)$ satisfies the equation $y'(x) = p(x)^{\frac{1}{2}}$. Show that the change of variables $u(x) = v(y(x))$ converts (1.1.1) to an equation with leading coefficient 1 and zero-order coefficient of the form $r_1(y) + r_0(x(y))$.

22. Suppose the coefficient p in (1.1.1) is positive. Show that there is a gauge transformation $u(x) = \varphi(x)v(x)$ so that the equation takes the form

$$p(x)v''(x) + r_1(x)v(x) = 0,$$

with no first-order term. (Assume that a first-order equation $p(x)f'(x) = g(x)f(x)$ has a nonvanishing solution f for any given function g.)

23. Eliminate the first-order term in (1.0.1) and in (1.0.2).

24. Show that the Wronskian of two solutions of (1.1.1) satisfies a homogeneous first-order linear differential equation. What are the possible solutions of this equation for (1.0.1)? For (1.0.2)?

25. Find the asymptotic series $\sum_{n=0}^{\infty} a_n x^{-n}$ for the function e^{-x}, $x \to \infty$. What does the result say about whether a function is determined by its asymptotic series expansion?

26. Use the method of Exercise 18 to determine the full asymptotic series expansion of the function $M(a,c;x)$ as $x \to \infty$, assuming that $\operatorname{Re} c > \operatorname{Re} a > 0$.

27. Verify that (1.4.2) follows from (1.4.1).

28. Verify that (1.4.3) follows from (1.4.2).

29. Show that the change of variables in the integral (1.4.4) given by the linear fractional transformation $w(z) = (kz + 1)/(z + k)$ converts (1.4.4) to the form (1.4.5) with a polynomial $P(w)$ of degree 3. What are the images $w(z)$ of the four roots $z = \pm 1$ and $z = \pm k$ of the original polynomial?

30. Suppose that P in (1.4.5) has degree 3, with no repeated roots. Show that there is a linear fractional transformation that converts (1.4.5) to the same form but with a polynomial of degree 4 instead.

1.6 Remarks

A comprehensive classical reference for the theory and history of special functions is Whittaker and Watson [435]. The theory is also treated in Andrews, Askey, and Roy [10], Hochstadt [189], Lebedev [249], Luke [264], Nikiforov and Uvarov [305], Rainville [339], Sneddon [371], and Temme [397]. Many historical references are found in [435] and in [10]. The connection with differential equations of type (1.0.1) and (1.0.2) is emphasized in [305]. There are a very large number of identities of type (1.1.9) and (1.2.1). Extensive lists of such identities and other basic formulas are found in the Bateman Manuscript Project [116, 117], Jahnke and Emde [202],

Magnus and Oberhettinger [269], Abramowitz and Stegun [4], Gradshteyn and Ryzhik [170], Magnus, Oberhettinger, and Soni [270] and Brychkov [56]. A revised and updated version of [4] is now available as Olver, Lozier, Clark, and Boisvert [312, 313]. For a short history and critical review of the handbooks, see Askey [20].

Some other organizing principles for approaching special functions are: integral equations, Courant and Hilbert [89]; differential-difference equations, Truesdell [404]; Lie theory, Miller [287]; group representations, Dieudonné [104], Müller [295], Talman [394], Varchenko [413], Vilenkin [414], Vilenkin and Klimyk [415], and Wawrzyńczyk [426]; generating functions, McBride [276]; Painlevé functions, Iwasaki, Komura, Shimomura, and Yoshida [196]; singularities, Slavyanov and Lay [370]; and zeta-functions, Kanemitsu and Tsukada [212].

With the exception of [312, 313], none of the previous references treat Painlevé transcendents. For these functions, see the references in the remarks for Chapter 15.

One of the principal uses of special functions is to develop in series the solutions of equations of mathematical physics. See Burkhardt [60] for an exhaustive historical account up to the beginning of the twentieth century, and Higgins [182] and Johnson and Johnson [208] for related questions.

Much of the theory of special functions was developed in the eighteenth and nineteenth centuries. For a general introduction to the history of mathematics in that period, see Dieudonné [103]. The recent book by Roy [347] contains a great deal of information about the historical and intellectual development of most of the topics touched on in this book.

A comment on terminology: the fact that a mathematician's name is attached to a particular equation or function is often an indication that the equation or function in question was first considered by someone else, e.g. by one of the Bernoullis or by Euler. (There are exceptions to this.) Nevertheless, we generally adhere to standard terminology.

2

Gamma, beta, zeta

The first two functions discussed in this chapter are due to Euler. The third is usually associated with Riemann, although it was also studied earlier by Euler. Collectively they are of great importance historically, theoretically, and for purposes of calculation.

Historically and theoretically, investigation of these functions and their properties has provided considerable impetus to the study and understanding of fundamental aspects of mathematical analysis, including limits, infinite products, and analytic continuation. They have also motivated advances in complex function theory, such as the theorems of Weierstrass and Mittag-Leffler on representations of entire and meromorphic functions. The zeta function and its generalizations are intimately connected with questions of number theory.

From the point of view of calculation, many of the explicit constants of mathematical analysis, especially those that come from definite integrals, can be evaluated in terms of the gamma and beta functions.

There is much to be said for proceeding historically in discussing these and other special functions, but we shall not make a point of doing so. In mathematics it is often, even usually, the case that later developments cast a new light on earlier ones. One result is that later expositions can often be made both more efficient and, one hopes, more transparent than the original derivations.

After introducing the gamma and beta function and their basic properties, we turn to a number of important identities and representations of the gamma function and its reciprocal. Two characterizations of the gamma function are established, one based on complex analytic properties, the other based on a geometric property. Asymptotic properties of the gamma function are considered in detail. The psi function and the incomplete gamma function are introduced.

The identity that evaluates the beta integral in terms of gamma functions has important modern generalizations due to Selberg and Aomoto. Aomoto's proof is sketched.

The zeta function, its functional equation, and Euler's evaluation of $\zeta(n)$ for $n = 2, 4, 6, \ldots$, are the subject of the final section.

2.1 The gamma and beta functions

The gamma function was introduced by Euler in 1729 [119] in answer to the question of finding a function that takes the value $n!$ at each nonnegative integer n. At that time, a "function" was understood as a formula expressed in terms of the standard operations of algebra and calculus, so the problem was not trivial. Euler's first solution was in the form of the limit of certain quotients of products, which we discuss below. For many purposes, the most useful version is a function that takes the value $(n-1)!$ at the positive integer n and is represented as an integral (also due to Euler [119]):

$$\Gamma(z) = \int_0^\infty e^{-t} t^z \frac{dt}{t}, \qquad \operatorname{Re} z > 0. \tag{2.1.1}$$

The integral is holomorphic as a function of z in the right half-plane. An integration by parts gives the *functional equation*

$$z\Gamma(z) = \Gamma(z+1), \qquad \operatorname{Re} z > 0. \tag{2.1.2}$$

This extends inductively to

$$(z)_n \Gamma(z) = \Gamma(z+n), \tag{2.1.3}$$

where, as in (1.1.8), the shifted factorial $(z)_n$ is

$$(z)_n = z(z+1) \cdots (z+n-1). \tag{2.1.4}$$

Since $\Gamma(1) = 1$, it follows that

$$\Gamma(n+1) = (1)_n = n!, \qquad n = 0, 1, 2, 3, \ldots \tag{2.1.5}$$

Theorem 2.1.1 *The gamma function extends to a meromorphic function on* **C**. *Its poles are simple poles at the nonpositive integers. The residue at* $-n$ *is* $(-1)^n/n!$. *The extension continues to satisfy the functional equations* (2.1.2) *and* (2.1.3) *for* $z \neq 0, -1, -2, -3, \ldots$

Proof The extension, and the calculation of the residues, can be accomplished by using the extended functional equation (2.1.3), which can be used to define $\Gamma(z)$ for $\operatorname{Re} z > -n$. Another way is to write the integral in (2.1.1) as the sum of two terms:

$$\Gamma(z) = \int_0^1 e^{-t} t^z \frac{dt}{t} + \int_1^\infty e^{-t} t^z \frac{dt}{t}.$$

In the first term, the power series representation of e^{-t} converges uniformly, so the series can be integrated term by term. Thus

$$\Gamma(z) = \int_0^1 \sum_{n=0}^{\infty} \frac{(-1)^n}{n!} t^{z+n} \frac{dt}{t} + \int_1^{\infty} e^{-t} t^z \frac{dt}{t}$$

$$= \sum_{n=0}^{\infty} \frac{(-1)^n}{n!(z+n)} + \int_1^{\infty} e^{-t} t^z \frac{dt}{t}.$$

The series in the last line converges for $z \neq 0, -1, -2, \ldots$, and defines a meromorphic function which has simple poles and residues $(-1)^n/n!$. The integral in the last line extends as an entire function of z. The functional equation represents a relationship between $\Gamma(z)$ and $\Gamma(z+1)$ that necessarily persists under analytic continuation. □

Euler's first definition of the gamma function started from the observation that for any positive integers k and n,

$$(n+k-1)! = (n-1)!(n)_k = (k-1)!(k)_n.$$

As $n \to \infty$ with k fixed, $(n)_k \sim n^k$, so

$$(k-1)! = \lim_{n \to \infty} \frac{(n-1)!(n)_k}{(k)_n} = \lim_{n \to \infty} \frac{(n-1)!\, n^k}{(k)_n}.$$

Thus, for k a positive integer,

$$\Gamma(k) = \lim_{n \to \infty} \frac{\Gamma(n)\, n^k}{(k)_n}. \tag{2.1.6}$$

As we shall see, the limit of this last expression exists for any complex k for which $(k)_n$ is never zero, i.e. $k \neq 0, -1, -2, \ldots$ Therefore there is another way to solve the original problem of extending the factorial function in a natural way: define the function for arbitrary k by (2.1.6). In the next section we show that this gives the same function Γ.

The beta function, or beta integral, occurs in many contexts. It is the function of two complex variables defined first for $\operatorname{Re} a > 0$ and $\operatorname{Re} b > 0$ by

$$\mathrm{B}(a,b) = \int_0^1 s^{a-1}(1-s)^{b-1}\, ds. \tag{2.1.7}$$

Taking $t = 1 - s$ as the variable of integration shows that B is symmetric:

$$\mathrm{B}(a,b) = \mathrm{B}(b,a).$$

Taking $u = s/(1-s)$ as the variable of integration gives the identity

$$\mathrm{B}(a,b) = \int_0^{\infty} u^a \left(\frac{1}{1+u}\right)^{a+b} \frac{du}{u}.$$

Both the beta integral and Euler's evaluation of it in terms of gamma functions [120] come about naturally when one seeks to evaluate the product $\Gamma(a)\Gamma(b)$:

$$\begin{aligned}
\Gamma(a)\Gamma(b) &= \int_0^\infty \int_0^\infty e^{-(s+t)} s^a t^b \frac{ds}{s} \frac{dt}{t} \\
&= \int_0^\infty \int_0^\infty e^{-t(1+u)} u^a t^{a+b} \frac{dt}{t} \frac{du}{u} \\
&= \int_0^\infty \int_0^\infty e^{-t} u^a \left(\frac{x}{1+u}\right)^{a+b} \frac{dx}{x} \frac{du}{u} \\
&= \Gamma(a+b) \int_0^\infty u^a \left(\frac{1}{1+u}\right)^{a+b} \frac{du}{u} \\
&= \Gamma(a+b) B(a,b).
\end{aligned}$$

Summarizing, we have shown that any one of three different expressions may be used to define or evaluate the beta function.

Theorem 2.1.2 *The beta function satisfies the following identities for* $\operatorname{Re} a > 0, \operatorname{Re} b > 0$:

$$\begin{aligned}
B(a,b) &= \int_0^1 s^{a-1}(1-s)^{b-1} ds \\
&= \int_0^\infty u^a \left(\frac{1}{1+u}\right)^{a+b} \frac{du}{u} \\
&= \frac{\Gamma(a)\Gamma(b)}{\Gamma(a+b)}.
\end{aligned} \tag{2.1.8}$$

The beta function has an analytic continuation to all complex values of a and b such that $a, b \neq 0, -1, -2, \dots$

Proof The identities were established above. The analytic continuation follows immediately from the continuation properties of the gamma function, since, as we show below, the gamma function has no zeros.   □

As noted above, for a fixed positive integer k,

$$\frac{(n+k-1)!}{(n-1)!} = n^k \left[1 + O(n^{-1})\right]$$

as $n \to \infty$. The beta integral allows us to extend this asymptotic result to noninteger values.

Proposition 2.1.3 *For any complex a,*

$$\frac{\Gamma(x+a)}{\Gamma(x)} = x^a \left[1 + O(x^{-1})\right] \tag{2.1.9}$$

as $x \to +\infty$.

Proof The extended functional equation (2.1.3) can be used to replace a by $a + n$, so we may assume that $\operatorname{Re} a > 0$. Then

$$\frac{\Gamma(x)}{\Gamma(x+a)} = \frac{B(x,a)}{\Gamma(a)} = \frac{1}{\Gamma(a)} \int_0^1 s^{a-1}(1-s)^{x-1}\, ds$$

$$= \frac{x^{-a}}{\Gamma(a)} \int_0^x t^{a-1} \left(1 - \frac{t}{x}\right)^x \left(1 - \frac{t}{x}\right)^{-1} dt$$

$$\sim \frac{x^{-a}}{\Gamma(a)} \int_0^\infty t^{a-1} e^{-t}\, dt$$

$$= x^{-a}.$$

This gives the principal part of (2.1.9). The error estimate is left as an exercise.

The extended functional equation (2.1.3) allows us to extend this result to the shifted factorials.

Corollary 2.1.4 *If $a, b \neq 0, -1, -2, \ldots$, then*

$$\lim_{n\to\infty} \frac{(a)_n}{(b)_n} n^{b-a} = \frac{\Gamma(b)}{\Gamma(a)}. \tag{2.1.10}$$

2.2 Euler's product and reflection formulas

The first result here is Euler's product formula [119].

Theorem 2.2.1 *The gamma function satisfies*

$$\Gamma(z) = \lim_{n\to\infty} \frac{(n-1)!\, n^z}{(z)_n} \tag{2.2.1}$$

for $\operatorname{Re} z > 0$ and

$$\Gamma(z) = \frac{1}{z} \prod_{n=1}^\infty \left(1 + \frac{1}{n}\right)^z \left(1 + \frac{z}{n}\right)^{-1} \tag{2.2.2}$$

for all complex $z \neq 0, -1, -2, \ldots$ In particular, the gamma function has no zeros.

Proof The identity (2.2.1) is essentially the case $a = 1$, $b = z$ of (2.1.10), since

$$(1)_n\, n^{z-1} = (n-1)!\, n^z.$$

To prove (2.2.2), we note that

$$\frac{(n-1)!\, n^z}{(z)_n} = \frac{n^z}{z} \prod_{j=1}^{n-1} \left(1 + \frac{z}{j}\right)^{-1}. \tag{2.2.3}$$

Writing

$$n = \frac{n}{n-1}\frac{n-1}{n-2}\cdots\frac{2}{1} = \left(1+\frac{1}{n-1}\right)\left(1+\frac{1}{n-2}\right)\cdots(1+1),$$

we can rewrite (2.2.3) as

$$\frac{1}{z}\prod_{j=1}^{n-1}\left(1+\frac{1}{j}\right)^{z}\left(1+\frac{z}{j}\right)^{-1}.$$

The logarithm of the jth factor is $O(j^{-2})$. It follows that the product converges uniformly in any compact set that excludes $z = 0, -1, -2, \ldots$ Therefore (first for $\mathrm{Re}\,z > 0$ and then by analytic continuation) taking the limit gives (2.2.2) for all $z \neq 0, -1, -2, \ldots$ □

The reciprocal of the gamma function is an entire function. Its product representation can be deduced from (2.2.1), since

$$\frac{(z)_n}{(n-1)!\,n^z} = z\exp\left(z\left[\sum_{k=1}^{n-1}\frac{1}{k} - \log n\right]\right)\prod_{k=1}^{n-1}\left(1+\frac{z}{k}\right)e^{-z/k}. \qquad (2.2.4)$$

The logarithm of the kth factor in the product (2.2.4) is $O(k^{-2})$, so the product converges uniformly in bounded sets. The coefficient of z in the exponential is

$$\sum_{k=1}^{n-1}\frac{1}{k} - \log n = \sum_{k=1}^{n-1}\int_k^{k+1}\left[\frac{1}{k}-\frac{1}{t}\right]dt = \sum_{k=1}^{n-1}\int_k^{k+1}\frac{t-k}{tk}\,dt.$$

The kth summand in the last sum is $O(k^{-2})$, so the sum converges as $n \to \infty$. The limit is known as *Euler's constant*:

$$\gamma = \lim_{n\to\infty}\left\{\sum_{k=1}^{n-1}\frac{1}{k} - \log n\right\} = \lim_{n\to\infty}\sum_{k=1}^{n-1}\int_k^{k+1}\left[\frac{1}{k}-\frac{1}{t}\right]dt. \qquad (2.2.5)$$

We have shown that (2.2.1) has the following consequence.

Corollary 2.2.2 *The reciprocal of the gamma function has the product representation*

$$\frac{1}{\Gamma(z)} = z\,e^{\gamma z}\prod_{n=1}^{\infty}\left(1+\frac{z}{n}\right)e^{-z/n}, \qquad (2.2.6)$$

where γ is Euler's constant (2.2.5).

This (or its reciprocal) is known as the "Weierstrass form" of the gamma function, although it was first proved by Schlömilch [356] and Newman [301] in 1848.

The next result is *Euler's reflection formula* [124].

Theorem 2.2.3 *For z not an integer*

$$\Gamma(z)\Gamma(1-z) = \frac{\pi}{\sin \pi z}. \tag{2.2.7}$$

Proof For $0 < \mathrm{Re}\, z < 1$,

$$\Gamma(z)\Gamma(1-z) = B(z, 1-z) = \int_0^\infty \frac{t^{z-1}\, dt}{1+t}.$$

The integrand is a holomorphic function of t for t in the complement of the positive real axis $[0, +\infty)$, choosing the argument of t in the interval $(0, 2\pi)$. Let C be the curve that comes from $+\infty$ to 0 along the "lower" side of the interval $[0, \infty)$ ($\arg t = 2\pi$) and returns to $+\infty$ along the "upper" side of the interval ($\arg t = 0$). Evaluating

$$\int_C \frac{t^{z-1}\, dt}{1+t}$$

by the residue calculus gives

$$(1 - e^{2\pi i z}) \int_0^\infty \frac{t^{z-1}\, dt}{1+t} = 2\pi i \operatorname{res} (t^{z-1})\big|_{t=-1} = -2\pi i e^{i\pi z};$$

see Appendix A. This proves (2.2.7) in the range $0 < \mathrm{Re}\, z < 1$. Once again, analytic continuation gives the result for all noninteger complex z. $\qquad \square$

Corollary 2.2.4 (Hankel's integral formula) *Let C be an oriented contour in the complex plane that begins at $-\infty$, continues on the real axis ($\arg t = -\pi$) to a point $-\delta$, follows the circle $\{|t| = \delta\}$ in the positive (counter-clockwise) direction, and returns to $-\infty$ along the real axis ($\arg t = \pi$). Then*

$$\frac{1}{\Gamma(z)} = \frac{1}{2\pi i} \int_C e^t\, t^{-z}\, dt. \tag{2.2.8}$$

Here t^{-z} takes its principal value where C crosses the positive real axis.

Proof The function defined by the integral in (2.2.8) is entire, so it suffices to prove the result for $0 < z < 1$. In this case we may take $\delta \to 0$. Setting $s = -t$, the right-hand side is

$$\frac{1}{2\pi i} \int_0^\infty e^{-s} \left\{ \left(se^{-i\pi}\right)^{-z} - \left(se^{i\pi}\right)^{-z} \right\} ds = \frac{\sin \pi z}{\pi} \int_0^\infty e^{-s} s^{-z}\, ds$$

$$= \frac{1}{\Gamma(z)\Gamma(1-z)} \cdot \Gamma(1-z),$$

where we have used (2.2.7). $\qquad \square$

The contour C described in Corollary 2.2.4 is often called a *Hankel loop*.

Corollary 2.2.5 (Euler's product for sine)

$$\frac{\sin \pi z}{\pi z} = \prod_{n=1}^{\infty}\left(1 - \frac{z^2}{n^2}\right). \tag{2.2.9}$$

Proof This follows from (2.2.7) together with (2.2.2). □

Corollary 2.2.6 $\Gamma(\frac{1}{2}) = \sqrt{\pi}$.

Proof Take $z = \frac{1}{2}$ in (2.2.7). □

This evaluation can be obtained in a number of other ways, for example:

$$\Gamma\left(\frac{1}{2}\right)^2 = B\left(\frac{1}{2},\frac{1}{2}\right) = \int_0^{\infty} \frac{t^{\frac{1}{2}}}{1+t}\frac{dt}{t}$$
$$= 2\int_0^{\infty}\frac{du}{1+u^2} = \tan^{-1} u\Big|_{-\infty}^{\infty} = \pi\,;$$
$$\Gamma\left(\frac{1}{2}\right) = \int_0^{\infty} e^{-t}\frac{dt}{\sqrt{t}} = 2\int_0^{\infty} e^{-u^2}\,du$$
$$= \left[\int_{-\infty}^{\infty}\int_{-\infty}^{\infty} e^{-(x^2+y^2)}\,dx\,dy\right]^{\frac{1}{2}} = \left[\int_0^{2\pi}\int_0^{\infty} e^{-r^2}\,r\,dr\,d\theta\right]^{\frac{1}{2}} = \sqrt{\pi}.$$

2.3 Formulas of Legendre and Gauss

Observe that

$$\Gamma(2n) = (2n-1)! = 2^{2n-1}\frac{1}{2}\cdot 1 \cdot \left(\frac{1}{2}+1\right)\cdot 2\cdots(n-1)\left(\frac{1}{2}+n-1\right)$$
$$= 2^{2n-1}(n-1)!\left(\frac{1}{2}\right)_n = 2^{2n-1}\Gamma(n)\frac{\Gamma(n+\frac{1}{2})}{\Gamma(\frac{1}{2})} = \frac{2^{2n-1}}{\sqrt{\pi}}\Gamma(n)\Gamma\left(n+\frac{1}{2}\right).$$

This is the positive integer case of *Legendre's duplication formula*:

$$\Gamma(2z) = \frac{2^{2z-1}}{\sqrt{\pi}}\Gamma(z)\Gamma\left(z+\frac{1}{2}\right), \qquad 2z \neq 0, -1, -2, \ldots \tag{2.3.1}$$

We give two proofs of (2.3.1). The second proof generalizes to give Gauss's formula for $\Gamma(mz)$. The first proof begins with $\mathrm{Re}\, z > 0$ and uses the change of variables $t = 4s(1-s)$ on the interval $0 \le s \le \frac{1}{2}$, as follows:

$$\frac{\Gamma(z)^2}{\Gamma(2z)} = B(z,z) = \int_0^1 [s(1-s)]^z \frac{ds}{s(1-s)}$$
$$= 2\int_0^{\frac{1}{2}} [s(1-s)]^z \frac{ds}{s(1-s)} = 2\int_0^1 \left(\frac{t}{4}\right)^z \frac{dt}{t\sqrt{1-t}}$$
$$= 2^{1-2z} B\left(z,\frac{1}{2}\right) = 2^{1-2z}\frac{\Gamma(z)\Gamma(\frac{1}{2})}{\Gamma(z+\frac{1}{2})}.$$

This gives (2.3.1) for $\operatorname{Re} z > 0$; analytic continuation gives the result for $2z \neq 0, -1, -2, \ldots$

The second proof uses (2.2.1), which implies

$$\Gamma(2z) = \lim_{n \to \infty} \frac{(2n)! (2n)^{2z-1}}{(2z)_{2n}}. \tag{2.3.2}$$

Now

$$(2n)! = 2^{2n} \left(\frac{1}{2}\right)_n (1)_n = 2^{2n} \frac{\Gamma(\frac{1}{2} + n) \Gamma(n+1)}{\Gamma(\frac{1}{2})} \tag{2.3.3}$$

and

$$(2z)_{2n} = 2^{2n} (z)_n \left(z + \frac{1}{2}\right)_n = 2^{2n} \frac{\Gamma(z+n) \Gamma(z + \frac{1}{2} + n)}{\Gamma(z) \Gamma(z + \frac{1}{2})}$$

$$= 2^{2n} \frac{\Gamma(n)}{B(z,n)} \cdot \frac{\Gamma(z) \Gamma(\frac{1}{2} + n)}{\Gamma(z + \frac{1}{2}) B(z, \frac{1}{2} + n)}. \tag{2.3.4}$$

According to the calculation at the beginning of Section 2.2,

$$B(z,n) \sim n^{-z} \Gamma(z)$$

as $n \to \infty$, so in taking the limit in (2.3.2), we may replace the expression in (2.3.4) with

$$2^{2n} n^{2z} \frac{\Gamma(n) \Gamma(\frac{1}{2} + n)}{\Gamma(z) \Gamma(z + \frac{1}{2})}. \tag{2.3.5}$$

Multiplying the quotient of (2.3.3) and (2.3.5) by $(2n)^{z-1}$ gives (2.3.1).

The previous proof can be adapted to prove the first part, (2.3.6), of the following result of Gauss [150]:

$$\Gamma(mz) = m^{mz-1} \frac{\Gamma(z) \Gamma(z + \frac{1}{m}) \Gamma(z + \frac{2}{m}) \cdots \Gamma(z + \frac{m-1}{m})}{\Gamma(\frac{1}{m}) \Gamma(\frac{2}{m}) \cdots \Gamma(\frac{m-1}{m})} \tag{2.3.6}$$

$$= m^{mz - \frac{1}{2}} \frac{\Gamma(z) \Gamma(z + \frac{1}{m}) \Gamma(z + \frac{2}{m}) \cdots \Gamma(z + \frac{m-1}{m})}{(2\pi)^{\frac{1}{2}(m-1)}}. \tag{2.3.7}$$

To prove the second part, (2.3.7), we evaluate the denominator in (2.3.6) using the reflection formula (2.2.7):

$$\left[\Gamma\left(\frac{1}{m}\right) \Gamma\left(\frac{2}{m}\right) \cdots \Gamma\left(\frac{m-1}{m}\right) \right]^2 = \prod_{k=1}^{m-1} \frac{\pi}{\sin\left(\frac{\pi k}{m}\right)}$$

$$= \pi^{m-1} \prod_{k=1}^{m-1} \frac{2i}{e^{\pi i k/m} (1 - e^{-2\pi i k/m})}.$$

Since $1 + 2 + \cdots + (m-2) + (m-1) = \frac{1}{2}m(m-1)$,

$$\prod_{k=1}^{m-1} \frac{2i}{e^{\pi i k/m}} = 2^{m-1}$$

and since $\omega = e^{2\pi i/m}$ is a primitive mth root of 1 it follows that

$$\prod_{k=1}^{m-1}(1 - \omega^k) = \lim_{t \to 1} \prod_{k=1}^{m-1}(t - \omega^k) = \lim_{t \to 1} \frac{t^m - 1}{t - 1} = m.$$

Therefore

$$\Gamma\left(\frac{1}{m}\right)\Gamma\left(\frac{2}{m}\right)\cdots\Gamma\left(\frac{m-1}{m}\right) = \frac{(2\pi)^{\frac{1}{2}(m-1)}}{m^{\frac{1}{2}}}, \tag{2.3.8}$$

and we obtain (2.3.7).

2.4 Two characterizations of the gamma function

The gamma function is not uniquely determined by the functional equation $\Gamma(z + 1) = z\Gamma(z)$; in fact if f is a function such as $\sin(2\pi x)$ that is periodic with period 1, then the product $f\Gamma$ also satisfies the functional equation. Here we give two characterizations, one as a function holomorphic on the half-plane $\{\text{Re}\, z > 0\}$ and one as a function on the positive reals.

Theorem 2.4.1 (Wielandt) *Suppose that G is holomorphic in the half-plane $\{\text{Re}\, z > 0\}$, bounded on the closed strip $\{1 \le \text{Re}\, z \le 2\}$, and satisfies the equations $G(1) = 1$ and $zG(z) = G(z+1)$ for $\text{Re}\, z > 0$. Then $G(z) \equiv \Gamma(z)$.*

Proof Let $F(z) = G(z) - \Gamma(z)$. Then F satisfies the functional equation and vanishes at $z = 1$, so it vanishes at the positive integers. This implies that F extends to an entire function. In fact, the functional equation allows us to extend by defining

$$F(z) = \frac{F(z+n)}{(z)_n}, \qquad -n < \text{Re}\, z \le 2 - n, \quad n = 1, 2, 3, \ldots$$

This is clearly holomorphic where $(z)_n \ne 0$, i.e. except for $z = 0, -1, \ldots, 1 - n$. These values of z are zeros of $F(z+n)$ and are simple zeros of $(z)_n$, so they are removable singularities for F. Therefore the extension of F is entire.

The functional equation and the fact that F is regular at 0 and bounded on the strip $\{1 \le \text{Re}\, z \le 2\}$ imply that F is bounded on the wider strip $S = \{0 \le \text{Re}\, z \le 2\}$. Therefore the function $f(z) = F(z)F(1 - z)$ is entire and bounded on S. Moreover,

$$f(z+1) = zF(z)F(-z) = -F(z)(-z)F(-z) = -f(z).$$

Thus $f(z+2) = f(z)$. Since f is bounded on a vertical strip of width 2, this implies that f is a bounded entire function, hence constant. But $f(1) = 0$, so $f \equiv 0$. It follows that $F \equiv 0$. $\qquad\square$

The next characterization starts from the observation that $\log \Gamma$ is a convex function on the interval $\{x > 0\}$. In fact, (2.2.6) implies that

$$\log \Gamma(x) = -\gamma x - \log x + \sum_{k=1}^{\infty} \left[\frac{x}{k} - \log \left(1 + \frac{x}{k} \right) \right], \qquad (2.4.1)$$

from which it follows that

$$(\log \Gamma)''(x) = \sum_{k=0}^{\infty} \frac{1}{(x+k)^2}, \qquad (2.4.2)$$

which is positive for $x > 0$. This implies convexity. The following theorem is the converse.

Theorem 2.4.2 (Bohr–Mollerup) *Suppose that $G(x)$ is defined and positive for $x > 0$ and satisfies the functional equation $xG(x) = G(x+1)$, and suppose that $\log G$ is convex and $G(1) = 1$. Then $G(x) = \Gamma(x)$ for all $x > 0$.*

Proof Let $f = \log G$. In view of the functional equation, it is enough to prove $G(x) = \Gamma(x)$ for $0 < x < 1$. The assumptions imply that $G(1) = G(2) = 1$ and $G(3) = 2$, so $f(1) = f(2) < f(3)$. The convexity of f implies that f is increasing on the interval $[2, \infty)$, and that for each integer $n \geq 2$,

$$f(n) - f(n-1) \leq \frac{f(n+x) - f(n)}{x} \leq f(n+1) - f(n).$$

By the functional equation and the definition of f, this is equivalent to

$$(n-1)^x \leq \frac{G(n+x)}{G(n)} \leq n^x.$$

The functional equation applied to the quotient leads to

$$\frac{(n-1)!\,(n-1)^x}{(x)_n} \leq \frac{G(x)}{G(1)} \leq \frac{(n-1)!\,n^x}{(x)_n}.$$

By (2.2.1), the expressions on the left and the right have limit $\Gamma(x)$. $\qquad\square$

Wielandt's theorem first appeared in Knopp [220]; see Remmert [340]. The theorem of Bohr and Mollerup is in [51].

2.5 Asymptotics of the gamma function

Suppose x is real and $x \to +\infty$. The integrand $e^{-t}t^x$ for $\Gamma(x+1)$ has its maximum $(x/e)^x$ at $t = x$ which suggests a change of variables $t = xu$:

$$\Gamma(x) = \frac{1}{x}\Gamma(x+1) = x^x \int_0^\infty \left(ue^{-u}\right)^x du$$

$$= \left(\frac{x}{e}\right)^x \int_0^\infty \left(ue^{1-u}\right)^x du.$$

Now ue^{1-u} attains its maximum 1 at $u = 1$, so the part of the last integral over a region $|u-1| > \delta > 0$ decays exponentially in x as $x \to \infty$. Note that ue^{1-u} agrees to second order at $u = 1$ with $e^{-(u-1)^2}$; it follows that we can change the variables $u \to s$ near $u = 1$ in such a way that

$$ue^{1-u} = e^{-\frac{1}{2}s^2}, \quad u'(s) = \sum_{n=0}^\infty a_n s^n,$$

with $a_0 = 1$. Combining these remarks, we obtain an asymptotic expansion

$$\int_0^\infty \left(ue^{1-u}\right)^x du \sim \sum_{n=0}^\infty a_n \int_{-\infty}^\infty e^{-\frac{1}{2}xs^2} s^n ds$$

$$= \sum_{m=0}^\infty a_{2m} x^{-m-\frac{1}{2}} \int_{-\infty}^\infty e^{-\frac{1}{2}s^2} s^{2m} ds.$$

(The odd terms vanish because their integrands are odd functions.) Since $a_0 = 1$, the first term in this expansion is $\sqrt{2\pi/x}$. The same considerations apply for complex z with $\operatorname{Re} z \to +\infty$. Thus we have *Stirling's formula* [381]:

$$\Gamma(z) = \frac{z^z}{e^z}\left\{\left(\frac{2\pi}{z}\right)^{\frac{1}{2}} + O(z^{-\frac{3}{2}})\right\} \quad \text{as} \quad \operatorname{Re} z \to +\infty. \tag{2.5.1}$$

The previous equation can be made more explicit. Binet [45] proved the following.

Theorem 2.5.1 *For* $\operatorname{Re} z > 0$,

$$\Gamma(z) = \frac{z^z}{e^z}\left(\frac{2\pi}{z}\right)^{\frac{1}{2}} e^{\theta(z)}, \tag{2.5.2}$$

where

$$\theta(z) = \int_0^\infty \left(\frac{1}{e^t - 1} - \frac{1}{t} + \frac{1}{2}\right) e^{-zt} \frac{dt}{t}. \tag{2.5.3}$$

Proof We follow Sasvari [353]. By definition, (2.5.2) holds with $\theta(z)$ replaced by

$$\varphi(z) = \log \Gamma(z) + z(1 - \log z) + \frac{1}{2} \log z - \frac{1}{2} \log(2\pi).$$

The functional equation implies that

$$\varphi(z) - \varphi(z+1) = \left(z + \frac{1}{2}\right) \log\left(\frac{z+1}{z}\right) - 1,$$

and (2.5.1) implies that $\varphi(z) \to 0$ as $\operatorname{Re} z \to +\infty$. Then

$$\varphi'(z) - \varphi'(z+1) = -\frac{1}{2}\left(\frac{1}{z} + \frac{1}{z+1}\right) + \log\left(1 + \frac{1}{z}\right);$$

$$\varphi''(z) - \varphi''(z+1) = -\frac{1}{2}\left(\frac{1}{z} + \frac{1}{z+1}\right)' + \frac{1}{z+1} - \frac{1}{z}.$$

With θ given by (2.5.3), we have

$$\theta(z) - \theta(z+1) = \int_0^\infty \left[e^{-t} + \left(\frac{1}{2} - \frac{1}{t}\right)(1 - e^{-t})\right] e^{-zt} \frac{dt}{t};$$

$$\theta'(z) - \theta'(z+1) = -\int_0^\infty \left[e^{-t} + \left(\frac{1}{2} - \frac{1}{t}\right)(1 - e^{-t})\right] e^{-zt} \, dt$$

$$= -\frac{1}{2}\left(\frac{1}{z} + \frac{1}{z+1}\right) + \int_0^\infty \frac{1}{t}(1 - e^{-t})e^{-zt} \, dt;$$

$$\theta''(z) - \theta''(z+1) = -\frac{1}{2}\left(\frac{1}{z} + \frac{1}{z+1}\right)' + \frac{1}{z+1} - \frac{1}{z}.$$

Since the various functions and derivatives have limit 0 as $\operatorname{Re} z \to \infty$, it follows that $\theta = \varphi$. □

The integrand in (2.5.3) includes the function

$$\frac{1}{e^t - 1} - \frac{1}{t} + \frac{1}{2} = \frac{1}{2} \frac{e^{t/2} + e^{-t/2}}{e^{t/2} - e^{-t/2}} - \frac{1}{t}.$$

This function is odd and is holomorphic for $|t| < 2\pi$, so it has the expansion

$$\frac{1}{e^t - 1} - \frac{1}{t} + \frac{1}{2} = \sum_{m=1}^\infty \frac{B_{2m}}{(2m)!} t^{2m-1}, \quad |t| < 2\pi. \tag{2.5.4}$$

The coefficients B_{2m} are known as the *Bernoulli numbers* [40]. They can be computed recursively; see Exercise 35. Putting partial sums of this expansion into (2.5.3) gives the asymptotic estimates

$$\theta(z) = \sum_{m=1}^N \frac{B_{2m}}{2m(2m-1)} z^{1-2m} + O(z^{-2N-1}), \qquad \operatorname{Re} z \to +\infty. \tag{2.5.5}$$

Stirling's formula (2.5.1) for positive integer z can be used in conjuction with the product formula (2.2.6) to extend (2.5.1) to the complement of the negative real axis. The method here is due to Stieltjes.

Theorem 2.5.2 *For* $|\arg z| < \pi$,

$$\Gamma(z) = \frac{z^z}{e^z}\left[\left(\frac{2\pi}{z}\right)^{\frac{1}{2}} + O(z^{-\frac{3}{2}})\right] \tag{2.5.6}$$

as $|z| \to \infty$, *uniformly for* $|\arg z| \le \pi - \delta < \pi$.

Proof Let $[s]$ denote the greatest integer function. Then

$$\int_k^{k+1} \frac{\frac{1}{2} - s + [s]}{s+z}\, ds = \int_k^{k+1} \left(\frac{\frac{1}{2} + k + z}{s+z} - 1\right) ds$$

$$= \left(k + \frac{1}{2} + z\right)\left[\log\left(k+1+z\right) - \log\left(k+z\right)\right] - 1$$

$$= \left[\left(k + \frac{1}{2} + z\right)\log\left(k+1+z\right)\right.$$

$$\left. - \left(k - \frac{1}{2} + z\right)\log\left(k+z\right)\right] - \log\left(k+z\right) - 1.$$

Summing,

$$\int_0^n \frac{\frac{1}{2} - s + [s]}{s+z}\, ds$$

$$= \left(n - \frac{1}{2} + z\right)\log\left(n+z\right) - \left(-\frac{1}{2} + z\right)\log z - \sum_{k=0}^{n-1}\log\left(k+z\right) - n.$$

Now

$$-\sum_{k=1}^{n-1}\log\left(k+z\right) = -\sum_{k=1}^{n-1}\log\left(1+\frac{z}{k}\right) - \log\Gamma(n)$$

$$= \sum_{k=1}^{n-1}\left[\frac{z}{k} - \log\left(1+\frac{z}{k}\right)\right] - z\sum_{k=1}^{n-1}\frac{1}{k} - \log\Gamma(n)$$

and

$$\log\left(n+z\right) = \log n + \log\left(1+\frac{z}{n}\right) = \log n + \frac{z}{n} + O(n^{-2}).$$

It follows from (2.2.5), (2.5.1), and (2.2.6), respectively, that

$$\sum_{k=1}^{n-1} \frac{1}{k} = \log n + \gamma + O(n^{-1});$$

$$\log \Gamma(n) = \left(n - \frac{1}{2}\right) \log n - n + \frac{1}{2} \log 2\pi + O(n^{-1});$$

$$\log \Gamma(z) = -z\gamma - \log z + \sum_{k=1}^{\infty} \left[\frac{z}{k} - \log\left(1 + \frac{z}{k}\right)\right].$$

Combining the previous formulas gives

$$\int_0^\infty \frac{\frac{1}{2} - s + [s]}{s + z} \, ds = \log \Gamma(z) - \left(z - \frac{1}{2}\right) \log z - \frac{1}{2} \log 2\pi + z. \qquad (2.5.7)$$

We estimate the integral by integrating by parts: let

$$f(s) = \int_0^s \left(\frac{1}{2} - t + [t]\right) dt.$$

This function has period 1 and is therefore bounded. Then

$$\int_0^\infty \frac{\frac{1}{2} - s + [s]}{s + z} \, ds = \int_0^\infty \frac{f(s)}{(s + z)^2} \, ds = O(|z|^{-1})$$

uniformly for $1 + \cos(\arg z) \geq \delta > 0$, since $z = re^{i\theta}$ implies

$$|s + z|^2 = s^2 + r^2 + 2sr \cos\theta \geq (s^2 + r^2) \min\{1, 1 + \cos\theta\}.$$

Therefore exponentiating (2.5.7) gives (2.5.6). $\qquad \square$

Corollary 2.5.3 *For real x and y,*

$$|\Gamma(x + iy)| = \sqrt{2\pi} \, |y|^{x - \frac{1}{2}} e^{-\frac{1}{2}\pi |y|} \left[1 + O(|y|^{-1})\right] \qquad (2.5.8)$$

as $|y| \to \infty$.

Another consequence of Theorem 2.5.2 is a sharpened form of (2.1.9).

Corollary 2.5.4 *For any complex a,*

$$\frac{\Gamma(a + z)}{\Gamma(z)} = z^a \left[1 + O(|z|^{-1})\right] \qquad (2.5.9)$$

as $z \to \infty$, *uniformly for* $|\arg z| \leq \pi - \delta < \pi$.

The proof is left as Exercise 13.

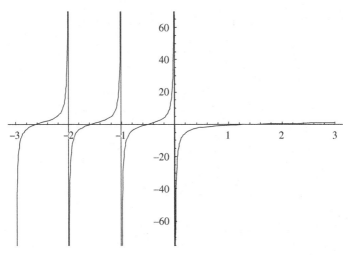

Figure 2.1. The ψ function.

2.6 The psi function and the incomplete gamma functions

The logarithmic derivative of the gamma function is denoted by $\psi(z)$:

$$\psi(z) = \frac{d}{dz} \log \Gamma(z) = \frac{\Gamma'(z)}{\Gamma(z)}.$$

Most of the properties of this function can be obtained directly from the corresponding properties of the gamma function. For instance, the only singularities of $\psi(z)$ are simple poles with residue -1 at the points $z = 0, -1, -2, \ldots$

The product formula (2.2.6) for $1/\Gamma(z)$ implies that

$$\psi(z) = -\gamma + \sum_{n=0}^{\infty} \left(\frac{1}{n+1} - \frac{1}{n+z} \right), \qquad z \neq 0, -1, -2, \ldots, \qquad (2.6.1)$$

and therefore, as noted in Section 2.4,

$$\psi'(z) = \sum_{n=0}^{\infty} \frac{1}{(z+n)^2}. \qquad (2.6.2)$$

The graph of $\psi(x)$ for real x is shown in Figure 2.1.

Using (2.6.1), the recurrence relation

$$\psi(z+1) = \psi(z) + \frac{1}{z}$$

is readily verified, and we have

$$\psi(1) = -\gamma, \quad \psi(k+1) = -\gamma + 1 + \frac{1}{2} + \frac{1}{3} + \cdots + \frac{1}{k}, \quad k = 1, 2, 3, \ldots$$

Taking the logarithmic derivative of (2.5.3) gives Binet's integral formula for ψ:

$$\psi(z) = \log z - \frac{1}{2z} - \int_0^\infty \left(\frac{1}{e^t - 1} - \frac{1}{t} + \frac{1}{2} \right) e^{-zt} \, dt, \quad \mathrm{Re}\, z > 0.$$

The *incomplete gamma function* $\gamma(\alpha, z)$ is defined by

$$\gamma(\alpha, z) = \int_0^z t^{\alpha-1} e^{-t} \, dt, \quad \mathrm{Re}\, \alpha > 0. \tag{2.6.3}$$

It is an analytic function of z in the right half-plane. We use the power series expansion of e^{-t} and integrate term by term to obtain

$$\gamma(\alpha, z) = z^\alpha \sum_{n=0}^\infty \frac{(-z)^n}{n!(n+\alpha)}, \quad \mathrm{Re}\, \alpha > 0.$$

The series converges for all z, so we may use this formula to extend the function in z and α for $\alpha \neq 0, -1, -2, \ldots$ If α is fixed, then the branch of $\gamma(\alpha, z)$ obtained after z encircles the origin m times is given by

$$\gamma(\alpha, z e^{2m\pi i}) = e^{2m\alpha\pi i} \gamma(\alpha, z), \quad \alpha \neq 0, -1, -2, \ldots \tag{2.6.4}$$

The *complementary incomplete gamma function* is defined by

$$\Gamma(\alpha, z) = \int_z^\infty t^{\alpha-1} e^{-t} \, dt. \tag{2.6.5}$$

In (2.6.5) there is no restriction on α if $z \neq 0$, and the principal branch is defined in the same manner as for $\gamma(\alpha, z)$. Combining with (2.6.3) gives

$$\gamma(\alpha, z) + \Gamma(\alpha, z) = \Gamma(\alpha). \tag{2.6.6}$$

It follows from (2.6.4) and (2.6.6) that

$$\Gamma(\alpha, z e^{2m\pi i}) = e^{2m\alpha\pi i} \Gamma(\alpha, z) + (1 - e^{2m\alpha\pi i}) \Gamma(\alpha), \quad m = 0, \pm 1, \pm 2, \ldots \tag{2.6.7}$$

The *error function* and *complementary error function* are defined, respectively, by

$$\mathrm{erf}\, z = \frac{2}{\sqrt{\pi}} \int_0^z e^{-t^2} \, dt, \quad \mathrm{erfc}\, z = \frac{2}{\sqrt{\pi}} \int_z^\infty e^{-t^2} \, dt.$$

Both are entire functions. Clearly,

$$\mathrm{erf}\, z + \mathrm{erfc}\, z = 1$$

and

$$\operatorname{erf} z = \frac{2}{\sqrt{\pi}} \sum_{n=0}^{\infty} \frac{(-1)^n z^{2n+1}}{n!(2n+1)}.$$

By a simple change of variable, one can show that

$$\operatorname{erf} z = \frac{1}{\sqrt{\pi}} \gamma\left(\frac{1}{2}, z^2\right), \quad \operatorname{erfc} z = \frac{1}{\sqrt{\pi}} \Gamma\left(\frac{1}{2}, z^2\right).$$

2.7 The Selberg integral

The Selberg integral is

$$S_n(a,b,c) = \int_0^1 \cdots \int_0^1 \prod_{i=1}^n x_i^{a-1} (1-x_i)^{b-1} \prod_{1 \le j < k \le n} |x_j - x_k|^{2c} \, dx_1 \cdots dx_n,$$

(2.7.1)

where convergence of the integral is assured by the conditions

$$\operatorname{Re} a > 0, \quad \operatorname{Re} b > 0, \quad \operatorname{Re} c > \max\left\{-\frac{1}{n}, -\frac{\operatorname{Re} a}{n-1}, -\frac{\operatorname{Re} b}{n-1}\right\}. \quad (2.7.2)$$

In particular, $S_n(a,b,0) = B(a,b)^n$. Thus Selberg's evaluation of S_n, stated in the following theorem, generalizes Euler's evaluation of the beta function in Theorem 2.1.2. Note that

$$S_n(a,b,c) = S_n(b,a,c).$$

Theorem 2.7.1 *Under the conditions* (2.7.2),

$$S_n(a,b,c) = \prod_{j=0}^{n-1} \frac{\Gamma(a+jc)\Gamma(b+jc)\Gamma(1+[j+1]c)}{\Gamma(a+b+[n-1+j]c)\Gamma(1+c)}. \quad (2.7.3)$$

For convenience, we let $\mathbf{x} = (x_1, x_2, \ldots, x_n)$, $d\mathbf{x} = dx_1 \cdots dx_n$, and denote the integrand in (2.7.1) by

$$w(\mathbf{x}) = \prod_{i=1}^n x_i^{a-1} (1-x_i)^{b-1} \prod_{1 \le j < k \le n} |x_j - x_k|^{2c}.$$

Note that $w(\mathbf{x})$ is symmetric in $x_1, \ldots, x_n$. Let C_n denote the n-dimensional cube $[0,1] \times \cdots \times [0,1]$. Then (2.7.1) is

$$S_n(a,b,c) = \int_{C_n} w(\mathbf{x}) \, d\mathbf{x}.$$

The proof of Theorem 2.7.1 outlined here is due to Aomoto [12]. It makes use of the related integrals

$$I_k = \int_{C_n} w(\mathbf{x}) \prod_{i=1}^k x_i \, d\mathbf{x}, \quad k = 0, 1, \ldots, n. \tag{2.7.4}$$

Note that $I_0 = S_n(a,b,c)$ and $I_n = S_n(a+1,b,c) = S_n(b,a+1,c)$. We may find a relation between I_k and I_{k-1} by integrating the identity

$$\frac{\partial}{\partial x_1}\left[(1-x_1)w(\mathbf{x})\prod_{j=1}^k x_j\right] = a(1-x_1)w(\mathbf{x})\prod_{j=2}^k x_j - b\,w(\mathbf{x})\prod_{j=1}^k x_j$$

$$+ (1-x_1)2c\,w(\mathbf{x})\left[\sum_{j=2}^n \frac{1}{x_1 - x_j}\right]\prod_{j=1}^k x_j. \tag{2.7.5}$$

The left side integrates to zero. The first two terms on the right integrate to $aI_{k-1} - (a+b)I_k$. The remaining terms can be integrated with the use of the following lemma, which is left as Exercise 30.

Lemma 2.7.2 *Let I_k be the integral* (2.7.4). *Then*

$$\int_{C_n} \frac{w(\mathbf{x})}{x_1 - x_j} \prod_{i=1}^k x_i \, d\mathbf{x} = \begin{cases} 0, & 2 \leq j \leq k; \\ \frac{1}{2}I_{k-1}, & k < j \leq n, \end{cases} \tag{2.7.6}$$

and

$$\int_{C_n} \frac{x_1 w(\mathbf{x})}{x_1 - x_j} \prod_{i=1}^k x_i \, d\mathbf{x} = \begin{cases} \frac{1}{2}I_k, & 2 \leq j \leq k; \\ I_k. & k < j \leq n. \end{cases} \tag{2.7.7}$$

Applying these identities to the integral of (2.7.5) gives the identity

$$I_k = \frac{a + (n-k)c}{a+b+(2n-k-1)c} I_{k-1}. \tag{2.7.8}$$

It follows that

$$S_n(a+1,b,c) = \prod_{j=0}^{n-1} \frac{a+jc}{a+b+(n-1+j)c} S_n(a,b,c).$$

Iterating this k times with a and b interchanged gives

$$S_n(a,b,c) = \prod_{j=0}^{n-1} \frac{(a+b+[n-1+j]c)_k}{(b+jc)_k} S_n(a,b+k,c).$$

We let $k \to \infty$ and use (2.1.10) to conclude that

$$S_n(a,b,c) = \prod_{j=0}^{n-1} \frac{\Gamma(b+jc)}{\Gamma(a+b+[n-1+j]c)} \cdot \lim_{k \to \infty} \left[k^{na+n(n-1)c} S_n(a,b+k,c) \right].$$

The expression in the limit can be rewritten as

$$k^{na+n(n-1)c} \int_0^k \cdots \int_0^k \prod_{i=1}^n \left(\frac{x_i}{k}\right)^{a-1} \left(1 - \frac{x_i}{k}\right)^{k+b-1} \prod_{i<j} \left|\frac{x_i}{k} - \frac{x_j}{k}\right|^{2c} \frac{d\mathbf{x}}{k^n}$$

$$= \int_0^k \cdots \int_0^k \prod_{i=1}^n x_i^{a-1} \left(1 - \frac{x_i}{k}\right)^{k+b-1} \prod_{i<j} |x_i - x_j|^{2c} \, d\mathbf{x}.$$

Taking the limit as $k \to \infty$, we obtain

$$S_n(a,b,c) = \prod_{j=0}^{n-1} \frac{\Gamma(b+jc)}{\Gamma(a+b+[n-1+j]c)}$$

$$\times \int_0^\infty \cdots \int_0^\infty \prod_{i=1}^n x_i^{a-1} e^{-x_i} \prod_{i<j} |x_i - x_j|^{2c} \, d\mathbf{x}. \qquad (2.7.9)$$

Denoting the last integral by $G_n(a,c)$ and using once again the symmetry in (a,b), we find that

$$\frac{G_n(a,c)}{\prod_{j=0}^{n-1} \Gamma(a+jc)} = \frac{G_n(b,c)}{\prod_{j=0}^{n-1} \Gamma(b+jc)}.$$

We denote the common value by $D_n(c)$ and note that $D_1(c) \equiv 1$. Returning to (2.7.9), we obtain

$$S_n(a,b,c) = \prod_{j=0}^{n-1} \frac{\Gamma(a+jc)\Gamma(b+jc)}{\Gamma(a+b+[n-1+j]c)} D_n(c). \qquad (2.7.10)$$

To complete the proof of Theorem 2.7.1 we need to evaluate $D_n(c)$. This will be done by using the following two lemmas, which are left as Exercises 31 and 32.

Lemma 2.7.3 *For any function f continuous on the interval $[0,1]$,*

$$\lim_{a \to 0+} \int_0^1 a\, t^{a-1} f(t)\, dt = f(0).$$

Lemma 2.7.4 *For any symmetric function $f(\mathbf{x})$,*

$$\int_{C_n} f(\mathbf{x})\, d\mathbf{x} = n! \int_0^1 \int_{x_n}^1 \cdots \int_{x_2}^1 f(\mathbf{x})\, dx_1\, dx_2 \cdots dx_n.$$

It follows that

$$\lim_{a\to0+}\frac{a}{n!}S_n(a,b,c)=\lim_{a\to0+}a\int_0^1 x_n^{a-1}(1-x_n)^{b-1}$$

$$\times\int_{x_n}^1\cdots\int_{x_2}^1\prod_{i=1}^{n-1}x_i^{a-1}(1-x_i)^{b-1}\prod_{i<j\le n}|x_i-x_j|^{2c}\,d\mathbf{x}$$

$$=\int_0^1\int_{x_{n-1}}^1\cdots\int_{x_2}^1\prod_{i=1}^{n-1}x_i^{2c-1}(1-x_i)^{b-1}\prod_{i<j<n}|x_1-x_j|^{2c}\,dx_1\cdots dx_n$$

$$=\frac{1}{(n-1)!}S_{n-1}(2c,b,c).\tag{2.7.11}$$

Since $\lim_{a\to0}a\Gamma(a)=1$, it follows from (2.7.11) and (2.7.10) that

$$D_n(c)=\frac{n\,\Gamma(nc)}{\Gamma(c)}D_{n-1}(c)=\frac{\Gamma(nc+1)}{\Gamma(c+1)}D_{n-1}(c).$$

But $D_1=1$, so

$$D_n(c)=\prod_{j=1}^n\frac{\Gamma(jc+1)}{\Gamma(c+1)}.$$

Combining this with (2.7.10) gives (2.7.3).

The relationship (2.7.8), together with (2.7.3) and the fact that $I_0=S_n(a,b,c)$, gives Aomoto's generalization of (2.7.3):

$$\int_0^1\cdots\int_0^1\prod_{i=1}^k x_i\prod_{i=1}^n x_i^{a-1}(1-x_i)^{b-1}\prod_{1\le i<j\le n}|x_i-x_j|^{2c}\,dx_1\cdots dx_n$$

$$=\prod_{i=1}^k\frac{a+(n-j)c}{a+b+(2n-j-1)c}S_n(a,b,c)$$

$$=\prod_{i=1}^k\frac{a+(n-j)c}{a+b+(2n-j-1)c}\prod_{j=0}^{n-1}\frac{\Gamma(a+jc)\Gamma(b+jc)\Gamma(1+[j+1]c)}{\Gamma(a+b+[n-1+j]c)\Gamma(1+c)}$$

if

$$\mathrm{Re}\,a>0,\quad \mathrm{Re}\,b>0,\quad \mathrm{Re}\,c>\max\left\{-\frac{1}{n},-\frac{\mathrm{Re}\,a}{n-1},-\frac{\mathrm{Re}\,b}{n-1}\right\}.$$

2.8 The zeta function

The zeta function is of particular importance in number theory. We mention it briefly here because its functional equation is closely connected with the gamma function.

For $\mathrm{Re}\, z > 1$, the *Riemann zeta function* is defined by

$$\zeta(z) = \sum_{n=1}^{\infty} \frac{1}{n^z}. \qquad (2.8.1)$$

The uniqueness of prime factorization implies that each n^{-z} occurs exactly once in the product over all primes p of the series

$$1 + \frac{1}{p^z} + \frac{1}{p^{2z}} + \cdots + \frac{1}{p^{mz}} + \cdots = \left(1 - \frac{1}{p^z}\right)^{-1},$$

so we obtain Euler's product formula

$$\zeta(z) = \prod_{p\ \mathrm{prime}} \left(1 - \frac{1}{p^z}\right)^{-1}, \quad \mathrm{Re}\, z > 1. \qquad (2.8.2)$$

Euler evaluated the zeta function at even positive integer z [126]:

$$\zeta(2m) = \frac{(-1)^{m-1}}{2} \frac{(2\pi)^{2m}}{(2m)!} B_{2m}, \qquad (2.8.3)$$

where the Bernoulli numbers are given by (2.5.4). In particular,

$$1 + \frac{1}{4} + \frac{1}{9} + \cdots + \frac{1}{n^2} + \cdots = \frac{\pi^2}{6}; \qquad (2.8.4)$$

$$1 + \frac{1}{16} + \frac{1}{81} + \cdots + \frac{1}{n^4} + \cdots = \frac{\pi^4}{90}; \qquad (2.8.5)$$

$$1 + \frac{1}{64} + \frac{1}{729} + \cdots + \frac{1}{n^6} + \cdots = \frac{\pi^6}{945}. \qquad (2.8.6)$$

The evaluation (2.8.3) follows from the product formula (2.2.9) for sine: taking the logarithm gives

$$\log(\sin \pi x) = \log(\pi x) + \sum_{n=1}^{\infty} \log\left(1 - \frac{x^2}{n^2}\right).$$

Differentiating both sides gives

$$\frac{\pi \cos \pi x}{\sin \pi x} = \frac{1}{x} + \sum_{n=1}^{\infty} \frac{2x}{x^2 - n^2} = \frac{1}{x} + \sum_{n=1}^{\infty} \left[\frac{1}{x+n} + \frac{1}{x-n}\right]. \qquad (2.8.7)$$

The function

$$f(x) = \frac{\pi \cos \pi x}{\sin \pi x} - \frac{1}{x}$$

is holomorphic near the origin. It follows from (2.8.7) that the derivative

$$f^{(k)}(x)\big|_{x=0} = \sum_{n=1}^{\infty} (-1)^k k! \left[\frac{1}{(x+n)^{k+1}} + \frac{1}{(x-n)^{k+1}} \right]\Big|_{x=0}$$

$$= \begin{cases} 0, & k = 2m; \\ -2\sum_{n=1}^{\infty} \dfrac{(2m-1)!}{n^{2m}}, & k = 2m-1. \end{cases}$$

Therefore the McLaurin expansion is

$$\frac{\pi \cos \pi x}{\sin \pi x} - \frac{1}{x} = -2 \sum_{m=1}^{\infty} \zeta(2m) x^{2m-1}. \tag{2.8.8}$$

On the other hand, expressing $\cos \pi x$ and $\sin \pi x$ in terms of exponentials and making use of (2.5.4) gives

$$\frac{\pi \cos \pi x}{\sin \pi x} - \frac{1}{x} = i\pi \frac{e^{i\pi x} + e^{-i\pi x}}{e^{i\pi x} - e^{-i\pi x}} - \frac{1}{x}$$

$$= 2i\pi \left[\frac{1}{e^{2\pi ix} - 1} + \frac{1}{2} - \frac{1}{2\pi ix} \right]$$

$$= 2i\pi \sum_{m=1}^{\infty} \frac{B_{2m}}{(2m)!} (2\pi ix)^{2m-1}. \tag{2.8.9}$$

Comparing coefficients of x^{2m-1} in the expansions (2.8.8) and (2.8.9) gives (2.8.3).

A change of variables in the integral defining $\Gamma(z)$ shows that

$$\frac{1}{n^z} = \frac{1}{\Gamma(z)} \int_0^\infty e^{-nt} t^z \frac{dt}{t}.$$

Therefore

$$\zeta(z) = \frac{1}{\Gamma(z)} \int_0^\infty \frac{e^{-t}}{1 - e^{-t}} t^z \frac{dt}{t} = \frac{1}{\Gamma(z)} \int_0^\infty \frac{t^{z-1} dt}{e^t - 1}.$$

The zeta function can be extended so as to be meromorphic in the plane. Set

$$f(z) = \int_C \frac{(-t)^{z-1} dt}{e^t - 1},$$

where C comes from $+\infty$ along the "lower" real axis ($\arg t = -2\pi$), circles zero in the negative (clockwise) direction at distance $\delta > 0$, and returns to $+\infty$ along the "upper" real axis ($\arg t = 0$); we choose the branch of $\log(-t)$ that is real for $t < 0$. The function f is entire. If $\mathrm{Re}\, z > 0$, we may let $\delta \to 0$ and evaluate the integral by the residue theorem to conclude that

$$f(z) = \left[e^{-i\pi z} - e^{i\pi z} \right] \Gamma(z)\zeta(z) = -2i\sin \pi z\, \Gamma(z)\zeta(z).$$

(See Appendix A.) Thus

$$\zeta(z) = -\frac{f(z)}{2i\,\Gamma(z)\sin\pi z}. \qquad (2.8.10)$$

This provides the analytic continuation to $\operatorname{Re} z \le 1$ and shows that the only pole of ζ in that half-plane, hence in $\mathbf{C}$, is a simple pole at $z = 1$.

The *functional equation* for the zeta function can be obtained from (2.8.10). We evaluate the function f by expanding the circle in the path of integration. The integrand has simple poles at $t = \pm 2n\pi i$, $n \in \mathbf{N}$, and the residue of $1/(e^t - 1)$ at each pole is 1. The residues of the integrand of f at $2n\pi i$ and $-2n\pi i$ sum to

$$(2n\pi)^{z-1} i (e^{-\frac{1}{2}\pi zi} - e^{\frac{1}{2}\pi zi}) = 2^z (n\pi)^{z-1}\sin\left(\frac{1}{2}\pi z\right).$$

Suppose now that $\operatorname{Re} z < 0$, so $\operatorname{Re}(z-1) < -1$. Letting the circle expand (between successive poles), we get

$$f(z) = (-2\pi i)\cdot 2^z \pi^{z-1}\sin\left(\frac{1}{2}\pi z\right)\sum_{n=1}^{\infty}\frac{1}{n^{1-z}}$$

$$= -i(2\pi)^z\, 2\sin\left(\frac{1}{2}\pi z\right)\zeta(1-z). \qquad (2.8.11)$$

This holds for other values by analytic continuation. Combining (2.8.10) and (2.8.11), we obtain the functional equation

$$\zeta(1-z) = \frac{2}{(2\pi)^z}\cos\left(\frac{1}{2}\pi z\right)\Gamma(z)\zeta(z). \qquad (2.8.12)$$

2.9 Exercises

1. For $\operatorname{Re} a > 0$ and $\operatorname{Re} b > 0$, show that

$$\int_0^1 t^{a-1}(1-t^2)^{b-1}\,dt = \frac{1}{2}\,\mathrm{B}\left(\frac{1}{2}a, b\right).$$

2. For $\operatorname{Re} a > 0$ and $\operatorname{Re} b > 0$, show that

$$\int_0^{\pi/2} \sin^{a-1}\theta\cos^{b-1}\theta\,d\theta = \frac{1}{2}\,\mathrm{B}\left(\frac{1}{2}a, \frac{1}{2}b\right).$$

3. Prove the functional relation

$$\mathrm{B}(a,b) = \frac{a+b}{b}\,\mathrm{B}(a, b+1).$$

4. Complete the proof of Proposition 2.1.3 by showing that the error in the appproximation is $O(x^{a-1})$.
5. Carry out the integration in the proof of Theorem 2.2.3.
6. Verify the contour integral representation of the beta function for $\mathrm{Re}\, a > 0$ and any complex b:

$$B(a,b)\frac{\sin \pi b}{b} = \frac{\Gamma(a)}{\Gamma(a+b)\Gamma(1-b)} = \frac{1}{2\pi i}\int_C s^{a-1}(s-1)^{b-1}\,ds.$$

Here the contour C is a counter-clockwise loop that passes through the origin and encloses the point $s = 1$. We take the arguments of s and $s-1$ to be zero for $s > 1$. Hint: assume first that $\mathrm{Re}\, b > 0$ and move the contour to run along the interval $[0,1]$ and back.

7. Use (2.2.7) to verify that for $\mathrm{Re}\, a < 1$ and $\mathrm{Re}\,(a+b) > 0$,

$$\frac{\Gamma(a+b)}{\Gamma(a)\Gamma(b)} = (a+b-1)\frac{e^{-i\pi a}}{2\pi i}\int_C t^{-a}(1+t)^{-b}\,dt,$$

where the curve C runs from $+\infty$ to 0 along the upper edge of the cut on $[0,\infty)$ and returns to $+\infty$ along the lower edge, and the principal branch of t^{-a} is taken along the upper edge.

8. Use (2.2.2) and the evaluation of $\Gamma(\frac{1}{2})$ to prove *Wallis's formula* [417]:

$$\frac{\pi}{4} = \frac{2}{3}\cdot\frac{4}{3}\cdot\frac{4}{5}\cdot\frac{6}{5}\cdots\frac{2n}{2n+1}\cdot\frac{2n+2}{2n+1}\cdots$$

9. One of the verifications of the evaluation of $\Gamma(\frac{1}{2})$ obtained the identity

$$\int_{-\infty}^{\infty} e^{-x^2}\,dx = \sqrt{\pi}$$

by evaluating

$$\int_{-\infty}^{\infty}\int_{-\infty}^{\infty} e^{-(x^2+y^2)}\,dx\,dy$$

in polar coordinates. Turned around, this can be viewed as a way of evaluating the length of the unit circle. Use a "polar coordinate" computation of the n-dimensional version of this integral to show that the $(n-1)$-volume (area) of the unit $(n-1)$-sphere is $2\pi^{n/2}/\Gamma(\frac{1}{2}n)$.

10. Use the Wielandt theorem or the Bohr–Mollerup theorem to prove Gauss's multiplication theorem, the first line of (2.3.6). Hint: replace z by z/m in the right side and define $G(z)$ as the result.

11. For real $y \neq 0$, show that

$$|\Gamma(iy)|^2 = \frac{\pi}{y\sinh \pi y}, \qquad \left|\Gamma\left(\frac{1}{2}+iy\right)\right|^2 = \frac{\pi}{\cosh \pi y}.$$

12. If x and y are real, prove that

$$\left|\frac{\Gamma(x)}{\Gamma(x+iy)}\right|^2 = \prod_{k=0}^{\infty}\left\{1+\frac{y^2}{(x+k)^2}\right\}, \quad x \neq 0,-1,-2,\ldots,$$

and hence $|\Gamma(x+iy)| \leq |\Gamma(x)|$.

13. If a and b are not negative integers, show that

$$\prod_{k=1}^{\infty}\frac{k(a+b+k)}{(a+k)(b+k)} = \frac{\Gamma(a+1)\Gamma(b+1)}{\Gamma(a+b+1)}.$$

14. Prove (2.5.5).

15. Prove (2.5.9).

16. Find the value

$$\int_0^{\infty} t^{x-1}e^{-\lambda t\cos\theta}\cos(\lambda t\sin\theta)\,dt,$$

where $\lambda > 0, x > 0$, and $-\frac{1}{2}\pi < \theta < \frac{1}{2}\pi$.

17. Let z be positive, and integrate $\pi w^{-z}\mathrm{cosec}\,\pi z/\Gamma(1-z)$ around the rectangle with vertices $c \pm iR, -R \pm iR$, where $c > 0$. Show that

$$\frac{1}{2\pi i}\int_{c-i\infty}^{c+i\infty} w^{-z}\Gamma(z)\,dz = e^{-w}.$$

Prove that this formula holds for $|\arg w| \leq \frac{1}{2}\pi - \delta$, where $\delta > 0$.

18. (Saalschütz) Prove that for $\mathrm{Re}\,z < 0$,

$$\Gamma(z) = \int_0^{\infty} t^{z-1}\left\{e^{-t} - 1 + t - \frac{t^2}{2!} + \cdots + (-1)^{k+1}\frac{t^k}{k!}\right\}dt,$$

where k is an integer between $\mathrm{Re}(-z)$ and $\mathrm{Re}(-z-1)$.

19. Show that for $s > 1$,

$$\log\zeta(s) = \sum_{p\text{ prime}}\sum_{m=1}^{\infty}\frac{1}{mp^{ms}}.$$

20. (a) The *Hurwitz zeta function* is defined by

$$\zeta(x,s) = \sum_{n=0}^{\infty}\frac{1}{(n+x)^s} \quad \text{for } x > 0.$$

Show that $\zeta(x+1,s) = \zeta(x,s) - x^{-s}$, and hence

$$\left(\frac{\partial\zeta(x+1,s)}{\partial s}\right)_{s=0} - \left(\frac{\partial\zeta(x,s)}{\partial s}\right)_{s=0} = \log x.$$

(b) Establish that for $\mathrm{Re}\,s > 1$,

$$\frac{\partial^2 \zeta(x,s)}{\partial x^2} = s(s+1)\sum_{n=0}^{\infty} \frac{1}{(n+x)^{s+2}}$$

and

$$\frac{d^2}{dx^2}\left(\frac{\partial \zeta(x,s)}{\partial s}\right)_{s=0} = \sum_{n=0}^{\infty} \frac{1}{(x+n)^2}.$$

(c) Show that the results in (a) and (b), together with (2.4.2), imply that

$$\left(\frac{\partial \zeta(x,s)}{\partial s}\right)_{s=0} = C + \log\Gamma(x).$$

(d) Prove that

$$\zeta'(0) = -\frac{1}{2}\log 2\pi.$$

Use this and the result in (c) to prove *Lerch's theorem*:

$$\left(\frac{\partial \zeta(x,s)}{\partial s}\right)_{s=0} = \log\frac{\Gamma(x)}{\sqrt{2\pi}}.$$

21. First prove that

$$\zeta(x,s) = \frac{1}{\Gamma(s)}\int_0^{\infty} \frac{t^{s-1}e^{-xt}}{1-e^{-t}}\,dt.$$

Then use the idea of Hankel's loop integral for the gamma function to derive the contour integral representation

$$\zeta(x,s) = \frac{e^{-i\pi s}\Gamma(1-s)}{2\pi i}\int_C \frac{t^{s-1}e^{-xt}}{1-e^{-t}}\,dt,$$

where C starts at infinity on the positive real axis, encircles the origin once in the positive direction, excluding the points $\pm 2n\pi i, n \geq 1$, and returns to positive infinity.

22. Deduce from (2.6.2) that for $x > 0$, $\Gamma(x)$ has a single minimum, which lies between 1 and 2.

23. Show that for $z \neq 0, \pm 1, \pm 2, \ldots$,

$$\psi(1-z) - \psi(z) = \pi\cot\pi z$$

and for $2z \neq 0, -1, -2, \ldots$

$$\psi(z) + \psi\left(z + \frac{1}{2}\right) + 2\log 2 = 2\psi(2z).$$

24. Prove that

$$\psi\left(\frac{1}{2}\right) = -\gamma - 2\log 2, \qquad \psi'\left(\frac{1}{2}\right) = \frac{1}{2}\pi^2.$$

25. Show that for $\operatorname{Re} z > 0$,

$$\psi(z) = \int_0^\infty \left(\frac{e^{-t}}{t} - \frac{e^{-zt}}{1 - e^{-t}} \right) dt.$$

This is known as Gauss's formula. Deduce that

$$\psi(z+1) = \frac{1}{2z} + \log z - \int_0^\infty \left[\frac{1}{2} \coth\left(\frac{1}{2}t\right) - \frac{1}{t} \right] e^{-zt} dt.$$

26. Show that $\gamma(a,z)/z^a \Gamma(a)$ is entire in both a and z. Furthermore, prove that

$$\frac{\gamma(a,z)}{z^a \Gamma(a)} = e^{-z} \sum_{n=0}^\infty \frac{z^n}{\Gamma(a+n+1)}.$$

27. Prove that

$$\frac{2}{\pi} \int_0^\infty \frac{e^{-zt^2}}{1+t^2} dt = e^z [1 - \operatorname{erf}(\sqrt{z})].$$

28. The generalized exponential integral is defined by

$$E_n(z) = \int_1^\infty \frac{e^{-zt}}{t^n} dt, \qquad n = 1, 2, \ldots$$

Show that

(a) $E_n(z) = z^{n-1} \Gamma(1-n, z);$

(b) $E_n(z) = \int_z^\infty E_{n-1}(t)\, dt = \cdots = \int_z^\infty \cdots \int_{t_{n-1}}^\infty \frac{e^{-t_n}}{t_n} dt_n \ldots dt_1;$

(c) $E_n(z) = \frac{e^{-z}}{(n-1)!} \int_0^\infty \frac{e^{-t} t^{n-1}}{z+t} dt.$

29. Let $0 < \lambda < 1$ and $k = 1, 2, \ldots$ Put

$$f(k) = \frac{\Gamma(k+\lambda)}{\Gamma(k+1)} (k+a)^{1-\lambda}$$

and

$$g(k) = \frac{f(k+1)}{f(k)}.$$

(a) Show that

$$\lim_{k\to\infty} g(k) = \lim_{k\to\infty} f(k) = 1$$

and

$$g(k) = \frac{k+\lambda}{k+1} \left(\frac{k+a+1}{k+a} \right)^{1-\lambda}.$$

Considering k as a continuous variable, show also that

$$g'(k) = \frac{A_k(\lambda; a)}{(k+a)^{2-\lambda}(k+1)^2(k+a+1)^\lambda},$$

where

$$A_k(\lambda;\alpha) = (1-\lambda)(-\lambda k + 2\alpha k - \lambda + \alpha^2 + \alpha).$$

(b) When $\alpha = 0$, prove (i) $g'(k)$ is negative for $0 < \lambda < 1$ and $k = 1, 2, \ldots$,
(ii) $g(k) > 1$ and $f(k) < 1$ for $k = 1, 2, \ldots$, and (iii)

$$\frac{\Gamma(k+\lambda)}{\Gamma(k+1)} < \frac{1}{k^{1-\lambda}}.$$

(c) When $\alpha = 1$, prove (i) $g'(k)$ is positive for $0 < \lambda < 1$ and $k = 1, 2, \ldots$,
(ii) $g(k) < 1$ and $f(k) > 1$ for $k = 1, 2, \ldots$, and (iii)

$$\frac{\Gamma(k+\lambda)}{\Gamma(k+1)} > \frac{1}{(k+1)^{1-\lambda}}.$$

These inequalities were first given by Gautschi [155], but the argument outlined here is taken from Laforgia [236].

30. Use the symmetry of the function $w(\mathbf{x})$ to prove Lemma 2.7.2.

31. Prove Lemma 2.7.3.

32. Prove Lemma 2.7.4. Hint: partition the domain C_n and use symmetry.

33. Use the product formula (2.2.9) to prove (2.8.4).

34. Use the product formula (2.2.9) to prove (2.8.5).

35. Multiply both sides of (2.5.4) by $t(e^t - 1)$ and show that B_{2m} satisfies

$$B_{2m} = -\left[\binom{2m}{2}\frac{B_{2m-2}}{3} + \binom{2m}{4}\frac{B_{2m-4}}{5} + \cdots + \frac{B_0}{2m+1}\right] + \frac{1}{2}$$

for $m = 1, 2, \ldots$, with $B_0 = 1$.

36. Compute the Bernoulli numbers B_2, B_4, and B_6, and use (2.8.3) to verify the equations (2.8.4), (2.8.5), and (2.8.6).

37. The *Bernoulli polynomials* $\{B_n(x)\}$ are defined by the identity

$$\frac{te^{xt}}{e^t - 1} = \sum_{n=0}^{\infty} \frac{B_n(x)}{n!} t^n, \quad |t| < 2\pi.$$

Thus $B_{2m}(0) = B_{2m}$. Set $B_{-1}(x) = 0$. Prove the identities

(a) $B_n'(x) = n B_{n-1}(x)$;

(b) $\displaystyle\int_0^1 B_0(x)\,dx = 1$;

(c) $\displaystyle\int_0^1 B_n(x)\,dx = 0, \quad n \neq 0$;

(d) $B_n(x+1) = B_n(x) + n x^{n-1}$;

(e) $1 + 2^n + 3^n + \cdots + m^n = \dfrac{B_{n+1}(m+1) - B_{n+1}(0)}{n+1}.$

38. Verify that the first six Bernoulli polynomials are

$$B_0(x) = 1;$$

$$B_1(x) = \frac{2x-1}{2};$$

$$B_2(x) = \frac{6x^2 - 6x + 1}{6};$$

$$B_3(x) = \frac{2x^3 - 3x^2 + x}{2};$$

$$B_4(x) = \frac{30x^4 - 60x^3 + 30x^2 - 1}{30};$$

$$B_5(x) = \frac{6x^5 - 15x^4 + 10x^3 - x}{6}.$$

39. The *Euler polynomials* $\{E_n(x)\}$ are defined by the identity

$$\frac{2e^{xt}}{e^t + 1} = \sum_{n=0}^{\infty} \frac{E_n(x)}{n!} t^n, \quad |t| < \pi.$$

Prove that

(a) $E'_n(x) = n E_{n-1}(x);$

(b) $E_n(x+1) + E_n(x) = 2x^n;$

(c) $1 - 2^n + 3^n - \cdots + (-1)^{m-1} m^n = \dfrac{E_n(0) + (-1)^{m-1} E_n(m+1)}{2}.$

40. Verify that the first six Euler polynomials are

$$E_0(x) = 1;$$

$$E_1(x) = \frac{2x-1}{2};$$

$$E_2(x) = x^2 - x;$$

$$E_3(x) = \frac{4x^3 - 6x^2 + 1}{4};$$

$$E_4(x) = x^4 - 2x^3 + x;$$

$$E_5(x) = \frac{2x^5 - 5x^4 + 5x^2 - 1}{2}.$$

41. The *Euler numbers* $\{E_n\}$ are defined by

$$E_n = 2^n E_n\left(\frac{1}{2}\right).$$

Prove that

$$\sum_{n=0}^{\infty} \frac{E_n}{n!} t^n = \frac{2}{e^t + e^{-t}}, \quad |t| < \frac{\pi}{2}.$$

42. Prove that $\zeta(-2m) = 0, \; m = 1, 2, 3, \ldots$

2.10 Remarks

The history and properties of the gamma function are discussed in detail in the book by Nielsen [303]; see also articles by Davis [95], Dutka [110], and Gautschi [157], and the books by Artin [17], Campbell [61], and Godefroy [166]. Copson's text [86] has an extensive list of exercises with identities involving the gamma function and Euler's constant.

Selberg's generalization of the beta integral, and its further elaborations, have been utilized in a number of areas, including random matrix theory, statistical mechanics, combinatorics, and integrable systems. See the chapter on the Selberg integral in [10] and the extensive survey by Forrester and Warnaar [139]. For a probabilistic proof of Selberg's formula, see the book by Mehta [280].

The literature on the zeta function and its generalizations is copious. See in particular the books of Titchmarsh [399], Edwards [113], Ivić [195], Patterson [319], and Motohashi [292]. The celebrated *Riemann hypothesis* is that all the nontrivial zeros of $\zeta(s)$ lie on the line $\{\mathrm{Re}\, s = \frac{1}{2}\}$. (The "trivial zeros" are at $s = -2, -4, -6, \ldots$) A number of consequences concerning analytic number theory would follow from the truth of the Riemann hypothesis.

3

Second-order differential equations

As noted in Chapter 1, most of the functions commonly known as "special functions" are solutions of second-order linear differential equations. These equations occur naturally in certain physical and mathematical contexts. In a sense there are exactly two (families of) equations in question: the confluent hypergeometric equation (Kummer's equation)

$$xu''(x) + (c - x)u'(x) - au(x) = 0, \tag{3.0.1}$$

with indices (a, c), and the hypergeometric equation

$$x(1 - x)u''(x) + [c - (a + b + 1)x]\, u'(x) - abu(x) = 0, \tag{3.0.2}$$

with indices (a, b, c), where a, b, c are constants. The various other equations (Bessel, Whittaker, Hermite, Legendre, ...) are obtained from these by specialization, by standard transformations, or by analytic continuation in the independent variable.

In this chapter, we give a brief general treatment of some questions concerning second-order linear differential equations, starting with gauge transformations, L^2 symmetry with respect to a weight, and the Liouville transformation.

The basic existence and uniqueness theorems are proved, followed by a discussion of the Wronskian, independence of solutions, comparison theorems, and zeros of solutions.

A natural classification question is treated: classifying symmetric problems whose eigenfunctions are polynomials. This question leads, up to certain normalizations, to the equations (3.0.1) and (3.0.2). General results on local maxima and minima of solutions of homogeneous equations are obtained and applied to some of these polynomials.

Another source of equations related to (3.0.1) and (3.0.2) is physics: problems involving the Laplace operator, when one seeks to find solutions by separating variables in special coordinate systems. We discuss the various equations that arise in this way and how they are related to (3.0.1) and (3.0.2).

3.1 Transformations and symmetry

Throughout this chapter, $p, q, r, f, u, v, \ldots$, will denote real functions defined on the finite or infinite open real interval $I = (a,b) = \{x : a < x < b\}$, and we assume that $p(x) > 0$, for all $x \in I$. All functions will be assumed to be continuous and to have continuous first and second derivatives as needed.

The general linear second-order differential equation on the interval I is

$$p(x)u''(x) + q(x)u'(x) + r(x)u(x) = f(x). \qquad (3.1.1)$$

The corresponding *homogeneous equation* is the equation with the right-hand side $f \equiv 0$:

$$p(x)u''(x) + q(x)u'(x) + r(x)u(x) = 0. \qquad (3.1.2)$$

The associated differential *operator* is

$$L = p(x)\frac{d^2}{dx^2} + q(x)\frac{d}{dx} + r(x). \qquad (3.1.3)$$

In (3.1.3), the functions p, q, and r are identified with the operations of multiplication by p, by q, and by r.

A *gauge transformation* of (3.1.1) is a transformation of the form

$$u(x) = \varphi(x)v(x), \quad \varphi(x) \neq 0. \qquad (3.1.4)$$

The function u satisfies (3.1.1) if and only if v satisfies

$$p(x)v''(x) + \left[2p(x)\frac{\varphi'(x)}{\varphi(x)} + q(x)\right]v'(x)$$
$$+ \left[p(x)\frac{\varphi''(x)}{\varphi(x)} + q(x)\frac{\varphi'(x)}{\varphi(x)} + r(x)\right]v(x) = \frac{f(x)}{\varphi(x)}. \qquad (3.1.5)$$

Note that the left-hand side of this equation is not changed if φ is replaced by $C\varphi$, C constant, $C \neq 0$. The corresponding transformed operator is

$$L_\varphi = p(x)\frac{d^2}{dx^2} + \left[2p(x)\frac{\varphi'(x)}{\varphi(x)} + q(x)\right]\frac{d}{dx}$$
$$+ \left[p(x)\frac{\varphi''(x)}{\varphi(x)} + q(x)\frac{\varphi'(x)}{\varphi(x)} + r(x)\right]. \qquad (3.1.6)$$

The usefulness of gauge transformations comes from the fact that the homogeneous linear first-order differential equation

$$\varphi'(x) = h(x)\varphi(x) \qquad (3.1.7)$$

always has a solution

$$\varphi(x) = \exp\left\{ \int_{x_0}^{x} h(y)\,dy \right\}, \tag{3.1.8}$$

where x_0 is any point of the interval I. This solution has no zeros in the interval. Note that if ψ is a second solution of (3.1.7), then the quotient ψ/φ has derivative 0 and is therefore constant.

In particular, a gauge transformation can be used to eliminate the first-order term qu' of (3.1.1) by taking φ such that $\varphi'/\varphi = -q/2p$. A second use is to *symmetrize* the operator (3.1.3). Suppose that $w > 0$ on I. The associated weighted L^2 space L_w^2 consists of all measurable real-valued functions f such that

$$\int_a^b f(x)^2\,w(x)\,dx < \infty.$$

The *inner product* (f, g) between two such functions is

$$(f, g) = (f, g)_w = \int_a^b f(x)\,g(x)\,w(x)\,dx.$$

The operator L of (3.1.3) is said to be *symmetric* with respect to the *weight function* w if

$$(Lu, v) = (u, Lv)$$

for every pair of twice continuously differentiable functions u, v that vanish outside some closed subinterval of I. The proofs of the following propositions are sketched in the exercises.

Proposition 3.1.1 *The operator L is symmetric with respect to the weight w if and only if it has the form*

$$L = p\frac{d^2}{dx^2} + \frac{(pw)'}{w}\frac{d}{dx} + r = \frac{1}{w}\frac{d}{dx}\left(pw\frac{d}{dx}\right) + r. \tag{3.1.9}$$

Proposition 3.1.2 *If L has the form (3.1.3) then there is a weight function w, unique up to a multiplicative constant, such that L is symmetric with respect to w.*

Proposition 3.1.3 *Given an operator (3.1.3) and a weight function w on the interval I, there is a gauge transformation (3.1.4) such that the corresponding operator L_φ is symmetric with respect to w.*

An invertible transformation T from an L^2 space with weight w_1 to an L^2 space with weight w_2 is said to be *unitary* if

$$(Tf, Tg)_{w_2} = (f, g)_{w_1}$$

for every pair f, g in $L^2_{w_1}$. Operators L_1 and L_2 in the respective spaces are said to be *unitarily equivalent* by T if

$$L_2 = TL_1 T^{-1}.$$

Proposition 3.1.4 *An operator symmetric with respect to a weight w on an interval I is unitarily equivalent, by a gauge transformation, to an operator that is symmetric with respect to the weight 1 on the interval I.*

A second useful method for transforming a differential equation like (3.1.1) is to make a change of the independent variable. If $y = y(x)$ and $u(x) = v(y(x))$, then

$$u'(x) = y'(x)v'(y(x)), \quad u''(x) = \left[y'(x)\right]^2 v''(y(x)) + y''(x)v'(y(x)).$$

In particular, we may eliminate the coefficient p by taking

$$y(x) = \int_{x_0}^x \frac{dt}{\sqrt{p(t)}}. \tag{3.1.10}$$

Then (3.1.1) becomes

$$v'' + \left[\frac{q}{\sqrt{p}} - \frac{p'}{2\sqrt{p}}\right] v' + rv = f.$$

This involves an abuse of notation: the primes on v refer to derivatives with respect to y, while the prime on p refers to the derivative with respect to x. To rectify this we consider p, q, r, and f as functions of $y = y(x)$ by taking $p(x) = p_1(y(x))$, etc. The previous equation becomes

$$v''(y) + \left[\frac{q_1(y)}{\sqrt{p_1(y)}} - \frac{p_1'(y)}{2\sqrt{p_1(y)}}\right] v'(y) + r_1(y)v(y) = f_1(y). \tag{3.1.11}$$

If we then eliminate the first-order term by a gauge transformation, the resulting composite transformation is known as the *Liouville transformation*.

3.2 Existence and uniqueness

The following standard fact about equations of the form (3.1.2) is crucial.

Theorem 3.2.1 *The set of solutions of the homogeneous equation*

$$p(x)u''(x) + q(x)u'(x) + r(x)u(x) = 0 \tag{3.2.1}$$

is a vector space of dimension 2.

Proof The set of solutions of (3.2.1) is a vector space, since a linear combination of solutions is again a solution (the "superposition principle").

To simplify notation, let us assume that the interval I contains the point $x = 0$. A gauge transformation is an invertible linear map, so it does not change the dimension. Therefore we may assume that $q \equiv 0$ and write the equation in the form

$$u''(x) = s(x)u(x), \qquad s(x) = -\frac{r(x)}{p(x)}. \tag{3.2.2}$$

We show first that there are two solutions u and v characterized by the conditions

$$u(0) = 1, \quad u'(0) = 0; \qquad v(0) = 0, \quad v'(0) = 1. \tag{3.2.3}$$

Solutions of (3.2.2) that satisfy these conditions would be solutions of the integral equations

$$u(x) = 1 + \int_0^x \int_0^y s(z)u(z)\,dz\,dy,$$
$$v(x) = \int_0^x \left\{ 1 + \int_0^y s(z)v(z)\,dz \right\} dy, \tag{3.2.4}$$

respectively. Conversely, solutions of these integral equations would be solutions of (3.2.2).

The equations (3.2.4) can be solved by the *method of successive approximations*. Let $u_0 = 1$, $v_0 = 0$, and define inductively

$$u_{n+1}(x) = 1 + \int_0^x \int_0^y s(z)u_n(z)\,dz\,dy;$$
$$v_{n+1}(x) = \int_0^x \left\{ 1 + \int_0^y s(z)v_n(z)\,dz \right\} dy, \qquad n \geq 0.$$

It is enough to show that each of the sequences $\{u_n\}$ and $\{v_n\}$ converges uniformly on each bounded closed subinterval $J \subset I$. We may assume $0 \in J$. Let

$$C = \sup_{x \in J} |s(x)|.$$

It is easily proved by induction that for $x \in J$,

$$|u_{n+1}(x) - u_n(x)| \leq \frac{C^n x^{2n}}{(2n)!};$$
$$|v_{n+1}(x) - v_n(x)| \leq \frac{C^n |x|^{2n+1}}{(2n+1)!}. \tag{3.2.5}$$

It follows that the sequences $\{u_n\}$ and $\{v_n\}$ are Cauchy sequences. They converge uniformly on J to the desired solutions u and v. The conditions (3.2.3) imply that u and v are linearly independent, so the dimension of the space of solutions of (3.2.2) is at least 2.

Suppose now that w is a solution of (3.2.2). Replacing w by

$$w(x) - w(0)u(x) - w'(0)v(x),$$

we may assume that $w(0) = w'(0) = 0$. The proof can be completed by showing that $w \equiv 0$ on each subinterval J as before. Let

$$M = \sup_{x \in J} |w(x)|.$$

Now

$$w(x) = \int_0^x \int_0^y s(z)w(z)\,dz\,dy.$$

It follows that for $x \in J$,

$$|w(x)| \leq \frac{CMx^2}{2}.$$

Inductively,

$$|w(x)| \leq \frac{C^n M x^{2n}}{(2n)!} \tag{3.2.6}$$

for all n. The right-hand side has limit 0 as $n \to \infty$, so $w(x) = 0$. $\qquad\square$

These arguments lead to a sharpened form of Theorem 3.2.2.

Theorem 3.2.2 *Given a point x_0 in the interval I and two constants c_0, c_1, there is a unique solution of the homogeneous equation*

$$p(x)u''(x) + q(x)u'(x) + r(x)u(x) = 0 \tag{3.2.7}$$

that satisfies the conditions

$$u(x_0) = c_0; \quad u'(x_0) = c_1. \tag{3.2.8}$$

In particular, if $u(x_0) = 0$, then either it is a simple zero, i.e. $u'(x_0) \neq 0$, or else $u \equiv 0$ on I. Moreover, if u is not identically zero, then wherever $r(x) \neq 0$, zeros of u' are simple: if $u'(x)$ and $u''(x)$ both vanish, then (3.2.7) implies that $u(x) = 0$. This proves the following.

Corollary 3.2.3 *If u is a solution of (3.2.7) that does not vanish identically, then any zero of u in I is a simple zero. Moreover, u' has only simple zeros wherever $r \neq 0$.*

In Chapter 12 we will want the higher-order versions of Theorems 3.2.1 and 3.2.2. Consider an equation of order n:

$$p_n(x)u^{(n)}(x) + p_{n-1}(x)u^{(n-1)}(x) + \cdots + p_0(x)u(x) = 0, \tag{3.2.9}$$

with coefficients continuous on an interval (a,b), and with $p_n(x) \neq 0$, $x \in (a,b)$.

Theorem 3.2.4 *The set of solutions of* (3.2.9) *is a vector space of dimension* n. *If* $x_0 \in (a,b)$, *there is a unique solution with specified values*

$$u^{(k)} = c_k, \qquad k = 0, 1, \ldots, n-1. \tag{3.2.10}$$

A proof of this result is outlined in Exercises 12 and 13. We note here that the coefficients may be allowed to have complex values.

3.3 Wronskians, Green's functions, and comparison

Suppose that u_1 and u_2 are two differentiable functions on the interval I. The *Wronskian* $W(u_1, u_2)$ is the function

$$W(u_1, u_2)(x) = \begin{vmatrix} u_1(x) & u_2(x) \\ u_1'(x) & u_2'(x) \end{vmatrix} = u_1(x)u_2'(x) - u_1'(x)u_2(x).$$

Proposition 3.3.1 *Suppose that* u_1 *and* u_2 *are solutions of the homogeneous equation* (3.1.2). *The Wronskian* $W(u_1, u_2)$ *is identically zero if* u_1 *and* u_2 *are linearly dependent, and nowhere zero if* u_1 *and* u_2 *are independent.*

Proof The assumption on u_1 and u_2 implies

$$pW' = p(u_1 u_2'' - u_1'' u_2) = -qW,$$

so W is the solution of the first-order homogeneous equation $W' = -qp^{-1}W$. It follows that W is either identically zero or never zero. Clearly $W \equiv 0$ is implied by linear dependence. Conversely, if $W \equiv 0$ then in any subinterval where $u_1 \neq 0$, we have

$$\left[\frac{u_2}{u_1} \right]' = \frac{W(u_1, u_2)}{u_1^2} = 0.$$

Therefore u_2/u_1 is constant.   □

Let us look for a solution of (3.1.1) that has the form

$$u(x) = \int_{x_0}^{x} G(x,y)f(y)\,dy. \tag{3.3.1}$$

Then

$$u'(x) = G(x,x)f(x) + \int_{x_0}^{x} G_x(x,y)f(y)\,dy.$$

To get rid of $f'(x)$ in taking the second derivative, we need $G(x,x) = 0$. Then

$$u''(x) = G_x(x,x)f(x) + \int_{x_0}^{x} G_{xx}(x,y)f(y)\,dy.$$

Therefore $Lu = f$, provided

$$LG(x, \cdot) = 0, \quad G(x,x) = 0; \quad p(x)G_x(x,x) = 1. \tag{3.3.2}$$

Suppose that u_1 and u_2 are linearly independent homogeneous solutions of $Lu = 0$ on the interval. The first equation in (3.3.2) implies that for each $y \in I$,

$$G(x,y) = v_1(y)u_1(x) + v_2(y)u_2(x).$$

The remaining two conditions in (3.3.2) give linear equations whose solution is

$$G(x,y) = \frac{u_1(y)u_2(x) - u_2(y)u_1(x)}{p(y)\,W(y)}, \tag{3.3.3}$$

where $W = W(u_1, u_2)$ is the Wronskian.

We may now generalize Theorem 3.2.2 to the inhomogeneous case.

Theorem 3.3.2 *Suppose that x_0 is a point of I. For any two real constants c_0, c_1, there is a unique solution u of* (3.1.1) *that satisfies the conditions*

$$u(x_0) = c_0, \quad u'(x_0) = c_1. \tag{3.3.4}$$

Proof   The solution (3.3.1), (3.3.3) satisfies the conditions $u(x_0) = 0$, $u'(x_0) = 0$. We may add to it any linear combination of u_1 and u_2. The Wronskian is not zero, so there is a unique linear combination that yields the conditions (3.3.4).   □

In order to satisfy more general boundary conditions, we look for a solution of the form

$$u(x) = \int_{y<x} G_+(x,y)f(y)\,dy + \int_{y>x} G_-(x,y)f(y)\,dy, \tag{3.3.5}$$

where $G_-(\cdot, y)$ satisfies a condition to the left and $G_+(\cdot, y)$ satisfies a condition to the right, as functions of the first variable. If $u_\pm$ are linearly independent solutions that satisfy such conditions, then $G_\pm(x,y) = v_\pm(y)u_\pm(x)$, and the previous argument shows that

$$G_-(x,y) = \frac{u_+(y)u_-(x)}{p(y)\,W(y)},$$

$$G_+(x,y) = \frac{u_-(y)u_+(x)}{p(y)\,W(y)}. \tag{3.3.6}$$

The Wronskian also plays a role in the proof of the following important result of Sturm [390].

Theorem 3.3.3 (Sturm comparison theorem) *Suppose that u_1 and u_2 are solutions of the equations*

$$p(x)u_j''(x) + q(x)u_j'(x) + r_j(x)u_j(x) = 0, \quad j = 1, 2, \tag{3.3.7}$$

on the interval I, neither u_1 nor u_2 is identically zero, and

$$r_1(x) < r_2(x) \quad \text{for all } x \in I.$$

Suppose that $u_1 = 0$ at points c, d in I, $c < d$. Then $u_2(x) = 0$ at some point x of the interval (c, d).

Proof The assumptions and the conclusion are unchanged under gauge transformations and under division of the equations by p, so we may assume for simplicity that $p \equiv 1$ and $q \equiv 0$. We may assume that u_1 has no zeros in (c, d); otherwise replace d by the first zero. If u_2 has no zeros in (c, d), then up to a change of sign we may assume that u_1 and u_2 are positive in the interval. The Wronskian $W = W(u_1, u_2)$ satisfies

$$W' = (r_1 - r_2) u_1 u_2,$$

so it is nonincreasing on the interval. Our assumptions to this point imply that $u_1'(c) > 0$ and $u_1'(d) < 0$, so

$$W(c) = -u_1'(c) u_2(c) \le 0, \quad W(d) = -u_1'(d) u_2(d) \ge 0.$$

It follows that $W \equiv 0$ on (c, d), so u_2/u_1 is constant. But this implies that u_2 satisfies both of the equations (3.3.7), $j = 1, 2$, which is incompatible with the assumptions that $r_1 < r_2$ and $u_2 \ne 0$ on the interval. $\square$

This proof also serves to prove the following generalization.

Theorem 3.3.4 *Suppose that u_1 and u_2 are solutions of the equations*

$$p(x) u_j''(x) + q(x) u_j'(x) + r_j(x) u_j(x) = 0, \quad j = 1, 2,$$

on an open interval I, neither is identically zero, and

$$r_1(x) < r_2(x) \quad \text{for all } x \in I.$$

Suppose that $u_1(c) = 0$ at a point c in I and that the Wronskian $u_1(x) u_2'(x) - u_2(x) u_1'(x)$ has limit 0 as x approaches one of the endpoints of I. Then $u_2(x) = 0$ at some point between c and that endpoint.

Corollary 3.3.5 *Suppose that $u(x, t)$, $0 \le t < T$, is a solution of the equation*

$$p(x) u''(x, t) + q(x) u'(x, t) + r(x, t) u(x, t) = 0$$

on an open interval I, where the primes denote derivatives with respect to x. Suppose that at a point $a \in I$

$$u(a, t) \equiv 0 \quad \text{or} \quad u'(a, t) \equiv 0.$$

Let $a < x_1(t) < x_2(t) < \cdots$ denote the zeros of $u(x, t)$ to the right of a. If $r(x, t)$ is continuous and increases with t, then $x_k(t)$ decreases as t increases.

Another useful result about zeros is the following.

Theorem 3.3.6 *Suppose that w is positive and r is negative on (c,d), and the real function u satisfies*

$$[w u']'(x) + r(x) u(x) = 0, \quad c < x < d,$$

and is not identically zero. Then u has at most one zero in (c,d).

If uu' is positive in a subinterval $(c, c + \varepsilon)$ or if $\lim_{x \to d} u(x) = 0$, then u has no zeros in (c,d).

Proof Suppose that $u(a) = 0$ for some a in the interval. Replacing u by its negative if necessary, we may assume that $u'(a) > 0$. Then $u(x)$ and $u'(x)$ are positive on some interval $a < x < b \leq d$. The equation shows that wu' is increasing on the interval, so $u' > 0$ on (a,b). Taking b to be maximal with respect to these properties, it is clear that $b = d$. It follows that u has at most one zero in (c,d). This argument also shows that either of the additional conditions implies that there are no zeros. $\qquad\square$

3.4 Polynomials as eigenfunctions

It is of interest to extend the symmetry condition of Section 3.1 to the largest possible "allowable" class of functions. In general, this requires the imposition of *boundary conditions*. Suppose that I is a bounded interval (a,b) and that w, w', p, p', q, and r extend as continuous functions on the closed interval $[a,b]$. Suppose that u and v are twice continuously differentiable on (a,b) and belong to L_w^2, and suppose that u, u', v, v' are continuous on the closed interval. Suppose also that L is symmetric. A previous calculation shows that

$$(Lu, v) - (u, Lv) = \left(p w u' v - p w u v' \right) \Big|_a^b.$$

If pw vanishes at both endpoints then we do not need additional constraints at the boundary; otherwise additional conditions must be imposed on the functions u, v. Similarly, if I is a semi-infinite interval (a, ∞), conditions must be imposed at $x = a$ unless $pw = 0$ at $x = a$.

Suppose that we have symmetry for such a maximal allowable class of functions. An allowable function u that is not identically zero is an *eigenfunction* for L with *eigenvalue* $-\lambda$ if $Lu + \lambda u = 0$.

If u_1 and u_2 are eigenfunctions with different eigenvalues $-\lambda_1$ and $-\lambda_2$, then

$$-\lambda_1(u_1, u_2) = (Lu_1, u_2) = (u_1, Lu_2) = -\lambda_2(u_1, u_2),$$

so $(u_1, u_2) = 0$: u_1 and u_2 are *orthogonal*.

In a variation on a question posed by Routh [346] and Bochner [50] we ask: under what conditions is it the case that the set of eigenfunctions of L includes polynomials of all degrees.

The symmetry condition implies that L has the form (3.1.9):

$$p\frac{d^2}{dx^2} + \frac{(pw)'}{w}\frac{d}{dx} + r.$$

Suppose that there are polynomials of degrees 0, 1, and 2 that are eigenfunctions of L. This is equivalent to assuming that the space of polynomials of degree $\leq k$ is in the domain of L and is taken into itself by L, $k = 0, 1, 2, \ldots$ In particular, polynomials belong to L_w^2, so

$$\int_a^b x^{2n} w(x)\, dx < \infty, \quad n = 0, 1, 2, \ldots \tag{3.4.1}$$

Applying L to the constant function $u_0(x) \equiv 1$ gives $Lu_0 = r$, so r must be constant, and (up to translating the eigenvalues by $-r$) we may take $r = 0$. Applying L to $u_1(x) = x$ gives

$$Lu_1 = \frac{(pw)'}{w} = p' + p\frac{w'}{w}, \tag{3.4.2}$$

so the last expression must be a polynomial of degree at most 1. Taking $u_2(x) = \frac{1}{2}x^2$,

$$Lu_2 = p + x\left(p' + p\frac{w'}{w}\right).$$

Since Lu_2 must be a polynomial of degree at most 2, it follows that p is a polynomial of degree at most 2.

The symmetry condition requires that

$$0 = (Lu, v) - (u, Lv) = \int_a^b \left[pw(u'v - uv')\right]'$$

for every u, v in the domain. As noted above, a necessary condition is that $pw \to 0$ at each endpoint of the interval.

By normalizations (affine maps of the line, multiplication of the weight, the operator, and/or the polynomials by constants), we reduce to five cases, of which two turn out to be vacuous.

Case I: p constant. We take $p(x) \equiv 1$. It follows from (3.4.2) that w'/w has degree at most 1, so we take $w = e^h$, where h is a real polynomial of degree at most 2. After another normalization, $w(x) = e^{-x}$ or $w(x) = e^{\pm x^2}$. In the former case, condition (3.4.1) requires that I be a proper subinterval, but then the condition that $pw = w$ vanish at finite boundary points cannot be met. In the latter case, the endpoint condition forces $I = \mathbf{R} = (-\infty, \infty)$ and condition (3.4.1) forces the sign choice $w(x) = e^{-x^2}$. Thus in this case (3.1.9) is the operator

$$L = \frac{d^2}{dx^2} - 2x\frac{d}{dx} \quad \text{in} \quad L_w^2(\mathbf{R}); \quad w(x) = e^{-x^2}. \tag{3.4.3}$$

This operator takes the space of polynomials of degree $\leq n$ to itself for each n, and it follows that for each n there is a polynomial ψ_n of degree n that is an eigenfunction. Consideration of the degree n terms of ψ_n and of $L\psi_n$ shows that the eigenvalue is $-2n$:

$$L\psi_n + 2n\,\psi_n = 0.$$

Since the eigenvalues are distinct, the ψ_n are orthogonal in $L^2(\mathbf{R}, e^{-x^2}\,dx)$. Up to normalization, the associated orthogonal polynomials ψ_n are the *Hermite polynomials*.

The functions

$$\psi_n(x)e^{-\frac{1}{2}x^2}$$

are an orthogonal basis for $L^2(\mathbf{R})$. They are eigenfunctions for the operator

$$\frac{d^2}{dx^2} + (1 - x^2). \tag{3.4.4}$$

Case II: p linear. We may normalize to $p(x) = x$. Then (3.4.2) implies that $w'/w = b + a/x$, so up to scaling, $w = x^a e^{bx}$. The endpoint condition implies that the interval is either $\mathbf{R}_- = (-\infty, 0)$ or $\mathbf{R}_+ = (0, \infty)$, and we may assume $I = \mathbf{R}_+$. Condition (3.4.1) implies $b < 0$ and $a > -1$. We may rescale to have $b = -1$. Thus in this case (3.1.9) is the operator

$$L = x\frac{d^2}{dx^2} + [(a+1) - x]\frac{d}{dx} \quad \text{in} \quad L_w^2(\mathbf{R}_+);$$
$$w(x) = x^a e^{-x}, \, a > -1. \tag{3.4.5}$$

Once again, the space of polynomials of degree $\leq n$ is mapped to itself, so there is a polynomial ψ_n of degree n that is an eigenfunction. Consideration of the degree n term shows that

$$L\psi_n + n\,\psi_n = 0.$$

Therefore the ψ_n are orthogonal. Up to normalization, the ψ_n are the *Laguerre polynomials*. The functions

$$\psi_n(x)x^{\frac{1}{2}a}e^{-\frac{1}{2}x}$$

are an orthogonal basis for $L^2(\mathbf{R}_+)$. They are eigenfunctions of the operator

$$x\frac{d^2}{dx^2} + \frac{d}{dx} - \frac{x}{4} - \frac{a^2}{4x} + \frac{a+1}{2}. \tag{3.4.6}$$

Case III: p quadratic, distinct real roots. We normalize to $p(x) = 1 - x^2$. Then (3.4.2) implies $w'/w = \beta(1 + x)^{-1} - \alpha(1 - x)^{-1}$ for some constants α and β, so the weight function $w(x) = (1 - x)^{\alpha}(1 + x)^{\beta}$. The endpoint condition forces $I = (-1, 1)$ and condition (3.4.1) forces $\alpha, \beta > -1$. Thus in this case (3.1.9) is the operator

$$L = (1 - x^2)\frac{d^2}{dx^2} + [\beta - \alpha - (\alpha + \beta + 2)x]\frac{d}{dx} \tag{3.4.7}$$

in $L_w^2((-1, 1))$; here $w(x) = (1 - x)^{\alpha}(1 + x)^{\beta}$, $\alpha, \beta > -1$.

Again, the space of polynomials of degree $\leq n$ is mapped to itself, so there is a polynomial ψ_n of degree n that is an eigenfunction. As before, consideration of the degree n term shows that

$$L\psi_n + n(n + \alpha + \beta + 1)\psi_n = 0.$$

Therefore the ψ_n are orthogonal.

Up to normalization, the associated orthogonal polynomials are the *Jacobi polynomials*.

We may rescale the interval by taking $\frac{1}{2}(1 - x)$ as the new x variable, so that the interval is $(0, 1)$. In the new coordinates, up to a constant factor, the weight is $x^{\alpha}(1 - x)^{\beta}$ and the operator is

$$L = x(1 - x)\frac{d^2}{dx^2} + [\alpha + 1 - (\alpha + \beta + 2)x]\frac{d}{dx}$$

$$\text{in}\quad L_w^2((0, 1)), \quad w(x) = x^{\alpha}(1 - x)^{\beta} \quad \alpha, \beta > -1.$$

This is the hypergeometric operator corresponding to (3.0.2) with indices $(a, b, c) = (\alpha + \beta + 1, 0, \alpha + 1)$.

Case IV: p quadratic, distinct complex roots. We normalize to $x^2 + 1$. Then (3.4.2), together with the assumption that $w > 0$, implies that w would have the form $w(x) = (1 + x^2)^{\alpha} e^{\beta \tan^{-1} x}$. The endpoint condition rules out bounded intervals I, and condition (3.4.1) rules out unbounded intervals.

Case V: p quadratic, double root. We may take $p(x) = x^2$. Then (3.4.2) implies that w would have the form $w(x) = x^{\alpha}\exp(\beta/x)$, and once again the endpoint condition and condition (3.4.1) cannot both be satisfied.

We have proved the following (somewhat informally stated) result.

Theorem 3.4.1 *Up to normalization, the classical orthogonal polynomials (Hermite, Laguerre, Jacobi) are the only ones that occur as the eigenfunctions for second-order differential operators that are symmetric with respect to a positive weight.*

(To account for other names: up to certain normalizations, *Gegenbauer* or *ultraspherical polynomials* are Jacobi polynomials with $\alpha = \beta$, *Legendre polynomials* are Jacobi polynomials with $\alpha = \beta = 0$, and *Chebyshev polynomials* are Jacobi polynomials in the two cases $\alpha = \beta = -\frac{1}{2}, \alpha = \beta = \frac{1}{2}$.)

There is a sense in which Case I reduces to two instances of Case II. The operator (3.4.3) and the weight $w(x) = e^{-x^2}$ are left unchanged by the reflection $x \to -x$. Therefore the even and odd parts of a function in L_w^2 also belong to L_w^2, and (3.4.3) maps even functions to even functions and odd functions to odd functions. An even function f can be written as $f(x) = g(x^2)$ and f belongs to L_w^2 if and only if g^2 is integrable on $(0, \infty)$ with respect to the weight $x^{-\frac{1}{2}} e^{-x}$. An odd function f can be written as $f(x) = xg(x^2)$, and f belongs to L^2 if and only if g^2 is integrable on $(0, \infty)$ with respect to the weight $x^{\frac{1}{2}} e^{-x}$. It follows from these considerations that, up to multiplicative constants, Hermite polynomials of even degree must be Laguerre polynomials in x^2, with index $\alpha = -\frac{1}{2}$, and Hermite polynomials of odd degree, when divided by x, must be Laguerre polynomials in x^2 with index $\alpha = \frac{1}{2}$.

These polynomials are related also by certain limiting relations. In Case III, we may normalize the weight function by taking

$$w_{\alpha,\beta}(x) = \frac{2^{-(\alpha+\beta+1)}}{B(\alpha+1, \beta+1)} (1-x)^\alpha (1+x)^\beta$$

$$= 2^{-(\alpha+\beta+1)} \frac{\Gamma(\alpha+\beta+2)}{\Gamma(\alpha+1)\Gamma(\beta+1)} (1-x)^\alpha (1+x)^\beta. \qquad (3.4.8)$$

The change of variables $x = 1 - 2y$ shows that

$$\int_{-1}^{1} w_{\alpha,\beta}(x)\,dx = 1.$$

Now take $\beta = \alpha > 0$ and let $x = y/\sqrt{\alpha}$, so that

$$w_{\alpha,\alpha}(x)\,dx = 2^{-(2\alpha+1)} \frac{\Gamma(2\alpha+2)}{\Gamma(\alpha+1)^2} \left(1 - \frac{y^2}{\alpha}\right)^\alpha \frac{dy}{\sqrt{\alpha}},$$

and the interval $-1 < x < 1$ corresponds to the interval $-\sqrt{\alpha} < y < \sqrt{\alpha}$. Taking into account the duplication formula (2.3.1) and (2.1.9), we see that as $\alpha \to +\infty$, the rescaled weight on the right converges to the normalized version of Case I:

$$w_H(x) = \frac{1}{\sqrt{\pi}} e^{-x^2}. \qquad (3.4.9)$$

It follows from a compactness argument, which we omit, that the orthonormal polynomials for the weight (3.4.8) with $\alpha = \beta$ (normalized to have positive leading coefficient) converge, under the change of variables $x = y/\sqrt{\alpha}$, to the orthonormal polynomials for the weight (3.4.9). Thus Hermite polynomials are limits of rescaled equal-index Jacobi polynomials.

To obtain Laguerre polynomials as limits of Jacobi polynomials, we take the version of Case III transferred to the interval $(0, 1)$:

$$\widetilde{w}_{\alpha,\beta} = \frac{1}{B(\alpha+1,\beta+1)} x^\alpha (1-x)^\beta. \qquad (3.4.10)$$

Assume $\beta > 0$ and make the change of variables $x = y/\beta$, so

$$\widetilde{w}_{\alpha,\beta}(x)\,dx = \frac{\Gamma(\alpha+\beta+2)}{\Gamma(\alpha+1)\Gamma(\beta+1)\beta^{\alpha+1}} y^\alpha \left(1 - \frac{y}{\beta}\right)^\beta dy.$$

Taking into account (2.1.9), as $\beta \to +\infty$, the rescaled weight on the right converges to the normalized version of Case II:

$$w_\alpha(x) = \frac{1}{\Gamma(\alpha+1)} x^\alpha e^{-x}. \qquad (3.4.11)$$

Again, the orthonormal polynomials for weight (3.4.10) converge to the orthonormal polynomials for weight (3.4.11) under the change of variables $x = y/\beta$.

3.5 Maxima, minima, and estimates

In order to study solutions of the eigenvalue equations

$$p(x)u''(x) + q(x)u'(x) + \lambda u(x) = 0, \qquad \lambda > 0, \qquad (3.5.1)$$

it is convenient to introduce the auxiliary function

$$V(x) = u(x)^2 + \frac{p(x)}{\lambda} u'(x)^2. \qquad (3.5.2)$$

Proposition 3.5.1 *The relative maxima of $|u(x)|$ are increasing as x increases on any interval where $p' - 2q > 0$, and decreasing on any interval where $p' - 2q < 0$.*

Proof It follows from earlier results that the zeros of u' are simple (u' satisfies an equation of the same form), so they determine relative extrema of u. At a relative extremum of $u(x)$, $V(x) = u(x)^2$. Equation (3.5.1) implies that

$$V'(x) = \frac{p'(x) - 2q(x)}{\lambda} u'(x)^2.$$

Thus V is increasing where the function $p' - 2q$ is positive and decreasing where it is negative. $\qquad \square$

A similar idea applies to equations in a somewhat different form.

Proposition 3.5.2 *Suppose that $w(x) > 0$, $r(x) > 0$, and*

$$[w\,u']'(x) + r(x)\,u(x) = 0.$$

Then, as x increases, the relative maxima of $|u(x)|$ are increasing in any interval where $(w\,r)' < 0$, and decreasing in any interval where $(w\,r)' > 0$.

Proof Here let

$$W(x) = u(x)^2 + \frac{\left[w(x)\,u'(x)\right]^2}{w(x)\,r(x)}.$$

Then

$$W'(x) = -[wr]'(x)\,\frac{u'(x)^2}{r(x)^2}.$$

At the relative extrema, $u(x)^2 = W(x)$. □

This argument proves the following.

Proposition 3.5.3 *Suppose that $w(x) > 0$, $r(x) > 0$, and*

$$[w\,u']'(x) + r(x)\,u(x) = 0$$

in an interval $a < x \le b$. If $(wr)' \le 0$ in this interval, then

$$u(x)^2 \le u(b)^2 + \frac{w(b)}{r(b)}\,u'(b)^2, \quad a < x \le b.$$

Let us apply these observations to the three cases in Section 3.4. In the Hermite case, $p' - 2q = 4x$, so the relative maxima of $|H_n(x)|$ increase as $|x|$ increases. In the Laguerre case, $p' - 2q = 1 - 2(\alpha + 1 - x) = 2x - (2\alpha + 1)$, so the relative maxima of $|L^{(\alpha)}(x)|$ decrease as x increases so long as $x \le \alpha + \frac{1}{2}$, and increase with x for $x \ge \alpha + \frac{1}{2}$. In the Jacobi case,

$$p'(x) - 2q(x) = -2x - 2[\beta - \alpha - (\alpha + \beta + 2)x]$$

$$= 2[\alpha - \beta + (\alpha + \beta + 1)x].$$

It follows that the relative maxima of $|P_n^{(\alpha,\beta)}|$ are either monotone, if one of α, β is $\ge -\frac{1}{2}$ and the other is $\le -\frac{1}{2}$, or decrease from $x = -1$ until $(\alpha + \beta + 1)x = \beta - \alpha$, and then increase to $x = 1$ if both α and β exceed $-\frac{1}{2}$. Thus

$$\sup_{|x| \le 1} |P_n^{(\alpha,\beta)}(x)| = \max\left\{|P_n^{(\alpha,\beta)}(-1)|, |P_n^{(\alpha,\beta)}(1)|\right\} \quad \text{if} \ \ \max\{\alpha, \beta\} \ge -\frac{1}{2}.$$

The results on relative maxima for Hermite and Laguerre polynomials can be sharpened by using Proposition 3.5.2 with $w(x) \equiv 1$. As noted in Section 3.4, the gauge transformations

$$H_n(x) = e^{x^2/2}\,u_n(x);$$

$$L_n^{(\alpha)}(x) = e^{x/2} x^{-(\alpha+1)/2} v_n(x)$$

lead to the equations

$$u_n''(x) + (2n + 1 - x^2) u_n(x) = 0;$$

$$v_n''(x) + \left[\frac{1 - \alpha^2}{4x^2} + \frac{2n + \alpha + 1}{2x} - \frac{1}{4} \right] v_n(x) = 0.$$

It follows from Proposition 3.5.2 that the relative maxima of $|u_n(x)|$ are increasing away from the origin, and that the relative maxima of $|v_n(x)|$ are increasing on the interval

$$x \geq \max \left\{ 0, \frac{\alpha^2 - 1}{2n + \alpha + 1} \right\}.$$

Moreover, Theorem 3.3.6 gives information about the zeros of u_n and v_n, which are the same as those of H_n and $L_n^{(\alpha)}$. Since H_n and $L_n^{(\alpha)}$ are polynomials, u_n and v_n have limit 0 as $x \to \infty$. The coefficient $r(x)$ for $u_n(x)$ is negative for $x > \sqrt{2n+1}$, so $H_n(x) = (-1)^n H_n(-x)$ has no zeros x with $x^2 > 2n + 1$. Similarly, the coefficient $r(x)$ for $v_n(x)$ is negative for

$$x > 2n + \alpha + 1 + \sqrt{(2n + \alpha + 1)^2 + 1 - \alpha^2},$$

so $L_n^{(\alpha)}$ has no zeros in this interval.

3.6 Some equations of mathematical physics

A number of second-order ordinary differential equations arise from partial differential equations of mathematical physics, by separation of variables in special coordinate systems. We illustrate this here for problems involving the *Laplacian* in $\mathbf{R}^3$. In Cartesian coordinates $\mathbf{x} = (x_1, x_2, x_3)$, the Laplacian has the form

$$\Delta = \frac{\partial^2}{\partial x_1^2} + \frac{\partial^2}{\partial x_2^2} + \frac{\partial^2}{\partial x_3^2}.$$

Because it is invariant under translations and rotations, it arises in many physical problems for isotropic media. Four examples are the *heat equation*, or *diffusion equation*

$$v_t(\mathbf{x}, t) = \Delta v(\mathbf{x}, t),$$

the *wave equation*

$$v_{tt} = (\mathbf{x}, t) = \Delta v(\mathbf{x}, t),$$

and the *Schrödinger equations*, for the quantized harmonic oscillator

$$iv_t = \Delta v - |\mathbf{x}|^2 v$$

and for the Coulomb potential

$$iv_t = \Delta v - \frac{a}{|\mathbf{x}|} v.$$

Separating variables, i.e. looking for a solution in the form $v(\mathbf{x},t) = \varphi(t)u(\mathbf{x})$, leads, after dividing by v, to the four equations

$$\frac{\varphi'(t)}{\varphi(t)} = \frac{\Delta u(\mathbf{x})}{u(\mathbf{x})};$$

$$\frac{\varphi''(t)}{\varphi(t)} = \frac{\Delta u(\mathbf{x})}{u(\mathbf{x})};$$

$$i\frac{\varphi'(t)}{\varphi(t)} = \frac{\Delta u(\mathbf{x})}{u(\mathbf{x})} - |\mathbf{x}|^2;$$

$$i\frac{\varphi'(t)}{\varphi(t)} = \frac{\Delta u(\mathbf{x})}{u(\mathbf{x})} - \frac{a}{|\mathbf{x}|}.$$

In each of these equations the left side is a function of t alone and the right side is a function of $\mathbf{x}$ alone, so each side is constant, say $-\lambda$. Thus we are led to three equations involving the Laplacian, the first of which is known as the *Helmholtz equation*:

$$\Delta u(\mathbf{x}) + \lambda u(\mathbf{x}) = 0; \tag{3.6.1}$$

$$\Delta u(\mathbf{x}) - |\mathbf{x}|^2 u(\mathbf{x}) + \lambda u(\mathbf{x}) = 0; \tag{3.6.2}$$

$$\Delta u(\mathbf{x}) - a\frac{u(\mathbf{x})}{|\mathbf{x}|} + \lambda u(\mathbf{x}) = 0. \tag{3.6.3}$$

The case $\lambda = 0$ of the Helmholtz equation is the *Laplace equation*:

$$\Delta u(x) = 0. \tag{3.6.4}$$

One approach to these equations is to choose a coordinate system and separate variables once more to find special solutions. One may then try to reconstruct general solutions from the special solutions. For this purpose, in addition to Cartesian coordinates we consider the following.
Spherical coordinates (r, φ, θ):

$$(x_1, x_2, x_3) = (r\cos\varphi\sin\theta, r\sin\varphi\sin\theta, r\cos\theta), \tag{3.6.5}$$

in which the operator Δ takes the form

$$\Delta = \frac{\partial^2}{\partial r^2} + \frac{2}{r}\frac{\partial}{\partial r} + \frac{1}{r^2\sin^2\theta}\frac{\partial^2}{\partial\varphi^2} + \frac{1}{r^2\sin\theta}\frac{\partial}{\partial\theta}\sin\theta\frac{\partial}{\partial\theta}.$$

(This is the notation for spherical coordinates commonly used in physics and often used in applied mathematics. It is common in the mathematical literature to reverse the roles of θ and φ here.)

Cylindrical coordinates (r, θ, z):

$$(x_1, x_2, x_3) = (r\cos\theta, r\sin\theta, z), \tag{3.6.6}$$

in which the operator Δ takes the form

$$\Delta = \frac{\partial^2}{\partial r^2} + \frac{1}{r}\frac{\partial}{\partial r} + \frac{1}{r^2}\frac{\partial^2}{\partial\theta^2} + \frac{\partial^2}{\partial z^2}.$$

Parabolic cylindrical coordinates (ξ, ζ, z):

$$(x_1, x_2, x_3) = (\tfrac{1}{2}[\xi^2 - \zeta^2], \xi\zeta, z), \tag{3.6.7}$$

in which the operator Δ takes the form

$$\Delta = \frac{1}{\xi^2 + \zeta^2}\left(\frac{\partial^2}{\partial\xi^2} + \frac{\partial^2}{\partial\zeta^2}\right) + \frac{\partial^2}{\partial z^2}.$$

Separating variables in the Helmholtz equation (3.6.1) in Cartesian coordinates by setting $u(\mathbf{x}) = u_1(x_1)u_2(x_2)u_3(x_3)$ leads to

$$\frac{u_1''}{u_1} + \frac{u_2''}{u_2} + \frac{u_3''}{u_3} + \lambda = 0.$$

If we rule out solutions with exponential growth, it follows that

$$\frac{u_j''}{u_j} = -k_j^2, \qquad k_1^2 + k_2^2 + k_3^2 = \lambda.$$

Therefore the solution is a linear combination of the complex exponentials

$$u(\mathbf{x}) = e^{i\mathbf{k}\cdot\mathbf{x}}, \qquad |\mathbf{k}|^2 = \lambda.$$

Separating variables in the Laplace equation (3.6.4) in spherical coordinates by setting $u(\mathbf{x}) = R(r)U(\varphi)V(\theta)$ leads to

$$\frac{r^2 R'' + 2rR'}{R} + \left\{\frac{U''}{\sin^2\theta\, U} + \frac{[\sin\theta\, V']'}{\sin\theta\, V}\right\} = 0. \tag{3.6.8}$$

Each of the summands must be constant. It follows that R is a linear combination of powers r^ν and $r^{-1-\nu}$ and

$$\frac{U''}{U} + \left\{\sin\theta\frac{[\sin\theta\, V']'}{V} + \nu(\nu+1)\sin^2\theta\right\} = 0.$$

Once again each summand is constant, and $U'' = -\mu^2 U$ leads to

$$\frac{1}{\sin\theta}\frac{d}{d\theta}\left\{\sin\theta\frac{dV}{d\theta}\right\} + \left[\nu(\nu+1) - \frac{\mu^2}{\sin^2\theta}\right]V = 0.$$

The change of variables $x = \cos\theta$ converts the preceding equation to the *spherical harmonic equation*:

$$[(1-x^2)u']' + \left[v(v+1) - \frac{\mu^2}{1-x^2}\right]u = 0, \quad 0 < x < 1. \tag{3.6.9}$$

The case $\mu = 0$ (solution invariant under rotation around the vertical axis) is *Legendre's equation*; the solutions are known as *Legendre functions*.

Separating variables in the Helmholtz equation (3.6.1) in cylindrical coordinates, $u(\mathbf{x}) = R(r)T(\theta)Z(z)$, leads to

$$\frac{r^2 R'' + rR'}{R} + \frac{T''}{T} + \frac{r^2 Z''}{Z} + \lambda r^2 = 0.$$

It follows that $Z''/Z + \lambda = \mu$ and T''/T are constant. Since T is periodic, $T''/T = -n^2$ for some integer n and

$$r^2 R''(r) + rR'(r) + \mu r^2 R(r) - n^2 R(r) = 0.$$

Assuming that $\mu = k^2$ is positive, we may set $R(r) = u(k^{-1}r)$ and obtain *Bessel's equation*

$$x^2 u''(x) + xu'(x) + [x^2 - n^2]u(x) = 0. \tag{3.6.10}$$

Solutions of (3.6.10) are known as *cylinder functions*.

Separating variables in the Helmholtz equation (3.6.1) in parabolic cylindrical coordinates, $u(\mathbf{x}) = X(\xi)Y(\zeta)Z(z)$, leads to the conclusion that $Z''/Z + \lambda = \mu$ is constant and

$$\frac{X''}{X} + \frac{Y''}{Y} + \mu(\xi^2 + \zeta^2) = 0.$$

It follows that there is a constant v such that

$$X'' + (\mu\xi^2 - v)X = 0 = Y'' + (\mu\zeta^2 + v)Y.$$

Assuming that $\mu = -k^2$ is negative, the changes of variables $\xi = k^{-1}x$ and $\zeta = k^{-1}x$ convert these to the standard forms

$$u''(x) - x^2 u(x) \pm v\, u(x) = 0. \tag{3.6.11}$$

Solutions are known as *parabolic cylinder functions* or *Weber functions*.

Separating variables in (3.6.2) in Cartesian coordinates leads again to (3.6.11). Separating variables in spherical or cylindrical coordinates leads to equations in the radial variable r which are not of classical type; in parabolic cylinder coordinates (3.6.2) does not separate.

Equation (3.6.3) separates in spherical coordinates, leading to an equation in the radial variable:

$$v''(r) + \frac{2}{r}v'(r) + \left(\lambda - \frac{a}{r} + \frac{\mu}{r^2}\right)v(r) = 0.$$

Taking $v(r) = r^{-1}w(r)$, followed by a change of scale, converts this to the *Coulomb wave equation*

$$u''(\rho) + \left[1 - \frac{2\eta}{\rho} - \frac{l(l+1)}{\rho^2}\right] u(\rho) = 0. \tag{3.6.12}$$

Solutions are known as *Coulomb wave functions*.

3.7 Equations and transformations

Let us begin with a list of the second-order equations encountered in this chapter. We claimed at the outset that all are related to the pair consisting of the confluent hypergeometric equation

$$xu''(x) + (c - x)u'(x) - au(x) = 0 \tag{3.7.1}$$

and the hypergeometric equation

$$x(1 - x)u''(x) + [c - (a + b + 1)x]u'(x) - abu(x) = 0, \tag{3.7.2}$$

in the sense that they can be reduced to one of these equations by gauge transformations and changes of variables.

Each of the equations has, up to multiplication by a function, the form

$$u'' + \frac{q_0}{p}u' + \frac{r_0}{p^2}u = 0, \tag{3.7.3}$$

where p and r_0 are polynomials of degree at most 2 and q_0 is a polynomial of degree at most 1. This general form is preserved under a gauge transformation $u(x) = \varphi(x)v(x)$,

$$\frac{\varphi'}{\varphi} = \frac{q_1}{p}, \tag{3.7.4}$$

where q_1 is any polynomial of degree at most 1. The polynomial q_1 can be chosen (not necessarily in a unique way) so that the equation for v has the canonical form

$$pv'' + qv' + \lambda v = 0, \tag{3.7.5}$$

where q has degree at most 1 and λ is constant. To accomplish this, one is led to three equations for the two coefficients of q_1 and the constant λ.

Classifying symmetric problems with polynomials as eigenfunctions led us to equations of the form

$$u'' - 2xu' + 2\lambda u = 0; \tag{3.7.6}$$

$$xu'' + (\alpha + 1 - x)u' + \lambda u = 0; \tag{3.7.7}$$

$$(1 - x^2)u'' + [\beta - \alpha - (\alpha + \beta + 2)x]\, u' + \lambda u = 0. \tag{3.7.8}$$

A unitary equivalence (to a symmetric problem with weight 1), mapped the first two of these three equations to

$$v'' + (2\lambda + 1 - x^2)v = 0; \tag{3.7.9}$$

$$x v'' + v' - \frac{x^2 + \alpha^2}{4x} v + \left(\lambda + \frac{\alpha + 1}{2}\right) v = 0. \tag{3.7.10}$$

As noted in Section 3.4, letting $x = 1 - 2y$ takes (3.7.8) to the form (3.7.2).

Separating variables in problems involving the Laplacian in special coordinate systems led us to the equations

$$(1 - x^2)u'' - 2xu' + \lambda u - \mu^2(1 - x^2)^{-1}u = 0; \tag{3.7.11}$$

$$x^2 u'' + xu' + (x^2 - v^2)u = 0; \tag{3.7.12}$$

$$u'' - x^2 u \pm v u = 0; \tag{3.7.13}$$

$$x^2 u'' + \left[x^2 - 2\eta x - l(l + 1)\right] u = 0. \tag{3.7.14}$$

The spherical harmonic equation (3.7.11) is in the general form (3.7.3). As noted above, it can be reduced to the canonical form by a gauge transformation, in this case

$$u(x) = (1 - x^2)^{\frac{1}{2}\mu}\, v(x),$$

leading to the equation

$$(1 - x^2)v''(x) - 2(\mu + 1)xv'(x) + \left[\lambda - \mu^2 - \mu\right] v(x) = 0,$$

which is the special case $\alpha = \beta = \mu$ of (3.7.8) and thus a particular case of (3.7.2).

Equation (3.7.7) is a particular case of the confluent hypergeometric equation (3.7.1), so the gauge-equivalent equation (3.7.10) is also.

Bessel's equation (3.7.12) is in the general form (3.7.3) with $p(x) = x$. Corresponding gauge transformations are

$$u(x) = e^{\pm ix} x^v v(x),$$

leading to the canonical form

$$x v'' + (2v + 1 \pm 2ix)\, v' \pm i(2v + 1)v = 0.$$

Letting $y = \mp 2ix$ converts this, in turn, to

$$y w'' + (2v + 1 - y)\, w' - \left(v + \frac{1}{2}\right) w = 0.$$

This is (3.7.1) with $c = 2v + 1$, $a = v + \frac{1}{2}$.

Up to the sign of the parameter, (3.7.9) and (3.7.13) are identical. Moreover, (3.7.9) is related to (3.7.6) by a gauge transformation. We can relate them to the confluent hypergeometric equation (3.7.1) by noting first that the even and odd parts of a solution $u(x)$ of (3.7.6) are also solutions. Writing an even solution as $u(x) = v(x^2)$ converts (3.7.6) to the form

$$xv'' + \left(\frac{1}{2} - x\right)v' + \frac{1}{2}\lambda v = 0, \tag{3.7.15}$$

the particular case of (3.7.1) with $c = \frac{1}{2}$, $a = -\frac{1}{2}\lambda$. Writing an odd solution as $u(x) = xv(x^2)$ converts (3.7.6) to the form

$$xv'' + \left(\frac{3}{2} - x\right)v' + \frac{1}{2}(\lambda - 1)v = 0, \tag{3.7.16}$$

the particular case of (3.7.1) with $c = \frac{3}{2}$, $a = \frac{1}{2}(1 - \lambda)$.

Finally, the gauge transformation $u(x) = e^{ix}x^{l+1}v(x)$ converts (3.7.14) to

$$xv''(x) + (2l + 2 + 2ix)v'(x) + [(2l + 2)i - 2\eta]\,v(x) = 0,$$

and the change of variables $v(x) = w(-2ix)$ converts this equation to

$$yw''(y) + (2l + 2 - y)w'(y) - (l + 1 + i\eta)w(y) = 0,$$

which is (3.7.1) with $c = 2l + 2$, $a = l + 1 + i\eta$.

3.8 Exercises

1. Verify that (3.1.1) and (3.1.5) are equivalent under the gauge transformation (3.1.4).
2. Show that if φ and ψ are two solutions of (3.1.7) and $\varphi(x) \neq 0$, $x \in I$, then $\psi(x) \equiv C\varphi(x)$, C constant.
3. Prove Proposition 3.1.1 by showing first that the symmetry condition is equivalent to

$$0 = (Lu, v) - (u, Lv) = \int_a^b \left[p(u'v - uv')' + q(u'v - uv')\right]w\,dx$$

$$= \int_a^b (u'v - uv')\left[qw - (pw)'\right]dx.$$

In particular, if $u \equiv 1$ wherever $v \neq 0$, then

$$0 = -\int_a^b v'\left[qw - (pw)'\right]dx = \int_a^b v\left[qw - (pw)'\right]'dx;$$

conclude that $qw - (pw)' = c$, constant. If $u(x) = x$ wherever $v \neq 0$, then

$$0 = c \int_a^b (v - xv') \, dx = 2c \int_a^b v \, dx$$

for all such v, so $c = 0$. Therefore symmetry implies $qw = (pw)'$. Conversely, show that the condition $qw = (pw)'$ implies symmetry.

4. Prove Proposition 3.1.2 by finding a first-order equation that characterizes w up to a constant.
5. Prove Proposition 3.1.3 by finding a first-order equation that characterizes φ up to a constant.
6. Prove Proposition 3.1.4 by finding a unitary map that has the form $Tf(x) = h(x)f(x)$.
7. Prove (3.1.11).
8. Complete the Liouville transformation, the reduction of (3.1.1) to an equation of the form

$$w''(y) + r_1(y) w(y) + s(y) w(y) = f_2(y),$$

by applying a gauge transformation to (3.1.11).
9. Deduce Theorem 3.2.2 from the proof of Theorem 3.2.1.
10. Prove the estimates (3.2.5) and (3.2.6).
11. Show that u is a solution of (3.2.7) and (3.2.8) if and only if $u = u_0$, where the vector $[u_0, u_1]'$ is a solution of

$$\begin{bmatrix} u_0' \\ u_1' \end{bmatrix} = \begin{bmatrix} 0 & 1 \\ -r/p & -q/p \end{bmatrix} \begin{bmatrix} u_0 \\ u_1 \end{bmatrix}, \qquad \begin{bmatrix} u_0(x_0) \\ u_1(x_0) \end{bmatrix} = \begin{bmatrix} c_0 \\ c_1 \end{bmatrix}. \tag{3.8.1}$$

12. Write (3.8.1) as an integral equation and use the method of successive approximations to show that it has a unique solution

$$\begin{bmatrix} u_0 \\ u_1 \end{bmatrix} = \sum_{k=0}^{\infty} \begin{bmatrix} u_{0k} \\ u_{1k} \end{bmatrix},$$

where

$$\begin{bmatrix} u_{00} \\ u_{10} \end{bmatrix} = \begin{bmatrix} c_0 \\ c_1 \end{bmatrix},$$

$$\begin{bmatrix} u_{0,k+1}(x) \\ u_{1,k+1}(x) \end{bmatrix} = \int_{x_0}^x \begin{bmatrix} 0 & 1 \\ -r(s)/p(s) & -q(s)/p(s) \end{bmatrix} \begin{bmatrix} u_{0k}(s) \\ u_{1k}(s) \end{bmatrix} ds.$$

13. Adapt the method of Exercise 12 to prove Theorem 3.2.4.
14. Find an integral formula for the solution of the equation

$$u''(x) + \lambda^2 u(x) = f(x), \qquad -\infty < x < \infty,$$

that satisfies the conditions $u(0) = u'(0) = 0$, where λ is a positive constant.

15. Find an integral formula for the solution of the equation

$$u''(x) + u(x) = f(x), \quad -\pi < x < \pi,$$

that satisfies the conditions $u(-\pi) = 0 = u(\pi)$.

16. Suppose that $I = (a, b)$ is a bounded interval of length $L = b - a$, and that $r(x)$ is continuous on I, and $|r(x)| \leq C$ for all $x \in I$. Suppose that $\lambda > 0$ and u is a nonzero real solution of

$$u''(x) + [r(x) + \lambda] u(x) = 0$$

on I. Let $N(\lambda)$ denote the number of zeros of u in the interval. Use Theorem 3.3.3 to prove that for sufficiently large λ,

$$\left| N(\lambda) - \frac{\sqrt{\lambda} L}{\pi} \right| < 2.$$

17. Prove Theorem 3.3.4.
18. Prove Corollary 3.3.5.
19. Verify that the gauge transformation

$$H_n(x) = e^{\frac{1}{2}x^2} h_n(x)$$

converts the equation for the Hermite function H_n to the equation

$$h_n''(x) + (2n+1)h_n(x) = x^2 h_n(x).$$

Deduce from this that h_n is the solution of the integral equation

$$h_n(x) = A_n \cos(\sqrt{2n+1}\, x + b_n)$$
$$+ \int_0^x \frac{\sin\left[\sqrt{2n+1}\,(x-y)\right]}{\sqrt{2n+1}} y^2 h_n(y)\, dy$$

for some choice of the constants A_n and b_n; determine these constants. As shown in Chapter 13, this implies the asymptotic result

$$h_n(x) = A_n \left[\cos(\sqrt{2n+1}\, x + b_n) + O(n^{-\frac{1}{2}}) \right]$$

as $n \to \infty$.

20. Determine the Liouville transformation that reduces (3.7.7) for Laguerre polynomials to an equation of the form

$$v''(y) + \lambda v(y) = r(y) v(y).$$

Find the form of an integral equation for the corresponding modified Laguerre functions $\ell_n^{(a)}(y)$ analogous to that for the modified Hermite functions $h_n(x)$. Can this be used to obtain an asymptotic result? (Specifically, can the constants A_n, b_n be determined?)

21. Determine the Liouville transformation that reduces (3.7.8) for Jacobi polynomials to an equation of the form

$$v''(y) + \lambda v(y) = r(y) v(y).$$

22. Verify that an equation in the general form described in connection with (3.7.3) can be reduced to the canonical form (3.7.5) by a gauge transformation as described in connection with (3.7.4).

23. Show that Riccati's equation [341]

$$u'(x) + u(x)^2 + x^m = 0$$

can be converted to a second-order linear equation by setting $u(x) = u_1'(x)/u_1(x)$. Show that the change of variables $u_1(x) = u_2(y)$ with

$$y = \frac{2 x^{(m+2)/2}}{m+2}$$

leads to the equation

$$u_2''(y) + \frac{m}{(m+2)y} u_2'(y) + u_2(y) = 0.$$

24. Show that eliminating the first-order term in the last equation in Exercise 23 leads to the equation

$$u_3''(y) + \left[1 + \frac{r - r^2}{y^2}\right] u_3(y) = 0, \qquad r = \frac{m}{2m+4}.$$

Show that the gauge transformation $u_3(y) = y^{\frac{1}{2}} v(y)$ converts this to Bessel's equation (3.6.10) with $n = 1/(m+2)$. Combined with results from Section 9.1, this proves a result of Daniel Bernoulli [37]: solutions of Riccati's equation can be expressed in terms of elementary functions when $2/(m+2)$ is an odd integer.

3.9 Remarks

The general theory of second-order equations is covered in most textbooks on ordinary differential equations. Three classic texts are Forsyth [140], Ince [193], and Coddington and Levinson [84]. The book by Ince has an extensive discussion of the classification of second-order linear equations with rational coefficients. The book by Hille [185] is concerned specifically with equations in the complex domain. For some indication of the modern ramifications of the study of equations in the complex domain, see the survey article by Varadarajan [411]. Explicit solutions for many second-order equations are collected in the handbooks by Kamke [209], Sachdev [349], and Zwillinger [449].

The idea of solving partial differential equations by separation of variables developed throughout the eighteenth century, e.g. in the work of D. Bernoulli, d'Alembert, and Euler. Fourier [141] was the first to put all the ingredients of the method in place. For a discussion of this, the Sturm–Liouville theory, and other developments, see Painlevé's survey [314] and Lützen [265].

Separation of variables for the Laplace, Helmholtz, and other equations is treated in detail by Miller [288]; see also Müller [295]. There are other coordinate systems in which one can separate variables for the Helmholtz or Laplace equations, but the functions that arise are not among those treated here: Mathieu functions, modified Mathieu functions, prolate spheroidal functions, etc.

The Wronskian appears in [190]. The concept of linear dependence of solutions and the connection with the Wronskian goes back to Christoffel [78]. The idea of the Green's function goes back to Green in 1828 [173]. The use of the method of approximation to prove existence of solutions originated with Liouville [260] and was developed in full generality by Picard [325].

The Liouville transform was introduced in [261]. Sturm's comparison theorem appeared in [390]. The systematic study of orthogonal functions is due to Murphy [297]. Techniques for estimating relative extrema were developed by Stieltjes [383] for Legendre polynomials, Sonine for Bessel functions [374], and others.

For extensive coverage of the nineteenth-century history, see [314] and part 2 of [224].

4

Orthogonal polynomials on an interval

It was shown in Chapter 3 that there are three cases in which the eigenfunctions of a second-order ordinary differential operator that is symmetric with respect to a weight are polynomials. The polynomials in the three cases are the classical orthogonal polynomials: the Hermite polynomials, Laguerre polynomials, and Jacobi polynomials.

Each of these sets of polynomials is an example of a family of polynomials that are orthogonal with respect to an inner product that is induced by a positive weight function w on an interval of the real line. The basic theory of general orthogonal polynomials of this type is covered in this chapter. This includes expressions as determinants, three-term recurrence relations, properties of the zeros, and basic asymptotics. It is shown that under a certain condition on the weight $w(x)$, which is satisfied in each of the three classical cases, each element of the space L_w^2 can be expanded in a series using the orthogonal polynomials, analogous to the Fourier series expansion. These results carry over to more general measures than those of the form $w(x)\,dx$.

Orthogonal polynomials occur naturally in connection with approximating the Stieltjes transform of the weight function or measure. This transform can also be viewed as a continued fraction.

The central role played by the three-term recurrence relations leads to the question: do such relations characterize orthogonal polynomials? The (positive) answer is known as Favard's theorem.

The chapter concludes with a brief discussion of the asymptotic distribution of zeros.

4.1 Weight functions and orthogonality

Let $w(x)$ be a positive weight function on an open interval $I = (a,b)$ and assume that the moments

$$A_n = \int_a^b x^n w(x)\,dx, \quad n = 0, 1, 2, \ldots,$$

are finite.

Let $\Delta_{-1} = 1$ and let Δ_n, $n \geq 0$, be the determinant

$$\Delta_n = \begin{vmatrix} A_0 & A_1 & \cdots & A_n \\ A_1 & A_2 & \cdots & A_{n+1} \\ & & \ddots & \\ A_n & A_{n+1} & \cdots & A_{2n} \end{vmatrix}. \tag{4.1.1}$$

The associated quadratic form

$$\sum_{j,k=0}^n A_{j+k}\, a_j a_k = \int_a^b \left[\sum_{j=0}^n a_j x^j \right]^2 w(x)\,dx$$

is positive definite, so the determinant Δ_n is positive.

Consider the Hilbert space L_w^2, with inner product

$$(f,g) = (f,g)_w = \int_a^b f(x) g(x) w(x)\,dx.$$

The polynomial

$$Q_n(x) = \begin{vmatrix} A_0 & A_1 & \cdots & A_{n-1} & 1 \\ A_1 & A_2 & \cdots & A_n & x \\ & & \ddots & & \\ A_n & A_{n+1} & \cdots & A_{2n-1} & x^n \end{vmatrix} \tag{4.1.2}$$

is orthogonal to x^m, $m < n$, while $(Q_n, x^n) = \Delta_n$. To see this, expand the determinant (4.1.2) along the last column. Computing the inner product of Q_n with x^m results in a determinant in which the last column of the determinant (4.1.1) has been replaced by column $m+1$ of the same determinant. Thus if $m < n$, there is a repeated column, while $m = n$ gives Δ_n. Now $Q_n(x) = \Delta_{n-1} x^n$ plus terms of lower degree, so

$$(Q_n, Q_n) = (Q_n, \Delta_{n-1} x^n) = \Delta_{n-1}(Q_n, x^n) = \Delta_{n-1}\Delta_n.$$

Therefore the polynomials

$$P_n(x) = \frac{1}{\sqrt{\Delta_{n-1}\Delta_n}} Q_n(x)$$

are orthonormal. They are uniquely determined by the requirement that the leading coefficient be positive. The leading coefficient of P_n is $h_n = \sqrt{\Delta_{n-1}/\Delta_n}$.

Note that $xP_n(x)$ has degree $n+1$ and is orthogonal to x^m, $m < n-1$. It follows that for some constants a_n, b_n, c_n,

$$xP_n(x) = a_n P_{n+1}(x) + b_n P_n(x) + c_n P_{n-1}(x). \qquad (4.1.3)$$

Comparing coefficients of x^{n+1}, we see that $a_n = h_n/h_{n+1}$. On the other hand, taking the inner product with P_{n-1}, we have

$$c_n = (xP_n, P_{n-1}) = (P_n, xP_{n-1}) = \frac{h_{n-1}}{h_n} = a_{n-1}.$$

For later use, we note that the existence of a three-term recurrence formula of the form (4.1.3) depends only on the orthogonality properties of the P_n, not on the fact that they have norm 1 (so long as we do not require that $c_n = a_{n-1}$, as we do in the following calculation).

It follows from the previous two equations that

$$(x-y)P_n(x)P_n(y) = a_n[P_{n+1}(x)P_n(y) - P_n(x)P_{n+1}(y)]$$
$$- a_{n-1}[P_n(x)P_{n-1}(y) - P_{n-1}(x)P_n(y)].$$

Iterating, summing, and dividing by $x - y$, we get the *Christoffel–Darboux formula*

$$a_n \left[\frac{P_{n+1}(x)P_n(y) - P_n(x)P_{n+1}(y)}{x-y} \right] = \sum_{j=0}^{n} P_j(x)P_j(y). \qquad (4.1.4)$$

Recall that the coefficient on the left is the ratio of the leading coefficient of P_n to the leading coefficient of P_{n+1}.

This has an interesting consequence. Suppose that q is any polynomial of degree $\leq n$. Then q is a linear combination of the P_k, $k \leq n$, and orthonormality implies that

$$q = \sum_{j=0}^{n} (q, P_j) P_j.$$

Thus (4.1.4) implies the following.

Proposition 4.1.1 *If q is a polynomial of degree $\leq n$, then*

$$q(x) = \int_a^b K_n(x,y) q(y) w(y) \, dy, \qquad (4.1.5)$$

where

$$K_n(x,y) = a_n \left[\frac{P_{n+1}(x)P_n(y) - P_n(x)P_{n+1}(y)}{x-y} \right],$$

and a_n is the ratio of the leading coefficient of P_n to the leading coefficient of P_{n+1}.

The kernel function K_n plays the same role with respect to expansions in orthogonal polynomials as the classical Dirichlet kernel plays with respect to the classical Fourier expansion, so we refer to it as the *Dirichlet kernel* for the polynomials $\{P_n\}$.

The kernel function K_n can be realized as a determinant:

$$K_n(x,y) = -\frac{1}{\Delta_n}\begin{vmatrix} 0 & 1 & x & x^2 & \dots & x^n \\ 1 & A_0 & A_1 & A_2 & \dots & A_n \\ y & A_1 & A_2 & A_3 & \dots & A_{n+1} \\ & & & \ddots & & \\ y^n & A_n & A_{n+1} & A_{n+2} & \dots & A_{2n} \end{vmatrix}. \tag{4.1.6}$$

See Exercise 2.

Taking the limit as $y \to x$ in (4.1.4) gives

$$a_n\left[P'_{n+1}(x)P_n(x) - P_{n+1}(x)P'_n(x)\right] = \sum_{j=0}^{n} P_j(x)^2. \tag{4.1.7}$$

A first consequence of (4.1.7) is that the real roots of the P_n are simple: if x_0 were a double root, the left side of (4.1.7) would vanish at $x = x_0$, but P_0 is a nonzero constant, so the right side of (4.1.7) is positive.

The next result locates the zeros.

Proposition 4.1.2 P_n *has n real zeros, all lying in the interval I.*

Proof This is trivial for $n = 0$. Suppose $n \geq 1$ and let $x_1, \dots, x_m$ be the real roots of P_n lying in I. Set $q(x) = \prod(x - x_j)$. The sign changes of P_n and q in the interval I occur precisely at the x_j. Therefore the product $q P_n$ has fixed sign in I, so $(P_n, q) \neq 0$. This implies that q has degree at least n, so $m = n$. $\square$

A second consequence of (4.1.7) is that the roots of successive polynomials P_{n-1} and P_n interlace.

Proposition 4.1.3 *Between each pair of zeros of P_n is a zero of P_{n-1}.*

Proof For $n < 2$ there is nothing to prove. For $n \geq 2$, (4.1.7) implies that

$$P'_n(x)P_{n-1}(x) - P_n(x)P'_{n-1}(x) > 0.$$

Suppose that $x_1 < x_2$ are two successive zeros of P_n. Then the previous inequality implies that

$$P'_n(x_j)P_{n-1}(x_j) > 0, \qquad j = 1, 2.$$

Since $P'_n(x_1)$ and $P'_n(x_2)$ must have different signs, it follows from the preceding inequality that the $P_{n-1}(x_j)$ have different signs. Therefore P_{n-1} has a zero in the interval (x_1, x_2). □

We turn to the question of *completeness*: can every element of the Hilbert space L^2_w be written as a linear combination of the P_n? Suppose that f belongs to L^2_w. Consider the question of finding an element in the span of $\{P_j\}_{j \leq n}$ that is closest to f, with respect to the distance in L^2_w:

$$d(f,g) = ||f - g|| = (f - g, f - g)^{\frac{1}{2}}.$$

Proposition 4.1.4 *Let*

$$f_n = \sum_{j=0}^{n} (f, P_j) P_j = \int_a^b K_n(x, y) f(y) w(y) \, dy. \qquad (4.1.8)$$

Then f_n is the closest function to f in the span of $\{P_0, P_1, \ldots P_n\}$.

Proof Write

$$f - g = (f - f_n) + (f_n - g).$$

Computing inner products shows that $f - f_n$ is orthogonal to every element of the span of $\{P_j\}_{j \leq n}$, so if g is also in the span, then $f - f_n$ and $f_n - g$ are orthogonal, so

$$||f - g||^2 = ||f - f_n||^2 + ||f_n - g||^2. \qquad (4.1.9)$$

The left-hand side is minimal exactly when $g = f_n$. □

Taking $g = 0$ in (4.1.9) and using orthonormality of the P_n, we obtain *Bessel's inequality*:

$$||f||^2 = \sum_{j=0}^{n} (f, P_j)^2 + ||f - f_n||^2 \geq \sum_{j=0}^{n} (f, P_j)^2. \qquad (4.1.10)$$

The following completeness theorem applies to all the cases we shall consider.

Theorem 4.1.5 *Suppose that w is a positive weight on the interval (a, b) and that for some $c > 0$,*

$$\int_a^b e^{2c|x|} w(x) \, dx < \infty. \qquad (4.1.11)$$

Let $\{P_n\}$ be the orthonormal polynomials for w. For any $f \in L^2_w$,

$$f = \sum_{n=0}^{\infty} (f, P_n) P_n$$

in the sense that the partial sums of the series on the right converge to f in norm in L_w^2. Moreover, one has Parseval's equality:

$$\|f\|^2 = \sum_{n=0}^{\infty} (f, P_n)^2. \tag{4.1.12}$$

If the interval (a, b) is bounded, then condition (4.1.11) is redundant and one may use the Weierstrass polynomial approximation theorem to show that polynomials are dense in L_w^2. A proof that remains valid in the general case is given in Appendix B.

Exercise 17 shows that a condition like (4.1.11) cannot be eliminated or weakened too much.

4.2 Stieltjes transform and Padé approximants

A natural object associated with the weight function w on the interval I is its *Stieltjes transform* $F = Sw$. For z not in the closure of I,

$$F(z) = \int_I \frac{w(s)\, ds}{z - s}. \tag{4.2.1}$$

It is convenient here to impose the normalization condition

$$\int_I w(x)\, dx = 1. \tag{4.2.2}$$

The weight can be recovered from the Stieltjes transform.

Proposition 4.2.1 *For $x \in I$,*

$$w(x) = -\frac{1}{2\pi i} \lim_{\varepsilon \to 0+} [F(x + i\varepsilon) - F(x - i\varepsilon)]. \tag{4.2.3}$$

The proof is left as Exercise 15 – or see the Cauchy transform in Appendix A.

Let us ask for a sequence of good *rational approximations* to F as $z \to \infty$, or at least for $|\mathrm{Im}\, z| \to \infty$. The normalization condition (4.2.2) implies that $F(z) \sim 1/z$ as $|\mathrm{Im}\, z| \to \infty$, so the rational approximations should have the form $P_n(z)/Q_n(z)$ for some monic polynomials P_n and Q_n of degree $n - 1$ and n, respectively. By definition, the *Padé approximant* to F of degree n is a rational function of this form which is optimal in the sense that

$$F(z) - \frac{P_n(z)}{Q_n(z)} = O(z^{-2n-1}) \tag{4.2.4}$$

as $z \to \infty$ along nonreal rays. In general, this is as close as such an approximation can be. In fact, (4.2.4) may be rewritten as

$$Q_n(z)F(z) = P_n(z) + O(z^{-n-1}), \tag{4.2.5}$$

which is equivalent to a system of n linear equations for the n coefficients of lower-order terms of Q_n.

It follows from (4.2.1) that for any polynomial Q of degree n,

$$Q(z)F(z) = P(z) + R(z),$$

where

$$P(z) = \int_I \frac{Q(z) - Q(x)}{z - x} w(x) dx$$

is a polynomial of degree $n - 1$ and along nonreal rays

$$R(z) = \int_I \frac{Q(x)w(x)dx}{z - x}$$

$$= \sum_{k=0}^{n-1} \left[\int_I Q(x) x^k w(x) dx \right] z^{-k-1} + O(z^{-n-1}); \tag{4.2.6}$$

see Exercise 5. Therefore the unique choice of a monic polynomial Q that satisfies condition (4.2.5) is the monic orthogonal polynomial Q_n of degree n. The associated numerator is

$$P_n(z) = \int_I \frac{Q_n(z) - Q_n(x)}{z - x} w(x) dx. \tag{4.2.7}$$

For these monic orthogonal polynomials Q_n, the general three-term recurrence relation (4.1.3) takes the form

$$xQ_{n-1}(x) = Q_n(x) + b_n Q_{n-1}(x) + a_n Q_{n-2}(x), \tag{4.2.8}$$

where $Q_{-1} = 0$, $Q_0 = 1$, and since

$$(xQ_{n-1}, Q_{n-2}) = (Q_{n-1}, xQ_{n-2}) = ||Q_{n-1}||^2,$$

we have, for $n \geq 2$,

$$a_n = \frac{(xQ_{n-1}, Q_{n-2})}{||Q_{n-2}||^2} = \frac{||Q_{n-1}||^2}{||Q_{n-2}||^2} > 0.$$

The numerators P_n satisfy the same three-term recurrence,

$$xP_{n-1}(x) = P_n(x) + b_n P_{n-1}(x) + a_n P_{n-2}(x), \tag{4.2.9}$$

with $P_0 = 0$, $P_1 = 1$; see Exercise 6.

The difference between successive Padé approximants is

$$\frac{P_n}{Q_n} - \frac{P_{n-1}}{Q_{n-1}} = \frac{P_n Q_{n-1} - P_{n-1} Q_n}{Q_n Q_{n-1}}.$$

It follows from the recurrence relations (4.2.8) and (4.2.9) that

$$S_n \equiv P_n Q_{n-1} - P_{n-1} Q_n = a_n S_{n-1}.$$

Since $S_1 = 1$, it follows that

$$\frac{P_n}{Q_n} - \frac{P_{n-1}}{Q_{n-1}} = \frac{a_n a_{n-1} \cdots a_2}{Q_n Q_{n-1}}. \tag{4.2.10}$$

Consider now the difference between the Stieltjes transform F and the Padé approximant P_n/Q_n:

$$F(z) - \frac{P_n(z)}{Q_n(z)} = \int_I \frac{w(x)\,dx}{z-x} - \frac{1}{Q_n(z)} \int_I \frac{Q_n(z) - Q_n(x)}{z-x} w(x)\,dx$$

$$= \frac{1}{Q_n(z)} \int_I \frac{Q_n(x)\,w(x)\,dx}{z-x}.$$

Let $\delta(z)$ denote the distance from z to the interval I. Since Q_n is monic and its roots are in I, it follows that

$$|Q_n(z)| \geq \delta(z)^n \geq |\mathrm{Im}\, z|^n.$$

It follows from this, the normalization (4.2.2), and the Cauchy–Schwarz inequality, that

$$\left| F(z) - \frac{P_n(z)}{Q_n(z)} \right| \leq \frac{1}{\delta(z)^{n+1}} \int_I |Q_n(x)|\, w(x)\,dx$$

$$\leq \frac{1}{\delta(z)^{n+1}} \|Q_n\|. \tag{4.2.11}$$

The monic polynomials have a useful minimality property: see Exercise 7.

Lemma 4.2.2 *The monic orthogonal polynomial Q_n has a minimal norm among all monic polynomials of degree n.*

This lemma and (4.2.11) imply the following (see Exercise 8).

Proposition 4.2.3

$$\left| F(z) - \frac{P_n(z)}{Q_n(z)} \right| \leq \frac{1}{\delta(z)^{n+1}} \left\{ \int_I x^{2n}\, w(x)\,dx \right\}^{1/2}.$$

If the interval $I = (a, b)$ is finite, then the preceding result can be improved; see Exercise 9.

Proposition 4.2.4 *If $I = (a, b)$, then*

$$\left| F(z) - \frac{P_n(z)}{Q_n(z)} \right| \leq \frac{1}{\delta(z)^{n+1}} \left\{ \frac{b-a}{2} \right\}^n.$$

4.3 Padé approximants and continued fractions

The construction of the Padé approximants P_n/Q_n to the Stieltjes transform F of the weight w can be formulated in a different way, which is entirely analogous to a classical method of approximating a given real number by rational numbers: the method of *continued fractions*. We illustrate this method first in the numerical case.

Given a real number x, there is a unique integer β_0 such that $0 \leq x - \beta_0 < 1$. If $x \neq \beta_0$, let $x_1 = 1/(x - \beta_0) > 1$, so that

$$x = \beta_0 + \frac{1}{x_1},$$

and choose the integer β_1 such that $x_1 - \beta_1$ is in $[0, 1)$. If $x_1 \neq \beta_1$, let $x_2 = 1/(x_1 - \beta_1)$, so that $x_1 = \beta_1 + 1/x_2$ and

$$x = \beta_0 + \cfrac{1}{\beta_1 + \cfrac{1}{x_2}}.$$

As long as this process continues, we have a sequence of rational approximations

$$\beta_0, \qquad \beta_0 + \frac{1}{\beta_1}, \qquad \beta_0 + \cfrac{1}{\beta_1 + \cfrac{1}{\beta_2}}, \quad \ldots \qquad (4.3.1)$$

More generally, consider sequences of the form

$$\beta_0, \qquad \beta_0 + \frac{\alpha_1}{\beta_1}, \qquad \beta_0 + \cfrac{\alpha_1}{\beta_1 + \cfrac{\alpha_2}{\beta_2}}, \quad \ldots, \qquad (4.3.2)$$

which can be written as successive quotients

$$\beta_0, \qquad \frac{\beta_1 \beta_0 + \alpha_1}{\beta_1}, \qquad \frac{\beta_2(\beta_1 \beta_0 + \alpha_1) + \alpha_2 \beta_0}{\beta_2 \beta_1 + \alpha_2}, \quad \ldots$$

Let us write these as

$$\frac{p_0}{q_0}, \quad \frac{p_1}{q_1}, \quad \frac{p_2}{q_2}, \quad \ldots$$

Of course the representation as a quotient is not unique, but calculating a few more terms suggests that there is a unique choice such that

$$p_{-1} = 1, \quad p_0 = \beta_0; \qquad q_{-1} = 0, \quad q_0 = 1, \qquad (4.3.3)$$

and the p_n and q_n satisfy the three-term recurrence

$$p_n = \beta_n p_{n-1} + \alpha_n p_{n-2}, \qquad q_n = \beta_n q_{n-1} + \alpha_n q_{n-2}, \qquad (4.3.4)$$

$n \geq 1$. This can be verified by induction; see Exercise 10. Conversely, (4.3.3) and (4.3.4) imply that the ratios p_n/q_n are the successive *convergents* of the continued fraction

$$\beta_0 + \cfrac{\alpha_1}{\beta_1 + \cfrac{\alpha_2}{\beta_2 + \cdots}}.$$

The difference between successive convergents is

$$\frac{p_n}{q_n} - \frac{p_{n-1}}{q_{n-1}} = \frac{p_n q_{n-1} - p_{n-1} q_n}{q_n q_{n-1}},$$

and a consequence of (4.3.4) and (4.3.3) is that

$$p_n q_{n-1} - p_{n-1} q_n = -\alpha_n(p_{n-1} q_{n-2} - p_{n-2} q_{n-1}) = \cdots$$
$$= (-1)^{n+1} \alpha_n \alpha_{n-1} \cdots \alpha_1. \qquad (4.3.5)$$

Thus

$$\frac{p_n}{q_n} - \frac{p_{n-1}}{q_{n-1}} = (-1)^{n+1} \frac{\alpha_1 \cdots \alpha_n}{q_n q_{n-1}}. \qquad (4.3.6)$$

In the case of the numerical approximation, the β_n are positive integers and the α_n are 1, $n \geq 1$, so long as both are defined. Also,

$$q_n = \beta_n q_{n-1} + q_{n-2} \geq q_{n-1} + q_{n-2} \geq 2q_{n-2},$$

so q_n is at least the nth Fibonacci number F_n and is certainly $\geq 2^{(n-1)/2}$. Then

$$\frac{p_n}{q_n} - \frac{p_{n-1}}{q_{n-1}} = \frac{(-1)^{n+1}}{q_n q_{n-1}}.$$

If the number x being approximated is rational, the process will terminate. Otherwise, p_n/q_n is the partial sum of an absolutely convergent alternating series. In fact,

$$\frac{p_2}{q_2} < \frac{p_4}{q_4} < \cdots < x < \cdots < \frac{p_3}{q_3} < \frac{p_1}{q_1}$$

by induction, so

$$\left| x - \frac{p_n}{q_n} \right| \leq \frac{1}{q_{n+1} q_n} \leq \frac{1}{2^{n-\frac{1}{2}}}.$$

Let us find an analogous approximation to the Stieltjes transform F. To simplify, we work with $z = is$ in the positive imaginary axis. Along this ray, the function F has an asymptotic expansion

$$F(z) = \frac{1}{z} + \frac{A_1}{z^2} + \frac{A_2}{z^3} + \cdots, \qquad A_n = \int_I x^n w(x)\, dx; \qquad (4.3.7)$$

see Exercise 11. We emphasize that the computations to follow are, to a point, purely formal. The function F has a formal reciprocal

$$\frac{1}{F(z)} = z - b_1 + O(1/z), \quad b_1 = A_1. \tag{4.3.8}$$

Subtracting $z - b_1$ from $1/F$ gives a function F_1, with $F_1 = O(1/z)$, or

$$F(z) = \frac{1}{(z - b_1) + F_1(z)}, \quad F_1(z) = O(1/z).$$

Assuming that $F_1 \sim c_2/z$ with $c_2 \neq 0$, we take b_2 such that $(c_2/F_1) - (z - b_2) = F_2$, with $F_2 = O(1/z)$. Then

$$F(z) = \frac{1}{(z - b_1) + \dfrac{c_2}{(z - b_2) + F_2(z)}}, \quad F_2(z) = O(1/z).$$

Continuing, we obtain a formal continued fraction expansion of the Stieltjes transform (4.2.1) in the form

$$F(z) = \cfrac{1}{z - b_1 + \cfrac{c_2}{z - b_2 + \cfrac{c_3}{z - b_3 + \cdots}}}. \tag{4.3.9}$$

The corresponding convergents are

$$\frac{P_1(z)}{Q_1(z)} = \frac{1}{z - b_1},$$

$$\frac{P_2(z)}{Q_2(z)} = \frac{1}{z - b_1 + \dfrac{c_2}{z - b_2}},$$

and so on. This fits the pattern in (4.3.2) with $\alpha_n = c_n$ and $\beta_n = z - b_n$, so we find

$$P_n(z) = (z - b_n)P_{n-1}(z) + c_n P_{n-2}(z);$$

$$Q_n(z) = (z - b_n)Q_{n-1}(z) + c_n Q_{n-2}(z).$$

It follows that the P_n and Q_n are monic polynomials. Moreover, it can be shown by induction that

$$F(z) = \frac{P_n(z)}{Q_n(z)} + O(z^{-2n-1}).$$

Therefore these are the same polynomials that occurred in Section 4.2, with $a_n = -c_n$, and we conclude that the Padé approximants are identical to the convergents of the continued fraction (4.3.9).

4.4 Generalization: measures

Although we took the weight function w on the interval I to be positive throughout the interval, most of the results of the previous sections carry over immediately to more general nonnegative weights, including discrete weights. For example, we could take the interval to be $[0, \infty)$ and take the inner product to be defined by

$$(f, g) = \frac{1}{e} \sum_{m=0}^{\infty} \frac{f(m) g(m)}{m!}. \tag{4.4.1}$$

The moments are

$$(p_n, 1) = \frac{1}{e} \sum_{m=0}^{\infty} \frac{m^n}{m!},$$

$p_n(x) = x^n$. The associated orthogonal polynomials, up to normalization, are the Charlier polynomials $C_n(x; 1)$; see Chapter 6.

More generally, we may take the inner product to be defined by a *positive measure* on the line. For our purposes, it is most convenient to characterize a positive measure as a positive linear functional λ on the space $C_0(\mathbf{R})$ of continous complex-valued functions that are defined on the line $\mathbf{R}$ and have limit 0 at infinity, i.e. the mapping $\lambda : C_0(\mathbf{R}) \to \mathbf{C}$ is linear and satisfies the positivity condition

$$\text{if } f > 0 \text{ on } \mathbf{R}, \text{ then } \lambda(f) > 0. \tag{4.4.2}$$

A limiting argument shows that this condition implies

$$\text{if } f \geq 0 \text{ on } \mathbf{R}, \text{ then } \lambda(f) \geq 0. \tag{4.4.3}$$

Such a linear functional is commonly written in the more suggestive form

$$\lambda(f) = \int_{-\infty}^{\infty} f(x) \, d\lambda(x), \tag{4.4.4}$$

though for the Charlier example (4.4.1) the "integral" is a sum:

$$\lambda(f) = \frac{1}{e} \sum_{m=0}^{\infty} \frac{f(m)}{m!}.$$

(For a justification of the notation (4.4.4), see Exercise 12.)

If λ is such a measure and f is any nonnegative continuous function on $\mathbf{R}$, we may find a sequence $\{f_n\}$ of functions in $C_0(\mathbf{R})$ such that

$$f_1 \leq f_2 \leq f_3 \cdots ; \quad \lim_{n \to \infty} f_n(x) = f(x) \tag{4.4.5}$$

for all x. It follows from positivity that $\lambda(f_n)$ is a nondecreasing sequence. The limit $\lambda(f)$ (which may be infinite) is independent of the choice of sequence; see

Exercise 13. This procedure allows one to extend λ to the space consisting of all continuous functions f such that $\lambda(|f|)$ is finite.

We consider only measures such that $\lambda(p)$ is finite for every nonnegative polynomial p. Such a measure determines an inner product on the space $\mathcal{P}$ of all polynomials:

$$\langle p, q \rangle = \lambda(p \bar{q}).$$

This inner product is not necessarily positive definite on $\mathcal{P}$. For example, if λ is given by a finite sum

$$\lambda(f) = \sum_{k=1}^{m} a_k f(x_k),$$

and $p(x) = \prod_{k=1}^{m} (x - x_k)$, then $\langle p, p \rangle = 0$. We rule out such cases and assume that the inner product is positive definite. Then sequences of orthogonal polynomials can be constructed exactly as in Section 4.1. The algebraic arguments carry over unchanged. In particular, the normalized polynomials $\{P_n\}$ satisfy the three-term recurrence

$$xP_{n-1}(x) = a_n P_n(x) + b_n P_{n-1}(x) + a_{n-1}(x)P_{n-2}(x), \quad n \geq 2, \qquad (4.4.6)$$

with $a_m > 0$, $m \geq 1$.

Results of Sections 4.2 and 4.3 carry over as well, with suitable modifications. We take the Stieltjes transform associated with a general measure λ to be

$$F(z) = \lambda(r_z), \quad r_z(x) = \frac{1}{z - x}. \qquad (4.4.7)$$

In the notation (4.4.4), this is

$$F(z) = \int_{-\infty}^{\infty} \frac{d\lambda(x)}{z - x}.$$

Then λ can be recovered from F by a weaker form of (4.2.3). Given $\varepsilon > 0$, let λ_ε be the measure that is determined by the weight function

$$w_\varepsilon(x) = -\frac{1}{2\pi i} [F(x + i\varepsilon) - F(x - i\varepsilon)].$$

This means that, for any $g \in C_0(\mathbf{R})$,

$$\lambda_\varepsilon(g) = \int_{-\infty}^{\infty} g(x) w_\varepsilon(x) dx.$$

Then

$$\lambda(g) = \lim_{\varepsilon \to 0} \lambda_\varepsilon(g). \qquad (4.4.8)$$

4.5 Favard's theorem and the moment problem

Three-term recurrence relations for orthogonal polynomials played a crucial role in each of the previous sections, leading to the Christoffel–Darboux formula, the relative location of zeros (see also Exercise 3), the Stieltjes transform, Padé approximants, and the continued fraction representation. It is reasonable to ask the following. Suppose that $\{Q_n\}$ is a family of real polynomials such that Q_n has degree n, has positive leading coefficient, and satisfies a three-term recurrence

$$xQ_{n-1}(x) = a_nQ_n(x) + b_nQ_{n-1}(x) + c_nQ_{n-2}(x), \quad n \geq 2. \tag{4.5.1}$$

Is there a positive measure λ that extends to the space $\mathcal{P}$ of polynomials, such that the Q_n are orthogonal with respect to λ:

$$\lambda(Q_nQ_m) = 0 \quad \text{if } n \neq m? \tag{4.5.2}$$

Remarks. 1. Any such recurrence formula can be extended to all n by taking $Q_m = 0$ for $m < 0$.

2. If there is such a measure, then comparing the leading coefficients of each side of (4.5.1) shows that a_n must be positive, while taking the inner product of each side against Q_{n-2} shows that each c_n is positive. Replacing Q_n by α_nQ_n, with $\alpha_0 = 1$ and $\alpha_n = (c_{n+1}/a_n)^{1/2}\alpha_{n-1}$, we may assume that $c_n = a_{n-1}$:

$$xQ_{n-1}(x) = a_nQ_n(x) + b_nQ_{n-1}(x) + a_{n-1}Q_{n-2}(x), \quad a_n > 0. \tag{4.5.3}$$

Question 1. Given real polynomials $\{Q_n\}$ of degree n that satisfy a three-term recurrence (4.5.3) with $a_n > 0$, is there a positive linear functional on the space $\mathcal{P}$ of complex polynomials, such that
   (a) $\lambda(Q_mQ_n) = 0$ if $m \neq n$ and $\lambda(Q_n^2) = 1$;
   (b) λ has a positive extension to all of $C_0(\mathbf{R})$?

Question 2. If λ exists, is it unique?

The answer to part (a) of Question 1 is elementary. Assume that the Q_n are normalized to have positive leading term and to satisfy the recurrence (4.5.3), with $Q_0 = 1$. We introduce a Hermitian inner product $\langle\,,\,\rangle$ in $\mathcal{P}$ by declaring $\{Q_n\}$ to be an orthonormal basis. The operation T of multiplication by x: $T(p) = xp$, is symmetric with respect to this inner product. In fact, (4.5.3) and the orthonormality of the Q_n imply

$$\left\langle T\sum_m \alpha_mQ_m, \sum_n \beta_nQ_n \right\rangle = \left\langle \sum_m \alpha_m(a_{m+1}Q_{m+1} + b_{m+1}Q_m + a_mQ_{m-1}), \sum_n \beta_nQ_n \right\rangle$$

$$= \sum_m \alpha_m(a_{m+1}\overline{\beta}_{m+1} + b_{m+1}\overline{\beta}_m + a_m\overline{\beta}_{m-1}).$$

A similar calculation, holding m fixed and expanding TQ_n, shows that

$$\left\langle \sum_m \alpha_m Q_m, T \sum_n \beta_n Q_n \right\rangle$$

has the same value. (The sums here are taken to be finite.)

Iteration of this result implies that multiplication by any real polynomial is symmetric. It follows that for complex polynomials we have the crucial identity

$$\langle pP, Q \rangle = \langle P, \bar{p} Q \rangle. \tag{4.5.4}$$

Define the functional λ on $\mathcal{P}$ by

$$\lambda(P) = \langle P, Q_0 \rangle = \langle P, 1 \rangle.$$

Positivity follows from (4.5.4). In fact, if the polynomial P is nonnegative on **R**, then the nonreal roots come in complex conjugate pairs and the real roots have even multiplicity, so P has a polynomial factorization $P = \overline{Q} Q$. Then

$$\lambda(P) = \langle \overline{Q} Q, 1 \rangle = \langle Q, Q \rangle \geq 0.$$

The interesting questions are whether λ has a positive extension to $C_0(\mathbf{R})$, and whether such an extension is unique. The answer to the first question is usually referred to as *Favard's theorem*, but see the remarks at the end of this chapter.

Theorem 4.5.1 *Given real polynomials Q_n of degree n, with positive leading coefficient, that satisfy a three-term recurrence (4.5.3) with $a_n > 0$, there is a positive measure λ such that the $\{Q_n\}$ are orthonormal with respect to λ.*

Proof Let $\mathcal{P}_n$ denote the space of complex polynomials of degree $\leq n$. Let π_n denote the orthogonal projection of $\mathcal{P}$ onto $\mathcal{P}_n$: $\pi_n(Q_k) = Q_k$, $k \leq n$, $\pi_n(Q_k) = 0$, $k > n$. The π_n are symmetric, so the operator $T_n = \pi_n T \pi_n$ is symmetric. Note that $T_n = T$ on $\mathcal{P}_{n-1}$. Let λ_n denote the functional

$$\lambda_n(P) = \langle P(T_n)Q_0, Q_0 \rangle.$$

This is a positive linear functional. The operator T_n is essentially acting in the n-dimensional space $\mathcal{P}_n$, and the standard finite-dimensional spectral theory implies that λ_n has the form

$$\lambda_n(P) = \sum_{j=1}^{n} a_j P(x_{j,n})$$

where the $x_{j,n}$ are the points of the spectrum of T_n restricted to $\mathcal{P}_n$, repeated according to multiplicity. In other words, λ_n is a discrete measure supported on these points. Moreover the total mass $\lambda_n(1)$ is 1, since $\langle Q_0, Q_0 \rangle \equiv 1$ by definition. It is a standard fact that such a sequence of measures has a subsequence $\{\lambda_{n_k}\}$ with a *weak-* limit* which is a measure λ_∞. This means that, for any continuous function f on **R** with limit 0 as $x \to \pm\infty$,

$$\lim_{n_k \to \infty} \lambda_{n_k}(f) = \lambda_\infty(f). \tag{4.5.5}$$

This is a general fact from functional analysis; see Exercise 14.

Finally, for any given polynomial Q, if N is sufficiently large, $Q(T_N) = Q(T)$, so $\lambda_N(Q) = \lambda(Q)$. Therefore λ_∞ is an extension of λ.   □

For convenience, we drop the subscript and denote the extension λ_∞ simply by λ. Consider the question of uniqueness. If the measure λ has *bounded support*, i.e. if there is a closed bounded interval I such that $\lambda(f) = 0$ whenever $f \equiv 0$ on I, then λ is unique. This is a consequence of the *Weierstrass approximation theorem*: *any continuous function on I is a uniform limit of polynomials*; see Appendix B.

To investigate uniqueness in general, we recall (4.4.8): λ can be recovered from its Stieltjes transform F_λ. We look at the Padé approximants to F_λ. The Q_n here are normalized to be orthonormal rather than monic, but we may construct the Padé approximants in the form $\{P_n/Q_n\}$, with P_n defined relative to Q_n as before:

$$P_n(z) = \int_{-\infty}^{\infty} \frac{Q_n(z) - Q_n(x)}{z - x} \, d\lambda(x). \tag{4.5.6}$$

Previously we considered the difference $F - P_n/Q_n$. Consider now

$$F_\lambda(z) Q_n(z) - P_n(z) = \int_{-\infty}^{\infty} \frac{Q_n(x) \, d\lambda(x)}{z - x}.$$

The term on the right can be considered as the inner product of the function $(z - x)^{-1}$ with Q_n. Since the Q_n are orthonormal in the space $L^2(d\lambda)$, Bessel's inequality gives

$$\sum_{n=0}^{\infty} |F_\lambda(z) Q_n(z) - P_n(z)|^2 \leq \int_{-\infty}^{\infty} \frac{d\lambda(x)}{|z - x|^2} \leq \frac{1}{|\mathrm{Im}\, z|^2}. \tag{4.5.7}$$

Suppose that there is a second measure $\mu \neq \lambda$ for which the Q_n are orthonormal. This measure must have the same moments as λ, and must therefore give rise to the same auxiliary polynomials P_n. However, its Stieltjes transform F_μ must differ from F_λ at some (in fact most) $z \notin \mathbf{R}$. It follows from (4.5.7), and the corresponding inequality involving F_μ, at a point z where F_μ and F_λ differ, that both the sequences $\{Q_n(z)\}$ and $\{P_n(z)\}$ are square summable:

$$\sum_{n=0}^{\infty} \left(|Q_n(z)|^2 + |P_n(z)|^2 \right) < \infty. \tag{4.5.8}$$

As in the previous construction, the auxiliary polynomials $\{P_n\}$ here satisfy the same three-term recurrence (4.5.3) as the $\{Q_n\}$ from which they are constructed. It is a simple exercise to verify that

$$
\begin{aligned}
P_n(z)Q_{n-1}(z) &- P_{n-1}(z)Q_n(z) \\
&= \frac{a_{n-1}}{a_n}\left[P_{n-1}(z)Q_{n-2}(z) - P_{n-2}(z)Q_{n-1}(z)\right] = \cdots \qquad (4.5.9) \\
&= \frac{a_1}{a_n}\left[P_1(z)Q_0(z) - P_0(z)Q_1(z)\right] = \frac{1}{a_n}.
\end{aligned}
$$

It follows from (4.5.8) that the sequence of terms in brackets in (4.5.9) has a finite sum. Thus a consequence of the existence of two distinct measures for which the Q_n are orthonormal is that $\sum_{n=1}^{\infty} 1/a_n$ is finite. We have derived a result due to Carleman [63].

Theorem 4.5.2 *If*

$$
\sum_{n=1}^{\infty} \frac{1}{a_n} = \infty,
$$

then there is a unique measure with respect to which the polynomials $\{Q_n\}$ are orthonormal.

Remark. The material in this section is closely related to the *moment problem*: given a real sequence $\{m_n\}_{n\geq 0}$, is there a measure μ such that $\{m_n\}$ is the sequence of moments

$$
m_n = \int_{-\infty}^{\infty} x^n \, d\mu(x), \quad n = 0, 1, 2, \ldots ?
$$

If so, is μ unique? See [7]. For an example of nonuniqueness, see Exercise 16. For a discussion, references, and further examples, see the article by Penson, Blasiak, Duchamp, Horzela, and Solomon [321].

4.6 Asymptotic distribution of zeros

Suppose that $\{P_n\}$ is a sequence of orthogonal polynomials on an interval I. Let $\{x_{n,k}\}_{k=1}^{n}$ be the zeros of P_n. Let λ_n be the associated normalized discrete measure

$$
\lambda_n(f) = \frac{1}{n}\sum_{k=1}^{n} f(x_{n,k}).
$$

This is a sequence of positive measures, each of which has total mass 1, so again some subsequence has a limit. The natural questions here are: is the limit unique, and, if so, what is it? These questions have been thoroughly investigated in recent decades: see the article [234] by Kuijlaars and Van

Assche and the books by Saff and Totik [350] and Van Assche [407]. Results for the classical orthogonal polynomials are given in the next chapter. Here we cite only a somewhat simplified version of the pioneering result of Erdős and Turán [118] for orthogonal polynomials on a bounded interval with weight w.

Theorem 4.6.1 (Erdős and Turán) *Suppose $I = (-1, 1)$ and the weight function w is positive on I; then for any continuous function f on the closed interval $[-1, 1]$,*

$$\lim_{n \to \infty} \lambda_n(f) = \frac{1}{\pi} \int_{-1}^{1} \frac{f(x) \, dx}{\sqrt{1 - x^2}}.$$

We return to this result in Section 5.7 for the special case of Jacobi polynomials, along with analogous results for (rescaled) Laguerre and Hermite polynomials. For the full result, see [118]. For generalizations, see [407] and [234].

4.7 Exercises

1. Show that the kernel K_n in Proposition 4.1.1 is uniquely determined by the following conditions: (i) it is a polynomial of degree n in x and in y and (ii) the identity (4.1.5) holds for every polynomial of degree $\leq n$.
2. Prove (4.1.6). Hint: show that the function on the right-hand side satisfies the conditions in Exercise 1.
3. Suppose that $\{Q_n\}$ is a sequence of monic orthogonal polynomials for the weight w on the bounded interval I. Let $\{P_n/Q_n\}$ be the Padé approximants to the Stieltjes transform F of w. Multiply the identity (4.2.10) by $Q_n Q_{n-1}$ and use the resulting identity to show that there is a zero of P_n between any two zeros of Q_n. Conclude that the zeros of P_n are simple.
4. Suppose that Q_n and P_n are as in Exercise 3. Let $\{x_{j,n}\}_{j=1}^{n}$ be the zeros of Q_n. Suppose that $p(z)$ is a polynomial of degree $\leq 2n - 1$. Let C be a curve that encloses the interval I. Note that

$$\frac{P_n(z)}{Q_n(z)} = \frac{\lambda_{1,n}}{z - x_{1,n}} + \frac{\lambda_{2,n}}{z - x_{2,n}} + \cdots + \frac{\lambda_{n,n}}{z - x_{n,n}}$$

for certain constants $\lambda_{j,n}$.

(a) Show that

$$\lambda_{j,n} = \frac{P_n(x_{j,n})}{Q'_n(x_{j,n})}.$$

(b) Integrate $p(z)/2\pi i$ against both sides of

$$F(z) = \frac{P_n(z)}{Q_n(z)} - R(z), \quad R(z) = O(z^{-2n-1})$$

over the curve C and show that

$$\int_I p(z)\, w(z)\, dz = \sum_{j=1}^n \lambda_{j,n} p(x_{j,n}) \tag{4.7.1}$$

for some constants $\lambda_{j,n}$. (Hint: $p(z) = q(z) Q_n(z) + r(z)$, where q, r are polynomials and r has degree $< n$. Then $p(x_{j,n}) = r(x_{j,n})$.)

(c) Let

$$p(x) = \left[\frac{Q_n(x)}{x - x_{j,n}} \right]^2,$$

a polynomial of degree $2n - 2$. Show that for this choice of p, the previous calculation gives

$$\int_I p(x)\, w(x)\, dx = \lambda_{j,n} \left[Q_n'(x_{j,n}) \right]^2.$$

The left-hand side is positive, so $\lambda_{j,n}$ is positive.

(d) Use the positivity of the $\lambda_{j,n}$ and the Weierstrass theorem on approximation by polynomials to show that for any continuous function f on the closed interval,

$$\int_I f(x)\, w(x)\, dx = \lim_{n \to \infty} \sum_{j=1}^n \lambda_{j,n} f(x_{j,n}).$$

(e) The identity (4.7.1), valid for every polynomial of degree $< 2n$, is known as the *Gauss quadrature formula*. It was proved by Gauss [151] for Legendre polynomials and by Stieltjes [380] for general orthogonal polynomials on a bounded interval. Note that the formula, and the proof, carry over to general measures $d\lambda(x)$; see Section 4.4.

5. Prove (4.2.6).

6. Prove (4.2.9).

7. Prove Lemma 4.2.2.

8. Prove Proposition 4.2.3.

9. Use Lemma 4.2.2 and a judicious choice of a comparison monic polynomial P of degree n to prove Proposition 4.2.4.

10. Prove (4.3.4).

11. Prove (4.3.7).

12. Justify the notation (4.4.4) as follows. Suppose that λ is a positive measure and suppose for convenience that $\lambda(1)$ is finite, say $\lambda(1) = 1$. At some risk of confusion, define the *function* λ by

$$\lambda(x) = \sup\{\lambda(g) : g \in C_0(\mathbf{R}),\ g \le 1,\ g(y) = 0 \text{ if } y \ge x\}.$$

Suppose that $f \in C_0(\mathbf{R})$ vanishes outside a bounded interval $[a,b]$. Given a partition P of $[a,b]$ defined by points $\{x_k\}$,

$$a = x_0 < x_1 < x_2 < \cdots < x_n = b,$$

define the Riemann sum

$$\lambda_P(f) = \sum_{k=1}^n f(x_k)\big[\lambda(x_k) - \lambda(x_{k-1})\big].$$

Show that $\lambda(f)$ is the limit of the $\lambda_P(f)$ as the mesh $\sup\{x_k - x_{k-1}\}$ converges to zero.

13. Prove that the limit in (4.4.5) is independent of the nondecreasing sequence $\{f_n\}$.

14. Prove the result used in the proof of the Favard theorem: if $\{\lambda_n\}$ is a sequence of positive measures with $\lambda_n(1) = 1$ for all n, then there is a positive measure λ and a subsequence $\{\lambda_{n_k}\}$ such that

$$\lim \lambda_{n_k}(f) = \lambda(f), \quad \text{all } f \in C_0(\mathbf{R}).$$

Hint: show that there is a countable subset of $C_0(\mathbf{R})$ that is dense with respect to uniform convergence. Then a diagonal process produces a subsequence and a limit such that (4.5.5) is true for each f in the countable dense subset. Uniform convergence and positivity carry the result over to limits of these functions.

15. Prove Proposition 4.2.1 and (4.4.8).

16. This exercise constructs an example of nonuniqueness for the moment problem.

(a) Show that for $\operatorname{Re}\omega > 0$ and $n = 0,1,2,\ldots$,

$$\int_0^\infty t^n e^{-\omega t}\, dt = \frac{\Gamma(n+1)}{\omega^{n+1}}. \tag{4.7.2}$$

(b) With $\omega = e^{\frac{1}{4}\pi i}$, show that the integral (4.7.2) is real when $n+1$ is divisible by 4.

(c) With ω as in part (b), show that

$$\int_0^\infty t^n \exp\left(-\frac{t}{\sqrt{2}}\right) \sin\left(\frac{t}{\sqrt{2}}\right) dt = 0$$

when $n+1$ is divisible by 4.

(d) Let $s(x) = e^{-x^{1/4}} \sin(x^{1/4})$, $x > 0$. Show that

$$\int_0^\infty x^n s(x)\, dx = 0, \qquad n = 0,1,2,\ldots$$

(e) Let

$$w_a(x) = e^{-x^{1/4}} - a s(x), \quad x > 0, \quad 0 \le a < 1.$$

Show that the w_a are positive weights on the half-line, and that they all have the same moments.

(f) Show that the moments are

$$\int_0^\infty x^n w_a(x)\,dx = 4(4n+3)!.$$

17. Use results from Exercise 16 to show that the conclusion of Theorem 4.1.5 fails for the weight $w(x) = e^{-x^{1/4}}$ on $(0, \infty)$: polynomials are not dense in L_w^2.

4.8 Remarks

General orthogonal polynomials are an integral part of the theory of moments, continued fractions, and spectral theory; see, for example, Akhiezer [7] and the various books cited below and at the end of the next chapter. They arose in studies by Chebyshev [74] and Stieltjes [384] of certain types of continued fractions.

The Christoffel–Darboux formula was found by Chebyshev [70] in 1855, then rediscovered in the case of Legendre polynomials by Christoffel [77] in 1858, and in the general case by Christoffel [79] and Darboux [93] in 1877–1878.

Versions of the result commonly known as Favard's theorem were found before Favard, e.g. by Stieltjes [384] in 1895, Perron [320] and Wintner [441] in 1929, Stone [389] in 1932, and Sherman [364] in 1933, as well as Favard [129] in 1935 and Shohat [365] in 1936. For a discussion of the history and further developments, see the article [272] by Marcellán and Álvarez-Nodarse. For a thorough discussion of connections with the classical moment problems and the general theory of symmetric operators in Hilbert space, see Akhiezer [7], Section X.4 of Stone's treatise [389], Simon's article [366] and Section 3.8 of Simon's monograph [367], Part 2.

Various nonclassical orthogonal polynomials are treated in the books by Freud [144], Krall [229], Macdonald [267], Nevai [299], Saff and Totik [350], Simon [367], and Stahl and Totik [375]. The book by Van Assche [407] is devoted to asymptotics of various classes of orthogonal polynomials. See also the books cited in the remarks at the end of the following chapter.

5

The classical orthogonal polynomials

This chapter relies on the first section of the preceding chapter: recurrence relations, the Christoffel–Darboux formula, and completeness. The chapter begins with an examination of some features that are common to the three types of classical orthogonal polynomials. These features include Rodrigues formulas and representations as integrals.

In succeeding sections, each of the three classical cases is considered in more detail, as are some special cases of Jacobi polynomials (Legendre and Chebyshev polynomials). The question of pointwise convergence of the expansion in orthogonal polynomials is addressed. Asymptotic distribution of zeros is obtained, as well as a characterization of zeros in terms of electrostatic equilibrium.

Finally, we return to integral representations and the construction of a second solution of each of the differential equations.

5.1 Classical polynomials: general properties, I

We return to the three cases corresponding to the classical polynomials, with interval I, weight w, and eigenvalue equation of the form

$$p(x)\,\psi_n''(x) + q(x)\,\psi_n'(x) + \lambda_n \psi_n(x) = 0, \quad q = \frac{(pw)'}{w}, \qquad (5.1.1)$$

or equivalently,

$$(pw\,\psi_n')' + \lambda_n w \psi_n = 0.$$

The cases are

$$I = \mathbf{R} = (-\infty, \infty), \quad w(x) = e^{-x^2}, \quad p(x) = 1, q(x) = -2x;$$

$$I = \mathbf{R}_+ = (0, \infty), \quad w(x) = x^\alpha e^{-x}, \ \alpha > -1,$$
$$p(x) = x, q(x) = \alpha + 1 - x;$$

94

$$I = (-1, 1), \quad w(x) = (1-x)^\alpha (1+x)^\beta, \ \alpha > -1, \ \beta > -1,$$

$$p(x) = 1 - x^2, \quad q(x) = \beta - \alpha - (\alpha + \beta + 2)x.$$

The derivative of a solution ψ_n of (5.1.1) satisfies a similar equation: differentiating (5.1.1) gives

$$p[\psi_n']'' + (q + p')[\psi_n']' + (q' + \lambda_n)\psi_n' = 0.$$

Now

$$q + p' = \frac{(pw)'}{w} + p' = \frac{p^2 w' + 2pp'w}{pw} = \frac{[p(pw)]'}{pw}.$$

Thus the function pw is also a weight. Since q' is a constant in each of the three cases, ψ_n' is an orthogonal polynomial of degree $n - 1$ for the weight pw. Continuing, ψ_n'' is an orthogonal polynomial of degree $n - 2$ for the weight $p^2 w$, with eigenvalue

$$-\lambda_n - q' - (p' + q)' = -\lambda_n - 2q' - p''.$$

By induction, the mth derivative $\psi_n^{(m)}$ corresponds to weight $p^m w$, with eigenvalue

$$-\lambda_n - mq' - \frac{1}{2} m(m-1) p''.$$

Since $\psi_n^{(n)}$ is constant, the corresponding eigenvalue is zero and we have the general formula

$$\lambda_n = -nq' - \frac{1}{2} n(n-1) p'',$$

which corresponds to the results obtained in the three cases above:

$$\lambda_n = 2n, \qquad \lambda_n = n, \qquad \lambda_n = n(n + \alpha + \beta + 1), \tag{5.1.2}$$

respectively.

Equation (5.1.1) can be rewritten as

$$w\psi_n = -\lambda_n^{-1}(pw\psi_n')'. \tag{5.1.3}$$

Since pw is the weight corresponding to ψ_n', this leads to

$$w\psi_n = [\lambda_n(\lambda_n + q')]^{-1}(p^2 w \psi_n'')'',$$

and finally to

$$w\psi_n = (-1)^n \prod_{m=0}^{n-1} \left[\lambda_n + mq' + \frac{1}{2} m(m-1) p'' \right]^{-1} (p^n w \psi_n^{(n)})^{(n)}. \tag{5.1.4}$$

Since $\psi_n^{(n)}$ is constant, we may normalize by taking

$$\psi_n(x) = w(x)^{-1} \frac{d^n}{dx^n} \{ p(x)^n w(x) \}. \tag{5.1.5}$$

This is known as the *Rodrigues formula*. In view of (5.1.4), with this choice of ψ_n, we have

$$\psi_n(x) = a_n x^n + \text{lower order}, \tag{5.1.6}$$

$$n!\, a_n = (-1)^n \prod_{m=0}^{n-1} \left[\lambda_n + mq' + \frac{1}{2}m(m-1)p'' \right].$$

In our three cases the product on the right is, respectively,

$$(-2)^n n!, \qquad (-1)^n n!, \qquad (-1)^n n!\,(\alpha + \beta + n + 1)_n,$$

so the leading coefficient a_n is, respectively,

$$(-2)^n, \qquad (-1)^n, \qquad (-1)^n (\alpha + \beta + n + 1)_n. \tag{5.1.7}$$

As a first application of the Rodrigues formula, we consider the calculation of weighted inner products of the form

$$(f, \psi_n) = (f, \psi_n)_w = \int_a^b f(x)\, \psi_n(x)\, w(x)\, dx.$$

By (5.1.5), $\psi_n w = (p^n w)^{(n)}$. The function $p^n w$ vanishes fairly rapidly at the endpoints of the interval I. Therefore, under rather mild conditions on the function f, we may integrate by parts n times without acquiring boundary terms, to obtain

$$\int_a^b \psi_n(x) f(x)\, w(x)\, dx = (-1)^n \int_a^b p(x)^n f^{(n)}(x)\, w(x)\, dx. \tag{5.1.8}$$

This idea can be used in conjunction with (5.1.6) to calculate the weighted L^2 norms:

$$\int_a^b \psi_n^2(x) w(x)\, dx = (-1)^n \int_a^b \psi_n^{(n)} p(x)^n w(x)\, dx$$

$$= \prod_{m=0}^{n-1} \left[\lambda_n + mq' + \frac{1}{2}m(m-1)p'' \right] \int_a^b p(x)^n w(x)\, dx. \tag{5.1.9}$$

In the three cases considered, the preceding integral is, respectively,

$$\int_{-\infty}^{\infty} e^{-x^2}\, dx = \sqrt{\pi};$$

$$\int_0^{\infty} x^{n+\alpha} e^{-x}\, dx = \Gamma(n + \alpha + 1);$$

$$\int_{-1}^1 (1 - x)^{n+\alpha} (1 + x)^{n+\beta}\, dx = 2^{2n+\alpha+\beta+1} \int_0^1 s^{n+\alpha}(1 - s)^{n+\beta}\, ds$$

$$= 2^{2n+\alpha+\beta+1} \mathrm{B}(n + \alpha + 1, n + \beta + 1).$$

Thus the squares of the weighted L^2 norm of ψ_n in the three cases are, respectively,

$$||\psi_n||^2 = 2^n n! \sqrt{\pi},$$

$$||\psi_n||^2 = n! \Gamma(n + \alpha + 1),$$

$$||\psi_n||^2 = 2^{2n+\alpha+\beta+1} n! \frac{\Gamma(n+\alpha+1)\Gamma(n+\beta+1)}{(2n+\alpha+\beta+1)\Gamma(n+\alpha+\beta+1)}.$$

The standard normalizations of the classical polynomials differ from the choice given by the Rodrigues formula (5.1.5). The Rodrigues formula for the Hermite, Laguerre, and Jacobi polynomials, respectively, is taken to be

$$H_n(x) = (-1)^n e^{x^2} \frac{d^n}{dx^n}(e^{-x^2}); \tag{5.1.10}$$

$$L_n^{(\alpha)}(x) = \frac{1}{n!} x^{-\alpha} e^x \frac{d^n}{dx^n}(e^{-x} x^{n+\alpha}); \tag{5.1.11}$$

$$P_n^{(\alpha,\beta)}(x) = \frac{(-1)^n}{n! 2^n}(1-x)^{-\alpha}(1+x)^{-\beta}$$
$$\times \frac{d^n}{dx^n}\{(1-x)^{n+\alpha}(1+x)^{n+\beta}\}. \tag{5.1.12}$$

In view of these normalizations, the previous calculation of weighted L^2 norms gives

$$||H_n||^2 = 2^n n! \sqrt{\pi}, \tag{5.1.13}$$

$$||L_n^{(\alpha)}||^2 = \frac{\Gamma(n+\alpha+1)}{n!}, \tag{5.1.14}$$

$$||P_n^{(\alpha,\beta)}||^2 = \frac{2^{\alpha+\beta+1}\Gamma(n+\alpha+1)\Gamma(n+\beta+1)}{n!(2n+\alpha+\beta+1)\Gamma(n+\alpha+\beta+1)}. \tag{5.1.15}$$

The normalizations, together with (5.1.7), imply that the leading coefficients of H_n, $L_n^{(\alpha)}$, and $P_n^{(\alpha,\beta)}$ are

$$2^n, \qquad \frac{(-1)^n}{n!}, \qquad \frac{(\alpha+\beta+n+1)_n}{2^n n!}, \tag{5.1.16}$$

respectively. The discussion at the end of Section 3.4 shows that Laguerre and Hermite polynomials can be obtained as certain limits of Jacobi polynomials. In view of that discussion and this calculation of leading coefficients, it follows that

$$H_n(x) = \lim_{\alpha \to +\infty} \frac{2^n n!}{\alpha^{n/2}} P_n^{(\alpha,\alpha)}\left(\frac{x}{\sqrt{\alpha}}\right);$$ (5.1.17)

$$L_n^{(\alpha)}(x) = \lim_{\beta \to +\infty} P_n^{(\alpha,\beta)}\left(1 - \frac{2x}{\beta}\right).$$ (5.1.18)

5.2 Classical polynomials: general properties, II

We may take advantage of the Cauchy integral formula for derivatives of $w p^n$ to derive an integral formula from (5.1.5); see Appendix A. Let Γ be a curve that encloses $x \in I$ but excludes the endpoints of I. Then (5.1.5) shows that

$$\frac{\psi_n(x)}{n!} = \frac{1}{2\pi i} \int_\Gamma \frac{w(z)}{w(x)} \frac{p(z)^n}{(z-x)^n} \frac{dz}{z-x}.$$ (5.2.1)

The *generating function* for the orthogonal polynomials $\{\psi_n/n!\}$ is defined to be

$$G(x,s) = \sum_{n=0}^{\infty} \frac{\psi_n(x)}{n!} s^n.$$

The integral formula (5.2.1) allows the evaluation of G:

$$\begin{aligned}
G(x,s) &= \frac{1}{2\pi i} \int_\Gamma \sum_{n=0}^{\infty} \frac{s^n p(z)^n}{(z-x)^n} \cdot \frac{w(z)}{w(x)} \cdot \frac{dz}{z-x} \\
&= \frac{1}{2\pi i} \int_\Gamma \frac{w(z)}{w(x)} \cdot \frac{dz}{z-x-sp(z)}.
\end{aligned}$$

We may assume that Γ encloses a single solution $z = \zeta(x,s)$ of $z - x = s p(z)$. Since the residue of the integrand at this point is $w(\zeta)w(x)^{-1}[1 - sp'(\zeta)]^{-1}$, we obtain

$$G(x,s) = \frac{w(\zeta)}{w(x)} \cdot \frac{1}{1 - sp'(\zeta)}, \qquad \zeta - sp(\zeta) = x.$$ (5.2.2)

In the following sections, we give the explicit evaluation in the case of the Jacobi, Hermite, and Laguerre functions. In each of the latter two cases we give a second derivation of the formula for the generating function.

The integral formula (5.2.1) can also be used to obtain recurrence relations of the type (4.1.3) for the polynomials ψ_n. It will be slightly more convenient to work with

$$\varphi_n(x) = \frac{\psi_n(x)}{n!} w(x) = \frac{1}{2\pi i} \int_C \frac{p(z)^n w(z) dz}{(z-x)^{n+1}}$$

and look for a three-term recurrence relation

$$a_n \varphi_{n+1}(x) = b_n(x) \varphi_n(x) + c_n \varphi_{n-1}(x) \quad b_n(x) = b_{n0} + b_{n1} x. \tag{5.2.3}$$

If a is constant, then an integration by parts gives

$$\int_C \left[\frac{a p(z)^{n+1} w(z)}{(z-x)^{n+2}} - \frac{b(z) p(z)^n w(z)}{(z-x)^{n+1}} - \frac{c p^{n-1}(z) w(z)}{(z-x)^n} \right] dz$$
$$= \int_C \left[\frac{\tilde{a} [p^{n+1} w]'(z) - b(z) p(z)^n w(z) - (z-x) c p(z)^{n-1} w(z)}{(z-x)^{n+1}} \right] dz,$$

where $\tilde{a} = a/(n+1)$. Constants a, b_0, b_1, and c can be chosen so that for fixed x the last integrand is a derivative:

$$\frac{d}{dz} \left\{ \frac{Q(z) p(z)^n w(z)}{(z-x)^n} \right\}, \quad Q(z) = Q_0 + Q_1(z-x),$$

where Q_0 and Q_1 are constants. Since $(pw)' = qw$, it follows that

$$[p^m w]' = [p^{m-1}(pw)]' = [(m-1)p' + q] p^{m-1} w.$$

Applying this to both previous expressions, we find that the condition on $\tilde{a}, b, c$, and Q is

$$\tilde{a}(np' + q)p - bp - c(z-x)$$
$$= -nQp + \{Q'p + Q[(n-1)p' + q]\}(z-x).$$

Expanding p, q, and Q in powers of $z - x$ leads to a system of four linear equations for the four unknowns $b/\tilde{a}$, $c/\tilde{a}$, Q_0, and Q_1. The results for the three cases we have been considering are the following, respectively:

$$a_n \varphi_{n+1}(x) = b_n(x) \varphi_n(x) + c_n \varphi_{n-1}(x); \tag{5.2.4}$$

$$\begin{cases} p(x) = 1, \quad w(x) = e^{-x^2} : \\ a_n = n+1, \quad b_n(x) = -2x, \quad c_n = -2; \end{cases}$$

$$\begin{cases} p(x) = x, \quad w(x) = x^\alpha e^{-x} : \\ a_n = n+1, \quad b_n(x) = (2n + \alpha + 1) - x, \quad c = -(n+\alpha); \end{cases}$$

$$\begin{cases} p(x) = 1 - x^2, \quad w(x) = (1-x)^\alpha (1+x)^\beta : \\ a_n = (n+1)(n+\alpha+\beta+1)(2n+\alpha+\beta); \\ b_n(x) = (2n+\alpha+\beta+1)[\beta^2 - \alpha^2 - (2n+\alpha+\beta)(2n+\alpha+\beta+2)x]; \\ c_n = -4(2n+\alpha+\beta+2)(n+\alpha)(n+\beta). \end{cases}$$

Taking into account the normalizations above, the three-term recurrence relations in the three classical cases are

$$H_{n+1}(x) = 2x H_n(x) - 2n H_{n-1}(x); \tag{5.2.5}$$

$$(n+1)L_{n+1}^{(\alpha)}(x) = (2n+\alpha+1-x)L_n^{(\alpha)}(x) - (n+\alpha)L_{n-1}^{(\alpha)}(x); \tag{5.2.6}$$

$$\frac{(2n+2)(n+\alpha+\beta+1)}{2n+\alpha+\beta+1} P_{n+1}^{(\alpha,\beta)}(x)$$
$$= \left[\frac{\alpha^2 - \beta^2}{2n+\alpha+\beta} + (2n+\alpha+\beta+2)x \right] P_n^{(\alpha,\beta)}(x)$$
$$- \frac{2(2n+\alpha+\beta+2)(n+\alpha)(n+\beta)}{(2n+\alpha+\beta)(2n+\alpha+\beta+1)} P_{n-1}^{(\alpha,\beta)}(x). \tag{5.2.7}$$

The starting point in the derivation of the Rodrigues formula (5.1.5) was that if ψ_n is the degree n polynomial for weight w, then the derivative ψ_n' is a multiple of the degree $n-1$ polynomial for weight pw. In the Hermite case, $pw = w$; in the other two cases going to pw raises the index or indices by 1. Taking into account the leading coefficients of the ψ_n given in (5.1.7) and the normalizations, we obtain

$$H_n'(x) = 2n H_{n-1}(x); \tag{5.2.8}$$

$$\left[L_n^{(\alpha)} \right]'(x) = -L_{n-1}^{(\alpha+1)}(x); \tag{5.2.9}$$

$$\left[P_n^{(\alpha,\beta)} \right]'(x) = \frac{1}{2}(n+\alpha+\beta+1) P_{n-1}^{(\alpha+1,\beta+1)}(x). \tag{5.2.10}$$

The method used to derive (5.2.3) can be used to obtain derivative formulas of a different form:

$$p(x)\varphi_n'(x) = a_n(x)\varphi_n(x) + b_n\varphi_{n-1}(x).$$

In fact,

$$p(x)\varphi_n'(x) = \frac{n+1}{2\pi i} \int_C \frac{p(x)p(z)^n w(z)\,dz}{(z-x)^{n+2}}.$$

For fixed x we may expand $p(x)$ in powers of $(z-x)$ and integrate one summand by parts to put the integrand into the form

$$\frac{\tilde{p}_0(p^n w)'(z) + p_1 p(z)^n w(z) + p_2(z-x)p(z)^n w(z)}{(z-x)^{n+1}},$$

where

$$\tilde{p}_0 = \frac{p_0}{n+1}, \; p_0 = p(x), \; p_1 = -p'(x), \; p_2 = \frac{1}{2}p''.$$

Using the integral forms of φ_n and φ_{n-1} as well, we may treat the equation

$$p(x)\varphi_n'(x) - a(x)\varphi_n(x) - b\varphi_{n-1}(x) = 0 \qquad (5.2.11)$$

in the same way as we treated (5.2.3). The resulting formulas are

$$H_n'(x) = 2n\,H_{n-1}(x); \qquad (5.2.12)$$

$$x\left[L_n^{(\alpha)}\right]'(x) = nL_n^{(\alpha)}(x) - (n+\alpha)L_{n-1}^{(\alpha)}(x); \qquad (5.2.13)$$

$$(1-x^2)\left[P_n^{(\alpha,\beta)}\right]'(x) = \left[\frac{n(\alpha-\beta)}{2n+\alpha+\beta} - nx\right]P_n^{(\alpha,\beta)}(x)$$

$$+ \frac{2(n+\alpha)(n+\beta)}{2n+\alpha+\beta}\,P_{n-1}^{(\alpha,\beta)}(x). \qquad (5.2.14)$$

The recurrence formulas and the calculation of the norms allow us to compute the associated Dirichlet kernels. To see this, suppose that we have the identities

$$x\varphi_n(x) = a_n\varphi_{n+1}(x) + b_n\varphi_n(x) + c_n\varphi_{n-1}(x) \qquad (5.2.15)$$

for a sequence of orthogonal polynomials $\{\varphi_n\}$. The associated orthonormal polynomials are

$$\widetilde{\varphi}_n = ||\varphi_n||^{-1}\varphi_n.$$

The Christoffel–Darboux formula (4.1.4) implies that the Dirichlet kernel

$$K_n(x,y) = \sum_{j=0}^n \widetilde{\varphi}_j(x)\widetilde{\varphi}_j(y)$$

$$= ||\varphi_n||^{-2}\varphi_n(x)\varphi_n(y) + K_{n-1}(x,y)$$

is given by

$$\alpha_n\left[\frac{\widetilde{\varphi}_{n+1}(x)\widetilde{\varphi}_n(y) - \widetilde{\varphi}_n(x)\widetilde{\varphi}_{n+1}(y)}{x-y}\right] = \beta_n\left[\frac{\varphi_{n+1}(x)\varphi_n(y) - \varphi_n(x)\varphi_{n+1}(y)}{x-y}\right]$$

for some constants α_n, β_n. It follows from these equations, together with (5.2.4), that the constant β_n is $a_n/||\varphi_n||^2$. These observations lead to the following evaluations of the Dirichlet kernels associated with the Hermite, Laguerre, and Jacobi polynomials respectively:

$$K_n^H(x,y) = \frac{1}{2^{n+1}n!\sqrt{\pi}}$$

$$\times \left[\frac{H_{n+1}(x)H_n(y) - H_n(x)H_{n+1}(y)}{x-y}\right]; \qquad (5.2.16)$$

$$K_n^{(\alpha)}(x,y) = -\frac{(n+1)!}{\Gamma(n+\alpha+1)}$$

$$\times \left[\frac{L_{n+1}^{(\alpha)}(x)L_n^{(\alpha)}(y) - L_n^{(\alpha)}(x)L_{n+1}^{(\alpha)}(y)}{x-y} \right]; \qquad (5.2.17)$$

$$K_n^{(\alpha,\beta)}(x,y) = \frac{2^{-\alpha-\beta}(n+1)!\,\Gamma(n+\alpha+\beta+2)}{(2n+\alpha+\beta+2)\Gamma(n+\alpha+1)\,\Gamma(n+\beta+1)}$$

$$\times \left[\frac{P_{n+1}^{(\alpha,\beta)}(x)P_n^{(\alpha,\beta)}(y) - P_n^{(\alpha,\beta)}(x)P_{n+1}^{(\alpha,\beta)}(y)}{x-y} \right]. \qquad (5.2.18)$$

The *discriminant* of a polynomial $P(z) = a \prod_{j=1}^{n} (z - z_j)$ is the polynomial

$$D(z) = a^{2n-2} \prod_{1 \le j < k \le n} (z_j - z_k)^2.$$

The discriminants of the Hermite, Laguerre, and Jacobi polynomials are, respectively,

$$D_n^H = 2^{3n(n-1)/2} \prod_{j=1}^{n} j^j;$$

$$D_n^{(\alpha)} = \prod_{j=1}^{n} j^{j-2n+2}(j+\alpha)^{j-1};$$

$$D_n^{(\alpha,\beta)} = \frac{1}{2^{n(n-1)}} \prod_{j=1}^{n} \frac{j^{j-2n+2}(j+\alpha)^{j-1}(j+\beta)^{j-1}}{(n+j+\alpha+\beta)^{j-n}}.$$

For a proof, see, for example, Section 6.71 of [392].

The next three sections contain additional results and some alternative derivations for these three classical cases.

5.3 Hermite polynomials

The Hermite polynomials $\{H_n\}$ are orthogonal polynomials associated with the weight e^{-x^2} on the line $\mathbf{R} = (-\infty, \infty)$. They are eigenfunctions

$$H_n''(x) - 2xH_n'(x) + 2nH_n(x) = 0, \qquad (5.3.1)$$

satisfy the derivative relation

$$H_n'(x) = 2nH_{n-1}(x),$$

and can be defined by the Rodrigues formula

$$H_n(x) = (-1)^n e^{x^2} \frac{d^n}{dx^n}\{e^{-x^2}\} = \left(2x - \frac{d}{dx}\right)^n \{1\}.$$

It follows that the leading coefficient is 2^n and that

$$H'_n(x) - 2x H_n(x) = -H_{n+1}(x). \tag{5.3.2}$$

They are limits:

$$H_n(x) = \lim_{a\to+\infty} \frac{2^n}{a^{n/2}} P_n^{(a,a)}\left(\frac{x}{\sqrt{a}}\right).$$

The three-term recurrence relation (5.2.5) may also be derived as follows. It is easily shown by induction that H_n is even if n is even and odd if n is odd:

$$H_n(-x) = (-1)^n H_n(x).$$

Therefore the relation must have the form

$$x H_n(x) = a_n H_{n+1}(x) + b_n H_{n-1}(x). \tag{5.3.3}$$

Identities (5.3.2) and (5.3.3) imply that $H'_n = 2b_n H_{n-1}$. Comparing leading coefficients, we see that $a_n = \frac{1}{2}$ and $b_n = n$:

$$x H_n(x) = \frac{1}{2}H_{n+1}(x) + n H_{n-1}(x). \tag{5.3.4}$$

If we write $H_n(x) = \sum_{k=0}^{n} a_k x^k$, (5.3.1) implies the relation

$$(k+2)(k+1)a_{k+2} = 2(k-n)a_k.$$

Since $a_n = 2^n$ and $a_{n-1} = 0$, this recursion gives

$$H_n(x) = \sum_{2j\le n}(-1)^j \frac{n!}{j!(n-2j)!}(2x)^{n-2j}. \tag{5.3.5}$$

The first six of the H_n are

$$H_0(x) = 1;$$
$$H_1(x) = 2x;$$
$$H_2(x) = 4x^2 - 2;$$
$$H_3(x) = 8x^3 - 12x;$$
$$H_4(x) = 16x^4 - 48x^2 + 12;$$
$$H_5(x) = 32x^5 - 160x^3 + 120x.$$

Taking into account the factor $(-1)^n$, the generating function

$$G(x,s) = \sum_{n=0}^{\infty} \frac{H_n(x)}{n!} s^n$$

is calculated from (5.2.2) with $p(x) = 1$, and s is replaced by $-s$, so $\zeta(x,s) = x - s$ and

$$\sum_{n=0}^{\infty} \frac{H_n(x)}{n!} s^n = \frac{e^{-(x-s)^2}}{e^{-x^2}} = e^{2xs-s^2}. \tag{5.3.6}$$

This can also be calculated from the three-term recurrence relation (5.3.4), which is equivalent to

$$2xG(x,s) = \frac{\partial G}{\partial s}(x,s) + 2sG(x,s).$$

Therefore

$$G(x,s) = c(x)e^{-(s-x)^2}.$$

Since $G(x,0) = 1$, we obtain (5.3.6).

The generating function (5.3.6) can be used to obtain two *addition formulas* for the Hermite polynomials:

$$H_n(x+y) = \sum_{j+k+2l=n} \frac{n!}{j!k!l!} H_j(x) H_k(y) \tag{5.3.7}$$

and

$$H_n(x+y) = 2^{-\frac{1}{2}n} \sum_{m=0}^{n} \binom{n}{m} H_m(\sqrt{2}x) H_{n-m}(\sqrt{2}y); \tag{5.3.8}$$

see Exercises 9 and 10.

The generating function may also be used to give an alternative calculation of the weighted L^2 norms (5.1.13):

$$\sum_{m,n=0}^{\infty} \frac{s^m t^n}{m!n!} \int_{-\infty}^{\infty} H_m(x) H_n(x) e^{-x^2} dx = \int_{-\infty}^{\infty} G(x,s) G(x,t) e^{-x^2} dx$$

$$= \int_{-\infty}^{\infty} e^{2st} e^{-(x-s-t)^2} dx$$

$$= e^{2st} \sqrt{\pi} = \sqrt{\pi} \sum_{n=0}^{\infty} \frac{(2st)^n}{n!}.$$

This confirms that the H_n are mutually orthogonal, and that

$$\int_{-\infty}^{\infty} H_n(x)^2 e^{-x^2} dx = n! 2^n \sqrt{\pi}. \tag{5.3.9}$$

Therefore the normalized polynomials are

$$\tilde{H}_n(x) = \frac{1}{\pi^{\frac{1}{4}}\sqrt{n!2^n}}H_n(x). \tag{5.3.10}$$

According to Theorem 4.1.5, a given function $f \in L^2(\mathbf{R}, e^{-x^2}dx)$ can be approximated in L^2 by the sequence

$$f_n(x) = \sum_{m=0}^{n}(f,\tilde{H}_m)\tilde{H}_m = \int K_n(x,y)f(y)e^{-y^2}dy. \tag{5.3.11}$$

To compute the coefficients $(f,\tilde{H}_n)$, or equivalently, (f,H_n), we may use the Rodrigues formula, as in (5.1.8). If the function f and its derivatives to order n are of at most exponential growth as $|x| \to \infty$, then

$$(f,H_n) = \int_{-\infty}^{\infty} f(x)H_n(x)e^{-x^2}dx = \int_{-\infty}^{\infty} e^{-x^2}f^{(n)}(x)dx.$$

For example, $(x^m, H_n) = 0$, unless m and n are both even or both odd, and unless $m \geq n$. If $m = n + 2k$, the previous calculation and a change of variable give

$$(x^m, H_n) = \frac{m!}{(m-n)!}\int_{-\infty}^{\infty} e^{-x^2}x^{m-n}dx$$

$$= \frac{m!}{(m-n)!}\int_{0}^{\infty} e^{-t}t^{\frac{1}{2}(m-n-1)}dt$$

$$= \frac{m!}{(m-n)!}\Gamma\left(\frac{1}{2}[m-n+1]\right). \tag{5.3.12}$$

Similarly,

$$(e^{ax}, H_n) = a^n \int_{-\infty}^{\infty} e^{ax-x^2}dx$$

$$= a^n \int_{-\infty}^{\infty} e^{-(x-\frac{1}{2}a)^2}e^{\frac{1}{4}a^2}dx = a^n e^{\frac{1}{4}a^2}\int_{-\infty}^{\infty} e^{-x^2}dx$$

$$= a^n e^{\frac{1}{4}a^2}\sqrt{\pi}. \tag{5.3.13}$$

It follows from Cauchy's theorem or by analytic continuation that the identity (5.3.13) remains valid for all complex a. In particular, we may take $a = \pm ib$ to calculate

$$(\cos bx, H_n) = \begin{cases} \sqrt{\pi}(ib)^n e^{-\frac{1}{4}b^2}, & n \text{ even,} \\ 0, & n \text{ odd;} \end{cases} \tag{5.3.14}$$

$$(\sin bx, H_n) = \begin{cases} -i\sqrt{\pi}(ib)^n e^{-\frac{1}{4}b^2}, & n \text{ odd,} \\ 0, & n \text{ even.} \end{cases} \tag{5.3.15}$$

A particular case of the calculation in (5.3.13) is the identity

$$e^{-x^2} = \frac{1}{\sqrt{\pi}} \int_{-\infty}^{\infty} e^{-2ixt - t^2} \, dt.$$

This identity and the Rodrigues formula imply the integral formula

$$H_n(x) = (-1)^n \frac{e^{x^2}}{\sqrt{\pi}} \int_{-\infty}^{\infty} (-2it)^n e^{-2ixt - t^2} \, dt. \qquad (5.3.16)$$

This formula can be used to find a generating function for the products

$$H_n(x)H_n(y) = \frac{e^{x^2 + y^2}}{\pi} \int_{-\infty}^{\infty} \int_{-\infty}^{\infty} (-4tu)^n e^{-2ixt - 2iyu - t^2 - u^2} \, dt \, du.$$

Indeed, for $|s| < 1$,

$$\sum_{n=0}^{\infty} \frac{H_n(x)H_n(y)}{2^n n!} s^n = \frac{e^{x^2 + y^2}}{\pi} \int_{-\infty}^{\infty} \int_{-\infty}^{\infty} e^{-2ixt - 2iyu - 2tus - t^2 - u^2} \, dt \, du$$

$$= \frac{e^{x^2 + y^2}}{\pi} \int_{-\infty}^{\infty} \left\{ \int_{-\infty}^{\infty} e^{2i(-x + ius)t - t^2} \, dt \right\} e^{-2iyu - u^2} \, du$$

$$= \frac{e^{x^2 + y^2}}{\pi} \int_{-\infty}^{\infty} e^{-(x - ius)^2 - 2iyu - u^2} \, du$$

$$= \frac{e^{y^2}}{\sqrt{\pi}} \int_{-\infty}^{\infty} e^{-2i(y - xs)u - (1 - s^2)u^2} \, du.$$

Taking $v = u\sqrt{1 - s^2}$ as a new variable of integration gives

$$\sum_{n=0}^{\infty} \frac{H_n(x)H_n(y)s^n}{2^n n!} = \frac{1}{\sqrt{1 - s^2}} \exp\left(\frac{2xys - s^2 x^2 - s^2 y^2}{1 - s^2} \right). \qquad (5.3.17)$$

The general result regarding zeros, Proposition 4.1.2, together with the results proved in Section 3.5, gives the following.

Theorem 5.3.1 *The Hermite polynomial $H_n(x)$ has n simple roots, lying in the interval*

$$-\sqrt{2n + 1} < x < \sqrt{2n + 1}.$$

The relative maxima of

$$|e^{-x^2/2} H_n(x)|$$

increase as $|x|$ increases.

The results in Section 3.5 can be used to give more detailed information about the zeros of $H_n(x)$; see Exercises 15 and 16.

Theorem 5.3.2 *The positive zeros $x_{1n} < x_{2n} < \cdots$ of $H_n(x)$ satisfy the following estimates. If $n = 2m$ is even, then*

$$\frac{(2k-1)\pi}{2\sqrt{2n+1}} < x_{kn} < \frac{4k+1}{\sqrt{2n+1}}, \quad k = 1, 2, \ldots, m. \tag{5.3.18}$$

If $n = 2m+1$ is odd, then

$$\frac{k\pi}{\sqrt{2n+1}} < x_{kn} < \frac{4k+3}{\sqrt{2n+1}}, \quad k = 1, 2, \ldots, m. \tag{5.3.19}$$

For the asymptotic distribution of zeros, see Section 5.7.

The following asymptotic result is proved in Section 13.1:

$$H_n(x) = 2^{\frac{1}{2}n} \frac{2^{\frac{1}{4}}(n!)^{\frac{1}{2}}}{(n\pi)^{\frac{1}{4}}} e^{\frac{1}{2}x^2} \left[\cos\left(\sqrt{2n+1}\,x - \frac{1}{2}n\pi \right) + O(n^{-\frac{1}{2}}) \right] \tag{5.3.20}$$

as $n \to \infty$, uniformly on any bounded interval. In view of (5.3.22) and (5.3.23) below, (5.3.20) also follows from Fejér's result for Laguerre polynomials [132]. For more comprehensive asymptotic results, see Chapter 7.

A different normalization $w(x) = e^{-\frac{1}{2}x^2}$ is sometimes used for the weight function. The corresponding orthogonal polynomials, denoted by $\{He_n\}$, are eigenfunctions

$$[He_n]''(x) - x[He_n]'(x) + nHe_n(x) = 0$$

and are given by the Rodrigues formula

$$He_n(x) = (-1)^n e^{\frac{1}{2}x^2} \frac{d^n}{dx^n} \{e^{-\frac{1}{2}x^2}\} = \left[x - \frac{d}{dx} \right]^n \{1\}.$$

Setting $y = x/\sqrt{2}$, it is clear that $H_n(y)$ must be a multiple of $He_n(x)$, and consideration of the leading coefficients shows that

$$He_n(x) = 2^{-n/2} H_n\left(\frac{x}{\sqrt{2}} \right). \tag{5.3.21}$$

As noted in Section 3.4, there is a close relationship between Hermite polynomials and certain Laguerre polynomials. Since the Hermite polynomials H_{2n} of even order are even functions, they are orthogonal with respect to the weight $w(x) = e^{-x^2}$ on the half-line $x > 0$. Let $y = x^2$; then the measure $w(x)\,dx$ becomes

$$\frac{1}{2} y^{-\frac{1}{2}} e^{-y}\,dy.$$

The polynomials $\{H_{2n}(\sqrt{y})\}$ are therefore multiples of the Laguerre polynomials $L_n^{(-\frac{1}{2})}$. Consideration of the leading coefficients (see the next section) shows that the relationship is

$$H_{2n}(x) = (-1)^n 2^{2n} n! L_n^{(-\frac{1}{2})}(x^2), \qquad n = 0, 1, 2, \ldots \tag{5.3.22}$$

Similarly, the polynomials $x^{-1}H_{2n+1}$ are even functions that are orthogonal with respect to the weight $x^2 e^{-x^2}$ on the half-line, so they must be multiples of the Laguerre polynomials $L_n^{(\frac{1}{2})}(x^2)$:

$$H_{2n+1}(x) = (-1)^n 2^{2n+1} n! x L_n^{(\frac{1}{2})}(x^2), \qquad n = 0, 1, 2, \ldots \tag{5.3.23}$$

5.4 Laguerre polynomials

The Laguerre polynomials $\{L_n^{(\alpha)}\}$ are orthogonal polynomials associated with the weight $w(x) = x^\alpha e^{-x}$ on the half-line $\mathbf{R}_+ = (0, \infty)$. For a given $\alpha > -1$ they are the eigenfunctions

$$x\left[L_n^{(\alpha)}\right]''(x) + (\alpha + 1 - x)\left[L_n^{(\alpha)}\right]'(x) + nL_n^{(\alpha)}(x) = 0; \tag{5.4.1}$$

see Case II of Section 3.4. They satisfy the derivative relation

$$\left[L_n^{(\alpha)}\right]'(x) = -L_{n-1}^{(\alpha+1)}(x)$$

and are given by the Rodrigues formula (5.1.11)

$$L_n^{(\alpha)}(x) = \frac{1}{n!} x^{-\alpha} e^x \frac{d^n}{dx^n} \{x^\alpha e^{-x} x^n\}$$

$$= \frac{1}{n!} \left[\frac{d}{dx} + \frac{\alpha}{x} - 1\right]^n \{x^n\}, \tag{5.4.2}$$

where the second version is obtained by using the gauge transformation $u = \varphi v$ with $\varphi = x^\alpha e^{-x}$. It follows that the leading coefficient is $(-1)^n/n!$. The Laguerre polynomials for $\alpha = 0$ are denoted also by L_n:

$$L_n(x) = L_n^{(0)}(x).$$

The Laguerre polynomials are limits:

$$L_n^{(\alpha)}(x) = \lim_{\beta \to +\infty} P_n^{(\alpha,\beta)}\left(1 - \frac{2x}{\beta}\right).$$

Set $L_n^{(\alpha)}(x) = \sum_{k=0}^n b_k x^k$. Equation (5.4.1) gives the recurrence relation

$$(k+1)(k+\alpha+1)b_{k+1} = -(n-k)b_k,$$

so

$$L_n^{(\alpha)}(x) = \sum_{k=0}^{n} (-1)^k \frac{(\alpha+1)_n}{k!(n-k)!(\alpha+1)_k} x^k$$

$$= \frac{(\alpha+1)_n}{n!} \sum_{k=0}^{n} \frac{(-n)_k}{(\alpha+1)_k k!} x^k. \tag{5.4.3}$$

The first four of the $L_n^{(\alpha)}$ are

$$L_0^{(\alpha)}(x) = 1;$$

$$L_1^{(\alpha)}(x) = \alpha + 1 - x;$$

$$L_2^{(\alpha)}(x) = \frac{(\alpha+1)(\alpha+2)}{2} - (\alpha+2)x + \frac{1}{2}x^2;$$

$$L_3^{(\alpha)}(x) = \frac{(\alpha+1)(\alpha+2)(\alpha+3)}{6} - \frac{(\alpha+2)(\alpha+3)}{2} x + \frac{(\alpha+3)}{2}x^2 - \frac{1}{6}x^3.$$

In particular,

$$L_n(x) = \sum_{k=0}^{n} (-1)^k \frac{n!}{k!(n-k)!k!} x^k.$$

Comparing coefficients shows that the general three-term recurrence relation (4.1.3) is

$$xL_n^{(\alpha)}(x) = -(n+1)L_{n+1}^{(\alpha)}(x) + (2n+\alpha+1)L_n^{(\alpha)}(x)$$
$$-(n+\alpha)L_{n-1}^{(\alpha)}(x). \tag{5.4.4}$$

By induction,

$$\frac{d^n}{dx^n}\{xf(x)\} = x\frac{d^n}{dx^n}\{f(x)\} + n\frac{d^{n-1}}{dx^{n-1}}\{f(x)\},$$

so the Rodrigues formula gives the recurrence relation

$$(n+1)L_{n+1}^{(\alpha)}(x) = \left[x\frac{d}{dx} + \alpha + n + 1 - x\right]\{L_n^{(\alpha)}(x)\}. \tag{5.4.5}$$

Taking into account the normalization (5.4.2), the generating function

$$G(x,s) = \sum_{n=0}^{\infty} L_n^{(\alpha)}(x)s^n$$

can be calculated from (5.2.2). Here $p(x) = x$, so $\zeta = x/(1-s)$, and therefore

$$G(x,s) = \frac{e^{-xs/(1-s)}}{(1-s)^{\alpha+1}}. \tag{5.4.6}$$

This can also be calculated from the three-term recurrence relation (5.4.4), which is equivalent to

$$\frac{\partial G}{\partial s}(x,s) = \frac{\alpha+1}{1-s}G(x,s) - \frac{x}{(1-s)^2}G(x,s).$$

Since $G(x,0) \equiv 1$, this implies (5.4.6).

As for Hermite polynomials, the generating function can be used to obtain an addition formula:

$$L_n^{(\alpha)}(x+y) = \sum_{j+k+l=n} (-1)^j \frac{(\alpha+2-j)_j}{j!} L_k^{(\alpha)}(x) L_l^{(\alpha)}(y); \qquad (5.4.7)$$

see Exercise 20.

The generating function can also be used to calculate L^2 norms:

$$\sum_{m,n=0}^{\infty} s^m t^n \int_0^\infty L_m^{(\alpha)}(x) L_n^{(\alpha)}(x) x^\alpha e^{-x}\,dx$$

$$= \int_0^\infty G(x,s)G(x,t) x^\alpha e^{-x}\,dx$$

$$= \frac{1}{(1-s)^{\alpha+1}(1-t)^{\alpha+1}} \int_0^\infty e^{-x(1-st)/(1-s)(1-t)} x^{\alpha+1} \frac{dx}{x}.$$

Letting $y = x(1-st)/(1-s)(1-t)$, the last integral is

$$\frac{\Gamma(\alpha+1)}{(1-st)^{\alpha+1}} = \sum_{n=0}^{\infty} \frac{\Gamma(\alpha+1+n)}{n!}(st)^n.$$

This confirms that the $L_n^{(\alpha)}$ are mutually orthogonal, and

$$\int_0^\infty \left[L_n^{(\alpha)}(x)\right]^2 x^\alpha e^{-x}\,dx = \frac{\Gamma(\alpha+n+1)}{n!}.$$

Therefore the normalized polynomials are

$$\tilde{L}_n^{(\alpha)}(x) = \frac{\sqrt{n!}}{\sqrt{\Gamma(\alpha+n+1)}} L_n^{(\alpha)}(x).$$

To compute the coefficients of the expansion

$$f = \sum_{n=0}^{\infty} (f, \tilde{L}_n^{(\alpha)}) \tilde{L}_n^{(\alpha)},$$

we may use (5.1.8). If f and its derivatives to order n are bounded as $x \to 0$ and of at most polynomial growth as $x \to +\infty$,

$$(f, L_n^{(\alpha)}) = \int_0^\infty f(x) L_n^{(\alpha)}(x) x^\alpha e^{-x}\,dx = \frac{(-1)^n}{n!} \int_0^\infty e^{-x} f^{(n)}(x) x^{n+\alpha}\,dx.$$

In particular, if $m \geq n$, then

$$\int_0^\infty x^m L_n^{(\alpha)}(x) x^\alpha e^{-x} \, dx = (-1)^n \binom{m}{n} \Gamma(\alpha + m + 1) \qquad (5.4.8)$$

and for $\operatorname{Re} a > -1$,

$$(e^{-ax}, L_n^{(\alpha)}) = \frac{a^n \Gamma(n + \alpha + 1)}{n! (a+1)^{n+\alpha+1}}. \qquad (5.4.9)$$

For general values of λ and for $c > 1$, the equation

$$xu''(x) + (c - x)u'(x) - \lambda u(x) = 0$$

has a unique solution that is regular at $x = 0$ with $u(0) = 1$. It is known as the *confluent hypergeometric function* or *Kummer function*, $_1F_1(\lambda, c; x) = M(\lambda, c; x)$; see Chapter 8. In view of this and (5.4.3),

$$L_n^{(\alpha)}(x) = \frac{(\alpha + 1)_n}{n!} \, _1F_1(-n, \alpha + 1; x)$$

$$= \frac{(\alpha + 1)_n}{n!} M(-n, \alpha + 1; x). \qquad (5.4.10)$$

The general result regarding zeros, Proposition 4.1.2, together with the results proved in Section 3.5, gives the following.

Theorem 5.4.1 *The Laguerre polynomial $L_n^{(\alpha)}(x)$ has n simple roots in the interval*

$$0 < x < 2n + \alpha + 1 + \sqrt{(2n + \alpha + 1)^2 + (1 - \alpha^2)}. \qquad (5.4.11)$$

The relative maxima of

$$|e^{-x/2} x^{(\alpha+1)/2} L_n^{(\alpha)}(x)|$$

increase as x increases.

For the asymptotic distribution of zeros, see Section 5.7.

The following asymptotic result of Fejér [132, 133] is proved in Section 13.2:

$$L_n^{(\alpha)}(x) = \frac{e^{\frac{1}{2}x} n^{\frac{1}{2}\alpha - \frac{1}{4}}}{\sqrt{\pi} \, x^{\frac{1}{2}\alpha + \frac{1}{4}}} \left[\cos\left(2\sqrt{nx} - \frac{1}{2}\left[\alpha + \frac{1}{2}\right]\pi\right) + O(n^{-\frac{1}{2}}) \right] \qquad (5.4.12)$$

as $n \to \infty$, uniformly on any subinterval $\delta \leq x \leq \delta^{-1}, \delta > 0$.

For more comprehensive asymptotic results, see Sections 7.1 and 7.2.

5.5 Jacobi polynomials

The Jacobi polynomials $\{P_n^{(\alpha,\beta)}\}$ with indices $\alpha, \beta > -1$ are orthogonal with respect to the weight $w(x) = (1 - x)^\alpha (1 + x)^\beta$ on the interval $(-1, 1)$. The norms are

$$||P_n^{(\alpha,\beta)}||^2 = \frac{2^{\alpha+\beta+1}\,\Gamma(n+\alpha+1)\,\Gamma(n+\beta+1)}{n!\,(2n+\alpha+\beta+1)\,\Gamma(n+\alpha+\beta+1)}.$$

Changing the sign of x,

$$P_n^{(\alpha,\beta)}(-x) = (-1)^n P_n^{(\beta,\alpha)}(x). \tag{5.5.1}$$

The $P_n^{(\alpha,\beta)}$ are eigenfunctions:

$$(1-x^2)\left[P_n^{(\alpha,\beta)}\right]'' + [\beta - \alpha - (\alpha+\beta+2)x]\left[P_n^{(\alpha,\beta)}\right]'$$
$$+ n(n+\alpha+\beta+1)P_n^{(\alpha,\beta)} = 0. \tag{5.5.2}$$

They satisfy the derivative relation

$$\left[P_n^{(\alpha,\beta)}\right]'(x) = \frac{1}{2}(n+\alpha+\beta+1)P_{n-1}^{(\alpha+1,\beta+1)}(x)$$

and can be defined by the Rodrigues formula (5.1.12)

$$P_n^{(\alpha,\beta)}(x) = \frac{(-1)^n}{n!\,2^n}(1-x)^{-\alpha}(1+x)^{-\beta}\frac{d^n}{dx^n}\{(1-x)^{\alpha+n}(1+x)^{\beta+n}\}.$$

It follows from the extended form of Leibniz's rule that

$$P_n^{(\alpha,\beta)}(x) = \frac{(-1)^n}{2^n}\sum_{k=0}^{n}(-1)^k\frac{1}{k!\,(n-k)!}$$

$$\times \frac{(\alpha+1)_n\,(\beta+1)_n}{(\alpha+1)_{n-k}\,(\beta+1)_k}(1-x)^{n-k}(1+x)^k, \tag{5.5.3}$$

which gives the endpoint values

$$P_n^{(\alpha,\beta)}(1) = \frac{(\alpha+1)_n}{n!}, \qquad P_n^{(\alpha,\beta)}(-1) = (-1)^n\frac{(\beta+1)_n}{n!}. \tag{5.5.4}$$

The three-term recurrence relation (5.2.7) is

$$\frac{(2n+2)(n+\alpha+\beta+1)}{2n+\alpha+\beta+1}P_{n+1}^{(\alpha,\beta)}(x)$$

$$= \left[\frac{\alpha^2-\beta^2}{2n+\alpha+\beta}+(2n+\alpha+\beta+2)x\right]P_n^{(\alpha,\beta)}(x)$$

$$-\frac{2(2n+\alpha+\beta+2)(n+\alpha)(n+\beta)}{(2n+\alpha+\beta)(2n+\alpha+\beta+1)}P_{n-1}^{(\alpha,\beta)}(x). \tag{5.5.5}$$

In view of the discussion in Section 3.5, we have the estimate

$$\sup_{|x|\leq 1}|P_n^{(\alpha,\beta)}(x)| = \max\left\{\frac{(\alpha+1)_n}{n!},\frac{(\beta+1)_n}{n!}\right\} \quad \text{if } \alpha \text{ or } \beta \geq -\frac{1}{2}. \tag{5.5.6}$$

Jacobi polynomials with either $\alpha = \pm\frac{1}{2}$ or $\beta = \pm\frac{1}{2}$ can be reduced to those with equal indices by using the following two identities and (5.5.1):

$$P_{2n}^{(\alpha,\alpha)}(x) = \frac{(n+\alpha+1)_n}{(n+1)_n} P_n^{(\alpha,-\frac{1}{2})}(2x^2-1);$$

$$P_{2n+1}^{(\alpha,\alpha)}(x) = \frac{(n+\alpha+1)_{n+1}}{(n+1)_{n+1}} x P_n^{(\alpha,\frac{1}{2})}(2x^2-1).$$

(5.5.7)

These identities follow from the relationship with hypergeometric functions, (5.5.13), together with two of the quadratic transformations, (10.5.2) and (10.5.22). For a direct proof, see Exercise 25.

The generating function can be calculated from (5.2.2). Taking into account the factor $(-\frac{1}{2})^n$ in the normalization (5.1.12), we replace s by $-\frac{1}{2}s$. Since $p(x) = 1 - x^2$, $y(x,s)$ is the solution of

$$y = x - \frac{s}{2}(1 - y^2),$$

so

$$y = s^{-1}\left[1 - \sqrt{1 - 2xs + s^2}\right];$$

$$\frac{1-y}{1+x} = \frac{2}{1 - s + \sqrt{1 - 2xs + s^2}};$$

$$\frac{1+y}{1+x} = \frac{2}{1 + s + \sqrt{1 - 2xs + s^2}}.$$

Thus

$$\sum_{n=0}^{\infty} P_n^{(\alpha,\beta)}(x) s^n = \frac{2^{\alpha+\beta}}{R(1 - s + R)^\alpha (1 + s + R)^\beta},$$

(5.5.8)

where $R = \sqrt{1 - 2xs + s^2}$.

The Liouville transformation for the Jacobi case starts with the change of variable

$$\theta(x) = \int_1^x \frac{dy}{\sqrt{p(y)}} = \int_1^x \frac{dy}{\sqrt{1 - y^2}} = \cos^{-1} x.$$

In the variable θ, the operator in (5.5.2) takes the form

$$\frac{d^2}{d\theta^2} + \frac{\alpha - \beta + (\alpha + \beta + 1)\cos\theta}{\sin\theta} \frac{d}{d\theta}.$$

The coefficient of the first-order term can be rewritten as

$$(2\alpha + 1)\frac{\cos\frac{1}{2}\theta}{2\sin\frac{1}{2}\theta} - (2\beta + 1)\frac{\sin\frac{1}{2}\theta}{2\cos\frac{1}{2}\theta}.$$

Therefore this coefficient can be eliminated by the gauge transformation $u = \varphi v$ with

$$\varphi(\theta) = \left(\sin\frac{1}{2}\theta\right)^{-\alpha-\frac{1}{2}} \left(\cos\frac{1}{2}\theta\right)^{-\beta-\frac{1}{2}}.$$

After this gauge transformation, the operator acting on the function v is

$$\frac{d^2}{d\theta^2} + \frac{(\alpha+\beta+1)^2}{4} + \frac{(1+2\alpha)(1-2\alpha)}{16\sin^2\frac{1}{2}\theta} + \frac{(1+2\beta)(1-2\beta)}{16\cos^2\frac{1}{2}\theta} \tag{5.5.9}$$

and the eigenvalue equation with $\lambda_n = n(n+\alpha+\beta+1)$ is

$$v''(\theta) + \left[\frac{(2n+\alpha+\beta+1)^2}{4} + \frac{(1+2\alpha)(1-2\alpha)}{16\sin^2\frac{1}{2}\theta} \right.$$

$$\left. + \frac{(1+2\beta)(1-2\beta)}{16\cos^2\frac{1}{2}\theta} \right] v(\theta) = 0. \tag{5.5.10}$$

In particular, if $2\alpha = \pm 1$ and $2\beta = \pm 1$, then (5.5.10) can be solved explicitly.

Equation (5.5.10) leads to estimates for the zeros of $P_n^{(\alpha,\beta)}$ for certain α, β; see Exercise 26.

Theorem 5.5.1 *Suppose $\alpha^2 \le \frac{1}{4}$ and $\beta^2 \le \frac{1}{4}$, and let $\cos\theta_{1n}, \ldots, \cos\theta_{nn}$ be the zeros of $P_n^{(\alpha,\beta)}$, $\theta_{1n} < \cdots < \theta_{nn}$. Then*

$$\frac{(k-1+\gamma)\pi}{n+\gamma} \le \theta_{kn} \le \frac{k\pi}{n+\gamma}, \quad \gamma = \frac{1}{2}(\alpha+\beta+1). \tag{5.5.11}$$

The inequalities are strict unless $\alpha^2 = \beta^2 = \frac{1}{4}$.

The following asymptotic result of Darboux [93] is proved in Section 13.3.

$$P_n^{(\alpha,\beta)}(\cos\theta) = \frac{\cos\left(n\theta + \frac{1}{2}[\alpha+\beta+1]\theta - \frac{1}{2}\alpha\pi - \frac{1}{4}\pi\right) + O(n^{-1})}{\sqrt{n\pi}\,(\sin\frac{1}{2}\theta)^{\alpha+\frac{1}{2}}(\cos\frac{1}{2}\theta)^{\beta+\frac{1}{2}}} \tag{5.5.12}$$

as $n \to \infty$, uniformly on any subinterval $\delta \le \theta \le \pi - \delta$, $\delta > 0$.

For more comprehensive asymptotic results, see Sections 7.1 and 7.2.

It is convenient for some purposes, such as making the connection to hypergeometric functions, to rescale the x-interval to $(0,1)$. Let $y = \frac{1}{2}(1-x)$. Up to a constant factor, the corresponding weight function is $w(y) = y^\alpha(1-y)^\beta$, while the rescaled polynomials are eigenfunctions for the operator

$$y(1-y)\frac{d^2}{dy^2} + [\alpha+1 - (\alpha+\beta+2)y]\frac{d}{dy},$$

with eigenvalues $-n(n + \alpha + \beta + 1)$; see Case III in Section 3.4. If we set

$$P_n^{(\alpha,\beta)}(1 - 2y) = \sum_{k=0}^{n} c_k y^k,$$

then the eigenvalue equation implies the identities

$$c_{k+1} = \frac{(n + \alpha + \beta + 1 + k)(-n + k)}{k(\alpha + 1 + k)} c_k.$$

By (5.5.4), $c_0 = (\alpha + 1)_n / n!$, so

$$P_n^{(\alpha,\beta)}(x) = \frac{(\alpha+1)_n}{n!} \sum_{k=0}^{n} \frac{(\alpha + \beta + 1 + n)_k(-n)_k}{(\alpha+1)_k k!} \left(\frac{1-x}{2}\right)^k$$

$$= \frac{(\alpha+1)_n}{n!} F\left(\alpha + \beta + 1 + n, -n, \alpha + 1; \frac{1}{2}(1-x)\right), \quad (5.5.13)$$

where F is the hypergeometric function associated with (3.7.2) with indices $\alpha + \beta + 1 + n, -n, \alpha + 1$: the solution that has value 1 at $y = 0$.

5.6 Legendre and Chebyshev polynomials

Up to normalization, these are Jacobi polynomials $P_n^{(\alpha,\alpha)}$ with a repeated index $\alpha = \beta$. Note that in any such case the weight function $(1 - x^2)^\alpha$ is an even function. It follows by induction that orthogonal polynomials of even degree are even functions, while those of odd degree are odd functions.

The *Legendre polynomials* $\{P_n\}$ are the case $\alpha = \beta = 0$:

$$P_n(x) = P_n^{(0,0)}(x).$$

The associated weight function is $w(x) \equiv 1$ and the eigenvalue equation is

$$(1 - x^2)P_n''(x) - 2xP_n'(x) + n(n+1)P_n(x) = 0. \quad (5.6.1)$$

The generating function is

$$\sum_{n=0}^{\infty} P_n(x)s^n = (1 - 2xs + s^2)^{-\frac{1}{2}}. \quad (5.6.2)$$

The recurrence and derivative formulas (5.2.7) and (5.2.14) specialize to

$$(n+1)P_{n+1}(x) = (2n+1)xP_n(x) - nP_{n-1}(x); \quad (5.6.3)$$

$$(1 - x^2)P_n'(x) = -nxP_n(x) + nP_{n-1}(x). \quad (5.6.4)$$

The recurrence relation (5.6.3) and the derivative identity

$$P'_n(x) - 2xP'_{n-1}(x) + P'_{n-2}(x) = P_{n-1}(x) \qquad (5.6.5)$$

can be derived from the generating function; see Exercises 29 and 30.

In Sections 11.2 and 10.6, we establish two integral representations:

$$P_n(\cos\theta) = \frac{1}{2\pi} \int_0^{2\pi} (\cos\theta + i\sin\theta \sin\alpha)^n \, d\alpha \qquad (5.6.6)$$

$$= \frac{1}{\pi} \int_0^1 \frac{\cos(s(n+\frac{1}{2})\theta)}{\cos\frac{1}{2}s\theta} \frac{ds}{\sqrt{s(1-s)}}. \qquad (5.6.7)$$

The general formula (5.2.18) for the Dirichlet kernel specializes to

$$K_n^{(0,0)}(x,y) = \frac{n+1}{2} \left[\frac{P_{n+1}(x)P_n(y) - P_n(x)P_{n+1}(y)}{x-y} \right]. \qquad (5.6.8)$$

The Liouville transformation takes the eigenvalue equation to the following equation for $u_n(\theta) = (\sin\theta)^{\frac{1}{2}} P_n(\cos\theta)$, the case $\alpha = \beta = 0$ of (5.5.10):

$$u''_n(\theta) + \left[\left(n + \frac{1}{2} \right)^2 + \frac{1}{4\sin^2\theta} \right] u_n(\theta) = 0. \qquad (5.6.9)$$

The first part of the following result is the specialization to $\alpha = \beta = 0$ of (5.5.6). The second part is the specialization of a result in Section 3.5. The remaining two statements follow from (5.6.9), together with Proposition 3.5.2 and Proposition 3.5.3.

Theorem 5.6.1 *The Legendre polynomials satisfy*

$$\sup_{|x|\leq 1} |P_n(x)| = 1.$$

The relative maxima of $|P_n(\cos\theta)|$ decrease as θ increases for $0 \leq \theta \leq \frac{1}{2}\pi$ and increase as θ increases for $\frac{1}{2}\pi \leq \theta \leq \pi$.

The relative maxima of $|(\sin\theta)^{\frac{1}{2}} P_n(\cos\theta)|$ increase with θ for $0 \leq \theta \leq \frac{1}{2}\pi$ and decrease as θ increases for $\frac{1}{2}\pi \leq \theta \leq \pi$. Moreover, for $0 \leq \theta \leq \pi$,

$$\left[(\sin\theta)^{\frac{1}{2}} P_n(\cos\theta) \right]^2 \leq P_n(0)^2 + \frac{P'_n(0)^2}{(n+\frac{1}{2})^2 + \frac{1}{4}}. \qquad (5.6.10)$$

A more explicit form of this last estimate can be derived:

$$|(\sin\theta)^{\frac{1}{2}} P_n(\cos\theta)| < \left(\frac{2}{n\pi} \right)^{\frac{1}{2}}; \qquad (5.6.11)$$

see Exercises 38 and 39. This is a sharp version, due to Bernstein [42], of an earlier result of Stieltjes [383].

Theorem 5.5.1 specializes to the following: let $\cos\theta_{1n}, \ldots, \cos\theta_{nn}$ be the zeros of $P_n(x)$, $\theta_{1n} < \cdots < \theta_{nn}$. Then

$$\frac{(k-1+\frac{1}{2})\pi}{n+\frac{1}{2}} < \theta_{kn} < \frac{k\pi}{n+\frac{1}{2}}. \tag{5.6.12}$$

The following inequality is due to Turán [405] (see also [391]):

$$P_n(x)^2 - P_{n-1}(x)P_{n+1}(x) \geq 0. \tag{5.6.13}$$

The *Chebyshev (or Tchebycheff) polynomials* $\{T_n\}$ and $\{U_n\}$ are the cases $\alpha = \beta = -\frac{1}{2}$ and $\alpha = \beta = \frac{1}{2}$, respectively:

$$T_n(x) = \frac{2\cdot4\cdot6\cdots(2n)}{1\cdot3\cdot5\cdots(2n-1)}\, P_n^{(-\frac{1}{2},-\frac{1}{2})}(x) = \frac{n!}{(\frac{1}{2})_n}\, P_n^{(-\frac{1}{2},-\frac{1}{2})}(x);$$

$$U_n(x) = \frac{4\cdot6\cdots(2n+2)}{3\cdot5\cdots(2n+1)}\, P_n^{(\frac{1}{2},\frac{1}{2})}(x) = \frac{(n+1)!}{(\frac{3}{2})_n}\, P_n^{(\frac{1}{2},\frac{1}{2})}(x).$$

Thus $T_n(1) = 1$, $U_n(1) = n+1$.

The *Gegenbauer polynomials* or *ultraspherical polynomials* $\{C_n^\lambda\}$ are the general case $\alpha = \beta$, normalized as follows:

$$C_n^\lambda(x) = \frac{(2\lambda)_n}{(\lambda+\frac{1}{2})_n}\, P_n^{(\lambda-\frac{1}{2},\lambda-\frac{1}{2})}(x).$$

In particular,

$$C_n^0(x) = T_n(x), \quad C_n^{\frac{1}{2}}(x) = P_n(x), \quad C_n^1(x) = U_n(x).$$

The Chebyshev polynomials simplify considerably under the Liouville transformation and gauge transformation considered above. With $\alpha = \beta = -\frac{1}{2}$, the operator (5.5.9) has a zero-order term and the eigenvalue $-\lambda_n$ is $-n^2$. Therefore the solutions of (5.5.2) that are even (resp. odd) functions of $\cos\theta$ when n is even (resp. odd) are multiples of $\cos(n\theta)$. The normalization gives the *Chebyshev polynomials of first kind*

$$T_n(\cos\theta) = \cos n\theta. \tag{5.6.14}$$

Since

$$\frac{dx}{\sqrt{1-x^2}} = d\theta,$$

the square of the L^2 norm is π if $n = 0$, and

$$\int_0^\pi \cos^2 n\theta \, d\theta = \frac{\pi}{2}, \quad n > 0.$$

The recurrence and derivative formulas are easily derived from the trigonometric identities

$$\cos(n\theta \pm \theta) = \cos n\theta \cos\theta \mp \sin n\theta \sin\theta.$$

These imply that

$$T_{n+1}(x) + T_{n-1}(x) = 2x T_n(x), \quad n \geq 1. \tag{5.6.15}$$

Also,

$$(1 - x^2) T_n'(x)\big|_{x=\cos\theta} = -\sin\theta \frac{d}{d\theta} \cos n\theta = n \sin n\theta \sin\theta,$$

so

$$(1 - x^2) T_n'(x) = -nx T_n(x) + n T_{n-1}(x). \tag{5.6.16}$$

For two other derivations of (5.6.15), see Exercises 33 and 34.

It follows from (5.6.15) and the norm calculation that the associated Dirichlet kernel is

$$
\begin{aligned}
K_n^T(x, y) &= \frac{2}{\pi} \sum_{k=0}^n T_k(x) T_k(y) \\
&= \frac{1}{\pi} \left[\frac{T_{n+1}(x) T_n(y) - T_n(x) T_{n+1}(y)}{x - y} \right]
\end{aligned}
\tag{5.6.17}
$$

for $n > 0$.

Note that $T_0 = 1$ and $T_1(x) = x$. Together with (5.6.15), this allows us to calculate an alternative generating function

$$G_T(x, s) = \sum_{n=0}^\infty T_n(x) s^n.$$

In fact,

$$s^{-1} [G_T(x, s) - xs - 1] + s G_T(x, s) = 2x [G_T(x, s) - 1],$$

so

$$\sum_{n=0}^\infty T_n(x) s^n = \frac{1 - xs}{1 - 2xs + s^2}. \tag{5.6.18}$$

With $\alpha = \beta = \frac{1}{2}$, the gauge transformation $u(\theta) = (\sin\theta)^{-1} v(\theta)$ reduces the eigenvalue equation to

$$v''(x) + [1 + n(n+2)] v(x) = v''(x) + (n+1)^2 v(x) = 0,$$

so the solutions u that are regular at $\theta = 0$ are multiples of $\sin(n+1)\theta / \sin\theta$. The normalization gives the *Chebyshev polynomials of the second kind*:

$$U_n(\cos\theta) = \frac{\sin(n+1)\theta}{\sin\theta}. \tag{5.6.19}$$

The weight function here is $\sin\theta$, so the square of the L^2 norm is

$$\int_0^\pi \sin^2(n+1)\theta \, d\theta = \frac{\pi}{2}.$$

The recurrence and derivation formulas are easily derived from the trigonometric identities

$$\sin(n\theta \pm \theta) = \sin n\theta \cos\theta \pm \cos n\theta \sin\theta.$$

These imply that

$$U_{n+1}(x) + U_{n-1}(x) = 2x\, U_n(x). \tag{5.6.20}$$

Also,

$$(1 - x^2)\, U_n'(x)\big|_{x=\cos\theta} = -\sin\theta \frac{d}{d\theta}\left\{ \frac{\sin(n+1)\theta}{\sin\theta} \right\}$$

$$= \frac{-(n+1)\cos(n+1)\theta \sin\theta + \sin(n+1)\theta \cos\theta}{\sin\theta},$$

so

$$(1 - x^2)U_n'(x) = -nx\, U_n(x) + (n+1)U_{n-1}(x). \tag{5.6.21}$$

For two other derivations of (5.6.20), see Exercises 33 and 36.

It follows from the formulas following (5.2.15) and the norm calculation that the associated Dirichlet kernel is

$$K_n^U(x,y) = \frac{2}{\pi} \sum_{k=0}^n U_k(x)U_k(y)$$

$$= \frac{1}{\pi}\left[\frac{U_{n+1}(x)U_n(y) - U_n(x)U_{n+1}(y)}{x - y} \right]. \tag{5.6.22}$$

Note that $U_0 = 1$ and $U_1(x) = 2x$. These facts and the recurrence relation allow us to compute an alternative generating function in analogy with G_T above:

$$\sum_{n=0}^\infty U_n(x)s^n = \frac{1}{1 - 2xs + s^2}. \tag{5.6.23}$$

The remaining cases where the eigenvalue equation (5.5.10) can be solved immediately are $\alpha = -\frac{1}{2}$, $\beta = \frac{1}{2}$ and $\alpha = \frac{1}{2}$, $\beta = -\frac{1}{2}$. In each case, the eigenvalue parameter for degree n is $\lambda_n = n(n+1)$, so after the gauge transformation the constant term is $\lambda_n + \frac{1}{4} = (n + \frac{1}{2})^2$. In the first case, the gauge function is $(\cos \frac{1}{2}\theta)^{-1}$ and the value at $\theta = 0$ should be $(\frac{1}{2})_n / n!$, so

$$P_n^{(-\frac{1}{2},\frac{1}{2})}(\cos\theta) = \frac{(\frac{1}{2})_n}{n!} \frac{\cos(n+\frac{1}{2})\theta}{\cos\frac{1}{2}\theta}. \tag{5.6.24}$$

In the second case, the gauge function is $(\sin\frac{1}{2}\theta)^{-1}$ and the value at $\theta = 0$ should be $(\frac{3}{2})_n / n!$, so

$$P_n^{(\frac{1}{2},-\frac{1}{2})}(\cos\theta) = \frac{(\frac{1}{2})_n}{n!} \frac{\sin(n+\frac{1}{2})\theta}{\sin\frac{1}{2}\theta}. \tag{5.6.25}$$

Combining the results of this section with (5.5.13) gives explicit evaluations of hypergeometric functions associated with the Jacobi indices $\alpha = \pm\frac{1}{2}$, $\beta = \pm\frac{1}{2}$:

$$F\left(n, -n, \frac{1}{2}; \frac{1}{2}[1 - \cos\theta]\right) = \cos(n\theta); \tag{5.6.26}$$

$$F\left(n+2, -n, \frac{3}{2}; \frac{1}{2}[1 - \cos\theta]\right) = \frac{\sin(n+1)\theta}{(n+1)\sin\theta}; \tag{5.6.27}$$

$$F\left(n+1, -n, \frac{1}{2}; \frac{1}{2}[1 - \cos\theta]\right) = \frac{\cos(n+\frac{1}{2})\theta}{\cos\frac{1}{2}\theta}; \tag{5.6.28}$$

$$F\left(n+1, -n, \frac{3}{2}; \frac{1}{2}[1 - \cos\theta]\right) = \frac{\sin(n+\frac{1}{2})\theta}{(2n+1)\sin\frac{1}{2}\theta}. \tag{5.6.29}$$

The previous arguments show that these identities are valid for all values of the parameter n. It will be shown in Section 10.6 that the integral representation (5.6.7) is a consequence of (5.6.28).

5.7 Distribution of zeros and electrostatics

We begin this section with a proof of a close relative of Theorem 4.6.1. The same method can be used to prove Theorem 4.6.1 itself in the case of Jacobi polynomials $\{P^{(\alpha,\beta)}\}$ for a fixed pair of indices α, β. Consider the sequence of positive measures $\{\lambda_n^{(\alpha,\beta)}\}$:

$$\lambda_n^{(\alpha,\beta)}(f) \equiv \frac{1}{n} \sum_{j=1}^{n} f(x_{j,n}), \tag{5.7.1}$$

where f is any continuous function on the interval $[-1, 1]$ and the $x_{j,n}$ are the zeros of $P_n^{(\alpha,\beta)}$. This can be extended to piecewise continuous functions whose points of discontinuity do not include any of the zeros. Taking f to be the characteristic function (indicator function) of a subinterval, $\lambda_n^{(\alpha,\beta)}(f)$ gives the fraction of the n zeros that lie in the given subinterval. If $\lambda_n^{(\alpha,\beta)}(f)$ has a limit, the limiting value tells us how the zeros are distributed asymptotically.

Theorem 5.7.1 (Erdős and Turán) *For any continuous function f on the interval $[-1, 1]$,*

$$\lim_{n \to \infty} \lambda_n^{(\alpha,\beta)}(f) = \frac{1}{\pi} \int_{-1}^{1} \frac{f(x)\, dx}{\sqrt{1 - x^2}}. \tag{5.7.2}$$

If one tries the same procedure as above for Laguerre or Hermite polynomials, the resulting limit is 0, because the zeros spread out. However, similar results are valid for rescaled versions of these polynomials.

For a Laguerre polynomial $L_n^{(\alpha)}$, the smallest interval that contains all the zeros is approximately $(0, \beta_n)$, where c is chosen so that $\beta_n = 4n + c$ dominates the numerical expression on the right in (5.4.11). Therefore it is natural to consider instead the rescaled monic Laguerre polynomials

$$l_n^{(\alpha)}(x) = (-1)^n \frac{n!}{(\beta_n)^n} L_n^{(\alpha)}(\beta_n x), \quad \beta_n = 4n + c, \tag{5.7.3}$$

whose roots lie in the interval $(0, 1)$, and let

$$\lambda_n^{(\alpha)}(f) \equiv \frac{1}{n} \sum_{j=1}^{n} f(x_{j,n}), \tag{5.7.4}$$

where f is any continuous function on the interval $[0, 1]$ and the $x_{j,n}$ are the zeros of $l_n^{(\alpha)}$.

The corresponding scaling in the Hermite case gives the monic polynomials

$$h_n(x) = \frac{1}{(2\alpha_n)^n} H_n(\alpha_n x), \quad \alpha_n = \sqrt{2n + 1}, \tag{5.7.5}$$

whose roots lie in the interval $(-1, 1)$. We let

$$\lambda_n(f) \equiv \frac{1}{n} \sum_{j=1}^{n} f(x_{j,n}), \tag{5.7.6}$$

where f is any continuous function on the interval $[-1, 1]$ and the $x_{j,n}$ are the zeros of h_n.

Theorem 5.7.2 *For any continuous function f on the interval $[0, 1]$,*

$$\lim_{n \to \infty} \lambda_n^{(\alpha)}(f) = \frac{4}{\pi} \int_{0}^{1} f(x) \sqrt{\frac{1}{x} - 1}\, dx. \tag{5.7.7}$$

Theorem 5.7.3 *For any continuous function f on the interval $[-1,1]$,*

$$\lim_{n\to\infty} \lambda_n(f) = \frac{2}{\pi} \int_{-1}^{1} f(x)\sqrt{1-x^2}\,dx. \tag{5.7.8}$$

The proofs of these three theorems follow the same pattern. Here we prove Theorem 5.7.3 and leave the other two proofs as Exercises 42 and 43.

In view of the Weierstrass approximation theorem (Appendix B), it is enough to prove (5.7.8) for polynomials f. The quotient h'_n/h_n has a simple pole with residue 1 at each zero $x_{j,n}$ of h_n:

$$\frac{h'_n(z)}{h_n(z)} = \frac{1}{z-x_{1,n}} + \frac{1}{z-x_{2,n}} + \cdots + \frac{1}{z-x_{n,n}}. \tag{5.7.9}$$

By the residue theorem (Appendix A), for polynomial f, we have

$$\lambda_n(f) = \frac{1}{2\pi i} \int_C \frac{h'_n(z)}{n\,h_n(z)} f(z)\,dz,$$

where C is any curve, oriented counter-clockwise, that encloses the interval $[-1,1]$. Thus we want to study the weighted ratio $\psi_n = h'_n/(nh_n)$. Now

$$\psi'_n = \frac{h''_n}{n\,h_n} - n\,\psi_n^2,$$

and the equation for H_n implies that

$$h''_n(z) = (2n+1)\big[2z\,h'_n(z) - 2n\,h_n(z)\big].$$

Combining the previous two equations,

$$n\,\psi_n(z)^2 - (2n+1)2z\,\psi_n(z) + (2n+1)2 = -\psi'_n(z),$$

or

$$\psi_n^2 - 4z\,\psi_n + 4 = \frac{1}{2n+1}\,\psi_n^2 - \frac{2}{2n+1}\,\psi'_n. \tag{5.7.10}$$

It follows from (5.7.9) that

$$|\psi_n(z)| \le \frac{1}{\delta(z)}, \qquad |\psi'_n(z)| \le \frac{1}{\delta(z)^2},$$

where $\delta(z)$ is the distance from z to the interval $[-1,1]$. Therefore, for any z not in the interval $[-1,1]$, $\psi_n(z)$ differs by $O(1/n)$ from one of the roots ψ of the equation $\psi^2 - 4z\psi + 4 = 0$. Since ψ_n is holomorphic outside the interval and is $\sim 1/z$ as $z\to\infty$, it follows that the root to choose is

$$\psi(z) = 2z - 2\sqrt{z^2-1}, \tag{5.7.11}$$

where the branch of the square root is taken so that $\psi(z) \sim 1/z$ at ∞. We have shown that

$$\psi_n(z) = \psi(z) + O(1/n).$$

Thus, for a fixed curve C enclosing the interval, and a fixed polynomial f,

$$\lim_{n \to \infty} \lambda_n(f) = \frac{1}{2\pi i} \int_C \psi(z) f(z) \, dz.$$

Letting the curve C shrink to the interval, we obtain the integral along $[-1,1]$ of the limit from the lower half-plane minus the limit from the upper half-plane. The result is (5.7.8). □

We turn next to a result of Stieltjes [382] that connects electrostatics and zeros of Jacobi polynomials. The general mathematical formulation involves a real interval $I = [a,b]$, a positive and differentiable density function w on I that vanishes at the endpoints, and a function defined for points $\{x_k\}$ in the open interval (a,b):

$$V(x_1, x_2, \ldots, x_n) = -\sum_{j \neq k} \log |x_j - x_k| - \sum_{j=1}^{n} \log w(x_j), \qquad (5.7.12)$$

with

$$x_1 < x_2 < \cdots < x_n, \quad x_j \in I.$$

This can be interpreted as the electrostatic potential energy of n equal charges, located at the points $\{x_k\}$, in the presence of an external field, represented by w, that confines the charges to the interval. One expects that the charges will arrange themselves so as to minimize V.

Lemma 5.7.4 *There is a configuration $\{x_j\}$ that minimizes the potential (5.7.12). The points in this configuration are the zeros of a polynomial Q with the property that $w(x)Q''(x) + w'(x)Q'(x)$ vanishes at each x_j.*

Proof If some distance $|x_j - x_k|$ is very small, or some distance to an endpoint is very small, then V is very large. Therefore there is a $\delta > 0$ such that the minimum attained in the compact set defined by

$$a + \delta \leq x_1 < x_2 < \cdots < x_n \leq b - \delta, \qquad x_{j+1} - x_j \geq \delta,$$

is minimal for the unconstrained problem. (This condition is formulated for the case of finite endpoints: replace $a + \delta$ by some finite lower bound if $a = -\infty$, and replace $b - \delta$ by some finite upper bound if $b = \infty$.) Now let $\{x_k\}$ be a minimizing configuration. Then it is a critical point. Vanishing of the derivative of V with respect to x_j implies that

$$2 \sum_{k \neq j} \frac{1}{x_j - x_k} = -\frac{w'(x_j)}{w(x_j)}. \qquad (5.7.13)$$

Let $Q(x) = \prod_{k=1}^{n}(x - x_k)$. Then for x distinct from the x_k,

$$Q'(x) = \sum_{k} \frac{Q(x)}{x - x_k} = \frac{Q(x)}{x - x_j} + O(x - x_j); \qquad (5.7.14)$$

$$Q''(x) = \sum_{k \neq j} \frac{Q(x)}{(x - x_j)(x - x_k)}$$

$$= 2\frac{Q(x)}{x - x_j} \sum_{j \neq k} \frac{1}{x_j - x_k} + O(x - x_j). \qquad (5.7.15)$$

It follows from (5.7.13), (5.7.14), and (5.7.15) that $wQ'' + w'Q'$ vanishes at each x_j. $\qquad\square$

The following three results are consequences of this lemma. The details, including uniqueness, are left as Exercises 44–46.

Theorem 5.7.5 (Stieltjes) *Let* $w(x) = (1 - x)^{\alpha+1}(1 + x)^{\beta+1}$, *with* $\alpha > -1$, $\beta > -1$, *and let* $I = [-1, 1]$. *There is a unique configuration* $\{x_j\}$ *that minimizes the potential* (5.7.12). *The* $\{x_j\}$ *are the roots of the Jacobi polynomial* $P_n^{(\alpha,\beta)}$.

Theorem 5.7.6 *Let* $w(x) = x^{\alpha+1}e^{-x}$, *with* $\alpha > -1$, *and let* $I = [0, \infty)$. *There is a unique configuration* $\{x_j\}$ *that minimizes the potential* (5.7.12). *The* $\{x_j\}$ *are the roots of the Laguerre polynomial* $L_n^{(\alpha)}$.

Theorem 5.7.7 *Let* $w(x) = e^{-x^2}$ *and let* $I = (-\infty, \infty)$. *There is a unique configuration* $\{x_j\}$ *that minimizes the potential* (5.7.12). *The* $\{x_j\}$ *are the roots of the Hermite polynomial* H_n.

These results are related to the idea of an *equilibrium measure*; see the extensive treatment in the book by Saff and Totik [350].

5.8 Expansion theorems

Suppose that w is one of the weights associated with the classical orthogonal polynomials,

$$w_H(x) = e^{-x^2}, \quad -\infty < x < \infty;$$

$$w_\alpha(x) = x^\alpha e^{-x}, \quad 0 < x < \infty;$$

$$w_{\alpha\beta}(x) = (1 - x)^\alpha(1 + x)^\beta, \quad -1 < x < 1,$$

and suppose that $\{\varphi_n\}_{n=0}^{\infty}$ is the corresponding set of orthonormal polynomials: the normalized version of the Hermite, Laguerre, or Jacobi polynomials associated with w. It follows from Theorem 4.1.5 that if f is a function in L_w^2, then the series

$$\sum_{n=0}^{\infty} (f, \varphi_n) \varphi_n(x) \tag{5.8.1}$$

converges to f in the L^2 sense:

$$\lim_{n\to\infty} ||f_n - f|| \to 0, \quad f_n(x) = \sum_{k=0}^{n} (f, \varphi_k) \varphi_k(x).$$

The partial sums are given by integration against the associated Dirichlet kernel:

$$f_n(x) = \int_I K_n(x, y) f(y) w(y) \, dy, \quad K_n(x, y) = \sum_{k=0}^{n} \varphi_k(x) \varphi_k(y).$$

The kernel K_n can be written in more compact form by using the Christoffel–Darboux formula (4.1.4). Taking $f \equiv 1$ and using orthogonality shows that

$$1 = \int_I K_n(x, y) w(y) \, dy. \tag{5.8.2}$$

In this section we consider pointwise convergence. In each case, if f belongs to L_w^2, then the series (5.8.1) converges to $f(x)$ at each point where f is differentiable. In fact, we may replace differentiability at x with the weaker condition that for some $\delta > 0$,

$$\left| \frac{f(y) - f(x)}{y - x} \right| \leq C \quad \text{for } 0 < |y - x| < \delta. \tag{5.8.3}$$

If f is piecewise continuously differentiable, then this condition is satisfied at every point of continuity.

We begin with the Hermite case. According to (5.1.13), the normalized polynomials can be taken to be

$$\varphi_n(x) = a_n H_n(x), \quad a_n = \frac{1}{\sqrt{2^n n! \pi^{\frac{1}{4}}}}.$$

Thus in this case the expansion (5.8.1) is

$$\sum_{n=0}^{\infty} c_n H_n(x), \quad c_n = \frac{1}{2^n n! \sqrt{\pi}} \int_{-\infty}^{\infty} f(x) H_n(x) e^{-x^2} \, dx. \tag{5.8.4}$$

Theorem 5.8.1 *Suppose that $f(x)$ is a real-valued function that satisfies*

$$\int_{-\infty}^{\infty} f(x)^2 e^{-x^2} \, dx < \infty, \tag{5.8.5}$$

and suppose that f satisfies the condition (5.8.3) at the point x. Then the series (5.8.4) converges to f(x).

It follows from (5.8.2) that

$$f(x) - f_n(x) = \int_{-\infty}^{\infty} K_n^H(x,y) \, [f(y) - f(x)] \, e^{-y^2} \, dy.$$

The Dirichlet kernel here is given by (5.2.16):

$$K_n^H(x,y) = \frac{1}{2^{n+1} n! \sqrt{\pi}} \left[\frac{H_{n+1}(x) H_n(y) - H_n(x) H_{n+1}(y)}{x - y} \right].$$

Therefore

$$\begin{aligned} f(x) - f_n(x) = & \frac{H_{n+1}(x)}{2^{n+1} n! \sqrt{\pi}} \int_{-\infty}^{\infty} H_n(y) g(x,y) e^{-y^2} \, dy \\ & - \frac{H_n(x)}{2^{n+1} n! \sqrt{\pi}} \int_{-\infty}^{\infty} H_{n+1}(y) g(x,y) e^{-y^2} \, dy, \end{aligned} \qquad (5.8.6)$$

where

$$g(x,y) = \frac{f(y) - f(x)}{y - x}. \qquad (5.8.7)$$

For $|y| \geq 2|x|$

$$y^2 g(x,y)^2 \leq 8[f(x)^2 + g(y)^2].$$

Together with assumptions (5.8.3) and (5.8.5), this implies that

$$\int_{-\infty}^{\infty} (1 + y^2) g(x,y)^2 \, e^{-y^2} \, dy \; < \; \infty. \qquad (5.8.8)$$

From (5.3.20) and Stirling's formula (2.5.1), we have the estimates

$$|H_n(x)| \leq A(x) n^{-\frac{1}{4}} (2^n n!)^{\frac{1}{2}}.$$

Thus it is enough to prove that both the integrals

$$\frac{n^{\frac{1}{4}}}{(2^n n!)^{\frac{1}{2}}} \int_{-\infty}^{\infty} H_n(y) g(x,y) e^{-y^2} \, dy, \qquad \frac{n^{-\frac{1}{4}}}{(2^n n!)^{\frac{1}{2}}} \int_{-\infty}^{\infty} H_{n+1}(y) g(x,y) e^{-y^2} \, dy$$

have limit 0 as $n \to \infty$. Changing n to $n-1$ in the second expression shows that the two expressions are essentially the same. Thus it is enough to prove the following lemma.

Lemma 5.8.2 *If*

$$\int_{-\infty}^{\infty} (1 + x^2) g(x)^2 e^{-x^2} \, dx \; < \; \infty,$$

then

$$\lim_{n \to \infty} \frac{n^{\frac{1}{4}}}{(2^n n!)^{\frac{1}{2}}} \int_{-\infty}^{\infty} H_n(x) g(x) e^{-x^2} \, dx = 0. \qquad (5.8.9)$$

We begin with an estimate for an integral involving H_n.

Lemma 5.8.3 *There is a constant B such that*

$$\int_{-\infty}^{\infty} \frac{H_n(x)^2}{1+x^2} e^{-x^2} dx \leq B 2^n n! n^{-\frac{1}{2}} \tag{5.8.10}$$

for all n.

Proof First,

$$
\begin{aligned}
J_n &= \frac{1}{2^n n!} \int_{-\infty}^{\infty} \frac{H_n(x)^2}{1+x^2} e^{-x^2} dx \\
&= 2 \int_0^{\infty} \left(\frac{1-x^2}{1+x^2}\right)^n \frac{e^{-x^2}}{1+x^2} dx;
\end{aligned} \tag{5.8.11}
$$

see Exercise 14. Making the change of variable $x \to 1/x$ in the integral over the interval $[1, \infty)$ converts the integral to

$$
\begin{aligned}
J_n &= 2 \int_0^1 \left(\frac{1-x^2}{1+x^2}\right)^n \frac{e^{-x^2} + (-1)^n e^{-1/x^2}}{1+x^2} dx \\
&\leq 4 \int_0^1 \left(\frac{1-x^2}{1+x^2}\right)^n \frac{dx}{1+x^2}.
\end{aligned}
$$

Let $t = 1 - (1-x^2)^2/(1+x^2)^2 = 4x^2/(1+x^2)^2$, so that

$$\sqrt{1-t} = \frac{2}{1+x^2} - 1, \qquad \frac{4dx}{1+x^2} = \frac{dt}{\sqrt{t(1-t)}}.$$

Then

$$4 \int_0^1 \left(\frac{1-x^2}{1+x^2}\right)^n \frac{dx}{1+x^2} = \int_0^1 (1-t)^{\frac{1}{2}n} \frac{dt}{\sqrt{t(1-t)}} = B\left(\frac{1}{2}, \frac{1}{2}n + \frac{1}{2}\right).$$

By (2.1.9), the last expression is $O(n^{-\frac{1}{2}})$, which gives (5.8.10). $\qquad \square$

It follows from this result and the Cauchy–Schwarz inequality that

$$\left(\int_{-\infty}^{\infty} H_n(x) g(x) e^{-x^2} dx\right)^2 \leq B 2^n n! n^{-\frac{1}{2}}$$

$$\times \int_{-\infty}^{\infty} g(x)^2 (1+x^2) e^{-x^2} dx \tag{5.8.12}$$

for all n, if the integral on the right is finite; here B is the constant in (5.8.10).

We can now prove the first lemma, and thus complete the proof of Theorem 5.8.1. Given $\varepsilon > 0$, we can choose $N = N(\varepsilon)$ so large that

$$\int_{|x|>N} g(x)^2 (1+x^2) e^{-x^2} dx < \frac{\varepsilon^2}{B}.$$

In view of (5.8.12), up to ε we only need to consider the integral (5.8.9) over the bounded interval $[-N, N]$. The asymptotic estimate (5.3.20) is uniform over such an interval. The product of the constant in (5.3.20) and the constant in (5.8.9) is independent of n. Thus it is enough to consider the integrals

$$\int_{-N}^{N} \cos\left(\sqrt{2n+1}\, x - \frac{1}{2}n\pi\right) g(x) e^{-x^2/2}\, dx, \qquad \int_{-N}^{N} n^{-\frac{1}{2}} g(x) e^{-x^2/2}\, dx.$$

Since N is fixed, the second integral is $O(n^{-\frac{1}{2}})$. The first integral tends to zero as $n \to \infty$ by the Riemann–Lebesgue lemma; see Appendix B. (Consider separately the cases n even, n odd.)

This result, and the ones to follow, can be extended to points of discontinuity: if $f \in L_w^2$ has one-sided limits $f(x\pm)$ at x, and one-sided estimates

$$\frac{|f(y) - f(x\pm)|}{|y - x|} \le C \quad \text{for } 0 < \pm(y - x) < \delta, \tag{5.8.13}$$

then

$$\lim_{n\to\infty} \sum_{k=0}^{n} (f, \varphi_k)\varphi_k(x) = \frac{1}{2}[f(x+) + f(x-)]; \tag{5.8.14}$$

see Exercise 41.

The Laguerre case is quite similar. According to (5.1.14), the normalized polynomials for the weight w_α on the interval $(0, \infty)$ can be taken to be

$$\varphi_n(x) = a_n L_n^{(\alpha)}(x), \quad a_n = \left[\frac{n!}{\Gamma(n + \alpha + 1)}\right]^{1/2}.$$

Therefore the corresponding expansion of a function $f \in L_{w_\alpha}^2$ is

$$\sum_{n=0}^{\infty} c_n L_n^{(\alpha)}(x), \quad c_n = \frac{n!}{\Gamma(n + \alpha + 1)} \int_0^{\infty} f(x) L_n^{(\alpha)}(x) x^\alpha e^{-x}\, dx. \tag{5.8.15}$$

Theorem 5.8.4 *Suppose that $f(x)$ is a real-valued function that satisfies*

$$\int_{-\infty}^{\infty} f(x)^2 x^\alpha e^{-x}\, dx < \infty, \tag{5.8.16}$$

and suppose that f satisfies the condition (5.8.3) at the point x, $0 < x < \infty$. Then the series (5.8.15) converges to $f(x)$.

The argument here is similar to, but somewhat more complicated than, the proof of Theorem (5.8.1). We refer to Uspensky's paper [406] for the details.

The proof of the corresponding result for Jacobi polynomials is simpler, because we have already established the necessary estimates. According to

(5.1.15), the orthonormal polynomials for the weight $w_{\alpha\beta}$ on the interval $(-1, 1)$ can be taken to be

$$\varphi_n(x) = a_n P_n^{(\alpha,\beta)}(x), \quad a_n = \left[\frac{n!(2n+\alpha+\beta+1)\Gamma(n+\alpha+\beta+1)}{2^{\alpha+\beta+1}\Gamma(n+\alpha+1)\Gamma(n+\beta+1)}\right]^{1/2}.$$

Therefore the corresponding expansion of a function f in $L^2_{w_{\alpha\beta}}$ is

$$\sum_{n=0}^{\infty} c_n P_n^{(\alpha,\beta)}(x), \quad c_n = \frac{n!(2n+\alpha+\beta+1)\Gamma(n+\alpha+\beta+1)}{2^{\alpha+\beta+1}\Gamma(n+\alpha+1)\Gamma(n+\beta+1)}$$

$$\times \int_{-1}^{1} f(x) P_n^{(\alpha,\beta)}(x)(1-x)^{\alpha}(1+x)^{\beta} \, dx. \quad (5.8.17)$$

Theorem 5.8.5 *Suppose that $f(x)$ is a real-valued function that satisfies*

$$\int_{-1}^{1} f(x)^2 (1-x)^{\alpha}(1+x)^{\beta} \, dx < \infty, \quad (5.8.18)$$

and suppose that f satisfies the condition (5.8.3) at the point x, $-1 < x < 1$. Then the series (5.8.17) converges to $f(x)$.

Proof It follows from (5.8.2) that the partial sums f_n of the series (5.8.17) satisfy

$$f_n(x) - f(x) = \int_{-1}^{1} K_n^{(\alpha,\beta)}(x,y) \, [f(y) - f(x)] \, (1-y)^{\alpha}(1+y)^{\beta} \, dy,$$

where the Dirichlet kernel here is given by (5.2.18):

$$K_n^{(\alpha,\beta)}(x,y) = \frac{2^{-\alpha-\beta}(n+1)!\,\Gamma(n+\alpha+\beta+2)}{(2n+\alpha+\beta+2)\Gamma(n+\alpha+1)\Gamma(n+\beta+1)}$$

$$\times \left[\frac{P_{n+1}^{(\alpha,\beta)}(x)P_n^{(\alpha,\beta)}(y) - P_n^{(\alpha,\beta)}(x)P_{n+1}^{(\alpha,\beta)}(y)}{x-y}\right].$$

It follows from (2.1.9) that the coefficient here is $O(n)$ as $n \to \infty$. The asymptotic result (5.5.12) implies that $P_n^{(\alpha,\beta)}(x)$ is $O(n^{-\frac{1}{2}})$ as $n \to \infty$. Thus to prove the result, we only need to show that the inner product

$$(P_n^{(\alpha,\beta)}, h) \equiv \int_{-1}^{1} P_n^{(\alpha,\beta)}(y) h(y)(1-y)^{\alpha}(1+y)^{\beta} \, dy = o(n^{-\frac{1}{2}})$$

as $n \to \infty$, where $h(y) = [f(y) - f(x)]/(y - x)$.

Let $\{Q_n\}$ be the normalized polynomials

$$Q_n(x) = ||P_n^{(\alpha,\beta)}||^{-1} P_n^{(\alpha,\beta)}(x).$$

By (5.1.15) and (2.1.9), $||P_n^{(\alpha,\beta)}||$ is $O(n^{-1/2})$. Therefore it is enough to show that

$$(Q_n, h) = o(1). \tag{5.8.19}$$

The assumptions (5.8.3) and (5.8.16) imply that h is square-integrable, so (5.8.19) follows from (4.1.12). $\qquad\square$

5.9 Functions of the second kind

A principal result of Section 5.2 was that, in the three cases considered, the function

$$u(x) = \frac{1}{w(x)} \int_C \frac{p^\nu(z)\, w(z)\, dz}{(z-x)^{\nu+1}} \tag{5.9.1}$$

is a solution of the equation

$$(pwu')' + \lambda_\nu wu = 0, \quad \lambda_\nu = -\nu q' - \frac{1}{2}\nu(\nu-1)p'' \tag{5.9.2}$$

for nonnegative integer values of ν. In each case, p was a polynomial of degree at most 2, q a polynomial of degree at most 1, $qw = (pw)'$, and the contour C enclosed x but excluded the zeros of p. This was derived as a consequence of the Rodrigues equation (5.1.5). We give now, under the assumptions just stated, a direct proof that (5.9.1) implies (5.9.2). We then note how similar integral representations of solutions can be obtained for arbitrary ν.

First,

$$p(x)w(x)u'(x) = (\nu+1)p(x) \int_C \frac{p(z)^\nu\, w(z)\, dz}{(z-x)^{\nu+2}}$$
$$- \frac{p(x)w'(x)}{w(x)} \int_C \frac{p(z)^\nu\, w(z)\, dz}{(z-x)^{\nu+1}}.$$

By assumption,

$$-pw' = -(pw)' + p'w = (p'-q)w,$$

and since p and q have degree at most 2 and 1, respectively,

$$p(x) = p(z) - p'(z)(z-x) + \frac{1}{2}p''(z)(z-x)^2,$$

$$p'(x) - q(x) = p'(z) - q(z) + \left[q'(z) - p''(z)\right](z-x)$$
$$= -\frac{p(z)w'(z)}{w(z)} + \left[q'(z) - p''(z)\right](z-x).$$

Combining the three preceding equations gives

$$(pwu')(x) = (v+1) \int_C \frac{p(z)^{v+1} w(z)\, dz}{(z-x)^{v+2}}$$
$$- \int_C \frac{(v+1)p(z)^v p'(z)w(z) + p(z)^{v+1} w'(z)\, dz}{(z-x)^{v+1}}$$
$$+ \left[q' + \frac{1}{2}(v-1)p'' \right] \int_C \frac{p(z)^v w(z)\, dz}{(z-x)^v}$$
$$= (v+1) \int_C \frac{p(z)^{v+1} w(z)\, dz}{(z-x)^{v+2}} - \int_C \frac{[p^{v+1} w]'(z)\, dz}{(z-x)^{v+1}}$$
$$+ \left[q' + \frac{1}{2}(v-1)p'' \right] \int_C \frac{p(z)^v w(z)\, dz}{(z-x)^v}.$$

Differentiating,

$$(pwu')'(x) = (v+1)(v+2) \int_C \frac{p(z)^{v+1} w(z)\, dz}{(z-x)^{v+3}}$$
$$- (v+1) \int_C \frac{[p^{v+1} w]'(z)\, dz}{(z-x)^{v+2}}$$
$$+ \left[vq' + \frac{1}{2}v(v-1)p'' \right] u(x)w(x). \tag{5.9.3}$$

Since

$$\frac{v+2}{(z-x)^{v+3}} = -\frac{d}{dz} \left\{ \frac{1}{(z-x)^{v+2}} \right\},$$

an integration by parts gives

$$(v+2) \int_C \frac{p(z)^{v+1}\, w(z)\, dz}{(z-x)^{v+3}} = \int_C \frac{(p^{v+1} w)'(z)\, dz}{(z-x)^{v+2}}. \tag{5.9.4}$$

Therefore (5.9.3) reduces to (5.9.2).

For the contour just discussed, the assumption that v is an integer was necessary in order for the contour to lie in one branch of $(s-x)^{-v-1}$; otherwise the integration by parts results in contributions from the points of C where one crosses from one branch to another. On the other hand, the argument would apply to a contour that lies (except possibly for its endpoints) in one branch of the power, provided that the endpoint contributions vanish. This provides a way to obtain a solution of (5.9.2) for more general values of v.

For each of three cases considered above, let $I = (a,b)$ be the associated real interval, and define a function of the *second kind*

$$u_v(x) = \frac{c_v}{w(x)} \int_a^b \frac{p(s)^v\, w(s)\, ds}{(s-x)^{v+1}}, \qquad \text{Re}\, v \geq 0,\ x \notin I. \tag{5.9.5}$$

In each case, $p^{\nu+1}w$ vanishes at a finite endpoint and vanishes exponentially at an infinite endpoint of the interval so long as $\mathrm{Re}\,\nu \geq 0$, so that once again (5.9.3) leads to (5.9.4). In addition, the argument that led to the recurrence relations and derivative formulas also carries over to the functions of the second kind. Summarizing, we have the following result.

Theorem 5.9.1 *For* $\mathrm{Re}\,\nu \geq 0$ *and x not real, the functions*

$$u_{\nu 1}(x) = (-1)^n n! e^{x^2} \int_{-\infty}^{\infty} \frac{e^{-s^2}\, ds}{(s-x)^{\nu+1}}; \qquad (5.9.6)$$

$$u_{\nu 2}(x) = x^{-\alpha} e^x \int_0^{\infty} \frac{s^{\nu+\alpha} e^{-s}\, ds}{(s-x)^{\nu+1}}; \qquad (5.9.7)$$

$$u_{\nu 3}(x) = \frac{1}{2^\nu}(1-x)^{-\alpha}(1+x)^{-\beta} \int_{-1}^{1} \frac{(1-s)^{\nu+\alpha}(1+s)^{\nu+\beta}\, ds}{(s-x)^{\nu+1}} \qquad (5.9.8)$$

satisfy the equations

$$u_{\nu 1}''(x) - 2x u_{\nu 1}'(x) + 2\nu u_{\nu 1}(x) = 0;$$

$$x u_{\nu 2}''(x) + (\alpha + 1 - x) u_{\nu 2}'(x) + \nu u_{\nu 2}(x) = 0;$$

$$(1 - x^2) u_{\nu 3}''(x) + [\beta - \alpha - (\alpha + \beta + 2)x]\, u_{\nu 3}'(x)$$
$$+ \nu(\nu + \alpha + \beta + 1) u_{\nu 3}(x) = 0,$$

respectively. Moreover, for $\mathrm{Re}\,\nu \geq 1$*, they satisfy the corresponding recurrence and derivative equations (5.2.5) and (5.2.8) for $u_{\nu 1}$, (5.2.6), (5.2.9), and (5.2.14) for $u_{\nu 2}$, and (5.2.7), (5.2.10), and (5.2.14) for $u_{\nu 3}$, with n replaced by ν.*

As the preceding proof shows, the basic result here holds in greater generality than the three specific cases considered above.

Theorem 5.9.2 *Suppose that p is a polynomial of degree at most 2, q a polynomial of degree at most 1, $pw'/w = q - p'$, and ν, x_0 are complex numbers. Suppose that C is an oriented contour in a region where p^ν and $(z - x_0)^\nu$ are holomorphic, and that the functions*

$$\frac{p(z)^\nu\, w(z)}{(z - x_0)^\nu}, \quad \frac{p(z)^{\nu+1}\, w(z)}{(z - x_0)^{\nu+1}}$$

are integrable on C, while the limit along the curve of

$$\frac{p(z)^{\nu+1}\, w(z)}{(z - x_0)^{\nu+2}}$$

as z approaches the (finite or infinite) endpoints a and b is 0. Then the function

$$u_\nu(x) = \frac{1}{w(x)} \int_C \frac{p(z)^\nu \, w(z) \, dz}{(z-x)^{\nu+1}}$$

is a solution of the equation

$$(pwu')' + \lambda_\nu wu = 0, \qquad \lambda_\nu = -\nu q' - \frac{1}{2}\nu(\nu-1)p''$$

in any region containing x_0 in which the assumptions continue to hold.

5.10 Exercises

1. Verify that the normalizations (5.1.10), (5.1.11), and (5.1.12) lead to (5.1.13), (5.1.14), and (5.1.15), respectively.
2. Verify the limit (5.1.17), at least for the leading coefficients.
3. Verify the limit (5.1.18), at least for the leading coefficients.
4. Verify one or more of the identities (5.2.4).
5. Use (5.2.4) and the normalizations (5.1.10), (5.1.11), and (5.1.12) to verify the identities (5.2.5), (5.2.6), and (5.2.7).
6. Use (5.1.7) and the normalizations (5.1.10), (5.1.11), and (5.1.12) to verify the identities (5.2.8), (5.2.9), and (5.2.10).
7. Use the method for (5.2.3) applied to (5.2.11) to derive one or more of the identities (5.2.12), (5.2.14), and (5.2.14).
8. Prove a converse to the result in Section 5.3: the recurrence relation (5.3.4) can be derived from the generating function formula (5.3.6).
9. Prove the addition formula (5.3.7).
10. Prove the addition formula (5.3.8).
11. Prove

$$\lim_{n\to\infty} \left(\frac{x}{n}\right)^n H_n\left(\frac{n}{2x}\right) = e^{-x^2}.$$

12. Prove that if $a^2 + b^2 = 1$, then

$$H_n(ax+by) = \sum_{k=0}^n \binom{n}{k} H_{n-k}(x) H_k(y) a^{n-k} b^k.$$

13. Let $u_n(x) = e^{-x^2/2} H_n(x)$. Prove that the Fourier transform of u_n,

$$\frac{1}{\sqrt{2\pi}} \int_{-\infty}^\infty e^{-ix\xi} u_n(x) \, dx,$$

is $(-i)^n u_n(\xi)$. Hint: use the identity $\exp(-ix\xi + \frac{1}{2}x^2) = e^{\xi^2/2}\exp(\frac{1}{2}[x - i\xi]^2)$.

14. Prove that

$$\frac{1}{2^n n!} \int_{-\infty}^{\infty} \frac{H_n(x)^2}{1+x^2} e^{-x^2} \, dx = \int_{-\infty}^{\infty} \left(\frac{1-x^2}{1+x^2} \right)^n \frac{e^{-x^2}}{1+x^2} \, dx.$$

Hint: (i) set $y = x$ in (5.3.17) and multiply both sides by $e^{-x^2}/(1+x^2)$; (ii) integrate the resulting equation on both sides with respect to x and evaluate the integral on the right by making the change of variables $x = \sqrt{(1+t)/(1-t)}\, y$.

15. Prove the lower bounds in Theorem 5.3.2 by using the gauge transformation $H_n(x) = e^{x^2/2} h_n(x)$ and noting that the zeros of H_n and h_n coincide.

16. Prove the upper bounds in Theorem 5.3.2 by using the gauge transformation $H_n(x) = e^{x^2/2} h_n(x)$ and then writing $h_n(x) = u_n(y)$, $y = x\sqrt{2n+1}$. Use Corollary 3.3.5 to relate the kth positive zero of u_n to that of u_{2k} (n even) or u_{2k+1} (n odd), and thus relate x_{kn} to $x_{k,2k} < \sqrt{4k+1}$ or to $x_{k,2k+1} < \sqrt{4k+3}$.

17. Prove

$$\lim_{a \to \infty} a^{-n} L_n^{(\alpha)}(ax) = \frac{(1-x)^n}{n!}.$$

18. Verify (5.4.4) by comparing coefficients, as in the derivation of (5.3.4).

19. Show that (5.4.4) can be derived from the generating function formula (5.4.6).

20. Prove the addition formula (5.4.7).

21. Prove the addition formula

$$L_n^{(\alpha+\beta+1)}(x+y) = \sum_{k=0}^{n} L_{n-k}^{(\alpha)}(x) L_k^{(\beta)}(y).$$

22. Expand the integrand in series to prove Koshlyakov's formula [228]:

$$L_n^{(\alpha+\beta)}(x) = \frac{\Gamma(n+\alpha+\beta+1)}{\Gamma(\beta)\Gamma(n+\alpha+1)} \int_{-1}^{1} t^\alpha (1-t)^{\beta-1} L_n^{(\alpha)}(xt) \, dt.$$

23. The *Laplace transform* of a function $f(x)$ defined for $x \geq 0$ is the function $\mathcal{L}f$ defined by

$$[\mathcal{L}f](s) = \int_0^\infty e^{-sx} f(x) \, dx$$

for all values of s for which the integral converges. Show that the Laplace transform of $x^\alpha L_n^{(\alpha)}$ is

$$\frac{\Gamma(n+\alpha+1)}{n!} \frac{(s-1)^n}{s^{n+\alpha+1}}, \qquad \operatorname{Re} s > 1.$$

24. Prove

$$\lim_{a \to \infty} a^{-n} P_n^{(\alpha,\beta)}(x) = \frac{1}{n!} \left(\frac{1+x}{2} \right)^n.$$

25. Prove (5.5.7) by showing that the functions on the right-hand side are orthogonal to polynomials in x of lower degree, with respect to the weight function $(1-x^2)^\alpha$, and comparing coefficients at $x=-1$.

26. Prove Theorem 5.5.1: use (5.5.10) and Theorem 3.3.3 to prove the lower bounds, then use (5.5.1) and the lower bounds to obtain the upper bounds. (Note that $x \to -x$ corresponds to $\theta \to \pi - \theta$.)

27. Suppose that f and its derivatives of order $\le n$ are bounded on the interval $(-1,1)$. Show that

$$\int_0^1 f(x) P_n^{(\alpha,\beta)}(x)(1-x)^\alpha(1+x)^\beta \, dx$$

$$= \frac{1}{2^n \, n!} \int_0^1 f^{(n)}(x)(1-x)^{n+\alpha}(1+x)^{n+\beta} \, dx.$$

28. Show that for integers $m \ge n$,

$$\int_0^1 (1+x)^m \, P_n^{(\alpha,\beta)}(x)(1-x)^\alpha(1+x)^\beta \, dx$$

$$= \binom{m}{n} \frac{\Gamma(n+\alpha+1)\Gamma(m+\beta+1)}{\Gamma(n+m+\alpha+\beta+2)} 2^{n+m+\alpha+\beta+1}.$$

29. Derive the recurrence relation (5.6.3) from the generating function (5.6.2) by differentiating both sides of (5.6.2) with respect to s, multiplying by $(1-2xs+x^2)$, and equating coefficients of s.

30. Derive the identity (5.6.5) from (5.6.2).

31. Derive the generating function (5.6.2) from the recurrence relation (5.6.3).

32. Prove

$$\int_{-1}^1 x^{n+2k} P_n(x) \, dx = \frac{(n+2k)!}{2^n \, (2k)!} \frac{\Gamma(k+\frac{1}{2})}{\Gamma(n+k+\frac{3}{2})}, \quad k=0,1,2,\ldots$$

33. (a) Show that the sequence of Chebyshev polynomials $\{T_n\}$ can be written as a linear combination of two geometric progressions.

   (b) Show that the space of linear combinations of two geometric progressions $\{r^n\}$, $\{s^n\}$, $r \ne s$, is precisely the space of solutions $\{u\}$ of a second-order difference equation

$$u_{n+2} + 2b u_{n+1} + c_n u_n = 0,$$

   and determine the coefficients b, c from r, s.

   (c) Use parts (a) and (b) to derive the three-term recurrence relation (5.6.15).

   (d) Adapt this method to derive (5.6.20) for the Chebyshev polynomials $\{U_n\}$.

34. Derive the recurrence relation (5.6.15) and the identity (5.6.16) from the generating function (5.6.18).
35. Derive (5.6.23) from (5.6.20) and the special cases U_0, U_1.
36. Derive (5.6.20) and (5.6.21) from (5.6.23).
37. Use the generating function to show that the Legendre polynomials satisfy the following: for even $n = 2m$, $P'_n(0) = 0$, and

$$P_n(0) = (-1)^m \frac{(\frac{1}{2})_m}{m!} = (-1)^m \frac{\Gamma(m+\frac{1}{2})}{\sqrt{\pi}\,\Gamma(m+1)},$$

while for odd $n = 2m+1$, $P_n(0) = 0$, and

$$P'_n(0) = (-1^m \frac{(\frac{3}{2})_m}{m!} = (-1)^m \frac{2\Gamma(m+\frac{3}{2})}{\sqrt{\pi}\,\Gamma(m+1)}.$$

38. Use the functional equation for the gamma function to show that the sequence

$$\frac{\sqrt{m}\,\Gamma(m+\frac{1}{2})}{\Gamma(m+1)}$$

is increasing, and use (2.1.9) to find the limit. Deduce from this and Exercise 37 that (5.6.11) is true for n even.
39. Prove (5.6.11) for n odd.
40. Use Exercise 28 to show that for any integer $m \geq 0$,

$$(1+x)^m = \sum_{n=0}^{m} c_n P_n^{(\alpha,\beta)}(x),$$

where

$$c_n = \frac{\Gamma(n+\alpha+\beta+1)}{\Gamma(2n+\alpha+\beta+1)} \frac{m!(n+\beta+1)_{m-n}}{(2n+\alpha+\beta+1)_{m-n}} 2^m.$$

41. Prove that if the assumption (5.8.3) in any of the theorems of Section 5.8 is replaced by the conditions (5.8.13), then the corresponding series converges to $[f(x_+) + f(x_-)]/2$. Hint: by subtracting a function that satisfies (5.8.3) at x, one can essentially reduce this to the fact that the integral, over an interval centered at x, of a function that is odd around x, is zero.
42. Prove Theorem 5.7.1: for a given choice of indices α, β, let

$$\psi_n(z) = \frac{[P_n^{(\alpha,\beta)}]'(z)}{n P_n^{(\alpha,\beta)}(z)}.$$

As in the proof of Theorem 5.7.3, use the differential equation to show that ψ_n is approximately the solution of a quadratic equation, and use the condition $\psi_n(z) \sim 1/z$ at infinity to determine the choice of solution:

$$\psi(z) = \frac{1}{\sqrt{z^2 - 1}}.$$

For a polynomial f, express $\lambda_n^{(\alpha,\beta)}(f)$ as a contour integral and collapse the contour to the interval $[-1, 1]$ to obtain the conclusion of Theorem 5.7.1.

43. Prove Theorem 5.7.2: follow the outline of Exercise 42. In this case, the approximation ψ is

$$\psi(z) = 2\left(1 - \sqrt{\frac{1}{z}}\right).$$

44. Prove Theorem 5.7.5.
45. Prove Theorem 5.7.6.
46. Prove Theorem 5.7.7.

5.11 Remarks

Legendre and Hermite polynomials were the first to be studied, and most of the general theory was first worked out in these cases. Legendre found recurrence relations and the generating function for Legendre polynomials in 1784–1785 [250, 251]; Laplace [242] found the orthogonality relation. Rodrigues proved the Rodrigues formula for the Legendre polynomials in 1816 [345]. Schläfli [355] gave the integral formula (5.2.1) for the Legendre polynomials in 1881. The series expansion (5.5.13) for Legendre polynomials was given by Murphy in 1833 [296].

Hermite polynomials occur in the work of Laplace on celestial mechanics [242] and on probability [243, 244], and in Chebyshev's 1859 paper [72], as well as in Hermite's 1864 paper [181]. Laguerre polynomials for $\alpha = 0$ were considered by Lagrange [237], Abel [2], Chebyshev [72], Laguerre [239], and, for general α, by Sonine [374].

Jacobi polynomials in full generality were introduced by Jacobi in 1859 [201]. The special case of Chebyshev polynomials was studied by Chebyshev in 1854 [69]. Gegenbauer polynomials were studied by Gegenbauer in [159, 160].

Further remarks on the history are contained in several of the books cited in Section 5.11, and in Szegő's article (and Askey's addendum) [393].

A common approach to classical orthogonal polynomials is to use the generating function formulas (5.3.6), (5.4.6), and (5.5.8) as definitions. The three-term recurrence relations and some other identities are easy consequences, as is orthogonality (once one has selected the correct weight) in the Hermite and Laguerre cases. The fact that the polynomials are the eigenfunctions of a symmetric second-order operator is easily established, but in the generating function approach this fact appears as something of a (fortunate) accident. It does not seem clear, from the generating function approach, why it is these polynomials and no others that have arisen as the "classical" orthogonal polynomials. In our view, the characterization theorem (3.4.1), the connection with the basic equations of mathematical physics in special coordinates (Section 3.6), and the characterization of "recursive" second-order equations (Section 1.1) provide natural explanations for the significance of precisely this set of polynomials.

The approach used in this chapter, deriving basic properties directly from the differential equation via the Rodrigues formula and the resulting complex integral representation, is the one used by Nikiforov and Uvarov [305].

The classical orthogonal polynomials are treated in every treatise or handbook of special functions. Some more comprehensive sources for the classical and general orthogonal polynomials are Szegő's classic treatise [392], as well as Askey [18], Chihara [76], Gautschi [156], Geronimus [164], Ismail [194], and Khrushchev [217]. An especially detailed and well-organized treatment is given in the book by Koekoek, Lesky, and Swarttouw [221].

The addition formulas for Hermite and Laguerre polynomials are easily derived from the generating function representation. There is a classical addition formula for Legendre polynomials that follows from the connection of these polynomials with spherical harmonics; see Section 11.1. Gegenbauer [161] obtained a generalization to all ultraspherical (Gegenbauer) polynomials ($\alpha = \beta$). More recently, Šapiro [352] found a formula valid for $\beta = 0$, and Koornwinder [225, 226] found a generalization valid for all Jacobi polynomials.

Discriminants of Jacobi polynomials were calculated by Hilbert [183] and Stieltjes [381, 382]. Zeros of Hermite and Laguerre polynomials are treated in the report by Hahn [175]. Hille, Shohat, and Walsh compiled an exhaustive bibliography up to 1940 [186].

As noted in Section 4.3, orthogonal polynomials are closely related to certain types of continued fractions. The book by Khrushchev [217] approaches the classical orthogonal polynomials from this point of view, noting that it was also Euler's point of view.

Expansion in orthogonal polynomials is treated in Sansone [351].

The subject of orthogonal polynomials continues to be a very active area of research, as evidenced, for example, by the books of Ismail [194] and Stahl and Totik [375]. Zeros of orthogonal polynomials are related to stationary points for electrostatic potentials, a fact that was used by Stieltjes in his calculation of the discriminants of Jacobi polynomials. These ideas have been extended considerably; see the books of Levin and Lubinsky [255] and Saff and Totik [350].

We have not touched here on the topic of "q-orthogonal polynomials," which are related to the q-difference operator

$$D_q f(x) = \frac{f(x) - f(qx)}{(1-q)x}$$

in ways similar to those in which the classical orthogonal polynomials are related to the derivative. For extensive treatments, see the books by Andrews, Askey, and Roy [10], Bailey [28], Gasper and Rahman [149], Ismail [194], and Slater [369].

6

Semi-classical orthogonal
polynomials

In Chapter 4 we discussed the question of polynomials that are orthogonal with respect to a weight function, which was assumed to be a positive continuous function on a real interval. This is an instance of a measure. Another example is a discrete measure, for example, one supported on the integers with masses w_m, $m = 0, \pm 1, \pm 2, \ldots$ Most of the results of Section 4.1 carry over to this case, although if w_m is positive at only a finite number $N + 1$ of points, the associated function space has dimension $N + 1$ and will be spanned by orthogonal polynomials of degree 0 through N.

In this context, the role of differential operators is played by difference operators. An analogue of the characterization in Theorem 3.4.1 is valid: up to normalization, the orthogonal polynomials that are eigenfunctions of a symmetric second-order difference operator are the "classical discrete polynomials," associated with the names Charlier, Krawtchouk, Meixner, and Hahn.

The theory of the classical discrete polynomials can be developed in a way that parallels the treatment of the classical polynomials in Chapter 5, using a discrete analogue of the formula of Rodrigues.

Working with functions of a complex variable and taking differences in the complex direction leads to two more characterizations: the Meixner–Pollaczek and continuous Hahn polynomials on the one hand, and the continuous dual Hahn and Wilson polynomials on the other. Except for the dual Hahn polynomials and the Racah polynomials, this completes the roster of polynomials in the "Askey scheme" – polynomials that can be expressed in terms of the generalized hypergeometric functions $_pF_q$, $q + 1 \leq p = 1, 2, 3, 4$.

6.1 Discrete weights and difference operators

Suppose that $w = \{w_n\}_{n=-\infty}^{\infty}$ is a two-sided sequence of nonnegative numbers. The corresponding inner product

$$(f,g) = (f,g)_w = \sum_{m=-\infty}^{\infty} f(m)g(m)w_m$$

is well-defined for all real functions f and g for which the norms $\|f\|_w$ and $\|g\|_w$ are finite, where

$$\|f\|_w^2 = (f,f)_w = \sum_{m=-\infty}^{\infty} f(m)^2 w_m.$$

The norm and inner product depend only on the values taken by functions on the integers, although it is convenient to continue to regard polynomials, for example, as being defined on the line. Polynomials have a finite norm if and only if the even moments

$$\sum_{m=-\infty}^{\infty} m^{2n} w_m$$

are finite. If so, then orthogonal polynomials ψ_n can be constructed exactly as in the case of a continuous weight function. Usually we normalize so that w is a probability distribution:

$$\sum_{m=-\infty}^{\infty} w_m = 1.$$

Again, there is a three-term recurrence relation. Suppose that

$$\psi_n(x) = a_n x^n + b_n x^{n-1} + \cdots$$

The polynomial

$$x\,\psi_n(x) - \frac{a_n}{a_{n+1}} \psi_{n+1} + \left(\frac{b_{n+1}}{a_{n+1}} - \frac{b_n}{a_n} \right) \psi_n(x)$$

has degree $n-1$ and is orthogonal to polynomials of degree $< n-1$, and is therefore a multiple $\gamma_n \psi_{n-1}$. Then

$$\gamma_n(\psi_{n-1},\psi_{n-1})_w = (x\psi_n,\psi_{n-1})_w = (\psi_n, x\psi_{n-1})_w = \frac{a_{n-1}}{a_n}(\psi_n,\psi_n)_w.$$

Thus

$$x\psi_n(x) = \alpha_n \psi_{n+1}(x) + \beta_n \psi_n(x) + \gamma_n \psi_{n-1}(x), \qquad (6.1.1)$$

$$\alpha_n = \frac{a_n}{a_{n+1}}, \quad \beta_n = \frac{b_n}{a_n} - \frac{b_{n+1}}{a_{n+1}}, \quad \gamma_n = \frac{a_{n-1}}{a_n} \frac{(\psi_n,\psi_n)_w}{(\psi_{n-1},\psi_{n-1})_w}.$$

As before, the three-term recurrence implies a Christoffel–Darboux formula.

If the weight w_m is positive at only finitely many integers, say at $m = 0, 1, \ldots, N$, then the L^2 space has dimension $N + 1$. The orthogonal polynomials have degrees $0, 1, \ldots, N$ and are a basis. In general, there is a completeness result analogous to Theorem 4.1.5, applicable to all the discrete cases to be considered in this chapter.

Theorem 6.1.1 *Suppose that w is a positive weight on the integers and suppose that for some $c > 0$,*

$$\sum_{-\infty}^{\infty} e^{2c|m|} w_m < \infty. \tag{6.1.2}$$

Let $\{P_n\}$ be the orthonormal polynomials for w. Then $\{P_n\}$ is complete: for any $f \in L^2_w$,

$$\lim_{n \to \infty} \|f - \sum_{k=0}^{n} (f, P_k) P_k\|_w = 0. \tag{6.1.3}$$

In the discrete case, it is easy to see that convergence in norm implies pointwise convergence at each point where $w_m > 0$.

Corollary 6.1.2 *Under the assumptions of Theorem 6.1.1, for any $f \in L^2_w$ and any m such that $w_m > 0$,*

$$f(m) = \sum_{n=0}^{\infty} (f, P_n) P_n(m). \tag{6.1.4}$$

Like an integrable function, a finite measure is determined by its Fourier transform. Therefore the proof of Theorem 4.1.5 in Appendix B also proves Theorem 6.1.1.

For functions defined on the integers, differentiation can be replaced by either the forward or backward difference operators

$$\Delta_+ f(m) = f(m+1) - f(m), \qquad \Delta_- f(m) = f(m) - f(m-1).$$

Each operator maps polynomials of degree d to polynomials of degree $d - 1$, so the product

$$[\Delta_+ \Delta_-] f(m) = [\Delta_- \Delta_+] f(m) = f(m+1) + f(m-1) - 2f(m)$$

decreases the degree of a polynomial by two and plays the role of a second-order derivative. It is convenient to express these operators in terms of the shift operators

$$S_\pm f(m) = f(m \pm 1).$$

Thus

$$\Delta_+ = S_+ - I, \qquad \Delta_- = I - S_-,$$

and

$$\Delta_+\Delta_- = \Delta_-\Delta_+ = S_+ + S_- - 2I = \Delta_+ - \Delta_-. \tag{6.1.5}$$

Therefore the general real second-order difference operator can be written as

$$L = p_+S_+ + p_-S_- + r,$$

where p_+, p_-, and r are real-valued functions. The condition for L to be symmetric with respect to w is that

$$\begin{aligned}
0 &= (Lf,g) - (f,Lg) \\
&= (p_+S_+f,g) - (f,p_-S_-g) + (p_-S_-f,g) - (f,p_+S_+g) \\
&= \sum_{m=-\infty}^{\infty} \left[p_+w - S_+(p_-w) \right](m)f(m+1)g(m) \\
&\quad + \sum_{m=-\infty}^{\infty} \left[p_-w - S_-(p_+w) \right](m)f(m-1)g(m)
\end{aligned}$$

for every f and g that vanish for all but finitely many integers. By choosing g to vanish except at one value m, and f to vanish except at $m+1$ or at $m-1$, we conclude that symmetry is equivalent to

$$S_-(p_+w) = p_-w, \qquad S_+(p_-w) = p_+w. \tag{6.1.6}$$

Note that S_+ and S_- are inverses, so these two conditions are mutually equivalent.

As for differential equations, symmetry implies that eigenfunctions that correspond to distinct eigenvalues are orthogonal: if $Lu_j = \lambda_j u_j, j = 1,2$, then

$$\lambda_1(u_1,u_2)_w = (Lu_1,u_2)_w = (u_1,Lu_2)_w = \lambda_2(u_1,u_2)_w.$$

We now ask: for which discrete weights w (with finite moments) and which symmetric operators L with coefficients $p_\pm(m)$ positive where $w_m > 0$ (with exceptions at endpoints) are the eigenfunctions of L polynomials? More precisely, when do the eigenfunctions of L include polynomials of degrees 0, 1, and 2? We assume that $w_m > 0$ for integers m in a certain interval that is either infinite or has $N+1$ points, and is zero otherwise, so we normalize to the cases

    (a) $w_m > 0$ if and only if $0 \le m \le N$;

    (b) $w_m > 0$ if and only if $m \ge 0$;

    (c) $w_m > 0$ for all m.

We shall see that case (c) does not occur.

The symmetry condition (6.1.6) implies that $p_-(0) = 0$ in cases (a) and (b) and that $p_+(N) = 0$ in case (a). We shall assume that otherwise $p_\pm$ is positive where w is positive.

With a change of notation for the zero-order term, we may write L as

$$L = p_+ \Delta_+ - p_- \Delta_- + r. \qquad (6.1.7)$$

Then $L(1) = r$ must be constant, and we may assume $r = 0$. Since $\Delta_\pm(x) = 1$, we have

$$L(x) = p_+ - p_-,$$

so $p_+ - p_-$ is a polynomial of degree 1. Next, $\Delta_\pm(x^2) = 2x \pm 1$, so

$$L(x^2) = 2x(p_+ - p_-) + (p_+ + p_-),$$

and it follows that $p_+ + p_-$ is a polynomial of degree at most 2. Therefore p_+ and p_- are both polynomials of degree at most 2, and at least one has positive degree. Moreover, if either has degree 2 then both do and they have the same leading coefficient.

The symmetry condition (6.1.6) and our positivity assumptions imply that

$$w_{m+1} = \frac{p_+(m)}{p_-(m+1)} w_m = \varphi(m) w_m, \qquad (6.1.8)$$

wherever $w_m > 0$. This allows us to compute all values w_m from w_0.

Suppose first that one of $p_\pm$ has degree 0, so the other has degree 1. Then (6.1.8) implies that $\varphi(m)$ is $O(|m|)$ in one direction or the other, so the moment condition rules out case (c). It follows that $p_-(0) = 0$, and p_+ is constant, which rules out case (a). We normalize by taking $p_+(x) = 1$. Then $p_-(x) = x/a$, $a > 0$, so

$$\frac{p_+(m)}{p_-(m+1)} = \frac{a}{m+1}.$$

To normalize w we introduce the factor e^{-a}:

$$w_m = e^{-a} \frac{a^m}{m!}, \quad m = 0, 1, 2, 3, \dots \qquad (6.1.9)$$

This is the probability distribution known as the Poisson distribution. The associated polynomials, suitably normalized, are the *Charlier polynomials*, also called the Poisson–Charlier polynomials.

Suppose next that both of $p_\pm$ have degree 1. If the leading coefficients are not the same, then $\varphi(m)$ grows in one direction, which rules out case (c). If the leading coefficients are the same, then asymptotically $\varphi(m) - 1 \sim b/m$ for some constant b, which implies that the products

$$\prod_{j=0}^{m} \varphi(j), \quad \prod_{j=0}^{m} \varphi(j)^{-1} \qquad (6.1.10)$$

are either identically 1 ($p_+ = p_-$), or one grows like m and the other decays like $1/m$ as $m \to \infty$. This rules out case (c). Thus if both have degree 1, then we have case (a) or (b) and $p_-(0) = 0$.

Continuing to assume that both of $p_\pm$ have degree 1, in the case of a finite interval we normalize by taking $p_+(x) = p(N - x)$ and $p_-(x) = qx$, where $p, q > 0$, $p + q = 1$. Then the normalized weight is

$$w_m = \binom{N}{m} p^m q^{N-m}, \quad m = 0, 1, 2, 3, \ldots, N, \tag{6.1.11}$$

the binomial probability distribution. Up to normalization, the associated polynomials are the *Krawtchouk polynomials*.

Suppose now that both $p_\pm$ have degree 1 and that the interval is infinite. We may normalize by taking $p_-(x) = x$ and $p_+(x) = c(x + b)$. Positivity implies $b, c > 0$ and finiteness implies $c < 1$. Then

$$\frac{p_+(m)}{p_-(m+1)} = \frac{c(m+b)}{(m+1)},$$

and the normalized weight is

$$w_m = (1 - c)^b \frac{(b)_m}{m!} c^m, \quad m = 0, 1, 2, 3, \ldots \tag{6.1.12}$$

The associated polynomials are the *Meixner polynomials*.

Suppose finally that one of $p_\pm$ has degree 2. Then both do, and the leading coefficients are the same, so

$$\frac{p_+(m)}{p_-(m+1)} = \frac{1 + am^{-1} + bm^{-2}}{1 + cm^{-1} + dm^{-2}} = 1 + \frac{a - c}{m} + O\left(\frac{1}{m^2}\right).$$

Arguing as before, we see that the moment condition rules out the possibility of an infinite interval. With the weight supported on $0 \leq m \leq N$, we have $p_-(0) = 0 = p_+(N)$ and there are two cases, which can be normalized to

$$p_-(x) = x(N + 1 + \beta - x), \quad p_+(x) = (N - x)(x + \alpha + 1), \quad \alpha, \beta > -1$$

or

$$p_-(x) = x(x - \beta - N - 1), \quad p_+(x) = (N - x)(-\alpha - 1 - x), \quad \alpha, \beta < -N.$$

It is convenient to give up the positivity assumption for $p_\pm$ and use the first formula for $p_\pm$ in either case. The weight function is

$$w_m = C \frac{(N - m + 1)_m (\alpha + 1)_m}{m!(N + \beta + 1 - m)_m} = C \binom{N}{m} \frac{(\alpha + 1)_m (\beta + 1)_{N-m}}{(\beta + 1)_N}. \tag{6.1.13}$$

The associated normalized polynomials are commonly known as the *Hahn polynomials*. We refer to them here as the *Chebyshev–Hahn polynomials*; see the remarks at the end of the chapter. In the case $\alpha = \beta = 0$, the polynomials are commonly known as the *discrete Chebyshev polynomials*.

We have proved the discrete analogue of Theorem 3.4.1, somewhat loosely stated.

Theorem 6.1.3 *Up to normalization, the Charlier, Krawtchouk, Meixner, and Chebyshev–Hahn polynomials are the only ones that occur as eigenfunctions of a second-order difference operator that is symmetric with respect to a positive weight.*

6.2 The discrete Rodrigues formula

Suppose that w is a weight on the integers and L is symmetric with respect to w and has polynomials as eigenfunctions. The eigenvalue equation for a polynomial ψ_n of degree n is

$$p_+ \Delta_+ \psi_n - p_- \Delta_- \psi_n + \lambda_n \psi_n = 0. \tag{6.2.1}$$

Applying Δ_+ to this equation gives an equation for the "derivative" $\psi_n^{(1)} = \Delta_+ \psi_n$. Using the discrete Leibniz identity

$$\Delta_+(fg) = (S_+ f)\Delta_+ g + g\,\Delta_+ f,$$

we obtain

$$p_+^{(1)} \Delta_+ \psi_n^{(1)} - p_-^{(1)} \Delta_- \psi_n^{(1)} + \lambda_n^{(1)} \psi_n^{(1)} = 0,$$

with

$$p_+^{(1)} = S_+ p_+, \quad p_-^{(1)} = p_-, \quad \lambda_n^{(1)} = \lambda_n + \Delta_+(p_+ - p_-).$$

The new operator here is symmetric with respect to the weight $w^{(1)} = p_+ w$:

$$S_+(w^{(1)} p_-^{(1)}) = (S_+ p_+) S_+(w p_-) = p_+^{(1)} w p_+ = w^{(1)} p_+^{(1)}.$$

Continuing, we see that the successive differences $\psi_n^{(k)} = (\Delta_+)^k \psi_n$ satisfy

$$p_+^{(k)} \Delta_+ \psi_n^{(k)} - p_- \Delta_- \psi_n^{(k)} + \lambda_n^{(k)} \psi_n^{(k)} = 0,$$

with

$$p_+^{(k)} = S_+^k p_+, \tag{6.2.2}$$

$$\lambda_n^{(k)} = \lambda_n + (I + S_+ + \cdots + S_+^{k-1})\Delta_+ p_+ - k\Delta_+ p_-,$$

and the corresponding operator is symmetric with respect to the weight

$$w^{(k)} = w \prod_{j=0}^{k-1} S_+^j p_+. \tag{6.2.3}$$

Now $\psi_n^{(k)}$ has degree $n - k$. In particular, $\psi_n^{(n)}$ is constant, so $\lambda_n^{(n)} = 0$, and we have proved that

$$\lambda_n = -(I + S_+ + \cdots + S_+^{n-1})\Delta_+ p_+ + n\Delta_+ p_-$$

$$= n\Delta_+(p_- - p_+) + \sum_{j=1}^{n-1}(I - S_+^j)\Delta_+ p_+. \tag{6.2.4}$$

The eigenvalue equation can be rewritten to obtain $\psi_n^{(k-1)}$ from $\psi_n^{(k)}$, which leads to a discrete Rodrigues formula for ψ_n. At the first stage, we rewrite the operator L, using the identities (6.1.6) and $S_-\Delta_+ = \Delta_-$:

$$wL = wp_+ \Delta_+ - S_-(wp_+)\Delta_- = wp_+ \Delta_+ - S_-(wp_+ \Delta_+)$$

$$= \Delta_-(wp_+ \Delta_+) = \Delta_-(w^{(1)}\Delta_+).$$

Therefore the eigenvalue equation can be solved to obtain

$$\psi_n = -\frac{1}{\lambda_n w}\Delta_-(w^{(1)}\psi_n^{(1)})$$

on the points where w is positive. We continue this process, and take the constant $\psi_n^{(n)}$ to be

$$\psi_n^{(n)} = \prod_{k=0}^{n-1}\left[\lambda_n + (S_+^k - I)p_+ - k\Delta_+ p_-\right] \equiv A_n. \tag{6.2.5}$$

The result is the discrete Rodrigues formula: where $w > 0$,

$$\psi_n = (-1)^n \frac{1}{w}\Delta_-^n(w^{(n)}), \quad w^{(n)} = w\prod_{k=0}^{n-1}S_+^k p_+. \tag{6.2.6}$$

Since $\Delta_- = I - S_-$, we may expand $\Delta_-^n = (I - S_-)^n$ and rewrite (6.2.6) as

$$\psi_n = \frac{1}{w}\sum_{k=1}^{n}(-1)^{n-k}\binom{n}{k}S_-^k(w^{(n)}). \tag{6.2.7}$$

As an application, we obtain a formula for the norm of ψ_n. Note that whenever the sums are absolutely convergent,

$$\sum_k \Delta_- f(k)\,g(k) = -\sum_k f(k)\,\Delta_+ g(k).$$

Therefore

$$\lambda_n ||\psi_n||_w^2 = -(L\psi_n, \psi_n)_w = -\sum_k (wL\psi_n)(k)\,\psi_n(k)$$

$$= -\sum_k [\Delta_-(wp_+\psi_n^{(1)})](k)\,\psi_n(k) = \sum_k [wp_+\psi_n^{(1)}](k)\,\psi_n^{(1)}(k)$$

$$= ||\psi_n^{(1)}||_{w^{(1)}}^2.$$

Continuing, we obtain

$$A_n||\psi_n||_w^2 = ||\psi_n^{(n)}||_{w^{(n)}}^2 = A_n^2||1||_{w^{(n)}}^2.$$

Therefore

$$||\psi_n||_w^2 = A_n \sum_k w^{(n)}(k). \tag{6.2.8}$$

It is useful to rewrite the operator in (6.2.6) as

$$\frac{1}{w}\Delta_-^n w^{(n)} = \left(\frac{1}{w}\Delta_- w^{(1)}\right)\left(\frac{1}{w^{(1)}}\Delta_- w^{(2)}\right)\cdots\left(\frac{1}{w^{(n-1)}}\Delta_- w^{(n)}\right).$$

It follows from the symmetry condition (6.1.6) that

$$\frac{1}{w}\Delta_-(w^{(1)}f) = p_+ f - \frac{S_-(p_+w)}{w}S_- f = p_+ f - p_- S_- f.$$

Then

$$\frac{1}{w}\Delta_- w^{(1)} \cdot \frac{1}{w^{(1)}}\Delta_- w^{(2)} = (p_+ - p_- S_-)(p_+^{(1)} - p_- S_-)$$

$$= p_+ p_+^{(1)} - 2p_+ p_- S_- + p_-(S_- p_-)S_-^2$$

and by induction,

$$(-1)^n \frac{1}{w}\Delta_-^n w^{(n)} = \sum_{k=0}^n (-1)^{n-k}\binom{n}{k}\prod_{j=0}^{n-k-1} p_+^{(j)}\prod_{j=0}^{k-1} p_-^{(j)} S_-^k,$$

where $p_-^{(j)} = S_-^j p_-$. Applying this operator to the constant function 1 gives

$$\psi_n = \sum_{k=0}^n (-1)^{n-k}\binom{n}{k}\prod_{j=0}^{n-k-1}(S_+^j p_+)\prod_{j=0}^{k-1}(S_-^j p_-). \tag{6.2.9}$$

A second approach to computing ψ_n is to use a discrete analogue of the series expansion, with the monomials x^k replaced by the polynomials

$$e_k(x) = (x - k + 1)_k = x(x-1)(x-2)\cdots(x-k+1) = (-1)^k(-x)_k.$$

Then

$$\Delta_+ e_k(x) = k e_{k-1}(x), \quad x\Delta_- e_k(x) = k e_k(x), \quad xe_k = e_{k+1} + k e_k.$$

In each of our cases, $p_-(x)$ is divisible by x, so applying the operator $L = p_+ \Delta_+ - p_- \Delta_-$ to the expansion

$$\psi_n(x) = \sum_{k=0}^{n} a_{nk} e_k(x) \qquad (6.2.10)$$

leads to recurrence relations for the coefficients a_{nk} that identify a_{nk} as a certain multiple of $a_{n,k-1}$.

Finally, we remark that both the coefficient β_n of the three-term recurrence relation (6.1.1) and the eigenvalue λ_n can be recovered directly from (6.2.1) by computing the coefficients of x^n and x^{n-1}:

$$
\begin{aligned}
0 &= L\psi_n(x) + \lambda_n \psi_n(x) \\
&= p_+ \Delta_+ (a_n x^n + b_n x^{n-1}) - p_- \Delta_- (a_n x^n + b_n x^{n-1}) \\
&\quad + \lambda_n a_n x^n + \lambda_n b_n x^{n-1} + \cdots \\
&= (p_+ - p_-)\left[n a_n x^{n-1} + (n-1) b_n x^{n-2} \right] \\
&\quad + (p_+ + p_-)\left[\binom{n}{2} a_n x^{n-2} + \binom{n-1}{2} b_n x^{n-3} \right] \\
&\quad + \lambda_n a_n x^n + \lambda_n b_n x^{n-1} + \cdots
\end{aligned}
\qquad (6.2.11)
$$

The coefficient of x^n on the right must vanish, and this determines λ_n. Using this value of λ_n in the coefficient of x^{n-1} determines the ratio b_n/a_n and therefore the term $\beta_n = b_n/a_n - b_{n+1}/a_{n+1}$ in the three-term recurrence.

6.3 Charlier polynomials

The interval is infinite, and

$$p_+(x) = 1, \quad p_-(x) = \frac{x}{a}, \quad w_m = e^{-a} \frac{a^m}{m!}.$$

Then $p_+^{(k)} = p_+$, $w^{(k)} = w$, and

$$\lambda_n = \frac{n}{a}.$$

Therefore the constant A_n in (6.2.5) is $n!/a^n$. From (6.2.8) we obtain the norm:

$$\|\psi_n\|_w^2 = \frac{n!}{a^n}. \qquad (6.3.1)$$

A standard normalization is $C_n(x;a) = (-1)^n \psi_n(x)$. Equation (6.2.9) gives

$$C_n(x;a) = \sum_{k=0}^{n} (-1)^k \binom{n}{k} a^{-k} (x-k+1)_k$$

$$= \sum_{k=0}^{n} \frac{(-n)_k (-x)_k}{k!} \left(-\frac{1}{a}\right)^k$$

$$= {}_2F_0 \left(\begin{matrix} -n, -x \\ - \end{matrix}; -\frac{1}{a}\right), \tag{6.3.2}$$

where ${}_2F_0$ is a generalized hypergeometric function; see Chapter 12. The leading coefficient is $(-a)^{-n}$. In this case, $\Delta_+ C_n$ is an orthogonal polynomial with respect to the same weight, and comparison of leading coefficients gives

$$\Delta_+ C_n(x;a) = -\frac{n}{a} C_{n-1}(x;a). \tag{6.3.3}$$

The associated difference equation is

$$\Delta_+ C_n(x;a) - \frac{x}{a} \Delta_- C_n(x;a) + \frac{n}{a} C_n(x;a) = 0. \tag{6.3.4}$$

The first four polynomials are

$$C_0(x;a) = 1;$$

$$C_1(x;a) = -\frac{x}{a} + 1;$$

$$C_2(x;a) = \frac{x(x-1)}{a^2} - \frac{2x}{a} + 1 = \frac{x^2}{a^2} - (1+2a)\frac{x}{a^2} + 1;$$

$$C_3(x;a) = -\frac{x(x-1)(x-2)}{a^3} + \frac{3x(x-1)}{a^2} - \frac{3x}{a} + 1$$

$$= -\frac{x^3}{a^3} + 3(a+1)\frac{x^2}{a^3} - (3a^2+3a+2)\frac{x}{a^3} + 1.$$

It is a simple matter to compute the generating function

$$G(x,t;a) \equiv \sum_{n=0}^{\infty} \frac{C_n(x;a)}{n!} t^n.$$

Note that the constant term of $C_n(x;a)$ is 1, so

$$G(x,0;a) = 1, \quad G(0,t;a) = e^t. \tag{6.3.5}$$

In general,

$$G(x,t;a) = \sum_{n=0}^{\infty} \sum_{k=0}^{n} t^n \frac{(-x)_k}{(n-k)!k!} \left(\frac{1}{a}\right)^k$$

$$= \sum_{m=0}^{\infty} \sum_{k=0}^{\infty} \frac{t^m}{m!} \frac{(-x)_k}{k!} \left(\frac{t}{a}\right)^k$$

$$= e^t \left(1 - \frac{t}{a}\right)^x. \tag{6.3.6}$$

The generating function allows another computation of the norms and inner products:

$$\sum_{n=0}^{\infty} \sum_{m=0}^{\infty} \frac{(C_n, C_m)_w}{n!m!} t^n s^m = \sum_{k=0}^{\infty} G(k,t;a) G(k,s;a) w_k = e^{st/a}. \tag{6.3.7}$$

The identity

$$\sum_{n=0}^{\infty} \frac{C_{n+1}(x;a)}{n!} t^n = \frac{\partial G}{\partial t}(x,t;a)$$

leads to the recurrence relation

$$C_{n+1}(x;a) = C_n(x;a) - \frac{x}{a} C_n(x-1;a). \tag{6.3.8}$$

The identity

$$G(x+1,t;a) = \left(1 - \frac{t}{a}\right) G(x,t;a)$$

leads to (6.3.3).

The general three-term recurrence (6.1.1) is easily computed. It follows from (6.3.2) that

$$(-1)^n C_n(x;a) = \frac{1}{a^n} x^n - \frac{\binom{n}{2} + na}{a^n} x^{n-1} + \cdots$$

Therefore

$$x C_n(x;a) = -a C_{n+1}(x;a) + (n+a) C_n(x;a) - n C_{n-1}(x;a). \tag{6.3.9}$$

The generating function can be used to obtain an addition formula:

$$C_n(x+y;a) = \sum_{j+k+l=n} (-1)^l \frac{n!}{j!k!l!} C_j(x;a) C_k(x;a); \tag{6.3.10}$$

see Exercise 6. The proof of the following addition formula is also left as Exercise 7:

$$C_n(x+y;a) = \sum_{k=0}^{n} \binom{n}{k} C_{n-k}(x;a) \frac{(-y)_k}{a^k}. \tag{6.3.11}$$

Charlier polynomials are connected with the Laguerre polynomials:

$$C_n(x;a) = (-1)^n \frac{n!}{a^n} L_n^{(x-n)}(a);$$

see [108]. Therefore $u(a) = a^n C_n(x;a)$ satisfies the confluent hypergeometric equation

$$au''(a) + (1 + x - n - a)u'(a) + nu(a) = 0.$$

Dunster [108] derived uniform asymptotic expansions in n for x in each of three intervals whose union is the real line $-\infty < x < \infty$. His results imply that for fixed real x,

$$C_n(x;a) \sim -\frac{n!e^a}{a^n} \frac{\sin \pi x}{\pi} \frac{\Gamma(1+x)}{n^{x+1}} = \frac{n!e^a}{a^n \Gamma(-x)n^{x+1}} \tag{6.3.12}$$

as $n \to \infty$. Therefore the zeros are asymptotically close to the positive integers. See Exercise 22 in Chapter 13.

6.4 Krawtchouk polynomials

The interval is finite and

$$p_+(x) = p(N-x), \quad p_-(x) = qx, \quad w_m = \binom{N}{m} p^m q^{N-m}, \tag{6.4.1}$$

where $p, q > 0$, $p + q = 1$. Then $\lambda_n = n$ and the constant A_n in (6.2.5) is $n!$.
Two standard normalizations here are

$$k_n^{(p)}(x;N) = \frac{1}{n!} \psi_n(x);$$

$$K_n(x;p,N) = (-1)^n \frac{(N-n)!}{N!p^n} \psi_n(x) = (-1)^n \binom{N}{n}^{-1} \frac{1}{p^n} k_n^{(p)}(x;N).$$

From (6.2.8) we obtain the norms

$$||k_n^{(p)}||_w^2 = \binom{N}{n}(pq)^n; \tag{6.4.2}$$

$$||K_n||_w^2 = \binom{N}{n}^{-1} \left(\frac{q}{p}\right)^n.$$

Most of the identities to follow have a simpler form in the version $k_n^{(p)}$.
Equation (6.2.9) gives

$$k_n^{(p)}(x;N) = \sum_{k=0}^{n} p^{n-k} q^k \frac{(x-N)_{n-k}(x-k+1)_k}{(n-k)!k!}. \tag{6.4.3}$$

The leading coefficient is

$$\frac{1}{n!}\sum_{k=0}^{n}\binom{n}{k}p^{n-k}q^{k} = \frac{(p+q)^{n}}{n!} = \frac{1}{n!}.$$

The polynomial $\Delta_{+}k_{n}$ is an eigenfunction for the weight $w^{(1)}$, which, after normalization, is the weight associated with p and $N-1$. Taking into account the leading coefficients, it follows that

$$\Delta_{+}k_{n}^{(p)}(x;N) = k_{n-1}^{(p)}(x;N-1). \tag{6.4.4}$$

The associated difference equation is

$$p(N-x)\,\Delta_{+}k_{n}^{(p)}(x;N) - qx\,\Delta_{-}k_{n}^{(p)}(x;N) + nk_{n}^{(p)}(x;N) = 0. \tag{6.4.5}$$

Using this equation and the expansion (6.2.10), we may derive a second form for the Krawtchouk polynomials:

$$k_{n}^{(p)}(x;N) = (-p)^{n}\binom{N}{n}\sum_{k=0}^{n}\frac{(-n)_{k}(-x)_{k}}{(-N)_{k}k!}p^{-k}$$

$$= (-p)^{n}\binom{N}{n}F\left(-n,-x,-N;\frac{1}{p}\right), \tag{6.4.6}$$

where F is the hypergeometric function (Chapter 10); see Exercise 11. The normalization of the alternate form K_{n} is chosen so that

$$K_{n}(x;p,N) = F\left(-n,-x,-N;\frac{1}{p}\right).$$

The first four polynomials are

$$k_{0}^{(p)}(x;N) = 1;$$

$$k_{1}^{(p)}(x;N) = x - Np;$$

$$k_{2}^{(p)}(x;N) = \frac{1}{2}\left[x^{2} + (2p - 1 - 2Np)x + N(N-1)p^{2}\right];$$

$$k_{3}^{(p)}(x;N) = \frac{1}{6}\{x^{3} + (6p - 3 - 3Np)x^{2}$$

$$+ \left[3Np(Np + 1 - 3p) + 2(3p^{2} - 3p + 1)\right]x$$

$$- N(N-1)(N-2)p^{3}\}.$$

We may consider $k_{n}^{(p)}(x;N)$ as being defined by (6.4.3) for all $n = 0,1,2,\ldots$ Note that for $n > N$ these polynomials vanish at the points $m = 0,1,2,\ldots,N$. The generating function is

$$G(x,t;N,p) \equiv \sum_{n=0}^{\infty}k_{n}^{(p)}(x;N)t^{n} = (1+qt)^{x}(1-pt)^{N-x}; \tag{6.4.7}$$

see Exercise 12. The identity

$$\sum_{n=0}^{\infty} (n+1) k_{n+1}^{(p)}(x;N) t^n = \frac{\partial G}{\partial t}(x,t;N,p)$$

leads to the recurrence relation

$$(n+1) k_{n+1}^{(p)}(x;N) = xq k_n^{(p)}(x-1;N-1) - (N-x)p k_n^{(p)}(x;N-1). \quad (6.4.8)$$

The identity

$$(1-pt) G(x+1,t;N,p) = (1+qt) G(x,t;N,p) \quad (6.4.9)$$

leads to

$$k_n^{(p)}(x+1;N) - k_n^{(p)}(x;N) = k_{n-1}^{(p)}(x;N). \quad (6.4.10)$$

The three-term recurrence

$$xk_n^{(p)}(x;N) = (n+1) k_{n+1}^{(p)}(x;N) + (pN+n-2pn) k_n^{(p)}(x;N)$$

$$+ pq(N-n+1) k_{n-1}^{(p)}(x;N) \quad (6.4.11)$$

can be computed using (6.2.11); see Exercise 16.

The generating function can be used to prove the addition formula

$$k_n^{(p)}(x+y;N) = \sum_{j+l+m=n} \frac{(N)_j p^j}{j!} k_l^{(p)}(x;N) k_m^{(p)}(y;N). \quad (6.4.12)$$

To describe the asymptotic behavior of $k_n^{(p)}(x;N)$ for fixed $x > 0$ and $p > 0$, we first let $N = n\mu$ for fixed $\mu \geq 1$, and set $q - 1 = p$. Following Qiu and Wong [337], we note first that there is a unique $\eta = \eta(\mu)$ such that

$$\eta - (\mu-1)\log\eta = (\mu-1)(1-\log q) - \mu\log\mu - \log p.$$

Let

$$t_0 = \frac{\mu-1}{\mu}, \quad s_0 = \frac{\mu-1}{\eta}, \quad \lambda = n\eta,$$

$$\gamma = -\mu\log\mu + (\mu-1)(1+\log\eta), \quad \zeta = \pm\sqrt{2(1-s_0+s_0\log s_0)},$$

where the positive sign is taken if and only if $s_0 \geq 1$. Then let

$$g_0(s_0) = \begin{cases} -\dfrac{1}{\sqrt{1-t_0}} \left(\dfrac{t_0-q}{s_0-1}\right)^x, & s_0 \neq 1; \\ -q^x p^{(x-1)/2}, & s_0 = 1. \end{cases}$$

Using an extension of the steepest descent method (see Chapter 13), Qiu and Wong showed that as $n \to \infty$,

(i) for $\mu \geq \frac{1}{p} + \varepsilon$,

$$k_n^{(p)}(x;n\mu) \sim (-1)^{n+1} e^{\lambda(s_0 - s_0 \log s_0)} \frac{(s_0 - 1)^x g(s_0)}{\sqrt{2\pi \lambda s_0}} \frac{p^{n-x}}{e^{n\gamma}};$$

(ii) for $\frac{1}{p} - \varepsilon < \mu < \frac{1}{p} + \varepsilon$,

$$k_n^{(p)} x;n\mu) \sim (-1)^{n+1} a_0 W(x, \sqrt{\lambda}\zeta) \frac{e^\lambda p^{n-x}}{e^{n\gamma} \lambda^{(x+1)/2}},$$

where

$$W(x,\zeta) = \frac{D_x(\zeta)}{\sqrt{2\pi} e^{\zeta^2/4}}, \qquad a_0 = \begin{cases} g_0(1) \left(\dfrac{\zeta}{s_0 - 1}\right)^{x+1}, & s_0 \neq 1; \\ g_0(1), & s_0 = 1, \end{cases}$$

and D_x is the parabolic cylinder function of Section 8.6;
(iii) for $1 \leq \mu \leq \frac{1}{p} - \varepsilon$,

$$k_n^{(p)}(x;n\mu) \sim (-1)^{n+1} \frac{g_0(1) e^\lambda p^{n-x}}{\Gamma(-x)(\lambda - \lambda s_0)^{x+1} e^{n\gamma}}.$$

6.5 Meixner polynomials

The interval is infinite and

$$p_+(x) = c(x+b), \quad p_-(x) = x, \quad w_m = (1-c)^b \frac{(b)_m}{m!} c^m, \tag{6.5.1}$$

where $b > 0$ and $0 < c < 1$. Therefore

$$\lambda_n = (1-c)n, \quad A_n = (1-c)^n n!.$$

Equation (6.2.9) gives

$$\psi_n(x) = \sum_{k=0}^n (-1)^{n-k} \binom{n}{k} (x+b)_{n-k} (x-k+1)_k c^{n-k},$$

with leading coefficient $(1-c)^n$. Two standard normalizations are

$$m_n(x;b,c) = (-c)^{-n} \psi_n(x)$$

$$= \sum_{k=0}^n \binom{n}{k} (x+b)_{n-k} (x-k+1)_k (-c)^{-k}; \tag{6.5.2}$$

$$M_n(x;b,c) = \frac{(-1)^n}{c^n (b)_n} \psi_n(x) = \frac{1}{(b)_n} m_n(x;b,c).$$

It follows that

$$||m_n||_w^2 = n! \frac{(b)_n}{(c)_n};$$

$$||M_n||_w^2 = \frac{n!}{(b)_n (c)_n}. \tag{6.5.3}$$

Most of the identities to follow have a simpler form in the version m_n.

The polynomial $\Delta_+ m_n$ is an eigenfunction for the weight $w^{(1)}$, which, after normalization, is the weight associated with $b+1$ and c. The leading coefficient of m_n is $(1 - 1/c)^n$, so

$$\Delta_+ m_n(x; b, c) = n \left(1 - \frac{1}{c}\right) m_{n-1}(x; b+1, c). \tag{6.5.4}$$

The associated difference equation is

$$c(x+b) \Delta_+ m_n(x; b, c) - x \Delta_- m_n(x; b, c) + (1 - c) n m_n(x; b, c) = 0. \tag{6.5.5}$$

Using this equation and the expansion (6.2.10) leads to a second form (normalized by taking into account the leading coefficient):

$$m_n(x; b, c) = (b)_n \sum_{k=0}^{n} \frac{(-n)_k (-x)_k}{(b)_k k!} \left(1 - \frac{1}{c}\right)^k$$

$$= (b)_n F\left(-n, -x, b; 1 - \frac{1}{c}\right), \tag{6.5.6}$$

where again F is the hypergeometric function. The normalization of M_n is chosen so that

$$M_n(x; b, c) = F\left(-n, -x, b; 1 - \frac{1}{c}\right).$$

The first four polynomials are

$m_0(x; b, c) = 1;$

$m_1(x; b, c) = \left(1 - \dfrac{1}{c}\right) x + b;$

$m_2(x; b, c) = \left(1 - \dfrac{1}{c}\right)^2 x^2 + \left(2b + 1 - \dfrac{2b}{c} - \dfrac{1}{c^2}\right) x + b(b+1);$

$m_3(x; b, c) = \left(1 - \dfrac{1}{c}\right)^3 x^3 + \left(3b + 3 - \dfrac{6b+3}{c} + \dfrac{3b-3}{c^2} + \dfrac{3}{c^3}\right) x^2$

$\qquad + \left(3b^2 + 6b + 2 - \dfrac{3b^2 + 3b}{c} - \dfrac{3b}{c^2} - \dfrac{2}{c^3}\right) x + b(b+1)(b+2).$

The generating function is

$$G(x,t;b,c) \equiv \sum_{n=0}^{\infty} \frac{m_n(x;b,c)}{n!} t^n = (1-t)^{-x-b} \left(1 - \frac{t}{c}\right)^x ; \qquad (6.5.7)$$

see Exercises 22 and 23.

The identity

$$\sum_{n=0}^{\infty} \frac{m_{n+1}(x;b,c)}{n!} t^n = \frac{\partial G}{\partial t}(x,t;b,c)$$

implies the recurrence relation

$$m_{n+1}(x;b,c) = (x+b) m_n(x;b+1,c) - \frac{x}{c} m_n(x-1;b+1,c). \qquad (6.5.8)$$

The identity

$$(1-t) G(x+1,t;b,c) = \left(1 - \frac{t}{c}\right) G(x,t;b,c)$$

implies

$$m_n(x+1;b,c) - m_n(x;b,c) = n m_{n-1}(x+1;b,c)$$
$$- \frac{n}{c} m_{n-1}(x;b,c). \qquad (6.5.9)$$

The three-term recurrence

$$(c-1)x m_n(x;b,c) = c m_{n+1}(x;b,c) - (bc + nc + n) m_n(x;b,c)$$
$$+ n(b+n-1) m_{n-1}(x;b,c) \qquad (6.5.10)$$

can be computed using (6.2.11); see Exercise 25.

The generating function can be used to prove the addition formula

$$m_n(x+y;b,c) = \sum_{j+k+l=n} \frac{n!(-b)_j}{j!k!l!} m_l(x;b,c) m_k(y;b,c); \qquad (6.5.11)$$

see Exercise 26.

It follows from the expansion (6.5.6) and the identity (5.4.10) that Laguerre polynomials are limits of Meixner polynomials: for $x \neq 0$,

$$m_n \left(\frac{cx}{1-c}; \alpha+1, c\right) \sim (\alpha+1)_n M(-n, \alpha+1; x) = n! L_n^{(\alpha)}(x) \qquad (6.5.12)$$

as $c \to 1-$; see Exercise 27.

Jin and Wong [206] used a modification of the steepest descent method due to Chester, Friedman, and Ursell [75] (see also Wong [443]), to derive an asymptotic expansion for $m_n(n\alpha; b, c)$ for $\alpha > 0$. When $n\alpha$ is bounded, they gave the simplified result

$$m_n(n\alpha; b, c) \sim -\frac{n!\,\Gamma(\alpha n + 1)}{c^n (1-c)^{n\alpha+b}\, n^{n\alpha+1}} \frac{\sin \pi n\alpha}{\pi} \tag{6.5.13}$$

as $n \to \infty$; see (3.13), (3.14), and (4.1) of [207] and Exercise 21 of Chapter 13.

Fix x and take $\alpha = x/n$. It follows from (6.5.13) that

$$m_n(x; b, c) \sim -\frac{\Gamma(b+n)\,\Gamma(x+1)}{c^n(1-c)^{x+b}\, n^{b+x}} \frac{\sin \pi x}{\pi}. \tag{6.5.14}$$

Thus the zeros of the Meixner polynomials are asymptotically close to the positive integers as $n \to \infty$.

6.6 Chebyshev–Hahn polynomials

The interval is finite, and for $\alpha, \beta > -1$ or $\alpha, \beta < -N$, we take

$$p_+(x) = (N-x)(x+\alpha+1),$$
$$p_-(x) = x(N+\beta+1-x). \tag{6.6.1}$$

In the case $\alpha, \beta < -N$, we have violated our condition that $p_\pm$ be positive at the interior points $\{1, 2, \dots, N-1\}$. In the following formulas this will only show up in the appearance of absolute values in the formulas for norms.

The weight is

$$w_m = C_0 \binom{N}{m} (\alpha+1)_m (\beta+1)_{N-m}, \tag{6.6.2}$$

where

$$C_0 = \frac{\Gamma(N+\alpha+\beta+2)}{\Gamma(\alpha+\beta+2)}.$$

With this choice of C_0, the total mass is

$$\sum_{m=0}^{N} w_m = 1; \tag{6.6.3}$$

see Exercise 28.

According to (6.2.9),

$$\psi_n(x) = \sum_{k=0}^{n} (-1)^{n-k} \binom{n}{k} \prod_{j=0}^{n-k-1} p_+(x+j) \prod_{j=0}^{k-1} p_-(x-j). \tag{6.6.4}$$

This appears to have degree $2n$ rather than n, but there is considerable cancellation. A more useful form can be obtained using (6.2.10), which leads to

$$\psi_n(x) = C \sum_{k=0}^{n} \frac{(-n)_k(-x)_k(n+\alpha+\beta+1)_k}{(-N)_k(\alpha+1)_k\, k!}$$

$$= C\,_3F_2 \left(\begin{matrix} -n, -x, n+\alpha+\beta+1 \\ -N, \alpha+1 \end{matrix} ; 1 \right),$$

where $_3F_2$ denotes a generalized hypergeometric function; see Chapter 12. To determine the constant C we note that the constant term in (6.6.4) comes from the summand with $k = 0$ and is therefore

$$(-1)^n \prod_{j=0}^{n-1} p_+(j) = (-1)^n (N+1-n)_n(\alpha+1)_n.$$

It follows that

$$\psi_n(x) = (-1)^n(N+1-n)_n(\alpha+1)_n$$
$$\times\,_3F_2 \left(\begin{matrix} -n, -x, n+\alpha+\beta+1 \\ -N, \alpha+1 \end{matrix} ; 1 \right).$$

One normalization is

$$Q_n(x;\alpha,\beta,N) = \,_3F_2 \left(\begin{matrix} -n, -x, n+\alpha+\beta+1 \\ -N, \alpha+1 \end{matrix} ; 1 \right)$$
$$= \sum_{k=0}^{n} \frac{(-n)_k(-x)_k(n+\alpha+\beta+1)_k}{(-N)_k(\alpha+1)_k\, k!}. \qquad (6.6.5)$$

A second normalization is

$$h_n^{(\alpha,\beta)}(x,N) = (-1)^n \frac{(N-n)_n(\beta+1)_n}{n!} Q_n(x;\beta,\alpha,N-1). \qquad (6.6.6)$$

Neither normalization results in particularly simple forms for the identities that follow; we use Q_n.

It follows from (6.2.4) that $\lambda_n = n(n+\alpha+\beta+1)$, so

$$\|\psi_n\|_w^2 = (n!)^2 \binom{N}{n} \left| \frac{(\alpha+1)_n(\beta+1)_n(\alpha+\beta+n+1)_{N+1}}{(\alpha+\beta+2)_N(\alpha+\beta+2n+1)} \right|.$$

Therefore

$$\|Q_n(x;\alpha,\beta,N)\|^2$$
$$= \left| \frac{n!(N-n)!(\beta+1)_n(\alpha+\beta+n+1)_{N+1}}{N!(\alpha+1)_n(\alpha+\beta+2)_N(\alpha+\beta+2n+1)} \right|. \qquad (6.6.7)$$

The weight $w^{(1)} = p_+w$ is a constant multiple of the weight associated with the indices $(\alpha+1,\beta+1,N-1)$. It follows from (6.6.5) that the leading coefficient of $Q_n(x; \alpha, \beta, N)$ is

$$\frac{(n+\alpha+\beta+1)_n}{(-N)_n (\alpha+1)_n},$$

so

$$\Delta_+ Q_n(x; \alpha, \beta, N)$$
$$= -\frac{(n+\alpha+\beta+1)}{N(\alpha+1)} Q_{n-1}(x; \alpha+1, \beta+1, N-1). \qquad (6.6.8)$$

The difference equation is

$$(N-x)(x+\alpha+1)\Delta_+ Q_n(x; \alpha, \beta, N)$$
$$- x(N+\beta+1-x)\Delta_- Q_n(x; \alpha, \beta, N)$$
$$+ n(n+\alpha+\beta+1)Q_n(x; \alpha, \beta, N) = 0. \qquad (6.6.9)$$

The first three polynomials are

$$Q_0(x; \alpha, \beta, N) = 1;$$
$$Q_1(x; \alpha, \beta, N) = -\frac{\alpha+\beta+2}{N(\alpha+1)} x + 1;$$
$$Q_2(x; \alpha, \beta, N) = \frac{(\alpha+\beta+3)(\alpha+\beta+4)}{N(N-1)(\alpha+1)(\alpha+2)} x^2$$
$$- \frac{(\alpha+\beta+3)[\alpha+\beta+4+2(N-1)(\alpha+2)]}{N(N-1)(\alpha+1)(\alpha+2)} x + 1.$$

A straightforward (but tedious) application of (6.2.11) yields

$$Q_n(x; \alpha, \beta, N) = a_n x^n + b_n x^{n-1} + \cdots,$$
$$\frac{b_n}{a_n} = -\frac{n\,[2N(\alpha+1)+(2N+\beta-\alpha)(n-1)]}{2(\alpha+\beta+2n)}.$$

The ratio of leading coefficients is

$$\frac{a_n}{a_{n+1}} = -\frac{(n+\alpha+\beta+1)(\alpha+n+1)(N-n)}{(2n+\alpha+\beta+1)(2n+\alpha+\beta+2)}.$$

Therefore, by (6.1.1), the three-term recurrence is

$$x Q_n(x; \alpha, \beta, N) = \alpha_n Q_{n+1}(x; \alpha, \beta, N) + \beta_n Q_n(x; \alpha, \beta, N)$$
$$+ \gamma_n Q_{n-1}(x; \alpha, \beta, N); \qquad (6.6.10)$$

$$\alpha_n = -\frac{(\alpha+\beta+n+1)(\alpha+n+1)(N-n)}{(\alpha+\beta+2n+1)(\alpha+\beta+2n+2)};$$
$$\gamma_n = -\frac{n(n+\beta)(\alpha+\beta+n+N+1)}{(\alpha+\beta+2n)(\alpha+\beta+2n+1)};$$
$$\beta_n = -(\alpha_n + \gamma_n).$$

It follows from the expansion (6.6.5) and the identity (5.5.13) that Jacobi polynomials are limits of Chebyshev–Hahn polynomials: for fixed $x \neq 0$,

$$Q_n(Nx; \alpha, \beta, N) \sim F(\alpha + \beta + 1 + n, -n, \alpha + 1; x)$$

$$= \frac{n!}{(\alpha + 1)_n} P_n^{(\alpha, \beta)}(1 - 2x) \qquad (6.6.11)$$

as $N \to \infty$; see Exercise 30. For refinements of this result, see Sharapudinov [363].

For fixed α and β and fixed ratio $n/N = c$, $0 < c < 1$, then for fixed real x,

$$Q_n(x; \alpha, \beta, N - 1) \sim -\frac{\Gamma(\alpha + 1)\Gamma(N - n)}{\Gamma(N) e^n n^{2x+2\alpha+2} \Gamma(-x)} (1 + c)^{n+N+a+\beta+\frac{1}{2}}$$

$$\times N^{x+n+\alpha+1} \qquad (6.6.12)$$

as $N \to \infty$; see Lin and Wong [258]. For $x > -\frac{1}{2}$, we use the reflection formula (2.2.3) to rewrite (6.6.12) in a form that exhibits the asymptotics of the zeros:

$$Q_n(x; \alpha, \beta, N - 1) \sim -\frac{\Gamma(\alpha + 1)\Gamma(N - n)}{\Gamma(N) e^n n^{2x+2\alpha+2}} (1 + c)^{n+N+a+\beta+\frac{1}{2}}$$

$$\times N^{x+n+\alpha+1} \Gamma(x + 1) \frac{\sin \pi x}{\pi} \qquad (6.6.13)$$

as $N \to \infty$. The polynomials that are commonly called "discrete Chebyshev polynomials" are the case $\alpha = \beta = 0$:

$$t_n(x, N) = (-1)^n (N - n)_n Q_n(x; 0, 0, N - 1)$$

$$= (-1)^n \sum_{k=0}^{n} \binom{n+k}{k} \frac{(N - n)_{n-k}(-x)_k(n - k + 1)_k}{k!}. \qquad (6.6.14)$$

If the ratio $n/N = c$ is fixed, $0 < c < 1$, then for fixed $x < 0$,

$$t_n(x, N + 1) = (-1)^{n+1} \frac{\Gamma(n + N + 2)N^x}{\Gamma(N + 1)\Gamma(-x)n^{2x+2}} [1 + O(1/N)] \qquad (6.6.15)$$

as $N \to \infty$, while for fixed $x > -\frac{1}{2}$,

$$t_n(x, N + 1) = (-1)^{n+1} \frac{\Gamma(n + N + 2)N^x \Gamma(x + 1)}{\Gamma(N + 1)n^{2x+2}}$$

$$\times \left\{ \frac{\sin \pi x}{\pi} [1 + O(1/N)] + O(e^{-\delta N}) \right\} \qquad (6.6.16)$$

as $N \to \infty$, where $\delta > 0$; see Pan and Wong [318].

6.7 Neo-classical polynomials

This section contains a brief derivation of the remaining members of what we have called "semi-classical orthogonal polynomials."

The analysis of second-order difference operators in Section 6.1 can be extended in a different (in fact orthogonal) direction, to differences in the imaginary direction. We replace the shift operators $S_\pm$ with the imaginary shifts $T_\pm$:

$$T_\pm f(x) = f(x \pm i). \tag{6.7.1}$$

Consider the operator

$$L = p_+ (T_+ - I) + p_- (T_- - I), \qquad p_+ p_- \neq 0. \tag{6.7.2}$$

Applying L to the functions $f_n(x) = x^n$ gives

$$Lf_0 = 0, \quad Lf_1 = i(p_+ - p_-), \quad Lf_2 = 2ix(p_+ - p_-) - (p_+ + p_-).$$

Let $\mathcal{P}_n$ denote the set of real polynomials of degree $\leq n$. We want L to map $\mathcal{P}_n$ into itself. Then necessarily

$$i(p_+ - p_-) \in \mathcal{P}_1, \qquad p_+ + p_- \in \mathcal{P}_2. \tag{6.7.3}$$

Thus $p_+(x) - p_-(x)$ is imaginary and $p_+(x) + p_-(x)$ is real for real x:

$$\overline{p_+(x) - p_-(x)} = p_-(x) - p_+(x), \quad \overline{p_+(x) + p_-(x)} = p_+(x) + p_-(x). \tag{6.7.4}$$

Let $p = p_+$. It follows from (6.7.4) that $p_- = \bar{p}$, where $\bar{p}(x) = \overline{p(x)}$ for real x, and in general

$$\bar{p}(z) = \overline{p(\bar{z})}.$$

The conditions (6.7.3) imply that, up to multiplication by a real constant, p has one of the two forms

(a) $\qquad p(x) = e^{i\psi}(\bar{a} - ix), \qquad -\pi/2 < \psi < \pi/2;$

(b) $\qquad p(x) = (\bar{a} - ix)(\bar{b} - ix), \qquad \mathrm{Re}\,(a + b) \neq 0.$

In either case, it is easily checked that L maps each $\mathcal{P}_n$ into itself.

Consider a weight function w that is meromorphic in the complex plane and positive on the real axis, which implies that $w = \bar{w}$. We begin with a formal calculation of the adjoint of pT_+ in the space $L^2(\mathbf{R}, w(x)\,dx)$. Suppose that f and g are real polynomials. A formal application of Cauchy's theorem gives

$$(p T_+ f, g) = \int_{-\infty}^{\infty} p(x) f(x + i) g(x) w(x)\, dx$$

$$= \int_{-\infty}^{\infty} p(x - i) f(x) g(x - i) w(x - i)\, dx$$

$$= \int_{-\infty}^{\infty} f(x)\overline{\bar{p}(x+i)}\overline{g(x+i)}\,\frac{\overline{w(x+i)}}{w(x)}\,w(x)\,dx$$

$$= (f, [T_+\bar{p}\,T_+w/w]T_+g). \tag{6.7.5}$$

This calculation is valid if pw has no poles in the strip $\{0 \le \mathrm{Im}\,z \le 1\}$ and satisfies the estimate

$$|p(z)w(z)| \le C\exp(-\varepsilon|z|) \tag{6.7.6}$$

uniformly in this strip, where $\varepsilon > 0$. Under this assumption, (6.7.5) shows that

$$(pT_+)^* = pT_+ \quad \text{if and only if} \quad \frac{p(x)}{\bar{p}(x+i)} = \frac{w(x+i)}{w(x)}. \tag{6.7.7}$$

Calculating $(\bar{p}T_-)^*$ leads to the same condition, so this condition and (6.7.6) imply symmetry of L on the space of polynomials. Since each $\mathcal{P}_n$ is invariant, it is spanned by an orthogonal basis of eigenfunctions. By Theorem 4.1.5, the estimate (6.7.6) implies that polynomials are dense in $L^2(\mathbf{R}, w(x)\,dx)$, and we conclude that the polynomial eigenfunctions of L are an orthogonal basis for this space.

To complete this discussion we need to solve (6.7.7) for w. Specifically, we need

(a) $$\frac{w(x+i)}{w(x)} = e^{2i\psi}\,\frac{\bar{a}-ix}{a+i(x+i)} = e^{2i\psi}\,\frac{\bar{a}-ix}{a+ix-1};$$

(b) $$\frac{w(x+i)}{w(x)} = \frac{(\bar{a}-ix)(\bar{b}-ix)}{(a+ix-1)(b+ix-1)}.$$

These can be solved factor by factor, using the identities

$$e^{2i\psi} = \frac{e^{2\psi(x+i)}}{e^{2\psi x}};$$

$$\bar{a}-ix = \frac{\Gamma(\bar{a}-ix+1)}{\Gamma(\bar{a}-ix)} = \frac{\Gamma(\bar{a}-i(x+i))}{\Gamma(\bar{a}-ix)};$$

$$a+ix-1 = \frac{\Gamma(a+ix)}{\Gamma(a+ix-1)} = \frac{\Gamma(a+ix)}{\Gamma(a+i(x+i))}.$$

Thus for case (a), the operator is

$$L = e^{i\psi}(\bar{a}-ix)(T_+ - I) + e^{-i\psi}(a+ix)(T_- - I), \tag{6.7.8}$$

and the weight function can be taken to be

$$w(x) = e^{2\psi x}\Gamma(\bar{a}-ix)\Gamma(a+ix) = e^{2\psi x}|\Gamma(a+ix)|^2, \quad \mathrm{Re}\,a > 0. \tag{6.7.9}$$

For a real, the eigenfunctions of L are, up to normalization, the *Meixner–Pollaczek polynomials*; see Meixner [285] and Pollaczek [333]. The associated parameters are usually taken to be $\lambda = a$ and $\phi = \psi + \frac{\pi}{2}$. With this choice of parameters, the Meixner–Pollaczek polynomials can be written in terms of the hypergeometric function:

$$P_n^{(\lambda)}(x;\phi) = \frac{(2\lambda)_n e^{in\phi}}{n!} F(-n, \lambda + ix, 2\lambda; 1 - e^{-2i\phi})$$

$$= \frac{(2\lambda)_n e^{in\phi}}{n!} \sum_{k=0}^{n} \frac{(-n)_k (\lambda + ix)_k}{(2\lambda)_k k!} (1 - e^{-2i\phi})^k. \quad (6.7.10)$$

The three-term recurrence relation is

$$(n+1)P_{n+1}^{(\lambda)}(x;\phi) - 2[x\sin\phi + (n+\lambda)\cos\phi] P_n^{(\lambda)}(x;\phi)$$

$$+ (n+2\lambda-1)P_{n-1}^{(\lambda)}(x;\phi) = 0. \quad (6.7.11)$$

See Exercises 31 and 32.

The asymptotics of these polynomials for large n has been studied by Li and Wong [256]. For the simplest case $\phi = \frac{1}{2}\pi$, i.e. the weight function $w(x) = |\Gamma(a+ix)|^2$,

$$P_n^{(\lambda)}\left(n\cos\theta; \frac{\pi}{2}\right) \sim \frac{(2\cos\theta)^{\frac{1}{2}-\lambda}}{\sqrt{n\pi \sin\theta}} \exp\left(\frac{n\pi|\cos\theta|}{2}\right)$$

$$\times \sin\left\{(n+\lambda)\theta - \frac{n\cos\theta}{2} \log\frac{1+\sin\theta}{1-\sin\theta} + \frac{\pi}{4}\right\}. \quad (6.7.12)$$

For case (b), the operator is

$$L = (\bar{a} - ix)(\bar{b} - ix)(T_+ - I) + (a+ix)(b+ix)(T_- - I), \quad (6.7.13)$$

and the weight function may be taken to be

$$w(x) = |\Gamma(a+ix)\Gamma(b+ix)|^2, \quad \text{Re}\,a > 0, \quad \text{Re}\,b > 0. \quad (6.7.14)$$

The eigenfunctions, up to normalization, are known as the *continuous Hahn polynomials*; see Atakishiyev and Suslov [24] and Askey [19], as well as [221]. In each case the estimate (6.7.6) is a consequence of Corollary 2.5.8.

With $c = \bar{a}$ and $d = \bar{b}$, the continuous Hahn polynomials have a representation in terms of the generalized hypergeometric function $_3F_2$:

$$p_n(x; a, b, c, d) = i^n \frac{(a+c)_n (a+d)_n}{n!} \tilde{p}_n(x)$$

$$= i^n \frac{(a+c)_n (a+d)_n}{n!} {}_3F_2\left(\begin{matrix} -n, n+a+b+c+d-1, a+ix \\ a+c, a+d \end{matrix}; 1\right)$$

$$= i^n \frac{(a+c)_n (a+d)_n}{n!} \sum_{k=0}^{n} \frac{(-n)_k (n+a+b+c+d-1)_k (a+ix)_k}{(a+c)_k (a+d)_k k!}.$$

$$(6.7.15)$$

The three-term recurrence relation is

$$(a + ix)\tilde{p}_n(x) = A_n \tilde{p}_{n+1}(x) - (A_n + C_n)\tilde{p}_n(x) + C_n \tilde{p}_{n-1}(x), \qquad (6.7.16)$$

where

$$A_n = -\frac{(n+a+b+c+d-1)(n+a+c)(n+a+d)}{(2n+a+b+c+d-1)(2n+a+b+c+d)},$$

$$C_n = \frac{n(n+b+c-1)(n+b+d-1)}{(2n+a+b+c+d-2)(2n+a+b+c+d-1)}.$$

See Exercises 33 and 34.

Let $u = \operatorname{Re}(a+b)$, $v = \operatorname{Im}(a+b)$, and

$$\pi_n(x) = \frac{n!}{(n+a+b+c+d-1)_n} p_n(x; a, b, c, d).$$

For t in any compact subset of the complement of $[-1/2, 1/2]$ in $\mathbf{C}$,

$$\pi_n(nt) \sim 2^{\frac{3}{2}-2u} \left(\frac{n}{4e}\right)^n \frac{t^{1-u}}{(4t^2-1)^{1/4}} \left(2t + \sqrt{4t^2-1}\right)^{n+u-\frac{1}{2}}$$

$$\times \exp\left\{(2nt+v)\sin^{-1}\frac{1}{2t}\right\} \qquad (6.7.17)$$

as $n \to \infty$; see Cao, Li, and Lin [62].

Consider, instead of $T_\pm$, the difference operators

$$(\nabla_\pm f)(y^2) = \frac{f((y\pm i)^2) - f(y^2)}{(y\pm i)^2 - y^2} \qquad (6.7.18)$$

and an operator

$$Lf(y^2) = p_+(y)(\nabla_+ f)(y^2) + p_-(y)(\nabla_- f)(y^2). \qquad (6.7.19)$$

Applying L to the functions $F_n(y) = f_n(y^2) = (y^2)^n$ gives

$$LF_0 = 0, \qquad LF_1 = p_+ + p_-,$$

$$LF_2 = (2y^2 - 1)(p_+ + p_-) + 2iy(p_+ - p_-).$$

Let $\mathcal{Q}_n$ denote the set of functions of the form

$$F(y) = f(y^2), \qquad f \in \mathcal{P}_n;$$

thus $\mathcal{Q}_n$ consists of the even polynomials in $\mathcal{P}_{2n}$. We want L to map $\mathcal{Q}_n$ to itself. Then necessarily,

$$p_+ + p_- \in \mathcal{Q}_1, \qquad iy(p_- - p_+) \in \mathcal{Q}_2. \qquad (6.7.20)$$

As above, these conditions imply that $p_- = \bar{p}_+$, and we set $p = p_+$. In addition, we want $p + \bar{p}$ to have order 2. This leads to two possible normal forms:

$$(c) \qquad p(y) = \frac{(a - iy)(\bar{b} - iy)(\bar{c} - iy)}{iy}, \qquad a \in \mathbf{R}; \qquad (6.7.21)$$

$$(d) \qquad p(y) = \frac{(\bar{a} - iy)(\bar{b} - iy)(\bar{c} - iy)(\bar{d} - iy)}{iy}. \qquad (6.7.22)$$

Additionally, in case (c), either both parameters b, c are real or $c = \bar{b}$; in case (d), nonreal parameters come in complex conjugate pairs.

We look now for conditions that guarantee that L be symmetric, acting on functions on the half-line $(0, \infty)$ with respect to an inner product

$$\int_0^\infty f(x)\overline{g(x)}\, w(x)\, dx = \int_0^\infty f(y^2)\overline{g(y^2)}\, w(y^2)\, 2y\, dy \qquad (6.7.23)$$

$$= \int_{-\infty}^\infty f(y^2)\overline{g(y^2)}\, \widetilde{w}(y)\, dy,$$

where $\widetilde{w}(y) = w(y^2)y$ for $y > 0$ and $\widetilde{w}(y) = \widetilde{w}(-y)$, $y < 0$. Now with $f(y^2) = F(y)$,

$$Lf(y^2) = q(y)(T_+ - I)F(y) + \bar{q}(y)(T_- - I)F(y), \qquad q(y) = \frac{p(y)}{2iy - 1}.$$

According to the argument above, the key step in establishing symmetry of L is to solve

$$\frac{\widetilde{w}(y + i)}{\widetilde{w}(y)} = \frac{q(y)}{\bar{q}(y + i)}. \qquad (6.7.24)$$

In addition to three or four factors of the form $(\bar{a} - iy)/(a + iy - 1)$, the right side of (6.7.24) has a factor

$$\frac{[-i(y + i)][-2i(y + i) - 1]}{iy(2iy - 1)} = -\frac{iy - 1}{-iy}$$

$$= \frac{2iy - 1}{1 - 2iy} \cdot \frac{2iy - 2}{-2iy}$$

$$= \frac{\Gamma(-2iy)}{\Gamma(-2i(y + i))} \cdot \frac{\Gamma(2iy)}{\Gamma(2i(y + i))}.$$

Thus in case (c) we may take

$$w(y^2) = \frac{1}{2y} \cdot \frac{|\Gamma(a + iy)\, \Gamma(b + iy)\, \Gamma(c + iy)|^2}{|\Gamma(2iy)|^2}, \qquad y > 0, \qquad (6.7.25)$$

where $a > 0$, b, and c have positive real parts, and either b and c are both real or $c = \bar{b}$. The eigenfunctions of L in this case, suitably normalized, are known as the *continuous dual Hahn polynomials*; see [22].

The continuous dual Hahn polynomials also have a representation in terms of the generalized hypergeometric function $_3F_2$:

$$\frac{S_n(y^2;a,b,c)}{(a+b)_n\,(a+c)_n} = \widetilde{S}_n(y^2)$$

$$= {}_3F_2\left(\begin{matrix}-n,a+iy,a-iy\\a+b,a+c\end{matrix};1\right)$$

$$= \sum_{k=0}^{n}\frac{(-n)_k\,(a+iy)_k\,(a-iy)_k}{(a+b)_k\,(a+c)_k\,k!}. \tag{6.7.26}$$

The three-term recurrence relation is

$$-(a^2+y^2)\widetilde{S}_n(y^2) = A_n\widetilde{S}_{n+1}(y^2) - (A_n+C_n)\widetilde{S}_n(y^2) + C_n\widetilde{S}_{n-1}(y^2), \tag{6.7.27}$$

where

$$A_n = (n+a+b)(n+a+c); \qquad C_n = n(n+b+c-1).$$

See Exercises 35 and 36.

In case (d) we may take

$$w(y^2) = \frac{1}{2y}\cdot\frac{|\Gamma(a+iy)\Gamma(b+iy)\Gamma(c+iy)\Gamma(d+iy)|^2}{|\Gamma(2iy)|^2}, \tag{6.7.28}$$

where a,b,c,d have positive real parts, and nonreal parameters come in complex conjugate pairs. The eigenfunctions of L in this case, suitably normalized, are known as the *Wilson polynomials*; see [436, 21].

The Wilson polynomials have a representation in terms of the generalized hypergeometric function $_4F_3$:

$$\frac{W_n(y^2;a,b,c,d)}{(a+b)_n(a+c)_n(a+d)_n} = \widetilde{W}_n(y^2)$$

$$= {}_4F_3\left(\begin{matrix}-n,n+a+b+c+d-1,a+iy,a-iy\\a+b,a+c,a+d\end{matrix};1\right)$$

$$= \sum_{k=0}^{n}\frac{(-n)_k\,(n+a+b+c+d-1)_k\,(a+iy)_k\,(a-iy)_k}{(a+b)_k\,(a+c)_k\,(a+d)_k\,k!}. \tag{6.7.29}$$

The three-term recurrence relation is

$$-(a^2+y^2)\widetilde{W}_n(y^2) = A_n\widetilde{W}_{n+1}(y^2) - (A_n+C_n)\widetilde{W}_n(y^2) + C_n\widetilde{W}_{n-1}(y^2), \tag{6.7.30}$$

where

$$A_n = \frac{(n+a+b+c+d-1)(n+a+b)(n+a+c)(n+a+d)}{(2n+a+b+c+d-1)(2n+a+b+c+d)},$$

$$C_n = \frac{n(n+b+c-1)(n+b+d-1)(n+c+d-1)}{(2n+a+b+c+d-2)(2n+a+b+c+d-1)}.$$

See Exercises 37 and 38.

If $2ix$ is not an integer, then

$$W_n(x^2; a, b, c, d) \sim \frac{n}{2ix} \left[\frac{(a+ix)_n(b+ix)_n(c+ix)_n(d+ix)_n}{(1+2ix)_n} \right.$$
$$\left. - \frac{(a-ix)_n(b-ix)_n(c-ix)_n(d-ix)_n}{(1-2ix)_n} \right]$$

$$(6.7.31)$$

as $n \to \infty$; see Wilson [439].

Remark. One can replace the function y^2 in (6.7.18) by $\lambda(y) = y(y + \alpha)$, $\alpha > 0$. The associated polynomials are the *dual Hahn polynomials* and the *Racah polynomials*; see [436]. They are orthogonal with respect to a weight supported on a finite set $\{0, 1, 2, \ldots, N\}$. Like the Wilson polynomials, they can be written in terms of $_4F_3$, where in this case one of the indices is $-N$. For asymptotics, see Wang and Wong [419].

6.8 Exercises

1. Use the expansion (6.2.10) and the difference equation for the Charlier polynomials to give another derivation of (6.3.2), assuming that the leading coefficient is $(-a)^{-n}$.
2. Verify (6.3.7) and show that it implies that the C_n are orthogonal and have norms given by (6.3.1).
3. Verify the recurrence relation (6.3.8).
4. Show that (6.3.3) and the identity $G(0, t; a) = e^t$ determine the Charlier generating function $G(x, t; a)$ uniquely (for integer x).
5. Show that the recurrence relation (6.3.8) and the identity $G(x, 0; a) = 1$ determine $G(x, t; a)$ uniquely (for integer x).
6. Prove the addition formula (6.3.10).
7. Prove (6.3.11).
8. Let $p_m(x) = (x - m + 1)_m = (x - m + 1)(x - m + 2) \cdots (x - 1)x$. Show that $\Delta_+ p_m = m p_{m-1}$ and conclude that $(\Delta_+)^k p_m(0) = m!$ if $k = m$ and 0 otherwise. Use this to conclude that if f is any polynomial, it has a discrete Taylor expansion

$$f(x+y) = \sum_{k \geq 0} \frac{(\Delta_+)^k f(x)}{k!} (y - k + 1)_k.$$

9. Use Exercise 8 to prove (6.3.11).
10. Show that

$$\lim_{a \to \infty} C_n(ax; a) = (1 - x)^n.$$

11. Derive (6.4.6).

12. Use the binomial expansion to verify (6.4.7).

13. Compute the constant term of $k_n^{(p)}(x;N)$, and deduce from this and the computation of the leading coefficient that the generating function must satisfy

$$G(x,0) = 1, \quad G(0,t) = (1 - pt)^N.$$

14. Use the result of Exercise 13 to show that the recurrence relation (6.4.8) determines the generating function (6.4.7) for integer x.

15. Verify (6.4.8) and (6.4.10).

16. Verify (6.4.11) by using (6.2.11).

17. Show that the coefficient b_n of x^{n-1} in $k_n^{(p)}(x;N)$ in (6.4.3) is

$$b_n = \frac{1}{n!}\left[-nNp + \binom{n}{2}(p-q)\right] = \frac{-Np + (n-1)(p-\frac{1}{2})}{(n-1)!}.$$

Hint: compute the coefficient of the kth summand in (6.4.3) and reduce the problem to computing sums like

$$\sum_{k=0}^{n}\binom{n}{k}p^k q^{n-k}k, \quad \sum_{k=0}^{n}\binom{n}{k}p^k q^{n-k}k(k-1).$$

Compare these sums to the partial derivatives F_p and F_{pp} of the function of two variables $F(p,q) = (p+q)^n$, evaluated at $q = 1 - p$.

18. Use Exercise 17 to give another proof of (6.4.11).

19. Prove the addition formula (6.4.12).

20. Show that as $N \to \infty$,

$$k_n^{(p)}(Nx;N) \sim \frac{N^n}{n!}(x-p)^n.$$

21. Use (6.5.5) and the expansion in (6.2.10) to prove (6.5.6).

22. Verify (6.5.7) using (6.5.2).

23. Verify (6.5.7) using (6.5.6).

24. Verify (6.5.8) and (6.5.9).

25. Verify (6.5.10) by using (6.2.11).

26. Prove the addition formula (6.5.11).

27. Verify (6.5.12).

28. Verify (6.6.3): show that

$$(\alpha+1)_m\,(\beta+1)_{N-m}$$
$$= \frac{\Gamma(N+\alpha+\beta+2)}{\Gamma(\alpha+1)\Gamma(\beta+1)}B(\alpha+1+m, \beta+1+N-m)$$

and use the integral representation (2.1.7) of the beta function to show that

$$\sum_{m=0}^{N} \binom{N}{m} B(\alpha + 1 + m, \beta + 1 + n - m) = B(\alpha + 1, \beta + 1).$$

29. Verify (6.6.10).
30. Verify (6.6.11).
31. Show that the polynomials $P_n^{(\lambda)}$ of (6.7.10) are eigenfunctions of L with L given by (6.7.8).
32. Show that the polynomials $P_n^{(\lambda)}$ of (6.7.10) satisfy the three-term recurrence (6.7.11).
33. Show that the polynomials p_n of (6.7.15) are eigenfunctions of L with L given by (6.7.13).
34. Show that the polynomials p_n of (6.7.15) satisfy the three-term recurrence (6.7.16).
35. Show that the polynomials $S_n(x)$ of (6.7.26) are eigenfunctions of L with L given by (6.7.19) and (6.7.21).
36. Show that the polynomials $S_n(x)$ of (6.7.26) satisfy the three-term recurrence (6.7.27).
37. Show that the polynomials $W_n(x)$ of (6.7.29) are eigenfunctions of L with L given by (6.7.19) and (6.7.22).
38. Show that the polynomials $W_n(x)$ of (6.7.29) satisfy the three-term recurrence (6.7.30).

6.9 Remarks

Discrete orthogonal polynomials are treated in the books by Chihara [76] and Ismail [194], who discuss the history and some of the classification results. Nikiforov, Suslov, and Uvarov [304] present the subject from the point of view of difference equations. Asymptotics are studied in the book by Baik, Kriecherbauer, McLaughlin, and Miller [27], using the Riemann–Hilbert method. Notation has not been completely standardized. The notation selected for [313] is such that each of these polynomials is a generalized hypergeometric series. This choice does not necessarily yield the simplest formulas.

The terminology here is more faithful to history than is often the case, in part because much of the history is relatively recent. Nevertheless, Chebyshev introduced the version of the "Hahn polynomials" treated above in 1858 [71]; see also [73]. Charlier introduced the Charlier polynomials in 1905 [66], Krawtchouk introduced the Krawtchouk polynomials in 1929 [229], Meixner introduced the Meixner polynomials in 1934 [285], and explicit formulas for the polynomials $h_n^{(\alpha,\beta)}$ were obtained by Weber and Erdélyi [429]. Dunkl [107] obtained addition formulas for Krawtchouk polynomials, analogous to those for Legendre and Jacobi polynomials, by group theoretic methods.

In 1949, Hahn [176] introduced a large class of "q-polynomials" that contain as a limiting case the discrete polynomials that had been introduced by Chebyshev. For q-polynomials in general, see the remarks and references at the end of Chapter 5.

Stanton [376] generalized Dunkl's addition formula to q-Krawtchouk polynomials. For other recent extensions of classical results to discrete and q-polynomials, see the books by Ismail [194] and Koekoek, Lesky, and Swarttouw [221].

The orthogonal polynomials introduced in Section 6.7, together with the classical orthogonal polynomials and the classical discrete orthogonal polynomials, complete the *Askey scheme*: a hierarchy of orthogonal polynomials expressible as generalized hypergeometric series (see Chapter 12). This hierarchy is closed under many limiting processes. For these polynomials and their q-generalizations, see Askey and Wilson [21, 22]. The book by Koekoek, Lesky, and Swarttouw [221] contains a detailed and well-organized treatment of both classical and semi-classical polynomials. For details and generalizations, see the survey article by Atakishiyev, Rahman, and Suslov [23]. For a combinatorial characterization, see Leonard [254]. For an extensive historical discussion, see Koelink [222]. For asymptotic relations among various semi-classical and classical polynomials, see Ferreira, López, and Sinusía [131] and Ferreira, López, and Pagola [130].

7

Asymptotics of orthogonal polynomials: two methods

Study of the asymptotics of orthogonal polynomials leads naturally to a division of the plane into three regions: the complement of the closed interval I that contains the zeros (perhaps after rescaling), the interior of the interval I, and the endpoints of I. For the classical polynomials, integral representations or differential equations can be used, as in Chapter 13. However, the discrete orthogonal polynomials do not satisfy a differential equation, and the Chebyshev–Hahn polynomials have no one-dimensional integral representation.

Each of the two methods illustrated in this chapter begins from the easiest part of the problem, a determination of the leading asymptotics in the exterior region, i.e. the complement of the interval. In the first section, we compute such an approximation in the case of Hermite polynomials. Corresponding results for Laguerre and Jacobi polynomials are stated, with the details left to the exercises.

The recently introduced *Riemann–Hilbert method* is a powerful tool for obtaining global asymptotics, i.e. asymptotic approximations in each of a few regions that cover the complex plane, for the orthogonal polynomials associated with certain types of weight function.

On the other hand, there are many orthogonal polynomials for which the weight function is not known or is not unique. In this case, the main tool has to be an associated three-term recurrence relation. We give an indication in the case of Jacobi polynomials. The three-term recurrence relation leads to a general form of the asymptotics on a complex neighborhood of the interval $(-1, 1)$. This form contains several functions that can be determined explicitly by matching to an approximation that is valid on the complement of the closed interval. Corresponding results are stated for Laguerre and Hermite polynomials.

In broad outline, the Riemann–Hilbert method consists in embedding a given object, such as an orthogonal polynomial, as part of the solution of a *Riemann–Hilbert problem*; the problem of determining a matrix-valued function of a complex variable that is holomorphic off a certain contour Σ and satisfies specified jump conditions across Σ as well as a normalization condition at infinity. Transforming such a problem into a more tractable form can lead to an explicit asymptotic expansion. We illustrate this in the case of Hermite polynomials. The ideas of Section 4.2 are used to set up the Riemann–Hilbert formulation in general. In succeeding sections, some transformations of the formulation for Hermite polynomials lead to a uniform asymptotic approximation.

7.1 Approximation away from the real line

A first asymptotic approximation to the quotient P'_n/nP_n for the Jacobi, rescaled Laguerre, and rescaled Hermite polynomials in the complement of the real axis was determined in Section 5.7. This quotient is the derivative of $\log(P_n^{1/n})$, so, up to a multiplication by a scalar c_n, we have a first asymptotic approximation to P_n. Carrying the argument of Section 5.7 a little further will determine c_n.

In the Hermite case, we worked with the Hermite polynomials rescaled so as to be monic as well, as in (5.7.5):

$$h_n(z) = \frac{1}{(2\alpha_n)^n} H_n(\alpha_n z), \quad \alpha_n = \sqrt{2n+1}. \tag{7.1.1}$$

We look for an approximation in the form

$$h_n(z) = \exp\{n\,g(z) + r(z) + O(1/n)\}, \quad z \notin \mathbf{R}.$$

Then

$$\psi_n(z) \equiv \frac{h'_n(z)}{n\,h_n(z)} = g'(z) + \frac{r'(z)}{n} + O\left(\frac{1}{n^2}\right).$$

The argument in Section 5.7 shows that we should take

$$g'(z) = \psi(z) = 2\left(z - \sqrt{z^2 - 1}\right). \tag{7.1.2}$$

We want $e^g \sim z$ at infinity, so we take

$$g(z) = z^2 - z\sqrt{z^2 - 1} - \frac{1}{2} + \log\left(\frac{z + \sqrt{z^2 - 1}}{2}\right). \tag{7.1.3}$$

Then $\psi_n = \psi + r'/n + O(1/n^2)$, and (5.7.10) is

$$\left(\psi(z) + \frac{r'(z)}{n}\right)^2 - 4z\left(\psi(z) + \frac{r'(z)}{n}\right) + 4$$

$$= \frac{1}{2n+1}\psi(z)^2 - \frac{2}{2n+1}\psi'(z) + O(1/n^2)$$

or

$$[2\psi(z) - 4z]\, r'(z) = \frac{\psi(z)^2}{2} - \psi'(z)$$

After some calculation, this gives

$$r'(z) = z - \sqrt{z^2 - 1} - \frac{z}{2(z^2 - 1)}.$$

We want $r \sim 0$ at infinity, so we take

$$r(z) = \frac{1}{2}g(z) - \frac{1}{4}\log(z^2 - 1). \tag{7.1.4}$$

This proves the following.

Theorem 7.1.1 *For $z \notin [-1, 1]$, the rescaled Hermite polynomials* (7.1.1) *satisfy*

$$h_n(z) = (z^2 - 1)^{-1/4} \exp\{(n + \tfrac{1}{2})g(z)\}[1 + O(1/n)] \tag{7.1.5}$$

as $n \to \infty$, where g is given by (7.1.3).

Similar arguments prove corresponding results for the monic rescaled Laguerre polynomials (5.7.3):

$$l_n^{(\alpha)}(z) = \frac{(-1)^n n!}{(4n)^n} L_n^{(\alpha)}(4nz).$$

(We have replaced $\beta_n = 4n + c$ in (5.7.3) by $4n$ to simplify the calculations slightly.)

Theorem 7.1.2 *For $z \notin [0, 1]$, the rescaled monic Laguerre polynomials satisfy*

$$l_n^{(\alpha)}(z) = \left(\frac{z}{z-1}\right)^{1/4} \left(\frac{2z - 1 + 2\sqrt{z^2 - z}}{4z}\right)^{\frac{1}{2}(\alpha+1)}$$

$$\times \exp\{ng(z)\}[1 + O(1/n)] \tag{7.1.6}$$

as $n \to \infty$, where

$$g(z) = (\sqrt{z} - \sqrt{z-1})^2 + 2\log\left(\frac{\sqrt{z} + \sqrt{z-1}}{2}\right).$$

See Exercise 1.

A similar result holds for the Jacobi polynomials $P_n^{(\alpha,\beta)}$. For convenience, we use the monic versions

$$p_n^{(\alpha,\beta)}(z) = \frac{2^n n!}{(\alpha + \beta + n + 1)_n} P_n^{(\alpha,\beta)}(z). \tag{7.1.7}$$

Theorem 7.1.3 *For* $z \notin [-1, 1]$, *the monic Jacobi polynomials (7.1.7) satisfy*

$$p_n^{(\alpha,\beta)}(z) = (z-1)^{-\frac{\alpha}{2}-\frac{1}{4}}(z+1)^{-\frac{\beta}{2}-\frac{1}{4}} \left(\frac{z + \sqrt{z^2 - 1}}{2} \right)^{n+\frac{1}{2}(\alpha+\beta+1)} [1 + O(1/n)]$$

(7.1.8)

as $n \to \infty$.

See Exercise 2.

7.2 Asymptotics by matching

The results of Section 7.1 can be combined with a general form for solutions of certain three-term recurrence relations to obtain asympotics of the classical polynomials in a complex neighborhood of the interval that contains the zeros.

For Jacobi polynomials, let us fix a pair of indices $\alpha > -1$, $\beta > -1$. The three-term recurrence (5.5.5) implies the corresponding three-term recurrence for the monic polynomials $p_n = p_n^{(\alpha,\beta)}$:

$$xp_n(x) = p_{n+1}(x) + a_n p_n(x) + b_n p_{n-1}(x), \tag{7.2.1}$$

where

$$a_n = \frac{\beta^2 - \alpha^2}{4(n+1+\gamma)(n+\gamma)} = O(1/n^2), \quad \gamma = \frac{\alpha+\beta}{2};$$

$$b_n = \frac{(n+\alpha)(n+\beta)(n+\alpha+\beta)n}{4(n+\gamma-\frac{1}{2})(n+\gamma)^2(n+\gamma+\frac{1}{2})} = \frac{1}{4} + O(1/n^2).$$

Wong and Li [444] have shown that any solution of a recurrence of this type has asymptotics, valid in a complex neighborhood of $(-1, 1)$, of the form

$$p_n(x) = n^a r^n [f(x) \cos n\varphi(x) + g(x) \sin n\varphi(x)][1 + O(1/n)], \tag{7.2.2}$$

where r and a are real constants, φ is real-valued on the interval, and f and g are complex-valued. We will use (7.2.1) to determine the constants r and a and the function φ, and then match the asymptotics (7.1.8) outside the interval to determine the functions f and g.

Combining the form (7.2.2) and the recurrence (7.2.1), and ignoring terms of order $1/n$, we factor out n^a and r^n, replace a_n by zero and b_n by 1/4, and compare coefficients of $\cos n\varphi$ and $\sin n\varphi$ to obtain

$$xf - \left(r + \frac{1}{4r} \right) f \cos \varphi - \left(r - \frac{1}{4r} \right) g \sin \varphi = 0; \tag{7.2.3}$$

$$xg + \left(r - \frac{1}{4r} \right) f \sin \varphi - \left(r + \frac{1}{4r} \right) g \cos \varphi = 0. \tag{7.2.4}$$

Since f and g are not both identically zero, the matrix of coefficients of this pair of homogeneous equations must vanish. This is equivalent to

$$x - \left(r + \frac{1}{4r}\right)\cos\varphi; \qquad \left(r - \frac{1}{4r}\right)\sin\varphi = 0.$$

Therefore $r = 1/2$ and $x = \cos\varphi$. We put $\theta = \theta(x) = \cos^{-1}x$ and choose $0 < \mathrm{Re}\,\theta < \pi$. Equation (7.2.2) simplifies to

$$p_n(x) = n^a 2^{-n}[f(x)\cos n\theta + g(x)\sin n\theta][1 + O(1/n)], \quad x = \cos\theta.$$

Now let us include terms of order $1/n$ in the recurrence (7.2.1). After division by n^a, the only new terms come from multiplying the p_{n+1} by a/n and the p_{n-1} term by $-a/n$. This leads to the condition $a = 0$ and the further simplification

$$p_n(x) \sim 2^{-n}[f(x)\cos n\theta + g(x)\sin n\theta]$$

$$= \frac{f(x) - ig(x)}{2}\left(\frac{e^{i\theta}}{2}\right)^n + \frac{f(x) + ig(x)}{2}\left(\frac{e^{-i\theta}}{2}\right)^n, \qquad (7.2.5)$$

where again $x = \cos\theta$. This is valid in a complex neighborhood of the interval $(-1,1)$, so we may compare it to (7.1.8), valid on the complement of the interval

$$A(z)p_n(z) \sim \left(\frac{z + \sqrt{z^2 - 1}}{2}\right)^{n + \frac{1}{2}a + \frac{1}{2}\beta + \frac{1}{2}}, \qquad (7.2.6)$$

where

$$A(z) = (z-1)^{\frac{1}{2}a + \frac{1}{4}}(z+1)^{\frac{1}{2}\beta + \frac{1}{4}}.$$

Taking $z = \cos\theta$ with $\mathrm{Re}\,\theta \in (0,\pi)$, it follows that

$$\mathrm{Im}\,z > 0 \quad \text{implies} \quad \mathrm{Im}\,\theta < 0, \quad \sqrt{z^2 - 1} = i\sin\theta;$$

$$\mathrm{Im}\,z < 0 \quad \text{implies} \quad \mathrm{Im}\,\theta > 0, \quad \sqrt{z^2 - 1} = -i\sin\theta.$$

In particular, $e^{-in\theta} \to 0$ for $\mathrm{Im}\,z > 0$ and $e^{in\theta} \to 0$ for $\mathrm{Im}\,z < 0$. Combining these considerations with (7.2.5) and (7.2.6),

$$A(z) \cdot \frac{f(z) - ig(z)}{2} = \left(\frac{e^{i\theta}}{2}\right)^{\frac{1}{2}(\alpha+\beta+1)}, \qquad \mathrm{Im}\,z > 0;$$

$$(7.2.7)$$

$$A(z) \cdot \frac{f(z) + ig(z)}{2} = \left(\frac{e^{-i\theta}}{2}\right)^{\frac{1}{2}(\alpha+\beta+1)}, \qquad \mathrm{Im}\,z < 0.$$

Now

$$z - 1 = \cos\theta - 1 = -2\sin^2(\tfrac{1}{2}\theta), \quad z + 1 = 2\cos^2(\tfrac{1}{2}\theta).$$

Therefore for z in a complex neighborhood of $(-1,1)$,

$$A(z) = 2^{\frac{1}{2}(\alpha+\beta+1)}(\sin^2\tfrac{1}{2}\theta)^{\frac{1}{2}\alpha+\frac{1}{4}}(\cos^2\tfrac{1}{2}\theta)^{\frac{1}{2}\beta+\frac{1}{4}}e^{i\pi(\frac{1}{2}\alpha+\frac{1}{4})}, \quad \mathrm{Im}\,z > 0;$$

$$A(z) = 2^{\frac{1}{2}(\alpha+\beta+1)}(\sin^2\tfrac{1}{2}\theta)^{\frac{1}{2}\alpha+\frac{1}{4}}(\cos^2\tfrac{1}{2}\theta)^{\frac{1}{2}\beta+\frac{1}{4}}e^{-i\pi(\frac{1}{2}\alpha+\frac{1}{4})}, \quad \mathrm{Im}\,z < 0.$$

$$(7.2.8)$$

Combining (7.2.7) and (7.2.8) yields

$$2^{\alpha+\beta}(\sin\tfrac{1}{2}\theta)^{\alpha+\frac{1}{2}}(\cos\tfrac{1}{2}\theta)^{\beta+\frac{1}{2}}f(z) = \cos\left(\frac{1}{2}[\alpha+\beta+1]\theta - \left[\frac{1}{2}\alpha + \frac{1}{4}\right]\pi\right);$$

$$2^{\alpha+\beta}(\sin\tfrac{1}{2}\theta)^{\alpha+\frac{1}{2}}(\cos\tfrac{1}{2}\theta)^{\beta+\frac{1}{2}}g(z) = -\sin\left(\frac{1}{2}[\alpha+\beta+1]\theta - \left[\frac{1}{2}\alpha + \frac{1}{4}\right]\pi\right).$$

This proves the following.

Theorem 7.2.1 *The monic Jacobi polynomials*

$$p_n^{(\alpha,\beta)}(x) = \frac{2^n n!}{(\alpha+\beta+n+1)_n} P_n^{(\alpha,\beta)}(x)$$

satisfy

$$p_n^{(\alpha,\beta)}(\cos\theta) \sim 2^{-\alpha-\beta} \frac{\cos\left(n\theta + \frac{1}{2}[\alpha+\beta+1]\theta - [\frac{1}{2}\alpha + \frac{1}{4}]\pi\right)}{\sin(\frac{1}{2}\theta)^{\alpha+\frac{1}{2}}\cos(\frac{1}{2}\theta)^{\beta+\frac{1}{2}}} \qquad (7.2.9)$$

as $n \to \infty$, for $x = \cos\theta$ in a complex neighborhood of the interval $(-1,1)$.

Similar arguments, but with more complications, using Theorems 7.1.2 and 7.1.1, lead to the following. For the Hermite case and a nonclassical case, see [435].

Theorem 7.2.2 *The Laguerre polynomials $L_n^{(\alpha)}$ satisfy*

$$L_n^{(\alpha)}(4nx) \sim 2(\cot\theta)^{1/2}\frac{(-n)^n}{n!}\left(\frac{1}{4x}\right)^{\frac{1}{2}(1+\alpha)} e^{n(2x-1)}$$

$$\times \cos\left(2n[\cos\theta\sin\theta - \theta] - [1+\alpha]\theta + \tfrac{1}{4}\pi\right) \quad (7.2.10)$$

as $n \to \infty$, for x in a complex neighborhood of the interval $(0,1)$, where $\theta = \cos^{-1}\sqrt{x} \in (0, \frac{1}{2}\pi)$.

Theorem 7.2.3 *The Hermite polynomials H_n satisfy*

$$H_n(\sqrt{2n+1}\,x) \sim \frac{\sqrt{2}(2n+1)^{n/2}}{(1-x^2)^{1/4}} e^{(n+\frac{1}{2})(x^2-\frac{1}{2})}\cos\left([n+\tfrac{1}{2}][\cos\theta\sin\theta - \theta + \tfrac{1}{2}\pi]\right)$$

$$(7.2.11)$$

as $n \to \infty$, for x in a complex neighborhood of the interval $(-1,1)$, where $\theta = \cos^{-1}x \in (0,\pi)$.

7.3 The Riemann–Hilbert formulation

Suppose that $\{Q_n\}$ is a family of monic polynomials, orthogonal with respect to a positive weight w on a real interval I. Recall from Section 4.2 that the quotients

$$\frac{P_n(z)}{Q_n(z)}, \quad P_n(z) = \int_I \frac{Q_n(z) - Q_n(x)}{z - x} w(x)\,dx$$

approximate the Stieltjes transform of the weight,

$$F(z) = \int_I \frac{w(x)\,dx}{z - x}, \quad z \notin I.$$

The remainder

$$R_n(z) = Q_n(z)F(z) - P_n(z)$$

is the Stieltjes transform of $Q_n w$:

$$R_n(z) = \int_I \frac{Q_n(x)\,w(x)\,dx}{z - x}.$$

As in the case of the weight itself (Proposition 4.2.1), this means that

$$-\frac{1}{2\pi i}\left[(R_n)_+(x) - (R_n)_-(x)\right] = Q_n(x)\,w(x) \tag{7.3.1}$$

on the interior of I.

Note that carrying (4.2.6) one step further and using the orthogonality property shows that

$$R_n(z) = \frac{c_n}{z^{n+1}} + O(z^{-n-2}); \tag{7.3.2}$$

$$c_n = \int_I Q_n(x)\,x^n\,w(x)\,dx = \int_I Q_n(x)^2 w(x)\,dx. \tag{7.3.3}$$

Let the 2×2 matrix-valued function $Y(z)$ be defined by

$$Y(z) = \begin{bmatrix} Q_n(z) & -\dfrac{1}{2\pi i}R_n(z) \\ a_n Q_{n-1}(z) & -\dfrac{a_n}{2\pi i}R_{n-1}(z) \end{bmatrix}. \tag{7.3.4}$$

Here a_n is a constant to be chosen. Then (7.3.1), for a fixed pair n, $n - 1$, is equivalent to the matrix equation

$$Y_+(x) = Y_-(x)\begin{bmatrix} 1 & w(x) \\ 0 & 1 \end{bmatrix}. \tag{7.3.5}$$

This formulation is due to Fokas, Its, and Kataev [138]. It is an example of a *matrix Riemann–Hilbert problem*: given certain matrix-valued data on an oriented contour in $\mathbf{C}$, find a matrix-valued function whose multiplicative jump across the contour is that data. The solution is not unique unless one specifies additional conditions, such as the behavior of the function as $z \to \infty$. In our case,

$$Y(z) = \begin{bmatrix} 1 + O(z^{-1}) & O(z^{-1}) \\ O(z^{-1}) & -\dfrac{a_n c_{n-1}}{2\pi i} + O(z^{-1}) \end{bmatrix} \begin{bmatrix} z^n & 0 \\ 0 & z^{-n} \end{bmatrix}. \tag{7.3.6}$$

We choose $a_n = -2\pi i / c_{n-1}$.

It is not obvious that the Riemann–Hilbert formulation (7.3.5), (7.3.6) has any advantage over the additive formulation (7.3.1). As we shall see, however, it can be a powerful tool for investigating asymptotics. As noted in the introduction, the basic idea is to renormalize (7.3.5), (7.3.6) in such a way that the problem can easily be solved asymptotically in n.

7.4 The Riemann–Hilbert problem in the Hermite case, I

The monic orthogonal polynomials in the Hermite case are

$$Q_n(x) = 2^{-n} H_n(x).$$

Note two symmetry properties: behavior under change of sign,

$$Q_n(-x) = (-1)^n Q_n(x), \tag{7.4.1}$$

and behavior under complex conjugaton,

$$Q_n(\bar{z}) = \overline{Q_n(z)}. \tag{7.4.2}$$

The scalar functions ψ, g, g_n, φ, ϕ, ρ, ξ_n, and the matrix-valued functions Y, U, T, L, $\hat{U}$, to be introduced below, all have the symmetry property (7.4.2). The matrix functions satisfy a more subtle version of (7.4.1); see (7.4.4).

According to (5.3.9), the corresponding constant

$$c_n = \int_{-\infty}^{\infty} Q_n(x)^2 e^{-x^2} \, dx = \frac{n! \sqrt{\pi}}{2^n}.$$

The Riemann–Hilbert formulation is

$$Y_+(x) = Y_-(x) \begin{bmatrix} 1 & e^{-x^2} \\ 0 & 1 \end{bmatrix},$$

with condition at ∞

$$Y(z) = \left[1 + O(|z|^{-1}) \right] \cdot \begin{bmatrix} z^n & 0 \\ 0 & z^{-n} \end{bmatrix}, \tag{7.4.3}$$

where **1** is the identity matrix and

$$
Y(z) = \begin{bmatrix} Q_n(z) & -\frac{1}{2\pi i} R_n(z) \\[2mm] -\frac{2\pi i}{c_{n-1}} Q_{n-1}(z) & \frac{1}{c_{n-1}} R_{n-1}(z) \end{bmatrix};
$$

$$
R_n(z) = \int_{-\infty}^{\infty} \frac{Q_n(x)}{z-x} e^{-x^2}\, dx = c_n z^{-n-1} + O(z^{-n-2}).
$$

More precisely, in (7.4.3) and similar formulas to follow, we mean that the condition holds uniformly in regions where $|\arg z \pm \frac{\pi}{2}| \le \frac{\pi}{2} - \delta$, i.e. the complement of the union of open wedges containing the positive and negative real axis. Throughout the discussion in this section, we assume implicitly the dependence of Y on the degree n, which is assumed to be large.

The symmetry condition (7.4.1) implies

$$
R_n(-z) = (-1)^{n+1} R_n(z),
$$

and therefore

$$
Y(-z) = (-1)^n J\, Y(z) J, \tag{7.4.4}
$$

where J is the matrix

$$
J = \begin{bmatrix} 1 & 0 \\ 0 & -1 \end{bmatrix}. \tag{7.4.5}
$$

As in Section 7.1, it is natural to rescale by setting

$$
U(z) = \begin{bmatrix} \alpha_n^{-n} & 0 \\ 0 & \alpha_n^n \end{bmatrix} Y(\alpha_n z), \qquad \alpha_n = \sqrt{2n+1}. \tag{7.4.6}
$$

The 11-element of U is then the rescaled monic polynomial $h_n(z)$. Equation (7.1.5) can be written as

$$
h_n(z) \sim e^{g_n(z)}, \tag{7.4.7}
$$

where g_n is

$$
g_n(z) = (n + \tfrac{1}{2})\left[z^2 - z\sqrt{z^2-1} - \frac{1}{2} + \log\left(\frac{\varphi(z)}{2}\right) \right] - \frac{1}{4}\log(z^2-1)
$$

$$
= (n + \tfrac{1}{2})\left[z^2 - \phi(z) - \frac{1}{2} - \log 2 \right] - \frac{1}{4}\log(z^2-1), \tag{7.4.8}
$$

where $\varphi(z) = z + \sqrt{z^2-1}$ and

$$
\phi(z) = z\sqrt{z^2-1} - \log\varphi(z) = z\sqrt{z^2-1} - \log(z + \sqrt{z^2-1}). \tag{7.4.9}
$$

The Riemann–Hilbert formulation for U is

$$U_+(x) = U_-(x) \begin{bmatrix} 1 & e^{-(2n+1)x^2} \\ 0 & 1 \end{bmatrix}, \quad x \in \mathbf{R};$$

$$U(z) = \left[\mathbf{1} + O(|z|^{-1}) \right] \cdot \begin{bmatrix} z^n & 0 \\ 0 & z^{-n} \end{bmatrix}. \tag{7.4.10}$$

The symmetry (7.4.4) carries over to U:

$$U(-z) = (-1)^n J\, U(z) J. \tag{7.4.11}$$

The next step is to regularize the problem at infinity. This can be done by multiplying the first column by z^{-n} and the second by z^n, but this introduces problems at the origin. Instead, we take advantage of (7.4.7) and the fact that $e^{g_n(z)} z^{-n} \sim 1$ as $z \to \infty$, and multiply by

$$e^{-g_n(z)J} = \begin{bmatrix} e^{-g_n(z)} & 0 \\ 0 & e^{g_n(z)} \end{bmatrix}.$$

It will be convenient to have an additional conjugation by a constant matrix. We take

$$T(z) = e^{-\frac{1}{2} l_n J} \left[U(z) e^{-g_n(z)J} \right] e^{\frac{1}{2} l_n J} = 1 + O(z^{-1}), \tag{7.4.12}$$

where l_n is a constant to be determined. The multiplicative jump matrix for T is

$$v_T(x) = T_-(x)^{-1} T_+(x)$$

$$= e^{-\frac{1}{2} l_n J} \left[e^{(g_n)_- J} U_-(x)^{-1} U_+(x) e^{-(g_n)_+ J} \right] e^{\frac{1}{2} l_n J}$$

$$= \begin{bmatrix} e^{(g_n)_- - (g_n)_+} & e^{-(2n+1)x^2 + (g_n)_+ + (g_n)_- - l_n} \\ 0 & e^{(g_n)_+ - (g_n)_-} \end{bmatrix}. \tag{7.4.13}$$

For $x \in (-1, 1)$,

$$g_n(x)_+ + g_n(x)_- = (2n+1)(x^2 - \tfrac{1}{2} - \log 2) - \log(\gamma_+ \gamma_-),$$

where $\gamma(z) = (z^2 - 1)^{1/4}$, so

$$\gamma_+(x)\,\gamma_-(x) = \sqrt{1 - x^2}.$$

We take

$$l_n = -(2n+1)(\tfrac{1}{2} + \log 2) + \log 2, \tag{7.4.14}$$

so that the 12-element of the jump matrix is $1/[2(\gamma)_+(\gamma)_-]$, independent of n.

Next, $x + i\sqrt{1-x^2}$ has modulus 1, so the function ϕ of (7.4.9) satisfies

$$\phi_+(x) = -\phi_-(x) = ix\sqrt{1-x^2} - \log(x + i\sqrt{1-x^2})$$

for $x \in (-1, 1)$. Therefore

$$g_n(x)_+ - g_n(x)_- = -(n + \tfrac{1}{2})[\phi_+(x) - \phi_-(x)] - \tfrac{1}{2}\pi i$$
$$= -(2n+1)\left[ix\sqrt{1-x^2} - \log(x + \sqrt{1-x^2})\right] - \tfrac{1}{2}\pi i.$$

Thus for $x \in [-1, 1]$, the jump matrix $v_T(x)$ is

$$v_T(x) = \begin{bmatrix} i(\omega_n)_+ & 1/(2\gamma_+\gamma_-) \\ 0 & -i(\omega_n)_- \end{bmatrix}, \tag{7.4.15}$$

where

$$\omega_n(z) = e^{(2n+1)\phi(z)}.$$

Similarly, for real x not in the interval $[-1, 1]$,

$$v_T(x) = \begin{bmatrix} 1 & 1/(2\gamma^2)e^{-(2n+1)\phi(x)} \\ 0 & 1 \end{bmatrix}. \tag{7.4.16}$$

Remark. For $\pm x > 1$,

$$\operatorname{Re}\phi(x) = \pm x\sqrt{x^2-1} - \log\left(\pm x + \sqrt{x^2-1}\right) > 0.$$

Therefore the off-diagonal term in v_T converges rapidly to zero as $n \to \infty$, uniformly on any pair of intervals $|x| \geq 1 + \delta > 1$.

The jump matrix (7.4.15) on the interval $[-1, 1]$ can be simplified by conjugation with a diagonal matrix:

$$\begin{bmatrix} (\gamma/\beta)_- & 0 \\ 0 & (\beta/\gamma)_- \end{bmatrix}\begin{bmatrix} i(\omega_n)_+ & 1/(2\gamma_-\gamma_+) \\ 0 & -i(\omega_n)_- \end{bmatrix}\begin{bmatrix} (\beta/\gamma)_+ & 0 \\ 0 & (\gamma/\beta)_+ \end{bmatrix}$$
$$= \begin{bmatrix} (\widehat{\omega}_n)_+ & 1 \\ 0 & (\widehat{\omega}_n)_- \end{bmatrix}. \tag{7.4.17}$$

This identity is valid so long as $\beta_+\beta_- = \tfrac{1}{2}$. We choose

$$\beta(z) = \left(\frac{z + \sqrt{z^2-1}}{2}\right)^{1/2}$$

so that the diagonal matrices are asymptotic to the identity matrix at ∞ and so that β is holomorphic on the complement of $[-1, 1]$. With this choice of β,

$$\widehat{\omega}_n(z) = \varphi(z)\omega_n(z) = \varphi(z)e^{(2n+1)\phi(z)}.$$

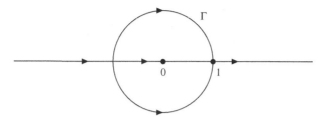

Figure 7.1. The contour $\Sigma = \mathbf{R} \cup \Gamma$.

This computation shows that we can get a simpler Riemann–Hilbert problem by modifying T. Set

$$\widehat{T}(z) = T(z) \begin{bmatrix} \beta/\gamma & 0 \\ 0 & \gamma/\beta \end{bmatrix}. \qquad (7.4.18)$$

The jump matrix for $\widehat{T}$ on $[-1, 1]$ is

$$v_{\widehat{T}}(x) = \begin{bmatrix} (\widehat{\omega}_n)_+ & 1 \\ 0 & (\widehat{\omega}_n)_- \end{bmatrix}$$

$$= \begin{bmatrix} 1 & 0 \\ (\widehat{\omega}_n)_- & 1 \end{bmatrix} \begin{bmatrix} 0 & 1 \\ -1 & 0 \end{bmatrix} \begin{bmatrix} 1 & 0 \\ (\widehat{\omega}_n)_+ & 1 \end{bmatrix}. \qquad (7.4.19)$$

The jump matrix for real x, $|x| > 1$, is

$$v_{\widehat{T}}(x) = \begin{bmatrix} 1 & \widehat{\omega}_n^{-1} \\ 0 & 1 \end{bmatrix}.$$

We take advantage of this in the following way. Consider the configuration $\Sigma = \mathbf{R} \cup \Gamma$, where Γ is the unit circle $\{|z| = 1\}$. Orient each branch of Σ in the direction of increasing $\mathrm{Re}\, z$; see Figure 7.1.

Define the matrix-valued function V for $z \notin \mathbf{R}$ by

$$V(z) = \widehat{T}(z), \quad |z| > 1,$$

$$V(z) = \widehat{T}(z) \begin{bmatrix} 1 & 0 \\ -\widehat{\omega}_n & 1 \end{bmatrix}, \quad |z| < 1, \ \mathrm{Im}\, z > 0;$$

$$V(z) = \widehat{T}(z) \begin{bmatrix} 1 & 0 \\ \widehat{\omega}_n & 1 \end{bmatrix}, \quad |z| < 1, \ \mathrm{Im}\, z < 0.$$

Then the Riemann–Hilbert problem for $\widehat{T}$ is equivalent to the following Riemann–Hilbert problem: $V(z)$ holomorphic, $V(z) = \mathbf{1} + O(|z|^{-1})$ as $z \to \infty$, with jump matrix $v_V(x) = v_{\widehat{T}}(x)$ for $|x| > 1$, and

$$v_V(z) = \begin{bmatrix} 1 & 0 \\ \widehat{\omega}_n & 1 \end{bmatrix}, \quad |z| = 1, \ \mathrm{Im}\, z > 0;$$

$$v_V(z) = \begin{bmatrix} 0 & 1 \\ -1 & 0 \end{bmatrix}, \quad -1 < z < 1;$$

$$v_V(z) = \begin{bmatrix} 1 & 0 \\ \widehat{\omega}_n & 1 \end{bmatrix}, \quad |z| = 1, \ \operatorname{Im} z < 0.$$

Formally, at least, this problem simplifies as $n \to \infty$.

Proposition 7.4.1 *The function $\widehat{\omega}_n(z)$ converges exponentially to zero as $n \to \infty$ at each point $z \in \Gamma$, $z \neq \pm 1$.*

See Exercise 4.

In view of this proposition and the remark, it appears that in the limit, the solution to the problem for V is holomorphic on the complement of $[-1, 1]$. We are thus led to examine the solution of the limiting Riemann–Hilbert problem: N holomorphic on the complement of $[-1, 1]$ and

$$N_+(x) = N_-(x) \begin{bmatrix} 0 & 1 \\ -1 & 0 \end{bmatrix}, \quad x \in [-1, 1]; \tag{7.4.20}$$

$$N(z) = 1 + O(|z|^{-1}).$$

Proposition 7.4.2 *The solution to the problem (7.4.20) is*

$$N(z) = \frac{1}{2} \begin{bmatrix} a(z) + a(z)^{-1} & i[a(z)^{-1} - a(z)] \\ i[a(z) - a(z)^{-1}] & a(z) + a(z)^{-1} \end{bmatrix},$$

where

$$a(z) = \left(\frac{z-1}{z+1} \right)^{1/4}.$$

See Exercise 5.

We use the identities

$$2\varphi(z)^{\pm 1} = 2z \pm 2\sqrt{z^2 - 1} = \left(\sqrt{z+1} \pm \sqrt{z-1} \right)^2$$

to rewrite

$$N(z) = \frac{1}{\sqrt{2}(z^2 - 1)^{1/4}} \begin{bmatrix} \varphi^{1/2} & i\varphi^{-1/2} \\ -i\varphi^{-1/2} & \varphi^{1/2} \end{bmatrix}. \tag{7.4.21}$$

Summarizing, we expect that in the limit, $\widehat{T} \sim N$. Now

$$U(z) = e^{\frac{1}{2} \ln J} \left[T(z) e^{g_n(z) J} \right] e^{-\frac{1}{2} \ln J},$$

and (7.4.18) gives

$$T(z) e^{g_n(z) J} = \widehat{T}(z) e^{\widehat{g}_n(z) J},$$

where

$$\widehat{g}_n(z) = g_n(z) + \tfrac{1}{4}\log(z^2 - 1) - \tfrac{1}{2}\log\left(\frac{\varphi(z)}{2}\right)$$

$$= (n + \tfrac{1}{2})[z^2 - \phi(z) - \tfrac{1}{2} - \log 2] - \tfrac{1}{2}\log\left(\frac{\varphi(z)}{2}\right).$$

Thus we expect

$$U(z) \sim L(z) \equiv e^{\frac{1}{2}\ln J}\left[N(z)e^{\widehat{g}_n(z)J}\right]e^{-\frac{1}{2}\ln J} \tag{7.4.22}$$

as $n \to \infty$.

It is tempting to go straight to the Riemann–Hilbert formulation for the function $L^{-1}U$. The jump matrices converge rapidly and uniformly to the identity matrix except at $x = \pm 1$, but at these points the jumps are singular because of the singularities of N at $x = \pm 1$. Overcoming this difficulty requires a further transformation of the problem.

7.5 The Riemann–Hilbert problem in the Hermite case, II

We begin with a close look at the function ϕ,

$$\phi(z) = z\sqrt{z^2 - 1} - \log\left(z + \sqrt{z^2 - 1}\right), \quad z \notin (-\infty, 1],$$

which is holomorphic on the complement of $(-\infty, 1]$.

Lemma 7.5.1 *The power $\phi^{2/3}$ has a holomorphic extension to the complement of $(-\infty, -1]$.*

Proof Let $z = 1 + \eta$, $\eta \notin (-\infty, 0)$. Near $z = 1$ the derivative

$$\phi'(z) = 2\sqrt{z^2 - 1} = 2\sqrt{2\eta + \eta^2} = 2\sqrt{2\eta}\left(1 + \frac{\eta}{2}\right)^{1/2}$$

$$= 2^{3/2}\eta^{1/2}\left[1 + \sum_{k=1}^{\infty} a_k\eta^k\right].$$

This expansion converges for $|\eta| < 2$. Integration gives a representation of $\phi(z)$ in the form $\frac{2}{3}(2\eta)^{3/2}(1 + \dots)$, so $\phi^{2/3}$ extends to $\{|z - 1| < 2\}$. The complement of $(-\infty, -1]$ is simply connected and ϕ has no zeros there, so $\phi^{2/3}$ extends to the entire region. $\qquad\square$

This lemma leads naturally to an anticipation of the asymptotic result to be proved. In view of (7.4.1), it is enough to consider asymptotics of $H_n(z)$ for $\operatorname{Re} z \geq 0$. Let

$$u_n(z) = h_n(z)e^{-(n+\frac{1}{2})z^2} = (2\alpha_n)^{-n}H_n(\alpha_n z)e^{-\frac{1}{2}(\alpha_n z)^2},$$

where again h_n is the rescaled monic polynomial (7.1.1). The function u_n satisfies the rescaled version of (5.3.1):

$$u_n''(z) = \alpha_n^4(z^2 - 1)u_n(z).$$

The most difficult regions to study are the neighborhoods of the *turning points* $z = \pm 1$; for real x, these are the places where u_n goes from being oscillatory to having a fixed sign. To study the behavior near $z = 1$, therefore, it is natural to consider u_n as a function of $\phi^{2/3} \sim 2(2/3)^{2/3}(z - 1)$. We also normalize by taking

$$u_n(z) = \rho(z)v_n(\xi_n(z)), \qquad \xi_n(z) = \beta_n \phi(z)^{2/3},$$

where β_n is a constant to be chosen, and $\rho = \rho_n$ is chosen so that the second-order equation for v_n has no first-order term. Thus

$$\xi_n'(z) = \frac{2\beta_n}{3} \phi(z)^{-1/3} \phi'(z) = \frac{4}{3}\beta_n^{3/2} \left[\frac{z^2 - 1}{\xi_n}\right]^{1/2}$$

and $2\rho'\xi_n' + \rho\xi_n'' = 0$, so

$$\rho(z) = \xi_n'(z)^{-1/2} = \sqrt{\frac{3}{4}}\beta_n^{-3/4} \left[\frac{\xi_n}{z^2 - 1}\right]^{1/4}.$$

Then

$$\alpha_n^4(z^2 - 1)u_n = u_n'' = \rho(\xi_n')^2 v_n'' + \rho'' v_n.$$

The choice $\beta_n = (3\alpha_n^2/4)^{2/3}$, i.e.

$$\xi_n(z) = \left[\frac{3}{2}\left(n + \frac{1}{2}\right)\phi(z)\right]^{2/3}, \tag{7.5.1}$$

leads to the simple form

$$v_n''(\xi_n) = \left[\xi_n - \frac{\rho''}{\rho(\xi_n')^2}\right]v_n(\xi_n).$$

With these choices,

$$u_n(z) = \frac{1}{\alpha_n}\left[\frac{\xi_n}{z^2 - 1}\right]^{1/4} v_n(\xi_n). \tag{7.5.2}$$

Now let $z = 1 + \eta$. For $|z - 1| < 2$, ξ_n is a power series in η with first term $2\beta_n(2/3)^{2/3}\eta$, so ξ_n' is a power series with leading term $2\beta_n(2/3)^{2/3}$. It follows that $\rho''/\rho(\xi_n')^2$ is $O(\beta_n^{-2})$ near $z = 1$:

$$v_n''(\xi_n) = \left[\xi_n + O(n^{-4/3})\right]v_n(\xi_n). \tag{7.5.3}$$

This is a small perturbation of Airy's equation (9.9.3), so we might expect v_n to be asymptotic in n to a solution of Airy's equation. We can determine which solution by looking at the asymptotics in the expected relation (7.4.22).

The 11-element U_{11} is the rescaled monic polynomial (7.1.1). According to the construction of U and to (7.4.22), we expect that

$$
\begin{aligned}
u_n(z) = h_n(z)e^{-(n+\frac{1}{2})z^2} &= U_{11}(z)e^{-(n+\frac{1}{2})z^2} \\
&\sim \left[N_{11}(z)e^{\widehat{g}_n(z)} \right] e^{-(n+\frac{1}{2})z^2} \\
&= N_{11}(z)e^{-(n+\frac{1}{2})[\phi(z)+\frac{1}{2}+\log 2]-\frac{1}{2}\log\frac{\varphi}{2}} \\
&= 2^{-n}e^{-\frac{1}{4}(2n+1)}\varphi(z)^{-1/2}N_{11}(z)e^{-(n+\frac{1}{2})\phi(z)} \\
&= 2^{-(n+\frac{1}{2})}e^{-\frac{1}{4}(2n+1)}(z^2-1)^{-1/4}e^{-(n+\frac{1}{2})\phi}.
\end{aligned}
\tag{7.5.4}
$$

Therefore we expect v_n to be asymptotic to a solution of Airy's equation that has exponential decay as $z \to +\infty$, hence to a multiple $c_n \mathrm{Ai}$.

The asymptotics of Ai are given in (9.9.7). Now

$$
\frac{2}{3}\zeta_n^{3/2} = \left(n+\frac{1}{2}\right)\phi,
$$

so (9.9.7) gives

$$
\mathrm{Ai}(\zeta_n) \sim \frac{1}{2\sqrt{\pi}\,\zeta_n^{1/4}}e^{-(n+\frac{1}{2})\phi}.
\tag{7.5.5}
$$

If we assume that $v_n \sim c_n \mathrm{Ai}$, then (7.5.2), (7.5.4), and (7.5.5) imply

$$
c_n = \frac{\sqrt{2\pi}\,\alpha_n}{2^n}e^{-\frac{1}{4}(2n+1)}.
$$

Tracing this all back, we expect that

$$
\begin{aligned}
H_n(\alpha_n z) = (2\alpha_n)^n h_n(z) &= (2\alpha_n)^n u_n(z)e^{(n+\frac{1}{2})z^2} \\
&\sim \sqrt{2\pi}\,(2n+1)^{n/2}\left(\frac{\zeta_n}{z^2-1}\right)^{1/4}e^{(n+\frac{1}{2})(z^2-\frac{1}{2})}\mathrm{Ai}(\zeta_n).
\end{aligned}
\tag{7.5.6}
$$

To prove this result, we return to the Riemann–Hilbert formulation. The mapping properties of ϕ are important.

Proposition 7.5.2 *The mapping properties of ϕ are as shown in Figure 7.2. The ray $[1,\infty)$ is mapped onto $[0,\infty)$, the interval $[-1,1]$ is mapped to the interval $[-\pi i, 0]$, the curve C_+ is mapped to the positive imaginary axis, and the curve C_- is mapped to the negative imaginary axis minus the interval $(-\pi i, 0)$. The curve C_- meets the real axis at angles $\pm\frac{2}{3}\pi$ and is asymptotic to the rays $\arg z = \pm\frac{3}{4}\pi$. The curve C_+ meets the real axis at angles $\pm\frac{1}{3}\pi$ and is asymptotic to the rays $\arg z = \pm\frac{1}{4}\pi$.*

See Exercise 6.

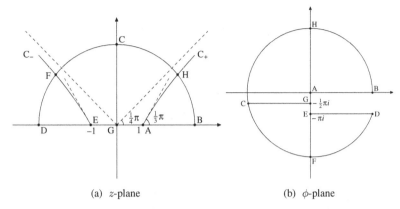

(a) z-plane   (b) ϕ-plane

Figure 7.2. Mapping properties of the function $\phi(z)$.

We return to the presumed asymptotic form of $U(z)$. Note that the choice of l_n implies that

$$\widehat{g}_n(z) - \tfrac{1}{2}l_n = (n + \tfrac{1}{2})[z^2 - \phi(z)] - \tfrac{1}{2}\log\varphi(z).$$

Therefore we expect $U(z)$ to be asymptotic to

$$
\begin{aligned}
L(z) &\equiv e^{\frac{1}{2}l_n J} N(z) e^{[(n+\frac{1}{2})(z^2-\phi)-\frac{1}{2}\log\varphi]J} \\
&= \frac{e^{\frac{1}{2}l_n J}}{\sqrt{2}(z^2-1)^{1/4}} \begin{bmatrix} \varphi^{1/2} & i\varphi^{-1/2} \\ -i\varphi^{-1/2} & \varphi^{1/2} \end{bmatrix} \begin{bmatrix} \varphi^{-1/2} & 0 \\ 0 & \varphi^{1/2} \end{bmatrix} e^{(n+\frac{1}{2})(z^2-\phi)J}.
\end{aligned} \tag{7.5.7}
$$

The symmetry property (7.4.11) is satisfied by L also.

Lemma 7.5.3 *For* $\mathrm{Re}\, z < 0$, $z \notin [-1, 0)$,

$$L(-z) = (-1)^n J L(z) J. \tag{7.5.8}$$

See Exercise 7; the key step is

$$
\begin{aligned}
e^{(n+\frac{1}{2})\phi(-z)} &= i(-1)^n e^{(n+\frac{1}{2})\phi(z)}; \\
(\varphi(-z))^{1/2} &= -i(\varphi(z))^{1/2}.
\end{aligned} \tag{7.5.9}
$$

It will be convenient to introduce into the product for L the identity matrix in the form

$$\mathbf{1} = \Phi \cdot \Phi^{-1} = \frac{1}{\sqrt{2}} \begin{bmatrix} 1 & -1 \\ -i & -i \end{bmatrix} \cdot \frac{1}{\sqrt{2}} \begin{bmatrix} 1 & i \\ -1 & i \end{bmatrix}.$$

Then

$$\begin{bmatrix} \varphi^{-1/2} & 0 \\ 0 & \varphi^{1/2} \end{bmatrix} \Phi = \frac{1}{\sqrt{2}} \begin{bmatrix} \varphi^{-1/2} & -\varphi^{-1/2} \\ -i\varphi^{1/2} & -i\varphi^{1/2} \end{bmatrix},$$

and

$$\frac{1}{\sqrt{2}} N(z) \begin{bmatrix} \varphi^{-1/2} & -\varphi^{-1/2} \\ -i\varphi^{1/2} & -i\varphi^{1/2} \end{bmatrix} = \frac{1}{(z^2-1)^{1/4}} \begin{bmatrix} 1 & 0 \\ -iz & -i\sqrt{z^2-1} \end{bmatrix}$$

$$= \begin{bmatrix} 1 & 0 \\ -iz & -i \end{bmatrix} \begin{bmatrix} (z^2-1)^{-1/4} & 0 \\ 0 & (z^2-1)^{1/4} \end{bmatrix}.$$

Combining these computations, and remembering the factor $[\xi_n/(z^2-1)]^{1/4}$ in (7.5.6), we write

$$L(z) = \frac{e^{\frac{1}{2} l_n J}}{\sqrt{2}} \begin{bmatrix} 1 & 0 \\ -iz & -i \end{bmatrix} \begin{bmatrix} \left[\dfrac{\xi_n}{z^2-1}\right]^{1/4} & 0 \\ 0 & \left[\dfrac{\xi_n}{z^2-1}\right]^{-1/4} \end{bmatrix} M(z) e^{(n+\frac{1}{2})z^2 J},$$

(7.5.10)

where

$$M(z) = \begin{bmatrix} \xi_n^{-1/4} e^{-(n+\frac{1}{2})\phi} & i\xi_n^{-1/4} e^{(n+\frac{1}{2})\phi} \\ -\xi_n^{1/4} e^{-(n+\frac{1}{2})\phi} & i\xi_n^{1/4} e^{(n+\frac{1}{2})\phi} \end{bmatrix}. \tag{7.5.11}$$

With this and (7.5.5) in mind, we look again at the asymptotics of $\mathrm{Ai}(z)$. The asymptotic formula in (9.9.7) is valid in the complement of $(-\infty, 0]$ and can be differentiated, giving

$$\mathrm{Ai}(\xi_n) \sim \frac{1}{2\sqrt{\pi}} \xi_n^{-1/4} e^{-(n+\frac{1}{2})\phi},$$

$$\mathrm{Ai}'(\xi_n) \sim -\frac{1}{2\sqrt{\pi}} \xi_n^{1/4} e^{-(n+\frac{1}{2})\phi}, \qquad -\pi < \arg \xi_n < \pi.$$

These account asymptotically for z in the first and fourth (open) quadrants Ω_1 and Ω_4.

Let $\omega = e^{2\pi i/3}$, so that $\omega^3 = 1$. In the first quadrant Ω_1,

$$\mathrm{Ai}(\omega^{-1}\xi_n) = \mathrm{Ai}(\omega^2 \xi_n) \sim \frac{e^{i\pi/6}}{2\sqrt{\pi}} \xi_n^{-1/4} e^{(n+\frac{1}{2})\phi};$$

$$\mathrm{Ai}'(\omega^{-1}\xi_n) = \mathrm{Ai}'(\omega^2 \xi_n) \sim -\frac{e^{-i\pi/6}}{2\sqrt{\pi}} \xi_n^{1/4} e^{(n+\frac{1}{2})\phi},$$

while in the fourth quadrant Ω_4,

$$\text{Ai}\,(\omega\xi_n) \sim \frac{e^{-i\pi/6}}{2\sqrt{\pi}}\,\xi_n^{-1/4}\,e^{(n+\frac{1}{2})\phi};$$

$$\text{Ai}\,'(\omega\xi_n) \sim -\frac{e^{i\pi/6}}{2\sqrt{\pi}}\,\xi_n^{1/4}\,e^{(n+\frac{1}{2})\phi}.$$

Since $-\omega^2 e^{i\pi/6} = i = \omega e^{-i\pi/6}$, it is reasonable to replace the matrix M in (7.5.11) in $\Omega_1 \cup \Omega_4$ with $2\sqrt{\pi}\,P(z)$, where

$$P(z) = \begin{bmatrix} \text{Ai}\,(\xi_n) & -\omega^2\text{Ai}\,(\omega^2\xi_n) \\ \text{Ai}\,'(\xi_n) & -\omega\text{Ai}\,'(\omega^2\xi_n) \end{bmatrix}, \qquad z \in \Omega_1; \qquad (7.5.12)$$

$$P(z) = \begin{bmatrix} \text{Ai}\,(\xi_n) & \omega\text{Ai}\,(\omega\xi_n) \\ \text{Ai}\,'(\xi_n) & \omega^2\text{Ai}\,'(\omega\xi_n) \end{bmatrix}, \qquad z \in \Omega_4. \qquad (7.5.13)$$

The relation (9.9.11) and its derivative,

$$\text{Ai}\,(\xi) + \omega\text{Ai}\,(\omega\xi) + \omega^2\text{Ai}\,(\omega^2\xi) = 0;$$

$$\text{Ai}\,'(\xi) + \omega^2\text{Ai}\,'(\omega\xi) + \omega\text{Ai}\,'(\omega^2\xi) = 0$$

imply the following jump relation for real $x > 0$:

$$P_+(x) = P_-(x)\begin{bmatrix} 1 & 1 \\ 0 & 1 \end{bmatrix}. \qquad (7.5.14)$$

Note also that the determinant of $P(z)$ is a Wronskian, hence constant, so from (7.5.11),

$$\det P(z) = \det\left\{\frac{1}{2\sqrt{\pi}}M(z)\right\} = \frac{i}{2\pi}. \qquad (7.5.15)$$

Up to this point, we have argued that a good candidate for the asymptotic form of $U(z)$ on $\Omega_1 \cup \Omega_4$ is

$$\widehat{U}(z) = E(z)P(z)\,e^{(n+\frac{1}{2})z^2 J}, \qquad (7.5.16)$$

where

$$E(z) = \sqrt{2\pi}\,e^{\frac{1}{2}\ln J}\begin{bmatrix} 1 & 0 \\ -iz & -i \end{bmatrix}\begin{bmatrix} \left[\frac{\xi_n}{z^2-1}\right]^{1/4} & 0 \\ 0 & \left[\frac{\xi_n}{z^2-1}\right]^{-1/4} \end{bmatrix}. \qquad (7.5.17)$$

We extend this to the remaining quadrants, Ω_2 and Ω_3, by using the symmetry (7.5.8) to define

$$\widehat{U}(z) = (-1)^n J\widehat{U}(-z)J, \qquad z \in \Omega_2 \cup \Omega_3. \qquad (7.5.18)$$

The next step is to compare the Riemann–Hilbert problem associated with $\widehat{U}$ to that associated with U.

Proposition 7.5.4 *For real* $x \neq 0$,

$$\widehat{U}_+(x) = \widehat{U}_-(x) \begin{bmatrix} 1 & e^{-(2n+1)x^2} \\ 0 & 1 \end{bmatrix}. \tag{7.5.19}$$

Proof Near $z = 1$, $\phi^{2/3} \sim 2(2/3)^{2/3}(z-1)$, so $\xi_n/(z^2-1)$ is holomorphic across $(-1, \infty)$, and the same is true for the matrix function E. Therefore, for $x > 0$, (7.5.14) implies that

$$\begin{aligned}
\widehat{U}_+(x) &= E(x)P_+(x)e^{(n+\frac{1}{2})x^2 J} \\
&= E(x)P_-(x)\begin{bmatrix} 1 & 1 \\ 0 & 1 \end{bmatrix}e^{(n+\frac{1}{2})x^2 J} \\
&= \widehat{U}_-(x)\begin{bmatrix} 1 & e^{-(2n+1)x^2} \\ 0 & 1 \end{bmatrix}.
\end{aligned}$$

For $x < 0$, (7.5.18) implies that

$$\begin{aligned}
\widehat{U}_+(x) &= (-1)^n J\,\widehat{U}_-(-x)J \\
&= (-1)^n J\,\widehat{U}_+(-x)\begin{bmatrix} 1 & -e^{-(2n+1)x^2} \\ 0 & 1 \end{bmatrix}J \\
&= \widehat{U}_-(x)\begin{bmatrix} 1 & e^{-(2n+1)x^2} \\ 0 & 1 \end{bmatrix}.
\end{aligned}$$

$\square$

Proposition 7.5.5 *As* $z \to \infty$,

$$\widehat{U}(z) = [\mathbf{1} + O(|z|^{-1})] \cdot \begin{bmatrix} z^n & 0 \\ 0 & z^{-n} \end{bmatrix}. \tag{7.5.20}$$

Proof Because of (7.5.18), it is enough to consider $z \in \Omega_1 \cup \Omega_4$. In this region, $P(z)$ was chosen so that $2\sqrt{\pi}P(z)$ has the same asymptotics as $M(z)$ in (7.5.10). Therefore $\widehat{U}$ has the same asymptotics here as

$$\begin{aligned}
L(z) &= e^{\frac{1}{2}\ln J}N(z)e^{[(n+\frac{1}{2})(z^2-\phi)-\frac{1}{2}\log\varphi]J} \\
&= e^{\frac{1}{2}\ln J}\left[N(z)e^{g_n J}\right]e^{-\frac{1}{2}\ln J}.
\end{aligned}$$

Now $N(z) = \mathbf{1} + O(|z|^{-1})$, and g_n was chosen so that $U(z)e^{-g_n J} = \mathbf{1} + O(|z|^{-1})$. The relation (7.5.20) follows. $\square$

7.6 Hermite asymptotics

In this section, we use the Riemann–Hilbert problem associated with

$$S(z) = e^{-\frac{1}{2}lnJ} U(z) \widehat{U}(z)^{-1} e^{\frac{1}{2}lnJ}$$

to prove uniform asymptotics for the Hermite polynomials.

We know from (7.5.19) and (7.5.20) that S is holomorphic across $\mathbf{R} \setminus \{0\}$ and $S(z) \sim 1$ as $z \to \infty$. It remains to consider the multiplicative jump across the imaginary axis, oriented in the direction of increasing imaginary part:

$$
\begin{aligned}
v_S(is) = S_-(is)^{-1} S_+(is) &= e^{-\frac{1}{2}lnJ} \widehat{U}_-(is)[\widehat{U}_+(is)]^{-1} e^{\frac{1}{2}lnJ} \\
&= e^{-\frac{1}{2}lnJ} \widehat{U}_-(is) \left\{ (-1)^n [J \widehat{U}_-(-is) J]^{-1} \right\} e^{\frac{1}{2}lnJ}, \quad s \in \mathbf{R}. \quad (7.6.1)
\end{aligned}
$$

Since

$$U(\bar{z}) = J\overline{U(z)}J, \qquad \widehat{U}(\bar{z}) = J\overline{\widehat{U}(z)}J \qquad (7.6.2)$$

(Exercise 8), we need only consider $s > 0$. Then

$$\widehat{U}_-(is) = \lim_{z \to is} \widehat{U}(z), \qquad z \in \Omega_1;$$

$$\widehat{U}_+(is) = (-1)^n \lim_{z \to is} J\,\widehat{U}(z)J, \qquad z \in \Omega_4.$$

We begin by rewriting (7.5.16) for $z \in \Omega_1$:

$$
\widehat{U}(z) = \frac{1}{\sqrt{2}} e^{\frac{1}{2}lnJ}
\begin{bmatrix} 1 & 0 \\ -iz & -i \end{bmatrix}
\begin{bmatrix} (z^2-1)^{-1/4} & 0 \\ 0 & (z^2-1)^{1/4} \end{bmatrix}
\widehat{P}(z) e^{(n+\frac{1}{2})z^2 J},
$$

where

$$
\widehat{P}(z) = 2\sqrt{\pi}
\begin{bmatrix}
\xi_n^{1/4} \mathrm{Ai}(\xi_n) & i(\omega^{-1}\xi_n)^{1/4} \mathrm{Ai}(\omega^{-1}\xi_n) \\
\xi_n^{-1/4} \mathrm{Ai}'(\xi_n) & -i(\omega^{-1}\xi_n)^{-1/4} \mathrm{Ai}'(\omega^{-1}\xi_n)
\end{bmatrix}.
$$

Let us write

$$2\sqrt{\pi}\, \xi^{1/4} \mathrm{Ai}(\xi) = u(\xi) e^{-\frac{2}{3}\xi^{3/2}};$$

$$-2\sqrt{\pi}\, \xi^{-1/4} \mathrm{Ai}'(\xi) = v(\xi) e^{-\frac{2}{3}\xi^{3/2}},$$

so that

$$
\widehat{P}(z) =
\begin{bmatrix} u(\xi_n) & iu(\omega^{-1}\xi_n) \\ -v(\xi_n) & iv(\omega^{-1}\xi_n) \end{bmatrix}
\begin{bmatrix} e^{-(n+\frac{1}{2})\phi} & 0 \\ 0 & e^{(n+\frac{1}{2})\phi} \end{bmatrix}, \quad z \in \Omega_1; \quad (7.6.3)
$$

$$
\widehat{P}(-z) =
\begin{bmatrix} u(\hat{\xi}_n) & iu(\omega\hat{\xi}_n) \\ -v(\hat{\xi}_n) & iv(\omega\hat{\xi}_n) \end{bmatrix}
\begin{bmatrix} e^{-(n+\frac{1}{2})\hat{\phi}} & 0 \\ 0 & e^{(n+\frac{1}{2})\hat{\phi}} \end{bmatrix}, \quad z \in \Omega_2, \quad (7.6.4)
$$

where $\hat{\xi}_n = \xi_n(-z)$, $\hat{\phi} = \phi(-z)$.

It follows from (7.5.15) that

$$\widehat{P}(-z)^{-1} = \frac{1}{2i}\begin{bmatrix} e^{(n+\frac{1}{2})\hat\phi} & 0 \\ 0 & e^{-(n+\frac{1}{2})\hat\phi} \end{bmatrix}\begin{bmatrix} iv(\omega\hat\xi_n) & -iu(\omega\hat\xi_n) \\ v(\hat\xi_n) & u(\hat\xi_n) \end{bmatrix}, \qquad z \in \Omega_2.$$

It follows from this and (7.5.9) that the last equation can be rewritten as

$$\widehat{P}(-z)^{-1} = \frac{(-1)^n}{2}J\begin{bmatrix} e^{(n+\frac{1}{2})\phi} & 0 \\ 0 & e^{-(n+\frac{1}{2})\phi} \end{bmatrix}\begin{bmatrix} iv(\omega\hat\xi_n) & -iu(\omega\hat\xi_n) \\ v(\hat\xi_n) & u(\hat\xi_n) \end{bmatrix}, \qquad z \in \Omega_2.$$

$$(7.6.5)$$

If z is in Ω_2, then $-z$ is in Ω_4 and $[(-z)^2 - 1]^{1/4} = -i[z^2 - 1]^{1/4}$, so

$$\widehat{U}(-z) = \frac{1}{\sqrt{2}}e^{\frac{1}{2}lnJ}\begin{bmatrix} 1 & 0 \\ iz & -i \end{bmatrix}\begin{bmatrix} i(z^2-1)^{-1/4} & 0 \\ 0 & -i(z^2-1)^{1/4} \end{bmatrix}\widehat{P}(-z)e^{(n+\frac{1}{2})z^2J}$$

$$= \frac{i}{\sqrt{2}}e^{\frac{1}{2}lnJ}J\begin{bmatrix} 1 & 0 \\ -iz & -i \end{bmatrix}\begin{bmatrix} (z^2-1)^{-1/4} & 0 \\ 0 & (z^2-1)^{1/4} \end{bmatrix}\widehat{P}(-z)e^{(n+\frac{1}{2})z^2J}.$$

Combining these computations, we obtain for $z \in \Omega_2$

$$\widehat{U}(z)^{-1} = (-1)^nJ\widehat{U}(-z)^{-1}J$$

$$= \frac{1}{i\sqrt{2}}e^{-(n+\frac{1}{2})z^2J}\begin{bmatrix} e^{(n+\frac{1}{2})\phi} & 0 \\ 0 & e^{-(n+\frac{1}{2})\phi} \end{bmatrix}\begin{bmatrix} iv(\omega\hat\xi_n) & -iu(\omega\hat\xi_n) \\ v(\hat\xi_n) & u(\hat\xi_n) \end{bmatrix}$$

$$\times \begin{bmatrix} (z^2-1)^{-1/4} & 0 \\ 0 & (z^2-1)^{1/4} \end{bmatrix}^{-1}\begin{bmatrix} 1 & 0 \\ -iz & -i \end{bmatrix}^{-1}e^{-\frac{1}{2}lnJ}.$$

For real s, $\xi_n(-is) = \overline{\xi_n(is)}$ and $u(\bar\xi) = \overline{u(\xi)}$, $v(\bar\xi) = \overline{v(\xi)}$. Therefore, for $s > 0$,

$$v_S(is) = \frac{1}{2i}K(is)\begin{bmatrix} u(\hat\xi_n) & iu(\omega^2\xi) \\ -v(\hat\xi_n) & iv(\omega^2\hat\xi_n) \end{bmatrix}\begin{bmatrix} \overline{iv(\omega^2\hat\xi_n)} & -\overline{iu(\omega^2\hat\xi_n)} \\ \overline{v(\hat\xi_n)} & \overline{u(\hat\xi_n)} \end{bmatrix}K(is)^{-1},$$

$$(7.6.6)$$

where

$$K(z) = \begin{bmatrix} 1 & 0 \\ -iz & -i \end{bmatrix}\begin{bmatrix} (z^2-1)^{-1/4} & 0 \\ 0 & (z^2-1)^{1/4} \end{bmatrix}$$

$$= \begin{bmatrix} (z^2-1)^{-1/4} & 0 \\ -iz(z^2-1)^{-1/4} & -i(z^2-1)^{1/4} \end{bmatrix}.$$

Because of the symmetries (7.6.2), taking the complex conjugate of (7.6.1) gives the jump on the remainder of the imaginary axis.

Now

$$u(\xi) \sim 1 + \sum_{k=1}^{\infty} \frac{u_k}{\zeta^k};$$

$$v(\xi) \sim 1 + \sum_{k=1}^{\infty} \frac{v_k}{\zeta^k},$$

$$(7.6.7)$$

where $\zeta = \frac{2}{3}\xi^{3/2}$, uniformly for $|\arg \xi| \le \pi - \delta$. Here u_k and v_k are numerical constants; see [312, 313]. In particular, if $\xi = \xi_n(z)$, then $\zeta = (n + \frac{1}{2})\phi(z)$. Moreover, for real s,

$$|\phi(is)| \ge \frac{\pi}{2}; \qquad \phi(is) \sim -s^2.$$

It follows from (7.6.6) and (7.6.7) that v_S has an asymptotic expansion

$$v_S(is) = 1 + \mathbf{J}(s)$$

$$\sim 1 + \sum_{k=1}^{\infty} \frac{1}{(n + \frac{1}{2})^k (1 + s^2)^k} J_k(s), \qquad (7.6.8)$$

where each matrix $J_k(s)$ is $O(|s|)$. Now

$$S(z) - 1 \sim 0 \quad \text{as } z \to \infty;$$

$$(S - 1)_+ - (S - 1)_- = S_- \mathbf{J},$$

so by the Sokhotski–Plemelj formula (Appendix A),

$$S(z) = 1 + \frac{1}{2\pi} \int_{-\infty}^{\infty} (is - z)^{-1} S_-(is) \mathbf{J}(s) \, ds$$

$$\sim 1 + \sum_{k=1}^{\infty} \frac{1}{(n + \frac{1}{2})^k} S_k(z) \qquad (7.6.9)$$

for $\operatorname{Re} z > 0$. It can be shown by induction that $S_k(z) = O(|z|^{-1})$ as $z \to \infty$. Explicit expressions for the coefficient functions $S_k(z)$ can be determined recursively; see the exercises.

The error term

$$E_p(z) = [S(z) - 1] - \sum_{k=1}^{p} \frac{1}{(n + \frac{1}{2})^k} S_k(z) \qquad (7.6.10)$$

can be estimated by successive approximations; see the exercises.

Remark. We used the imaginary axis as the jump contour for convenience of exposition. With this choice, we can obtain asymptotics that are uniform in regions $\operatorname{Re} z \ge \delta > 0$. The components that make up $\widehat{P}$ are holomorphic in the complement of $(-\infty, -1]$ and the asymptotics we used are valid to the right of the curve C_- in Figure 7.2, so there are many other choices of contour. In particular, we could choose the vertical line $\operatorname{Re} z = -\frac{1}{2}$ and get asymptotics that are uniform in a neighborhood of the imaginary axis as well.

Finally we are in a position to state and prove the main result, using the identity

$$U(z) = e^{\frac{1}{2}l_n J} S(z) e^{-\frac{1}{2}l_n J} \widehat{U}(z). \tag{7.6.11}$$

Theorem 7.6.1 *The asymptotic expansion of the rescaled Hermite polynomials is given by*

$$H_n(\sqrt{2n+1}z) = \sqrt{2\pi}\, 2^n (2n+1)^{n/2} \exp\left\{\frac{1}{2}[(2n+1)z^2 + l_n]\right\}$$
$$\times \left[\operatorname{Ai}(\xi_n)A(z,n) - \operatorname{Ai}'(\xi_n)B(z,n)\right], \tag{7.6.12}$$

where

$$A(z,n) \sim \left(\frac{\xi_n}{z^2-1}\right)^{1/4}\left[1 + \sum_{k=1}^{\infty}\frac{A_k(z)}{(n+\frac{1}{2})^k}\right];$$
$$B(z,n) \sim \left(\frac{z^2-1}{\xi_n}\right)^{1/4}\sum_{k=1}^{\infty}\frac{B_k(z)}{(n+\frac{1}{2})^k}. \tag{7.6.13}$$

Here ξ_n and l_n are defined by (7.5.1) and (7.4.14). The coefficients $A_k(z)$ and $B_k(z)$ are analytic functions for $\operatorname{Re} z \neq 0$. This expansion holds uniformly for $\operatorname{Re} z \geq 0$.

A corresponding asymptotic expansion can be obtained for $\operatorname{Re} z \leq 0$ by using the reflection formula $H_n(-z) = (-1)^n H_n(z)$.

Proof We operate with the previous construction, with a jump on the imaginary axis, bearing in mind that we could also use $\operatorname{Re} z = -\frac{1}{2}$ in order to get a result uniform up to the imaginary axis.

Denoting matrix entries of U, S, and $\widehat{U}$ by subscripts, we have from (7.6.11) that

$$H_n(\sqrt{2n+1}z) = 2^n Q_n(\sqrt{2n+1}z) = 2^n(2n+1)^{n/2}U_{11}(z)$$
$$= 2^n(2n+1)^{n/2}\left[S_{11}(z)\widehat{U}_{11}(z) + e^{l_n}S_{12}(z)\widehat{U}_{21}(z)\right]. \tag{7.6.14}$$

According to (7.5.16),

$$\widehat{U}_{11}(z) = e^{\frac{1}{2}l_n}c_n(z)\operatorname{Ai}(\xi_n);$$
$$\widehat{U}_{21}(z) = -ize^{-\frac{1}{2}l_n}c_n(z)\operatorname{Ai}(\xi_n) - ie^{-\frac{1}{2}l_n}d_n(z)\operatorname{Ai}'(\xi_n), \tag{7.6.15}$$

where

$$c_n(z) = \sqrt{2\pi}\, e^{(n+\frac{1}{2})z^2} \left(\frac{\zeta_n}{z^2-1}\right)^{1/4};$$

$$d_n(z) = \sqrt{2\pi}\, e^{(n+\frac{1}{2})z^2} \left(\frac{\zeta_n}{z^2-1}\right)^{-1/4}. \tag{7.6.16}$$

Combining (7.6.14), (7.6.15), and (7.6.16), and with the asymptotic expansion (7.6.9), we obtain (7.6.12) for $\mathrm{Re}\, z > 0$. □

7.7 Exercises

1. Prove Theorem 7.1.2: for a given index α, look for an approximation to $l_n = l_n^{(\alpha)}$ in the form $\exp(ng + r + O(1/n))$. Use Exercise 43 of Chapter 5 to find g: $g' = \psi$, $g = \log z + o(1)$ at infinity. Use the equation

$$z\, l_n''(z) + (\alpha + 1 - 4nz) l_n'(z) + 4n^2 l_n = 0$$

to derive the conditions

$$g'(z) = \psi(z) = 2 - 2\left(1 - \frac{1}{z}\right)^{1/2};$$

$$r'(z) = \frac{-1}{4(z^2 - z)} - \frac{\alpha + 1}{2z} + \frac{\alpha + 1}{2\sqrt{z^2 - z}}.$$

Use the conditions $g \sim \log z$, $r \sim 0$ at infinity to determine g and r and thus complete the proof.

2. Prove Theorem 7.1.3: for a given pair of indices (α, β), approximate $p_n = p_n^{(\alpha,\beta)}$ as in Exercise 1, which leads to

$$g'(z) = \psi(z) = \frac{1}{\sqrt{z^2 - 1}};$$

$$r'(z) = \frac{-a - cz}{2z^2 - 2z} + \frac{c}{2\sqrt{z^2 - 1}},$$

where $a = \alpha - \beta$ and $c = \alpha + \beta + 1$. Use the conditions $g \sim \log z$, $r \sim 0$ at infinity to determine g and r and thus complete the proof.

3. Verify (7.4.11).
4. Prove Proposition 7.4.1. (Hint: differentiate along vertical lines.)
5. Derive the result in Proposition 7.4.2. (Hint: find a matrix Φ such that

$$\begin{bmatrix} 0 & 1 \\ -1 & 0 \end{bmatrix} \Phi = \Phi \begin{bmatrix} -i & 0 \\ 0 & i \end{bmatrix}$$

and solve the scalar problems with multiplicative jumps $-i$ and i and asymptotics 1.)

6. Prove Proposition 7.5.2. In particular, prove that the curves $C_\pm$ characterized by $\operatorname{Re}\phi = 0$ have the stated properties at ± 1 and at ∞. Show also that the picture is qualitatively correct, in that in each quadrant the corresponding branch of $C_\pm$ is always either rising or falling, i.e. along $C_\pm$, parameterized by $s \in \mathbf{R}$,

$$\operatorname{sgn}\left\{\frac{\dot{y}}{\dot{x}}\right\} = \operatorname{sgn} xy,$$

where $z = x(s) + iy(s)$, and the dot denotes differentiation with respect to s.

7. Prove Lemma 7.5.3.
8. Prove (7.6.2).
9. Use (7.6.8) and (7.6.9) to show that the coefficient functions $S_k(z)$ in (7.6.9) satisfy the recursive formula

$$S_k(z) = \frac{1}{2\pi} \int_{-\infty}^{\infty} \left[\sum_{j=1}^{k} (S_{k-j})_-(is)J_j(is)\right] \frac{ds}{is-z}, \qquad k = 1, 2, \ldots$$

In particular, derive the explicit formulas: $S_0(z) = I$,

$$S_1(z) = \frac{1}{2\pi} \int_{-\infty}^{\infty} \frac{J_1(is)}{is-z}\, ds,$$

$$S_2(z) = \frac{1}{2\pi} \int_{-\infty}^{\infty} \left[(S_1)_-(is)J_1(is) + J_2(is)\right] \frac{ds}{is-z}.$$

10. (a) Use (7.6.1) and (7.6.8) to show that the error term $E_p(z)$ in (7.6.10) satisfies

$$(E_p)_+(is) = (E_p)_-(is)\,[I + \mathbf{J}(is)] + F_p(is)$$

for $s \in \mathbf{R}$, where $F_p(z)$ is defined by

$$F_p(z) = \left[I + \sum_{k=1}^{p} \frac{(S_k)_-(z)}{(n+\frac{1}{2})^k}\right] [\mathbf{J}(z) - \mathbf{1}]$$

$$- \sum_{k=1}^{p} \frac{1}{(n+\frac{1}{2})^k}\left[\sum_{j=1}^{k}(S_{k-j})_-(z)J_j(z)\right].$$

(b) Rewrite $F_p(z)$ in (a) as

$$F_p(z) = \left[I + \sum_{k=1}^{p} \frac{(S_k)_-(z)}{(n+\frac{1}{2})^k} \right] \left[J^*(z) - \sum_{k=1}^{p} \frac{J_k(z)}{(n+\frac{1}{2})^k} \right]$$

$$+ \sum_{k=p+1}^{2p} \frac{1}{(n+\frac{1}{2})^k} \left[\sum_{l=k-p}^{p} (S_{k-l})_-(z)J_l(z) \right],$$

and show that the norm of $F_p(z)$ is bounded by $(n+\frac{1}{2})^{-(p+1)}$ as $n \to \infty$, uniformly for $z \in i\mathbf{R}$.

(c) From (a), use the Sokhotski–Plemelj formula to derive the equation

$$E_p(z) = \frac{1}{2\pi} \int_{-\infty}^{\infty} \frac{(E_p)_-(is)J^*(is)}{is - z} ds + \frac{1}{2\pi i} \int_{-\infty}^{\infty} \frac{F_p(is)}{is - z} ds.$$

(d) Define the sequence $\{E_p^{(l)}(z) : l = 0,1,2,\dots\}$ successively by $E_p^{(0)}(z) = 0$ and

$$E_p^{(l)}(z) = \frac{1}{2\pi} \int_{-\infty}^{\infty} \frac{(E_p^{(l-1)})_-(is)J^*(is)}{is - z} ds + \frac{1}{2\pi} \int_{-\infty}^{\infty} \frac{F_p(is)}{is - z} ds$$

for $l = 1,2,\dots$ Show that

$$\lim_{l \to \infty} E_p^{(l)}(z) = E_p(z)$$

is the unique solution of the equation in (c), and that

$$\| E_p^{(l)}(z) \| \leq C_p \left(n + \frac{1}{2} \right)^{-(p+1)}$$

for all z bounded away from $i\mathbf{R}$, where C_p is a positive constant and $\| \cdot \|$ denotes any matrix norm.

7.8 Remarks

The matching method is due to Wang and Wong [418]. It may be applied to any case in which the asymptotics outside the critical interval can be computed, and for which the three-term recurrence takes a suitable (very general) form. Since it does not rely on having a differential equation or an integral representation, or on knowing the weight function, this method has very wide potential applicability. For examples of this type, see [418] and Dai, Ismail, and Wang [91]. A three-term recurrence is a second-order linear difference equation. An exhaustive treatment was given by Birkhoff [46] and Birkhoff and Trjitzinsky [47], but [444] is considerably simpler.

Riemann–Hilbert problems were introduced by Riemann in 1857 [342] (and included by Hilbert in his famous list of problems in 1900), in connection with differential equations in the complex plane. The question in its original form was investigated by Riemann [343], Hilbert [184], and others, and was settled by Plemelj in 1908 [329]; see [196]. For a general modern treatment of this type of problem, see Clancey and Gohberg [81]. Similar problems come up again in the context of inverse scattering theory; see the remarks at the end of Chapter 15.

The Riemann–Hilbert approach to asymptotics was developed by Deift and Zhou [100], in connection with inverse scattering problems. It was applied to orthogonal polynomials by Deift, Kriecherbauer, McLaughlin, Venakides, and Zhou in a series of papers; [99] and [98] give clear presentations of the motivations and techniques, in general, with Hermite polynomials as one of the examples. The treatment of Hermite polynomials in this chapter follows Wong and Zhang [445]. See also the exposition and application to generalized Jacobi polynomials by Kuijlaars [233].

The asymptotic result (5.3.20) is valid only in real intervals that have length $O(n^{-1/2})$ under the scaling (7.1.1). Asymptotics for the Hermite polynomials similar to (7.6.12) were proved by Plancherel and Rotach [328]; see also [392]. The Plancherel–Rotach results are stated separately for several nonoverlapping regions of the complex plane, unlike the uniform, overlapping results here.

Classical turning-point problems for differential equations are approached by making a change of variables to reduce to a standard case, e.g. Airy's equation, as in Section 7.5, Bessel's equation, as in Section 13.2, or Weber's equation, as in Section 15.8. A second method for such problems, applicable when there is an integral representation, is an extension of the steepest descent method introduced by Chester, Friedman, and Ursell [75] in 1957: the phase function in the integral is mapped to a cubic polynomial, leading to an Airy function. Finite difference equations require new methods, which have been developed recently by Wang and Wong [420, 421].

8

Confluent hypergeometric functions

The confluent hypergeometric equation

$$x u''(x) + (c - x) u'(x) - a u(x) = 0$$

has one solution, the Kummer function $M(a, c; x)$, with value 1 at the origin, and a second solution, $x^{1-c} M(a + 1 - c, 2 - c; x)$, which is $\sim x^{1-c}$ at the origin, provided that c is not an integer. A particular linear combination of the two gives a solution $U(a, c; x) \sim x^{-a}$ as $x \to +\infty$. The Laguerre polynomials are particular cases, corresponding to particular values of the parameters. Like the Laguerre polynomials, the general solutions satisfy a number of linear relations involving derivatives and different values of the parameters a and c. Special consideration is required when c is an integer.

In addition to the Laguerre polynomials, functions that can be expressed in terms of Kummer functions include the exponential function, error function, incomplete gamma function, complementary incomplete gamma function, Fresnel integrals, exponential integral, and the sine integral and cosine integral functions.

The closely related parabolic cylinder functions are solutions of *Weber's equation*

$$u''(x) + \left[\mp \frac{x^2}{4} + v + \frac{1}{2} \right] u(x) = 0.$$

Three solutions are obtained by utilizing the three solutions of the confluent hypergeometric equation mentioned above.

A gauge transformation removes the first-order term of the confluent hypergeometric equation and converts it to *Whittaker's equation*:

$$u''(x) + \left[-\frac{1}{4} + \frac{\kappa}{x} + \frac{1 - 4\mu^2}{4x^2} \right] u(x) = 0.$$

Again there are three solutions, related to the three solutions of the confluent hypergeometric equation, which satisfy a number of relations. The Coulomb wave functions are special cases.

8.1 Kummer functions

The confluent hypergeometric equation is the equation

$$xu''(x) + (c - x)u'(x) - au(x) = 0. \tag{8.1.1}$$

As noted in Chapter 1, it has a solution represented by the power series

$$M(a, c; x) = \sum_{n=0}^{\infty} \frac{(a)_n}{(c)_n n!} x^n. \tag{8.1.2}$$

This is only defined for $c \neq 0, -1, -2, \ldots$, since otherwise the denominators vanish for $n > c$. The function (8.1.2) is known as the *Kummer function*. The notation $\Phi(a, c; x)$ is common in the Russian literature. Another notation is ${}_1F_1(a, c; x)$, a special case of the class of functions ${}_pF_q$ introduced in Chapter 12. Thus

$$M(a, c; x) = \Phi(a, c; x) = {}_1F_1(a, c; x), \quad c \neq 0, -1, -2, \ldots$$

If $a = -m$ is an integer ≤ 0, then $(a)_n = 0$ for $n > m$: $M(-m, c; x)$ is a polynomial of degree m. If $c > 0$, it is a multiple of the Laguerre polynomial $L_m^{(c-1)}(x)$; see the next section. In general, the ratio test shows that $M(a, c; x)$ is an entire function of x. (The adjective "confluent" here is explained in Exercise 1 of Chapter 10.)

Assuming that $\operatorname{Re} c > \operatorname{Re} a > 0$, we obtain an integral representation, using the identity

$$\frac{(a)_n}{(c)_n} = \frac{\Gamma(a+n)}{\Gamma(a)} \frac{\Gamma(c)}{\Gamma(c+n)} = \frac{\Gamma(c)}{\Gamma(a)\Gamma(c-a)} \mathrm{B}(a+n, c-a)$$

$$= \frac{\Gamma(c)}{\Gamma(a)\Gamma(c-a)} \int_0^1 s^{n+a-1}(1-s)^{c-a-1} ds.$$

As in Section 1.2, this leads to the computation

$$M(a, c; x) = \frac{\Gamma(c)}{\Gamma(a)\Gamma(c-a)} \int_0^1 \left\{ s^{a-1}(1-s)^{c-a-1} \sum_{n=0}^{\infty} \frac{(sx)^n}{n!} \right\} ds$$

$$= \frac{\Gamma(c)}{\Gamma(a)\Gamma(c-a)} \int_0^1 s^{a-1}(1-s)^{c-a-1} e^{sx} ds, \quad \operatorname{Re} c > \operatorname{Re} a > 0. \tag{8.1.3}$$

For another integral representation, due to Barnes, see (12.3.14).

Let D denote the operator

$$D = x\frac{d}{dx}.$$

After multiplication by x, the confluent hypergeometric equation (8.1.1) has the form

$$D(D+c-1)u - x(D+a)u = 0. \tag{8.1.4}$$

Under the gauge transformation $u(x) = x^b v(x)$, the operator D acting on the function u becomes the operator $D+b$ acting on v:

$$x^{-b} D\{x^b v(x)\} = (D+b)v(x). \tag{8.1.5}$$

Letting $b = 1 - c$, (8.1.4) becomes

$$D(D+1-c)v - x(D+a+1-c)v = 0.$$

Therefore, if $c \neq 2, 3, 4, \ldots$,

$$x^{1-c} M(a+1-c, 2-c; x) \tag{8.1.6}$$

is a second solution of (8.1.1). (This is easily checked directly.) If c is an integer $\neq 1$, then one of the two functions (8.1.2), (8.1.6) is not defined. If $c = 1$, the functions coincide. We consider these cases in Section 8.3.

Denote the solution (8.1.2) by $M_1(x)$ and the solution (8.1.6) by $M_2(x)$. Equation (8.1.1) implies that the Wronskian

$$W(x) = W(M_1, M_2)(x) \equiv M_1(x)M_2'(x) - M_2(x)M_1'(x)$$

satisfies the first-order equation

$$x W'(x) = (x - c) W(x),$$

so $W(x) = Ax^{-c}e^x$ for some constant A. We may evaluate A by looking at the behavior of M_1 and M_2 as $x \to 0$:

$$M_1(x) = 1 + O(x), \qquad M_2(x) = x^{1-c}[1 + O(x)].$$

The result is

$$W(M_1, M_2)(x) = (1 - c)x^{-c}e^x. \tag{8.1.7}$$

Write (8.1.1) for $x \neq 0$ as

$$u''(x) - u'(x) + \frac{c}{x}u'(x) - \frac{a}{x}u(x) = 0.$$

As $x \to \infty$, we may expect any solution of (8.1.1) to resemble a solution of $v'' - v' = 0$ – a linear combination of e^x and 1. In fact, the change of variables $(1 - s)x = t$ converts the integral in (8.1.3) to

$$x^{a-c}e^x \int_0^x e^{-t}\left(1 - \frac{t}{x}\right)^{a-1} t^{c-a-1}\, dt.$$

This last integral has the limit

$$\int_0^\infty e^{-t} t^{c-a-1} \, dt = \Gamma(c-a)$$

as $\mathrm{Re}\, x \to +\infty$, giving the asymptotics

$$M(a,c;x) \sim \frac{\Gamma(c)}{\Gamma(a)} x^{a-c} e^x \quad \text{as} \quad \mathrm{Re}\, x \to +\infty, \tag{8.1.8}$$

when $\mathrm{Re}\, c > \mathrm{Re}\, a > 0$. Expanding $(1 - t/x)^{a-1}$ gives the asymptotic expansion

$$M(a,c;x) \sim \frac{\Gamma(c)}{\Gamma(a)\Gamma(c-a)} x^{a-c} e^x \sum_{n=0}^\infty \int_0^\infty e^{-t} \frac{(1-a)_n}{n!} \left(\frac{t}{x}\right)^n t^{c-a-1} \, dt$$

$$= \frac{\Gamma(c)}{\Gamma(a)} x^{a-c} e^x \sum_{n=0}^\infty \frac{(1-a)_n (c-a)_n}{n!} \frac{1}{x^n} \tag{8.1.9}$$

for $\mathrm{Re}\, c > \mathrm{Re}\, a > 0$. The results in Section 8.5 can be used to show that this asymptotic result is valid for all indices (a,c) with $c \neq 0, -1, -2, \ldots$

It is easily seen that if $v(x)$ is a solution of (8.1.1), with the index a replaced by $c - a$,

$$x v''(x) + (c-x) v'(x) - (c-a) v(x) = 0,$$

then $u(x) = e^x v(-x)$ is a solution of (8.1.1). Comparing values at $x = 0$ establishes *Kummer's identity*:

$$e^x M(c-a,c;-x) = M(a,c;x). \tag{8.1.10}$$

A second identity due to Kummer is

$$M(a,2a;4x) = e^{2x} {}_0F_1\left(a+\frac{1}{2};x^2\right) \equiv e^{2x} \sum_{n=0}^\infty \frac{x^{2n}}{(a+\frac{1}{2})_n n!}; \tag{8.1.11}$$

see Exercise 16 of Chapter 10.

It will be shown in Chapter 13 that $M(a,c;x)$ has the following asymptotic behavior as $a \to -\infty$:

$$M(a,c;x) = \frac{\Gamma(c)}{\sqrt{\pi}} \left(\frac{1}{2}cx - ax\right)^{\frac{1}{4}-\frac{1}{2}c} e^{\frac{1}{2}x}$$

$$\times \left[\cos\left(\sqrt{2cx-4ax} - \frac{1}{2}c\pi + \frac{1}{4}\pi\right) + O\left(|a|^{-\frac{1}{2}}\right)\right].$$

This was proved by Erdélyi [114] and Schmidt [358], generalizing Fejér's result for Laguerre polynomials (5.4.12).

8.2 Kummer functions of the second kind

To find a solution u of the confluent hypergeometric equation (8.1.1) that does not have exponential growth as $\operatorname{Re} x \to +\infty$, we note that such a solution should be expressible as a Laplace transform (see Exercise 23 of Chapter 5). Thus we look for an integral representation

$$u(x) = [\mathcal{L}\varphi](x) = \int_0^\infty e^{-xt} \varphi(t)\, dt$$

for some integrable function $\varphi(t)$. If a function $u(x)$ has this form, then

$$xu''(x) + (c-x)u'(x) - au(x) = \int_0^\infty e^{-xt}\left[xt^2 + (x-c)t - a\right]\varphi(t)\,dt$$

$$= \int_0^\infty \left[xe^{-xt}\right](t^2+t)\varphi(t)\,dt - \int_0^\infty e^{-xt}(a+ct)\varphi(t)\,dt.$$

Suppose that $t\varphi(t)$ has limit 0 as $t \to 0$ and that φ' is integrable. Then integration by parts of the first integral in the last line leads to

$$xu''(x) + (c-x)u'(x) - au(x)$$

$$= \int_0^\infty e^{-xt}\left\{\left[(t^2+t)\varphi(t)\right]' - (a+ct)\varphi(t)\right\}dt.$$

It follows that u is a solution of (8.1.1) if the expression in braces vanishes. This condition is equivalent to

$$\frac{\varphi'(t)}{\varphi(t)} = \frac{(c-2)t + a - 1}{t^2+t} = \frac{c-a-1}{1+t} + \frac{a-1}{t},$$

or $\varphi(t) = A t^{a-1}(1+t)^{c-a-1}$. Then $t\varphi(t)$ has limit 0 as $t \to 0$ if $\operatorname{Re} a > 0$. This argument shows that we may obtain a solution $U(a,c;\cdot)$ of (3.0.1) by taking

$$U(a,c;x) = \frac{1}{\Gamma(a)} \int_0^\infty e^{-xt} t^{a-1}(1+t)^{c-a-1}\, dt$$

$$= \frac{x^{-a}}{\Gamma(a)} \int_0^\infty e^{-s} s^{a-1}\left(1+\frac{s}{x}\right)^{c-a-1} ds, \quad \operatorname{Re} a > 0. \quad (8.2.1)$$

The second form leads to a full asymptotic series expansion for U as $\operatorname{Re} x \to +\infty$. The first term gives

$$U(a,c;x) \sim x^{-a} \quad \text{as } \operatorname{Re} x \to +\infty, \quad \operatorname{Re} a > 0. \quad (8.2.2)$$

The full expansion is

$$U(a,c;x) \sim x^{-a} \sum_{n=0}^\infty \frac{(a)_n (a+1-c)_n}{n!} \frac{(-1)^n}{x^n}, \quad \operatorname{Re} a > 0. \quad (8.2.3)$$

As noted below, U can be extended to all index pairs (a, c) with c not an integer. The results in Section 8.5, together with (8.2.5) below, make it possible to show that the asymptotic results (8.2.2) and (8.2.3) extend to all such index pairs.

The function U is called the *confluent hypergeometric function of the second kind*. It is sometimes denoted by Ψ:

$$\Psi(a, c; x) = U(a, c; x).$$

A different integral representation is given in (12.3.15).

The solution (8.2.1) must be a linear combination of the solutions (8.1.2) and (8.1.6),

$$U(a, c; x) = A(a, c) M(a, c; x)$$
$$+ B(a, c) x^{1-c} M(a + 1 - c, 2 - c; x), \qquad (8.2.4)$$

with coefficients $A(a, c)$ and $B(a, c)$ that are meromorphic functions of a and c. The coefficients can be determined by considering the integral representation in two special cases.

If $1 - c > 0$, then the second summand on the right in (8.2.4) vanishes at $x = 0$, while the value of (8.2.1) at $x = 0$ is

$$\frac{1}{\Gamma(a)} \int_0^\infty t^a (1 + t)^{c-a-1} \frac{dt}{t} = \frac{B(a, 1 - c)}{\Gamma(a)} = \frac{\Gamma(1 - c)}{\Gamma(a + 1 - c)}.$$

Therefore $A(a, c) = \Gamma(1 - c) / \Gamma(a + 1 - c)$ for $\operatorname{Re} a > 0$, $1 - c > 0$, and, by analytic continuation, for all noninteger values of c.

If $c - 1 > 0$, then $x^{c-1} M(a, c; x)$ vanishes at $x = 0$, while

$$x^{c-1} U(a, c; x) = x^{c-1} \frac{x^{-a}}{\Gamma(a)} \int_0^\infty e^{-t} t^{a-1} \left(1 + \frac{t}{x}\right)^{c-a-1} dt$$
$$= \frac{1}{\Gamma(a)} \int_0^\infty e^{-t} t^{a-1} (x + t)^{c-a-1} dt$$
$$\rightarrow \frac{\Gamma(c - 1)}{\Gamma(a)} \quad \text{as } x \rightarrow 0+.$$

Therefore $B(a, c) = \Gamma(c - 1) / \Gamma(a)$ for $\operatorname{Re} a > 0$ and for all noninteger values of c.

We have proved

$$U(a, c; x) = \frac{\Gamma(1 - c)}{\Gamma(a + 1 - c)} M(a, c; x)$$
$$+ \frac{\Gamma(c - 1)}{\Gamma(a)} x^{1-c} M(a + 1 - c, 2 - c; x) \qquad (8.2.5)$$

for all values of $\operatorname{Re} a > 0$ and c not an integer. Conversely, we may use (8.2.5) to remove the limitation $\operatorname{Re} a > 0$ and *define* U for all a and for all noninteger values of c. It follows from (8.2.5) and (8.1.10) that

$$U(a,c;x) = x^{1-c} U(a+1-c, 2-c; x); \tag{8.2.6}$$

see Exercise 5.

The Wronskian of the solutions M and U may be computed from (8.1.8) and (8.2.2), or from (8.2.5) and (8.1.7). The result is

$$W\left(M(a,c;\cdot), U(a,c;\cdot)\right)(x) = -\frac{\Gamma(c)}{\Gamma(a)} x^{-c} e^x; \tag{8.2.7}$$

see Exercise 6.

It follows from (8.1.10) that a solution of (8.1.1) that decays exponentially as $x \to -\infty$ is

$$\widetilde{U}(a,c;x) = e^x U(c-a,c;-x)$$
$$= \frac{1}{\Gamma(c-a)} \int_0^\infty e^{x(1+t)} t^{c-a-1} (1+t)^{a-1} \, dt$$
$$= \frac{(-x)^{a-c} e^x}{\Gamma(c-a)} \int_0^\infty e^{-s} s^{c-a-1} \left(1 - \frac{s}{x}\right)^{a-1} \, ds \tag{8.2.8}$$

for $\operatorname{Re} c > \operatorname{Re} a > 0$. Then $\widetilde{U}(a,c;x) = e^x(-x)^{a-c}(1 + O(-1/x))$ as $x \to -\infty$. Again, the results in Section 8.5 show that this asymptotic result extends to all index pairs (a,c), c not an integer.

In terms of the solutions (8.1.2) and (8.1.6), we use (8.2.5) and (8.1.10) to obtain

$$\widetilde{U}(a,c;x) = \frac{\Gamma(1-c)}{\Gamma(1-a)} e^x M(c-a,c;-x)$$
$$+ \frac{\Gamma(c-1)}{\Gamma(c-a)} e^x (-x)^{1-c} M(1-a, 2-c; -x)$$
$$= \frac{\Gamma(1-c)}{\Gamma(1-a)} M(a,c;x)$$
$$+ \frac{\Gamma(c-1)}{\Gamma(c-a)} (-x)^{1-c} M(1+a-c, 2-c; x). \tag{8.2.9}$$

It will be shown in Chapter 13 that a consequence of (8.1.12) and (8.2.5) is that U has the following asymptotic behavior as $a \to -\infty$:

$$U(a,c;x) = \frac{\Gamma(\frac{1}{2}c - a + \frac{1}{4})}{\sqrt{\pi}} x^{\frac{1}{4} - \frac{1}{2}c} e^{\frac{1}{2}x}$$
$$\times \left[\cos\left(\sqrt{2cx - 4ax} - \frac{1}{2}c\pi + a\pi + \frac{1}{4}\pi\right) + O\left(|a|^{-\frac{1}{2}}\right)\right]. \tag{8.2.10}$$

8.3 Solutions when c is an integer

As noted above, for $c \neq 1$ an integer, only one of the two functions (8.1.2) and (8.1.6) is defined, while if $c = 1$ they coincide. When c is an integer, a second solution can be obtained from the solution of the second kind U. In view of (8.2.6), it is enough to consider the case $c = m$ a positive integer.

Assume first that a is not an integer. We begin by modifying the solution M so that it is defined for all c. Assuming first that $c \neq 0, -1, -2, \dots$, let

$$N(a,c;x) \equiv \frac{\Gamma(a)}{\Gamma(c)} M(a,c;x) = \sum_{n=0}^{\infty} \frac{\Gamma(a+n)}{\Gamma(c+n)n!} x^n.$$

The series expansion is well-defined for all values of the parameters, except for a a nonpositive integer. Note that if $c = -k$ is an integer ≤ 0, the first $k+1$ terms of the series vanish. In particular, if $c = m$ is a positive integer,

$$N(a+1-m, 2-m; x) = \sum_{n=m-1}^{\infty} \frac{\Gamma(a+1-m+n)}{\Gamma(2-m+n)n!} x^n$$

$$= x^{m-1} \sum_{k=0}^{\infty} \frac{\Gamma(a+k)}{\Gamma(m+k)k!} x^k$$

$$= x^{m-1} N(a,m;x). \tag{8.3.1}$$

For noninteger c and a, we use the reflection formula (2.2.7) to rewrite (8.2.5) as

$$U(a,c;x) = \frac{\pi}{\sin \pi c \, \Gamma(a) \Gamma(a+1-c)}$$

$$\times \left[N(a,c;x) - x^{1-c} N(a+1-c, 2-c; x) \right]. \tag{8.3.2}$$

In view of (8.3.1), the difference in brackets has limit 0 as $c \to m$, m a positive integer. It follows that $U(a,c;x)$ has a limit as $c \to m$, and the limit is given by l'Hôpital's rule: for a not an integer,

$$U(a,m;x) = \frac{(-1)^m}{\Gamma(a)\Gamma(a+1-m)}$$

$$\times \frac{\partial}{\partial c}\Big|_{c=m} \left[N(a,c;x) - x^{1-c} N(a+1-c, 2-c; x) \right].$$

Therefore, for noninteger values of a and positive integer values of m, calculating the derivative shows that

$$U(a,m;x) = \frac{(-1)^m}{\Gamma(a+1-m)(m-1)!}\left\{ M(a,m;x)\log x \right.$$

$$\left. + \sum_{n=0}^{\infty} \frac{(a)_n}{(m)_n n!} [\psi(a+n) - \psi(n+1) - \psi(m+n)]\, x^n \right\}$$

$$+ \frac{(m-2)!}{\Gamma(a)} x^{1-m} \sum_{n=0}^{m-2} \frac{(a+1-m)_n}{(2-m)_n n!} x^n, \qquad (8.3.3)$$

where $\psi(b) = \Gamma'(b)/\Gamma(b)$ and the last sum is taken to be zero if $m = 1$.

The function in (8.3.3) is well-defined for all values of a. By a continuity argument, it is a solution of (8.1.1) for all values of a and c, with $c = m$, and all values of x not in the interval $(-\infty, 0]$. If a is not an integer less than m, $U(a,m;x)$ has a logarithmic singularity at $x = 0$ and is therefore independent of the solution $M(a,m;x)$. If a is a positive integer less than m, then the coefficient of the term in braces vanishes and $U(a,m;x)$ is the finite sum, which is a rational function that is again independent of $M(a,m;x)$.

If a is a nonpositive integer, then $U(a,m;x) \equiv 0$. To obtain a solution in this case we start with noninteger a and multiply (8.3.3) by $\Gamma(a)$. Since

$$\frac{\Gamma(a)}{\Gamma(a+1-m)} = (a+1-m)_{m-1}$$

has a nonzero value for $a = 0, -1, -2, \ldots$, the limiting value of $\Gamma(a)\,U(a,c;x)$ is a solution of (8.1.1) that has a logarithmic singularity at $x = 0$.

8.4 Special cases

The exponential function

$$e^x = \sum_{n=0}^{\infty} \frac{(c)_n}{(c)_n n!} x^n = M(c,c;x), \quad c \neq 0, -1, -2, \ldots, \qquad (8.4.1)$$

is one example of a confluent hypergeometric function.

The Laguerre polynomial $L_n^{(\alpha)}(x)$ satisfies (8.1.1) with $c = \alpha + 1$ and $a = -n$. Since the constant term is $(\alpha+1)_n/n!$, it follows that

$$L_n^{(\alpha)}(x) = \frac{(\alpha+1)_n}{n!} M(-n, \alpha+1; x). \qquad (8.4.2)$$

Combining this with the identities (5.3.22) and (5.3.23), that relate Hermite and Laguerre polynomials, gives

$$H_{2m}(x) = (-1)^m 2^{2m} \left(\frac{1}{2}\right)_m M\left(-m, \frac{1}{2}; x^2\right);$$

$$H_{2m+1}(x) = (-1)^m 2^{2m+1} \left(\frac{3}{2}\right)_m x M\left(-m, \frac{3}{2}; x^2\right).$$

(8.4.3)

These identities, together with the identity (5.3.21) relating the Hermite polynomials $\{H_n\}$ and the modified version $\{He_n\}$, give

$$He_{2m}(x) = (-1)^m 2^m \left(\frac{1}{2}\right)_m M\left(-m, \frac{1}{2}; \frac{1}{2}x^2\right);$$

$$He_{2m+1}(x) = (-1)^m 2^{m+\frac{1}{2}} \left(\frac{3}{2}\right)_m \frac{x}{\sqrt{2}} M\left(-m, \frac{3}{2}; \frac{1}{2}x^2\right).$$

(8.4.4)

Other special cases of Kummer functions include the error function, the incomplete gamma function, and the complementary incomplete gamma function of Section 2.6:

$$\operatorname{erf} x \equiv \frac{2}{\sqrt{\pi}} \int_0^x e^{-t^2} dt = \frac{2x}{\sqrt{\pi}} M\left(\frac{1}{2}, \frac{3}{2}; -x^2\right);$$

(8.4.5)

$$\gamma(a, x) \equiv \int_0^x e^{-t} t^{a-1} dt = \frac{x^a}{a} M(a, a+1; -x);$$

(8.4.6)

$$\Gamma(a, x) \equiv \int_x^\infty e^{-t} t^{a-1} dt = x^a e^{-x} U(1, a+1; x).$$

(8.4.7)

Among other such functions are the Fresnel integrals

$$C(x) \equiv \int_0^x \cos\left(\tfrac{1}{2}t^2\pi\right) dt;$$

$$S(x) \equiv \int_0^x \sin\left(\tfrac{1}{2}t^2\pi\right) dt,$$

and the exponential integral, cosine integral, and sine integral functions

$$\operatorname{Ei}(z) \equiv \int_{-\infty}^z \frac{e^t \, dt}{t}, \quad z \notin [0, \infty);$$

$$\operatorname{Ci}(z) \equiv \int_\infty^z \frac{\cos t \, dt}{t}, \quad z \notin (-\infty, 0];$$

$$\operatorname{Si}(z) \equiv \int_0^z \frac{\sin t \, dt}{t}, \quad z \notin (-\infty, 0].$$

These functions are related to the Kummer functions for special values of the indices:

$$C(x) = \frac{x}{2} \left[M\left(\frac{1}{2}, \frac{3}{2}; \frac{1}{2}ix^2\pi\right) + M\left(\frac{1}{2}, \frac{3}{2}; -\frac{1}{2}ix^2\pi\right) \right]; \qquad (8.4.8)$$

$$S(x) = \frac{x}{2i} \left[M\left(\frac{1}{2}, \frac{3}{2}; \frac{1}{2}ix^2\pi\right) - M\left(\frac{1}{2}, \frac{3}{2}; -\frac{1}{2}ix^2\pi\right) \right], \qquad (8.4.9)$$

and

$$\mathrm{Ei}(-z) = -e^{-z} U(1, 1; z); \qquad (8.4.10)$$

$$\mathrm{Ci}(x) = -\frac{1}{2} \left[e^{-ix} U(1, 1; ix) + e^{ix} U(1, 1, -ix) \right]; \qquad (8.4.11)$$

$$\mathrm{Si}(x) = \frac{1}{2i} \left[e^{-ix} U(1, 1; ix) - e^{ix} U(1, 1, -ix) \right] + \frac{\pi}{2}. \qquad (8.4.12)$$

See Exercises 10–17 for the identities (8.4.5)–(8.4.12).

8.5 Contiguous functions

Two Kummer functions are said to be *contiguous* if the first of the two indices is the same for each function and the second indices differ by ± 1, or if the second indices are the same and the first indices differ by ± 1. Any triple of contiguous Kummer functions satisfies a linear relationship. For convenience, fix a, c, x, and let

$$M = M(a, c; x); \quad M(a\pm) = M(a\pm 1, c; x); \quad M(c\pm) = M(a, c\pm 1; x).$$

There are six basic linear relations: the relations between M and any of the six pairs chosen from the four contiguous functions $M(a\pm)$, $M(c\pm)$.

Denote the coefficient of x^n in the expansion of M by

$$\varepsilon_n = \frac{(a)_n}{(c)_n n!}.$$

As above, let $D = x(d/dx)$. The coefficients of x^n in the expansions of DM, $M(a+)$, and $M(c-)$, respectively, are

$$n\varepsilon_n, \qquad \frac{a+n}{a}\varepsilon_n, \qquad \frac{c-1+n}{c-1}\varepsilon_n.$$

Therefore

$$DM = a[M(a+) - M] = (c-1)[M(c-) - M], \qquad (8.5.1)$$

and

$$(a - c + 1)M = aM(a+) - (c-1)M(c-). \qquad (8.5.2)$$

The coefficient of x^n in the expansion of xM is

$$\frac{(c+n-1)n}{a+n-1}\varepsilon_n = \left[n + (c-a) - (c-a)\frac{a-1}{a+n-1}\right]\varepsilon_n$$

so

$$xM = DM + (c-a)M - (c-a)M(a-).$$

Combining this with (8.5.1) gives

$$(2a - c + x)M = aM(a+) - (c-a)M(a-); \tag{8.5.3}$$

$$(a - 1 + x)M = (c-1)M(c-) - (c-a)M(a-). \tag{8.5.4}$$

The coefficient of x^n in the expansion of M' is

$$\frac{a+n}{c+n}\varepsilon_n = \left[1 + \frac{a-c}{c}\frac{c}{c+n}\right]\varepsilon_n,$$

so

$$cM' = cM - (c-a)M(c+).$$

Multiplying by x gives

$$cDM = cxM - (c-a)xM(c+),$$

and combining this with (8.5.1) gives

$$c(a + x)M = acM(a+) + (c-a)xM(c+); \tag{8.5.5}$$

$$c(c - 1 + x)M = c(c-1)M(c-) + (c-a)xM(c+). \tag{8.5.6}$$

Eliminating $M(c-)$ from (8.5.4) and (8.5.6) gives

$$cM = cM(a-) + xM(c+). \tag{8.5.7}$$

The relations (8.5.2)–(8.5.7) are the six basic relations mentioned above.

Contiguous relations for the solution $U(a,c;x)$ can be derived from those for $M(a,c;x)$, using (8.2.5). However, it is simpler to start with $\mathrm{Re}\,a > 0$ and use the integral representation (8.2.1). The identities extend to general values of the parameter by analytic continuation. We use the same notational conventions as before:

$$U = U(a,c;x); \quad U(a\pm) = U(a\pm 1,c;x); \quad U(c\pm) = U(a,c\pm 1;x).$$

Differentiating the first integral in (8.2.1) with respect to x gives the two identities

$$U'(a,c;x) = -aU(a+1,c+1;x); \qquad (8.5.8)$$

$$U'(a,c;x) = U(a,c;x) - U(a,c+1;x). \qquad (8.5.9)$$

Replacing c by $c-1$ and combining these two identities gives

$$U = aU(a+) + U(c-). \qquad (8.5.10)$$

Integrating the identity

$$\frac{d}{dt}\left[e^{-xt}t^{a-1}(1+t)^{c-a}\right] = e^{-xt}\left[-xt^{a-1}(1+t)^{c-a} + (a-1)t^{a-2}(1+t)^{c-a}\right.$$
$$\left. + (c-a)t^{a-1}(1+t)^{c-a-1}\right]$$

for $0 < t < \infty$ gives

$$(c-a)U = xU(c+) - U(a-). \qquad (8.5.11)$$

The two identities (8.5.8) and (8.5.9) give

$$aU(a+1,c+1;x) = U(c+) - U.$$

Combining this with (8.5.11), with $a+1$ in place of a, gives

$$(x+a)U = xU(c+) + a(a+1-c)U(a+). \qquad (8.5.12)$$

Eliminating $U(a+)$ from (8.5.10) and (8.5.12) gives

$$(x+c-1)U = xU(c+) + (c-a-1)U(c-). \qquad (8.5.13)$$

Eliminating $U(c+)$ from (8.5.11) and (8.5.12) gives

$$(x+2a-c)U = U(a-) + a(1+a-c)U(a+). \qquad (8.5.14)$$

Finally, eliminating $U(a+)$ from (8.5.10) and (8.5.14) gives

$$(a+x-1)U = U(a-) + (c-a-1)U(c-). \qquad (8.5.15)$$

8.6 Parabolic cylinder functions

A parabolic cylinder function, or Weber function, is a solution of one of the Weber equations

$$u''(x) - \frac{x^2}{4}u(x) + \left(v + \frac{1}{2}\right)u(x) = 0; \qquad (8.6.1)$$

$$u''(x) + \frac{x^2}{4}u(x) + \left(v + \frac{1}{2}\right)u(x) = 0. \qquad (8.6.2)$$

If u is a solution of (8.6.1), then $v(x) = u(e^{\frac{1}{4}i\pi}x)$ is a solution of (8.6.2) with $v + \frac{1}{2}$ replaced by $i(v + \frac{1}{2})$, and vice versa. We shall consider only (8.6.1).

Suppose that u is a solution of (8.6.1) that is holomorphic near $x = 0$. The coefficients of the expansion $u(x) = \sum b_n x^n$ can be computed from the first two terms b_0, b_1 by setting $b_{-2} = b_{-1} = 0$ and using the three-term recurrence

$$(n+2)(n+1)b_{n+2} = -\left(v + \frac{1}{2}\right)b_n + \frac{1}{4}b_{n-2}. \tag{8.6.3}$$

In particular, taking $b_0 \neq 0$, $b_1 = 0$ determines an even solution, while $b_0 = 0$, $b_1 \neq 0$ determines an odd solution.

The gauge transformation $u(x) = e^{-\frac{1}{4}x^2}v(x)$ converts (8.6.1) to

$$v''(x) - x v'(x) + v\, v(x) = 0. \tag{8.6.4}$$

For $v = n$ a nonnegative integer, one solution of this equation is the modified Hermite polynomial He_n.

Each of the equations (8.6.1), (8.6.2), and (8.6.4) is unchanged under the coordinate change $x \to -x$. Therefore each has an even solution and an odd solution. In the case of (8.6.4), we know that the polynomial solutions are even or odd according to whether v is an even or odd nonnegative integer. Let us look for an even solution of the modified equation (8.6.4) in the form $v(x) = w(x^2/2)$. Then (8.6.4) is equivalent to

$$y w''(y) + \left(\frac{1}{2} - y\right)w'(y) + \frac{1}{2}v\, w(y) = 0. \tag{8.6.5}$$

This is the confluent hypergeometric equation (8.1.1) with $c = \frac{1}{2}$, $a = -\frac{1}{2}v$. Thus the even solutions of (8.6.1) are multiples of

$$Y_{v1}(x) = e^{-\frac{1}{4}x^2}M\left(-\frac{1}{2}v, \frac{1}{2}; \frac{1}{2}x^2\right).$$

According to (8.1.6), a second solution of (8.6.5) is

$$\sqrt{y}M\left(-\frac{1}{2}v + \frac{1}{2}, \frac{3}{2}; y\right).$$

Therefore the odd solutions of (8.6.1) are multiples of

$$Y_{v2}(x) = e^{-\frac{1}{4}x^2}\frac{x}{\sqrt{2}}M\left(-\frac{1}{2}v + \frac{1}{2}, \frac{3}{2}; \frac{1}{2}x^2\right).$$

It follows from (8.1.8) that the solutions Y_{v1} and Y_{v2} of (8.6.5) grow like $e^{\frac{1}{4}x^2}$ as $|x| \to \infty$. To obtain a solution with decay at ∞, we use the Kummer function of the second kind $U(-\frac{1}{2}v, \frac{1}{2}; \cdot)$ instead. The standard normalized solution is

$$D_\nu(x) = 2^{\frac{1}{2}\nu} e^{-\frac{1}{4}x^2} U\left(-\frac{1}{2}\nu, \frac{1}{2}; \frac{1}{2}x^2\right)$$

$$= 2^{\frac{1}{2}\nu} \left\{ \frac{\Gamma(\frac{1}{2})}{\Gamma(\frac{1}{2} - \frac{1}{2}\nu)} Y_{\nu 1}(x) + \frac{\Gamma(-\frac{1}{2})}{\Gamma(-\frac{1}{2}\nu)} Y_{\nu 2}(x) \right\}. \qquad (8.6.6)$$

For $\operatorname{Re}\nu < 0$, (8.2.1) implies the integral representation

$$D_\nu(x) = \frac{2^{\frac{1}{2}\nu} e^{-\frac{1}{4}x^2}}{\Gamma\left(-\frac{1}{2}\nu\right)} \int_0^\infty e^{-\frac{1}{2}tx^2} t^{-\frac{1}{2}\nu-1}(1+t)^{\frac{1}{2}\nu-\frac{1}{2}} dt. \qquad (8.6.7)$$

Another integral representation valid for $\operatorname{Re}\nu < 0$ is

$$D_\nu(x) = \frac{2^{\frac{1}{2}\nu} \Gamma\left(\frac{1}{2}\right) e^{-\frac{1}{4}x^2}}{\Gamma\left(\frac{1}{2} - \frac{1}{2}\nu\right) \Gamma(-\frac{1}{2}\nu)} \int_0^\infty e^{-t - \sqrt{2}tx} t^{-\frac{1}{2}\nu-1} dt; \qquad (8.6.8)$$

see Exercise 8.

If $\nu = 2m$ is an even nonnegative integer, then the second summand on the right in (8.6.6) vanishes and the coefficient of the first summand is

$$2^m \frac{\Gamma(\frac{1}{2})}{\Gamma(\frac{1}{2} - m)} = (-1)^m 2^m \left(\frac{1}{2}\right)_m.$$

If $\nu = 2m + 1$ is an odd positive integer, then the first summand on the right in (8.6.6) vanishes and the coefficient of the second summand is

$$2^{m+\frac{1}{2}} \frac{\Gamma(-\frac{1}{2})}{\Gamma(-\frac{1}{2} - m)} = (-1)^m 2^{m+\frac{1}{2}} \left(\frac{3}{2}\right)_m.$$

In view of (8.4.4), therefore,

$$D_n(x) = e^{-\frac{1}{4}x^2} He_n(x), \quad n = 0, 1, 2, \ldots \qquad (8.6.9)$$

The behavior as $x \to 0$ is given by

$$D_\nu(x) = 2^{\frac{1}{2}\nu} \sqrt{\pi} \left\{ \frac{1}{\Gamma(\frac{1}{2} - \frac{1}{2}\nu)} - \frac{2^{\frac{1}{2}} x}{\Gamma(-\frac{1}{2}\nu)} \right\} + O(x^2); \qquad (8.6.10)$$

the remaining coefficients can be computed from the recursion (8.6.3). The definition (8.6.6) and the identity (8.2.6) imply the identity

$$D_\nu(x) = 2^{\frac{1}{2}\nu-\frac{1}{2}} e^{-\frac{1}{4}x^2} x U\left(-\frac{1}{2}\nu + \frac{1}{2}, \frac{3}{2}; \frac{1}{2}x^2\right). \qquad (8.6.11)$$

As noted above, (8.6.1) is unchanged under the coordinate change $x \to -x$. It is also unchanged if we replace the pair $(x, \nu + \frac{1}{2})$ with the pair $(\pm ix, -\nu - \frac{1}{2})$.

Therefore four solutions of (8.6.1) are

$$D_\nu(x), \quad D_\nu(-x), \quad D_{-\nu-1}(ix), \quad D_{-\nu-1}(-ix). \tag{8.6.12}$$

The behavior of each of these solutions as $x \to 0$ follows from (8.6.10). This allows one to express any one of the solutions (8.6.12) as a linear combination of any two of these solutions. In particular, it follows from (8.6.10), using (2.2.7) and (2.3.1), that

$$D_\nu(x) = \frac{\Gamma(\nu+1)}{\sqrt{2\pi}} \left\{ e^{\frac{1}{2}\nu\pi i} D_{-\nu-1}(ix) + e^{-\frac{1}{2}\nu\pi i} D_{-\nu-1}(-ix) \right\};$$

$$D_{-\nu-1}(ix) = \frac{\Gamma(-\nu)}{\sqrt{2\pi}} \left\{ i e^{\frac{1}{2}\nu\pi i} D_\nu(x) - i e^{-\frac{1}{2}\nu\pi i} D_\nu(-x) \right\}. \tag{8.6.13}$$

The Wronskian of two solutions of (8.6.1) is constant. It follows from (8.6.10) and the duplication formula (2.3.1) that

$$W(D_\nu(x), D_\nu(-x)) = \frac{2^{\nu+\frac{3}{2}}\pi}{\Gamma(\frac{1}{2} - \frac{1}{2}\nu)\Gamma(-\frac{1}{2}\nu)} = \frac{\sqrt{2\pi}}{\Gamma(-\nu)}. \tag{8.6.14}$$

The reflection formula (2.2.7) and (8.6.10) imply that

$$W(D_\nu(x), D_{-\nu-1}(ix)) = -i e^{-\frac{1}{2}\pi\nu i} = e^{-\frac{1}{2}\pi(\nu+1)i};$$

$$W(D_\nu(x), D_{-\nu-1}(-ix)) = i e^{\frac{1}{2}\nu\pi i} = e^{\frac{1}{2}\pi(\nu+1)i}. \tag{8.6.15}$$

The remaining Wronskians of the four solutions (8.6.12) can be deduced from these. In particular, $D_\nu(x)$ and $D_\nu(-x)$ are independent if and only if ν is not a nonnegative integer, while $D_\nu(x)$ and $D_{-\nu-1}(ix)$ are always independent.

The identities (8.6.6), (8.5.8), and (8.6.11) imply

$$D_\nu'(x) + \frac{x}{2} D_\nu(x) - \nu D_{\nu-1}(x) = 0. \tag{8.6.16}$$

Similarly, the identities (8.6.6), (8.5.9), and (8.6.11) imply

$$D_\nu'(x) - \frac{x}{2} D_\nu(x) + D_{\nu+1}(x) = 0. \tag{8.6.17}$$

Eliminating D_ν' from these identities gives the recurrence identity

$$D_{\nu+1}(x) - x D_\nu(x) + \nu D_{\nu-1}(x) = 0. \tag{8.6.18}$$

The asymptotic result

$$D_\nu(x) \sim \frac{2^{\frac{1}{2}\nu}}{\sqrt{\pi}} \Gamma\left(\frac{1}{2}\nu + \frac{1}{2}\right) \cos\left(\sqrt{\nu + \frac{1}{2}}x - \frac{1}{2}\pi\nu\right) \tag{8.6.19}$$

as $\nu \to +\infty$ is a consequence of (8.6.6) and the asymptotic result (8.2.10). A direct proof will be given in Chapter 13.

8.7 Whittaker functions

The first-order term in the confluent hypergeometric equation (8.1.1) can be eliminated by a gauge transformation $u(x) = \varphi(x)v(x)$ with $\varphi'/\varphi = \frac{1}{2}(1-c/x)$, i.e. $\varphi(x) = e^{\frac{1}{2}x}x^{-\frac{1}{2}c}$. The resulting equation for v is *Whittaker's equation*:

$$v''(x) + \left[-\frac{1}{4} + \frac{\kappa}{x} + \frac{1-4\mu^2}{4x^2} \right] v(x) = 0, \tag{8.7.1}$$

$$\kappa = \frac{c}{2} - a, \quad \mu = \frac{c-1}{2}.$$

Conversely, solutions of (8.7.1) have the form $x^{\frac{1}{2}c}e^{-\frac{1}{2}x}V(x)$, where V is a solution of the confluent hypergeometric equation (8.1.1) with indices a, c given by

$$a = \mu - \kappa + \frac{1}{2}, \quad c = 1 + 2\mu.$$

Since also

$$a + 1 - c = -\mu - \kappa + \frac{1}{2}, \quad 2 - c = 1 - 2\mu, \quad x^{\frac{1}{2}c}x^{1-c} = x^{1-\frac{1}{2}c} = x^{-\mu+\frac{1}{2}},$$

so long as 2μ is not an integer, there are independent solutions: the Whittaker functions

$$M_{\kappa,\mu}(x) = e^{-\frac{1}{2}x}x^{\mu+\frac{1}{2}}M\left(\mu - \kappa + \frac{1}{2}, 1 + 2\mu; x \right);$$

$$M_{\kappa,-\mu}(x) = e^{-\frac{1}{2}x}x^{-\mu+\frac{1}{2}}M\left(-\mu - \kappa + \frac{1}{2}, 1 - 2\mu; x \right), \tag{8.7.2}$$

that correspond to (8.1.2) and (8.1.6), respectively.

The Kummer functions are entire when defined. Both the functions in (8.7.2) are defined so long as 2μ is not a nonzero integer. The Whittaker functions are multiple-valued unless 2μ is an odd integer, in which case whichever of the functions (8.7.2) is defined is a single-valued function.

The asymptotics of the solutions (8.7.2) follow from (8.1.8):

$$M_{\kappa,\mu} \sim \frac{\Gamma(1+2\mu)}{\Gamma(\mu - \kappa + \frac{1}{2})} x^{-\kappa}e^{\frac{1}{2}x} \quad \text{as } \operatorname{Re} x \to +\infty. \tag{8.7.3}$$

The Wronskian of two solutions of (8.7.1) is constant. It follows from this and the behavior at zero,

$$M_{\kappa,\mu}(x) \sim x^{\mu+\frac{1}{2}},$$

that the Wronskian of the solutions (8.7.2) is

$$W(M_{\kappa,\mu}, M_{\kappa,-\mu})(x) \equiv -2\mu. \tag{8.7.4}$$

In view of (8.2.2), there is a solution that is exponentially decreasing as $x \to +\infty$,

$$W_{\kappa,\mu}(x) = e^{-\frac{1}{2}x} x^{\mu+\frac{1}{2}} U\left(\mu - \kappa + \frac{1}{2}, 1 + 2\mu; x\right)$$

$$= \frac{\Gamma(-2\mu)}{\Gamma(-\mu - \kappa + \frac{1}{2})} M_{\kappa,\mu}(x) + \frac{\Gamma(2\mu)}{\Gamma(\mu - \kappa + \frac{1}{2})} M_{\kappa,-\mu}(x)$$

$$= W_{\kappa,-\mu}(x), \tag{8.7.5}$$

provided 2μ is not an integer. It follows from (8.2.2) that

$$W_{\kappa,\mu}(x) \sim x^\kappa e^{-\frac{1}{2}x} \quad \text{as } \mathrm{Re}\, x \to +\infty. \tag{8.7.6}$$

The Wronskian of $M_{\kappa,\mu}$ and $W_{\kappa,\mu}$ can be computed from (8.7.5) and (8.7.4), or from the asymptotics (8.7.3) and (8.7.6):

$$W(M_{\kappa,\mu}, W_{\kappa,\mu})(x) \equiv -\frac{\Gamma(1+2\mu)}{\Gamma(\mu - \kappa + \frac{1}{2})}. \tag{8.7.7}$$

Since (8.7.1) is unchanged under $(x, \kappa) \to (-x, -\kappa)$, it follows that a solution exponentially decreasing at $-\infty$ is

$$W_{-\kappa,\mu}(x) = \frac{\Gamma(-2\mu)}{\Gamma(-\mu + k + \frac{1}{2})} M_{-k,\mu}(-x)$$

$$+ \frac{\Gamma(2\mu)}{\Gamma(\mu + \kappa + \frac{1}{2})} M_{-k,-\mu}(-x). \tag{8.7.8}$$

It follows from (8.7.5) and (8.2.10) that

$$W_{\kappa,\mu}(x) \sim \frac{\Gamma(\kappa + \frac{1}{4}) x^{\frac{1}{4}}}{\sqrt{\pi}} \cos\left(2\sqrt{\kappa x} - \kappa\pi + \frac{1}{4}\pi\right) \tag{8.7.9}$$

as $\kappa \to +\infty$.

The Coulomb wave equation (3.6.12) is

$$u''(\rho) + \left[1 - \frac{2\eta}{\rho} - \frac{l(l+1)}{\rho^2}\right] u(\rho) = 0.$$

Let $u(\rho) = v(2i\rho)$. Then the equation becomes

$$v''(x) + \left[-\frac{1}{4} + \frac{i\eta}{x} - \frac{l(l+1)}{x^2}\right] v(x) = 0.$$

This is Whittaker's equation with $\kappa = i\eta$ and $\mu = l + \frac{1}{2}$. When the equation is obtained by separating variables in spherical coordinates, the parameter l is a nonnegative integer. In this case, in addition to a solution that is regular at $x = 0$, there is a solution with a logarithmic singularity. The normalization of the regular solution in [313] is

$$
F_l(\eta, \rho) = \frac{C_l(\eta)}{(\pm 2i)^{l+1}} M_{\pm i\eta, l+\frac{1}{2}}(\pm 2i\rho)
$$
$$
= C_l(\eta) \rho^{l+1} e^{\mp i\rho} M(l + 1 \mp i\eta, 2l + 2; \pm 2i\rho), \qquad (8.7.10)
$$

where the normalizing constant is

$$
C_l(\eta) = 2^l e^{-\pi \eta/2} \frac{|\Gamma(l + 1 + i\eta)|}{\Gamma(2l + 2)}.
$$

(By (8.1.10), the choice of sign does not matter.) The reason for this choice of C_l is so that

$$
F_l(\eta, \rho) \sim \sin \theta_l(\eta, \rho) \qquad (8.7.11)
$$

as $\rho \to +\infty$, where θ_l is given by (8.7.14) below; see Exercise 26.

Irregular solutions are defined by

$$
H_l^{\pm}(\eta, \rho) = (\pm i)^l e^{\pm i\sigma_l(\eta)} e^{\frac{1}{2}\pi\eta} W_{\mp i\eta, l+\frac{1}{2}}(\mp 2i\rho)
$$
$$
= e^{\pm i\theta_l(\eta, \rho)} (\mp 2i\rho)^{l+1\pm i\eta} U(l + 1 \pm i\eta, 2l + 2; \mp 2i\rho). \quad (8.7.12)
$$

Here the normalizing phases are the *Coulomb wave shift*

$$
\sigma_l(\eta) = \arg \Gamma(l + 1 + i\eta) \qquad (8.7.13)
$$

and

$$
\theta_l(\eta, \rho) = \rho - \eta \log(2\rho) - \frac{1}{2}l\pi + \sigma_l(\eta). \qquad (8.7.14)
$$

The real and imaginary parts are

$$
H_l^{\pm}(\eta, \rho) = G_\lambda(\eta, \rho) \pm i F_l(\eta, \rho), \qquad (8.7.15)
$$

where F_l is the regular solution (8.7.10) and G_l is defined by (8.7.15):

$$
G_l(\eta, \rho) = \frac{1}{2} \left[H_l^+(\eta, \rho) + H_l^-(\eta, \rho) \right]. \qquad (8.7.16)
$$

When l is a nonnegative integer, G_l has a logarithmic singularity at $\rho = 0$; see (8.3.3).

8.8 Exercises

1. Show that the expansion of $e^{\lambda x}$ as a sum of Jacobi polynomials is

$$e^{\lambda x} = \sum_{n=0}^{\infty} C_n M(n+\beta+1, 2n+\alpha+\beta+2; 2\lambda) P_n^{(\alpha,\beta)}(x),$$

where

$$C_n = \frac{\Gamma(n+\alpha+\beta+1)}{\Gamma(2n+\alpha+\beta+1)} (2\lambda)^n e^{-\lambda}.$$

Hint: write $e^{\lambda x} = e^{-\lambda} e^{\lambda(1+x)}$ and use Exercise 40 of Chapter 5.

2. Show that integration and summation can be interchanged to calculate the Laplace transform (see Exercise 23 of Chapter 5) of $f(x) = x^{b-1} M(a,c;x)$:

$$[\mathcal{L}f](s) = \frac{\Gamma(b)}{s^b} F\left(a, b, c; \frac{1}{s}\right),$$

where F is the hypergeometric function (see Exercise 7 of Chapter 1, or Chapter 10).

3. Verify the integral representation

$$M(a,c;x) = \frac{\Gamma(c)}{2\pi i} \int_C \left(1 - \frac{x}{t}\right)^{-a} t^{-c} e^t \, dt,$$

where C is the Hankel loop of Corollary 2.2.4.

4. The asymptotic result (8.1.8) implies that for most values of nonzero constants A and B, the linear combination

$$u(x) = A M(a,c;x) + B x^{1-c} M(a+1-c, 2-c; x)$$

will have exponential growth as $x \to +\infty$. Determine a necessary condition on the ratio A/B to prevent this from happening. Compare this to (8.2.5).

5. Prove the identity (8.2.6).

6. Verify the evaluation (8.2.7).

7. Prove that

$$U\left(a, \frac{1}{2}; x^2\right) = \frac{\Gamma(\frac{1}{2})}{\Gamma(a+\frac{1}{2})\Gamma(a)} \sum_{n=0}^{\infty} \frac{\Gamma(a+\frac{1}{2}n)}{n!} (-2x)^n.$$

8. Use Exercise 7 to prove that for $\operatorname{Re} a > 0$,

$$U\left(a, \frac{1}{2}; x^2\right)$$

$$= \frac{\Gamma(\frac{1}{2})}{\Gamma(a+\frac{1}{2})\Gamma(a)} \int_0^{\infty} \exp(-t - 2\sqrt{t}x) t^{a-1} \, dt.$$

9. Verify the identity for the Laguerre polynomials

$$L_n^{(a)}(x) = \frac{(-1)^n}{n!} U(-n, a+1; x).$$

10. Prove the identity (8.4.5) for the error function. Hint: $1/(2n+1) = \left(\frac{1}{2}\right)_n / \left(\frac{3}{2}\right)_n$.

11. Prove the identity (8.4.6) for the incomplete gamma function. Hint: $a/(a+n) = (a)_n/(a+1)_n$.

12. Use (8.2.1) to prove the identity (8.4.7) for the complementary incomplete gamma function.

13. Let $\omega = e^{i\pi/4} = (1+i)/\sqrt{2}$. Show that

$$\frac{1}{1+i} \operatorname{erf}\left(\omega x \sqrt{\frac{\pi}{2}}\right) = \int_0^x e^{-it^2\pi/2} \, dt = C(x) - iS(x),$$

where $C(x)$ and $S(x)$ are the Fresnel integrals of Section 8.4.

14. Use Exercise 13 and (8.4.5) to prove (8.4.8) and (8.4.9).

15. Use (8.2.1) to prove (8.4.10).

16. Use Cauchy's theorem to prove that

$$\int_0^\infty \frac{\sin t \, dt}{t} = \frac{1}{2} \int_{-\infty}^\infty \frac{\sin t \, dt}{t}$$

$$= \frac{1}{2} \operatorname{Im} \left[\lim_{\varepsilon \to 0+} \int_{|t|>\varepsilon} \frac{e^{it} \, dt}{t} \right] = \frac{\pi}{2}.$$

Hint: Integrate e^{it}/t over the boundary of the region $\{|z| > \varepsilon, 0 < \operatorname{Im} z < R, |\operatorname{Re} z| < S\}$, and let first S, then R go to ∞.

Thus for $x > 0$,

$$\int_x^\infty \frac{\sin t \, dt}{t} = \frac{\pi}{2} - \int_0^x \frac{\sin t \, dt}{t}.$$

17. Use (8.4.10) and Exercise 16 to prove (8.4.11) and (8.4.12).

18. Use the results of Section 8.5 to show that the asymptotic results (8.1.8) and (8.1.9) are valid for all indices (a, c), $c \neq 0, 1, 2, \ldots$

19. Use the results of Section 8.5 to show that the asymptotic results (8.2.2) and (8.2.3) are valid for all indices (a, c), c not an integer.

20. Suppose $c > 0$. The operator

$$L = x \frac{d^2}{dx^2} + (c-x) \frac{d}{dx}$$

is symmetric in L_w^2, $w(x) = x^{c-1} e^{-x}$, $x > 0$, with eigenvalues $0, 1, 2, \ldots$; the Laguerre polynomials with index $c - 1$ are eigenfunctions.

Given $\lambda > 0$ and $f \in L_w^2$, the equation $Lu + \lambda u = f$ has a unique solution $u \in L_w^2$, expressible in the form

$$u(x) = \int_0^\infty G_\lambda(x, y) f(y) \, dy.$$

Compute the Green's function G_λ. Hint: see Section 3.3. The appropriate boundary conditions here are: regular at $x = 0$, having at most polynomial growth as $x \to +\infty$.

21. Prove (8.6.13).
22. Express $D_{-\nu-1}(ix)$ as a linear combination of $D_\nu(x)$ and $D_\nu(-x)$.
23. Prove the Wronskian formulas (8.6.14) and (8.6.15).
24. Compute the Wronskians

$$W(D_\nu(x), D_{-\nu-1}(ix));$$

$$W(D_\nu(x), D_{-\nu-1}(-ix));$$

$$W(D_{-\nu-1}(ix), D_{-\nu-1}(-ix)).$$

25. Verify the identities (8.6.16), (8.6.17), and (8.6.18).
26. Show that the asymptotic result (8.2.2) is also valid for imaginary values of the argument x. Deduce that the Coulomb wave functions satisfy

$$H_l^\pm(\eta, \rho) \sim e^{\pm i\theta_l(\eta, \rho)}$$

as $\rho \to +\infty$, and thus verify (8.7.11).

8.9 Remarks

The Kummer functions were introduced by Kummer [235] in 1836, although the series (8.1.2) in the case $c = 2m$ had been investigated by Lagrange [237] in 1762–1765. Weber introduced the parabolic cylinder functions in 1869 [427]. Whittaker [434] introduced the Whittaker functions in 1903 and showed that many known functions, including the parabolic cylinder functions and the functions in Section 8.4, can be expressed in terms of the $W_{\kappa,\mu}$. Coulomb wave functions were studied in 1928 by Gordon [167] and Mott [293]. The current normalization and notation are due to Yost, Wheeler, and Breit [446]; see also Seaton [361]. For an application of Kummer and Whittaker functions to the study of singular and degenerate hyperbolic equations, see Beals and Kannai [34].

Three monographs devoted to confluent hypergeometric functions are Buchholz [57], Slater [368], and Tricomi [402]. Buchholz puts particular emphasis on the Whittaker functions and on applications, with many references to the applied literature, while Tricomi emphasizes the Kummer functions.

As noted above, there are several standard notations for the Kummer functions. We have chosen to use the notation M and U found in the handbooks of Abramowitz and Stegun [4], Jahnke and Emde [202], and Olver, Lozier, Clark, and Boisvert [312, 313]. Tricomi [402] uses Φ and Ψ.

9

Cylinder functions

A cylinder function of order v is a solution of Bessel's equation

$$x^2 u''(x) + x u'(x) + (x^2 - v^2) u(x) = 0. \tag{9.0.1}$$

As before, we write $D = x(d/dx)$, so that Bessel's equation takes the form

$$(D^2 - v^2) u(x) + x^2 u(x) = 0.$$

For $x \sim 0$, this can be viewed as a perturbation of the equation $(D^2 - v^2)u = 0$, which has solutions $u(x) = x^{\pm v}$, so we expect to find solutions that have the form

$$x^{\pm v} f_v(x), \tag{9.0.2}$$

with f holomorphic near $x = 0$. Suitably normalized, solutions that have this form are the Bessel functions of the first kind, or simply "Bessel functions." For v not an integer, one obtains two independent solutions of this form. For v an integer, there is one solution of this form, and a second solution normalized at $x = 0$ known as a Bessel function of the second kind.

For x large, (9.0.1) can be viewed as a perturbation of

$$u''(x) + \frac{1}{x} u'(x) + u(x) = 0.$$

The gauge transformation $u(x) = x^{-\frac{1}{2}} v(x)$ converts this to

$$v''(x) + \left(1 + \frac{1}{4x^2}\right) v(x) = 0,$$

which can be viewed as a perturbation of $v'' + v = 0$. Therefore we expect to find solutions of (9.0.1) that have the asymptotic form

$$v(x) \sim x^{-\frac{1}{2}} e^{\pm i x} g_v(x), \qquad x \to +\infty,$$

where g_v has algebraic growth or decay. Bessel functions of the third kind, or Hankel functions, are a basis for such solutions.

We may remove the first-order term from (9.0.1) by the gauge transformation $u(x) = x^{-\frac{1}{2}} v(x)$. The equation for v is then

$$x^2 v''(x) + \left(x^2 - v^2 + \frac{1}{4} \right) v(x) = 0.$$

Therefore when $v = \pm \frac{1}{2}$, the solutions of (9.0.1) are linear combinations of

$$\frac{\cos x}{\sqrt{x}}, \quad \frac{\sin x}{\sqrt{x}}.$$

Because of this and the recurrence relations for Bessel functions, Bessel functions are elementary functions whenever v is a half-integer.

Replacing x by ix in (9.0.1) gives the equation

$$x^2 u''(x) + x u'(x) - (x^2 + v^2) u(x) = 0. \tag{9.0.3}$$

Solutions are known as modified Bessel functions. The normalized solutions are real and have specified asymptotics as $x \to +\infty$.

The Airy equation

$$u''(x) - x u(x) = 0$$

is related to (9.0.3) with $v^2 = \frac{1}{9}$ by a simple transformation, so its solutions can be obtained from the modified Bessel functions.

In this chapter, we establish various representations of these functions in order to determine the recurrence and derivative formulas, and to determine the relations among the various normalized solutions.

9.1 Bessel functions

The simplest way to obtain solutions of Bessel's equation (9.0.1) that have the form (9.0.2) is to use the gauge transformation $u(x) = x^v v(x)$, so that the equation becomes

$$D^2 v(x) + 2v D v(x) + x^2 v(x) = 0,$$

and to determine the power series expansion $v(x) = \sum_{n=0}^{\infty} a_n x^n$. The coefficients must satisfy

$$n(n + 2v) a_n + a_{n-2} = 0.$$

It follows that $a_1 = 0$ and thus all odd terms vanish. We normalize by setting $a_0 = 1$ and obtain for the even terms

$$a_{2m} = (-1)^m \frac{1}{2m(2m-2) \cdots 2(2m+2v)(2m+2v-2) \cdots (2+2v)}$$

$$= (-1)^m \frac{1}{4^m m! (v+1)_m}, \quad v+1 \neq 0, -1, -2, \ldots, 1-m.$$

The corresponding solution of Bessel's equation is defined for v not a negative integer:

$$x^v \sum_{m=0}^{\infty} \frac{(-1)^m}{(v+1)_m \, m!} \left(\frac{x}{2}\right)^{2m}. \tag{9.1.1}$$

As a function of v for given $x > 0$, the function (9.1.1) has a simple pole at each negative integer. Division by $\Gamma(v+1)$ removes the pole. A further slight renormalization gives the *Bessel function of the first kind*:

$$J_v(x) = \sum_{m=0}^{\infty} \frac{(-1)^m}{\Gamma(v+1+m) \, m!} \left(\frac{x}{2}\right)^{v+2m}. \tag{9.1.2}$$

The series is convergent for all complex x. Taking the principal branch of x^v gives a function that is holomorphic on the complement of the real interval $(-\infty, 0]$.

It follows from the expansion (9.1.2) that

$$[x^v J_v]' = x^v J_{v-1}, \quad [x^{-v} J_v]' = -x^{-v} J_{v+1}, \tag{9.1.3}$$

which implies that

$$\left(\frac{1}{x}\frac{d}{dx}\right)^n \left[x^v J_v(x)\right] = x^{v-n} J_{v-n}(x);$$

$$\left(\frac{1}{x}\frac{d}{dx}\right)^n \left[x^{-v} J_v(x)\right] = (-1)^n x^{-v-n} J_{v+n}(x). \tag{9.1.4}$$

Now

$$x^{1-v} [x^v J_v]'(x) = x J_v'(x) + v J_v(x),$$
$$x^{1+v} [x^{-v} J_v]'(x) = x J_v'(x) - v J_v(x).$$

Eliminating $J_v'(x)$ and $J_v(x)$, respectively, from the resulting pairs of equations gives the recurrence and derivative relations

$$J_{v-1}(x) + J_{v+1}(x) = \frac{2v}{x} J_v(x); \tag{9.1.5}$$

$$J_{v-1}(x) - J_{v+1}(x) = 2 J_v'(x). \tag{9.1.6}$$

As remarked in the introduction to this chapter, the Bessel function $J_{\pm\frac{1}{2}}$ must be linear combinations of $x^{-\frac{1}{2}} \cos x$ and $x^{-\frac{1}{2}} \sin x$. In general, as $x \to 0+$,

$$J_{\pm v}(x) \sim \frac{1}{\Gamma(\pm v + 1)} \left(\frac{x}{2}\right)^{\pm v}, \quad J_{\pm v}'(x) \sim \pm \frac{v}{2} \frac{1}{\Gamma(\pm v + 1)} \left(\frac{x}{2}\right)^{\pm v - 1}. \tag{9.1.7}$$

It follows that

$$J_{\frac{1}{2}}(x) = \frac{\sqrt{2}\sin x}{\sqrt{\pi x}}; \quad J_{-\frac{1}{2}}(x) = \frac{\sqrt{2}\cos x}{\sqrt{\pi x}}. \tag{9.1.8}$$

These can also be derived from the series expansion (9.1.2) by using the duplication formula (2.3.1). It follows from (9.1.8) and (9.1.5) that $J_\nu(x)$ is expressible in terms of trigonometric functions and powers of x whenever $\nu + \frac{1}{2}$ is an integer.

For ν not an integer, the two solutions J_ν and $J_{-\nu}$ behave differently as $x \to 0$, so they are clearly independent. To compute the Wronskian, we note first that for any two solutions $u_1(x)$, $u_2(x)$ of (9.0.1), the Wronskian $W(x) = W(u_1, u_2)(x)$ must satisfy $x^2 W'(x) = -x W(x)$. Therefore $W = c/x$ for some constant c. The constant is easily determined, using the identities (9.1.7) and (2.2.7), which give

$$\Gamma(\nu + 1)\Gamma(-\nu + 1) = \nu\,\Gamma(\nu)\Gamma(1 - \nu) = \frac{\nu\pi}{\sin\nu\pi}.$$

It follows that

$$W(J_\nu, J_{-\nu})(x) = \frac{\sin\nu\pi}{\nu\pi} \begin{vmatrix} x^\nu & x^{-\nu} \\ \nu x^{\nu-1} & -\nu x^{-\nu-1} \end{vmatrix} = -\frac{2\sin\nu\pi}{\pi x}, \tag{9.1.9}$$

confirming that these solutions are independent for ν not an integer. If $\nu = -n$ is a negative integer, examination of the expansion (9.1.2) shows that the first nonvanishing term is the term with $m = n$. Setting $m = n + k$,

$$J_{-n}(x) = (-1)^n \sum_{k=0}^{\infty} \frac{(-1)^k}{k!(n+k)!} \left(\frac{x}{2}\right)^{n+2k} = (-1)^n J_n(x)$$

$$= \cos n\pi\, J_n(x). \tag{9.1.10}$$

These considerations lead to the choice of the *Bessel function of the second kind*

$$Y_\nu(x) = \frac{\cos\nu\pi\, J_\nu(x) - J_{-\nu}(x)}{\sin\nu\pi}. \tag{9.1.11}$$

In particular,

$$Y_{\frac{1}{2}}(x) = -\frac{\sqrt{2}\cos x}{\sqrt{\pi x}}; \quad Y_{-\frac{1}{2}}(x) = \frac{\sqrt{2}\sin x}{\sqrt{\pi x}}. \tag{9.1.12}$$

This solution of Bessel's equation (9.0.1) is first defined for ν not an integer. In view of (9.1.10), both the numerator and the denominator have simple zeros at integer ν, so the singularity is removable and Y_ν can be considered as a solution for all ν. The Wronskian is

$$W(J_\nu, Y_\nu)(x) = -\frac{W(J_\nu, J_{-\nu})(x)}{\sin\nu\pi} = \frac{2}{\pi x},$$

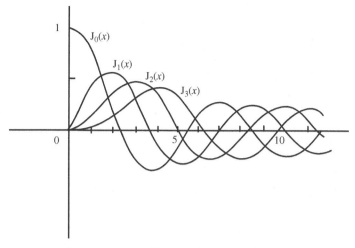

Figure 9.1. Bessel function $J_\nu(x)$, $\quad \nu = 0,1,2,3$.

so $J_\nu(x)$ and $Y_\nu(x)$ are independent solutions for all ν. It follows from (9.1.10) and (9.1.11) that

$$Y_{-n}(x) = (-1)^n Y_n(x), \qquad n = 0,1,2,\ldots \tag{9.1.13}$$

The identities (9.1.5) and (9.1.6) yield the corresponding identities for $Y_\nu(x)$:

$$Y_{\nu-1}(x) + Y_{\nu+1}(x) = \frac{2\nu}{x} Y_\nu(x); \tag{9.1.14}$$

$$Y_{\nu-1}(x) - Y_{\nu+1}(x) = 2 Y_\nu'(x). \tag{9.1.15}$$

Figures 9.1 and 9.2 show the graphs of these functions for some nonnegative values of the argument x and the parameter ν.

The series expansion (9.1.2) can be used to find the asymptotic behavior of $J_\nu(x)$ for large values of ν so long as $y^2 = x^2/\nu$ is bounded. A formal calculation gives

$$\frac{(\sqrt{\nu}\, y)^\nu}{2^\nu \Gamma(\nu+1)} \sum_{m=0}^\infty \frac{(-1)^m}{m!} \left(\frac{y}{2}\right)^{2m}$$

as the limiting value of the series. It is not difficult to show that

$$J_\nu(\sqrt{\nu}\, y) \sim \frac{(\sqrt{\nu}\, y)^\nu}{2^\nu \Gamma(\nu+1)} e^{-\frac{1}{4}y^2} \tag{9.1.16}$$

as $\nu \to +\infty$, uniformly on bounded intervals. For asymptotic results when x is allowed to be comparable to ν, see Watson [425], Chapter 8.

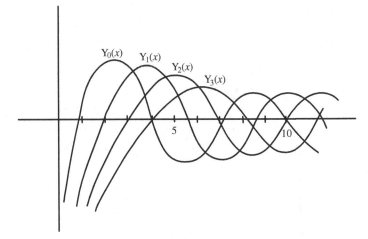

Figure 9.2. Bessel function $Y_\nu(x)$, $\nu = 0, 1, 2, 3$.

9.2 Zeros of real cylinder functions

Any real nonzero cylinder function with index ν has the form

$$u(x) = A J_\nu(x) + B Y_\nu(x), \tag{9.2.1}$$

where A and B are real constants. As noted in the introduction to this chapter, the gauge transformation $u(x) = x^{-\frac{1}{2}} v(x)$ converts Bessel's equation to a perturbation of the equation $w'' + w = 0$. Therefore we might expect any such function u to be oscillatory, that is, it has an infinite number of zeros in the half-line $(0, \infty)$, tending to infinity. This was proved for J_ν by Lommel [263]. Moreover, we might expect the spacing between zeros to be asymptotic to π.

Theorem 9.2.1 *A real nonzero cylinder function $u(x)$ has a countable number of positive zeros*

$$0 < x_1 < x_2 < \cdots < x_n < \cdots$$

The spacing $x_{n+1} - x_n$ is $\geq \pi$ if $|\nu| \geq \frac{1}{2}$, and $\leq \pi$ if $|\nu| \leq \frac{1}{2}$. As $n \to \infty$,

$$x_{n+1} - x_n = \pi + O(n^{-2}).$$

Proof As noted earlier, the gauge transformation $u(x) = x^{-\frac{1}{2}} v(x)$ converts Bessel's equation to

$$v''(x) + \left(1 - \frac{\nu^2 - \frac{1}{4}}{x^2}\right) v(x) = 0. \tag{9.2.2}$$

The cylinder function u itself is a linear combination of solutions with behavior x^ν and $x^{-\nu}$ as $x \to 0$, so v has no zeros in some interval $(0, \varepsilon]$. We make use of Sturm's comparison theorem, Theorem 3.3.3, by choosing comparison functions of the form

$$w(x) = \cos(ax+b), \quad a > 0. \tag{9.2.3}$$

Note that $w''(x) + a^2 w(x) = 0$ and the gap between the zeros of $w(x)$ is π/a. If a is chosen so that a^2 is a lower bound for the zero-order coefficient

$$1 - \frac{v^2 - \frac{1}{4}}{x^2}, \quad \varepsilon \le x < \infty, \tag{9.2.4}$$

then the comparison theorem implies that for $x \ge \varepsilon$, there is a zero of v between each pair of zeros of w. This proves that v, hence u, has countably many zeros. Moreover, by choosing b in (9.2.3) so that w vanishes at a given zero of v, we may conclude that the distance to the next zero of v is at most π/a.

Similarly, by choosing a so that a^2 is an upper bound for (9.2.4) and choosing b so that w vanishes at a zero of v, we conclude that the distance to the next zero of v is at least π/a. In particular, $a^2 = 1$ is a lower bound when $v^2 \le \frac{1}{4}$ and an upper bound when $v^2 \ge \frac{1}{4}$, so π is a lower bound for the gaps in the first case and an upper bound in the second case.

This argument shows that the nth zero x_n has magnitude comparable to n. On the interval $[x_n, \infty)$, the best lower and upper bounds of (9.2.4) differ from 1 by an amount that is $O(n^{-2})$, so the gaps differ from π by a corresponding amount. $\qquad\square$

Between any two positive zeros of a real cylinder function u, there is a zero of u'. It follows from (9.0.1) that such a zero x is simple if $x \ne v$, so $u(x)$ is a local extremum for u.

Theorem 9.2.2 *Suppose that u is a nonzero real cylinder function and that the zeros of u' in the interval (v, ∞) are*

$$y_1 < y_2 < \cdots < y_n < \cdots$$

Then

$$|u(y_1)| > |u(y_2)| > \cdots > |u(y_n)| > \cdots, \tag{9.2.5}$$

and

$$(y_1^2 - v^2)^{\frac{1}{4}} |u(y_1)| < (y_2^2 - v^2)^{\frac{1}{4}} |u(y_2)| < \cdots < (y_n^2 - v^2)^{\frac{1}{4}} |u(y_n)| < \cdots \tag{9.2.6}$$

Proof Bessel's equation may be written in the form

$$[xu']'(x) + \left[x - \frac{v^2}{x}\right] u(x) = 0.$$

Therefore the inequalities (9.2.5) are a consequence of Proposition 3.5.2. The inequalities (9.2.6) are left as an exercise. □

The inequalities (9.2.5) are a special case of the results of Sturm [390]. The inequalities (9.2.6) are due to Watson [424].

The asymptotic results (9.4.9) and (9.4.10), together with the representation (9.2.1), can be used to show that the sequence in (9.2.6) has a limit and to evaluate the limit; see Exercise 15.

Suppose $\nu > 0$. Since $J_\nu(x) \sim (x/2)^\nu / \Gamma(\nu + 1)$ as $x \to 0+$, it follows that $J_\nu J_\nu' > 0$ for small x. By Theorem 3.3.6, J_ν is positive throughout the interval $(0, \nu]$. Therefore the previous theorem can be sharpened for J_ν itself.

Corollary 9.2.3 *Suppose* $\nu > 0$. *The local extrema of* $J_\nu(x)$ *decrease in absolute value as x increases,* $x > 0$.

A standard notation for the positive zeros of J_ν is

$$0 < j_{\nu,1} < j_{\nu,2} < \cdots < j_{\nu,n} < \cdots \tag{9.2.7}$$

Theorem 9.2.4 *The zeros of J_ν and $J_{\nu+1}$ interlace:*

$$0 < j_{\nu,1} < j_{\nu+1,1} < j_{\nu,2} < j_{\nu+1,2} < j_{\nu,3} < \cdots$$

Proof It follows from (9.1.3) that

$$x^{-\nu} J_{\nu+1}(x) = -\frac{d}{dx}[x^{-\nu} J_\nu(x)],$$

$$x^{\nu+1} J_\nu(x) = \frac{d}{dx}[x^{\nu+1} J_{\nu+1}(x)]. \tag{9.2.8}$$

The consecutive zeros $j_{\nu,k}$ and $j_{\nu,k+1}$ of $J_\nu(x)$ are also zeros of $x^{-\nu} J_\nu(x)$. Therefore there is at least one zero of the derivative of $x^{-\nu} J_\nu$, and hence of $J_{\nu+1}(x)$, between $j_{\nu,k}$ and $j_{\nu,k+1}$. Conversely, suppose that $\mu_1 < \mu_2$ are zeros of $J_{\nu+1}(x)$. Then they are also zeros of $x^{\nu+1} J_{\nu+1}(x)$, and by (9.2.8) there is at least one zero of $J_\nu(x)$ between μ_1 and μ_2. This shows that the zeros interlace. It follows from Theorem 3.3.4 that the first zero of J_ν is less than the first zero of $J_{\nu+1}$, and it follows from the remark preceding Corollary 9.2.3 that $j_{\nu,1} > \nu$. □

We conclude this section with a result that was proved by Fourier [141] for J_0 and by Lommel [263] for general real ν.

Theorem 9.2.5 *All zeros of $J_\nu(z)$ are real when $\nu > -1$. All zeros of $J_\nu'(z)$ are real when $\nu \geq 0$. When $-1 < \nu < 0$, $J_\nu'(z)$ has two imaginary zeros and all other zeros are real.*

Proof From the power series representations of $J_\nu(z)$ and $J_\nu'(z)$, it is readily seen that these functions do not have purely imaginary zeros. Bessel's equation implies that

$$(\alpha^2 - \beta^2) \int_0^z t J_\nu(\alpha t) J_\nu(\beta t) \, dt = z \left[\beta J_\nu(\alpha z) J_\nu'(\beta z) - \alpha J_\nu(\beta z) J_\nu'(\alpha z) \right] \quad (9.2.9)$$

for $\nu > -1$. Indeed, $J_\nu(\alpha z)$ satisfies the differential equation

$$\frac{1}{z} \left[z w'(z) \right]' (z) + \left(\alpha^2 - \frac{\nu^2}{z^2} \right) w(z) = 0.$$

Multiply this equation by $J_\nu(\beta z)$, and multiply the corresponding equation for $J_\nu(\beta z)$ by $J_\nu(\alpha z)$. Subtracting the two gives

$$J_\nu(\beta z) \left[\alpha z J_\nu'(\alpha z) \right]' - J_\nu(\alpha z) \left[\beta z J_\nu'(\beta z) \right]' = (\beta^2 - \alpha^2) z J_\nu(\alpha z) J_\nu(\beta z),$$

or equivalently,

$$\left[\alpha z J_\nu(\beta z) J_\nu'(\alpha z) - \beta z J_\nu(\alpha z) J_\nu'(\beta z) \right]' = (\beta^2 - \alpha^2) z J_\nu(\alpha z) J_\nu(\beta z).$$

Formula (9.2.9) follows from an integration of the last equation. Now let α be a nonreal zero and $\alpha \notin i\mathbb{R}$. Then $z = \overline{\alpha}$ is also a zero. Put $z = 1$ and $\beta = \overline{\alpha}$ in (9.2.9), so that the equation becomes

$$(\alpha^2 - \overline{\alpha}^2) \int_0^1 t J_\nu(\alpha t) J_\nu(\overline{\alpha} t) \, dt = [\overline{\alpha} J_\nu(\alpha) J_\nu'(\overline{\alpha}) - \alpha J_\nu(\overline{\alpha}) J_\nu'(\alpha)]. \quad (9.2.10)$$

Since $J_\nu(\alpha) = J_\nu(\overline{\alpha}) = 0$, the right-hand side of (9.2.10) vanishes. Also, since $\operatorname{Re} \alpha \neq 0$ and $\operatorname{Im} \alpha \neq 0$, we deduce from (9.2.10) that

$$\int_0^1 t J_\nu(\alpha t) J_\nu(\overline{\alpha} t) \, dt = 0,$$

which is impossible since the integrand is positive.

If $J_\nu'(\alpha) = J_\nu'(\overline{\alpha}) = 0$, then the right-hand side of (9.2.10) again vanishes. Therefore the same argument can be used for $J_\nu'(z)$, except that now there is a pair of imaginary zeros when $-1 < \nu < 0$. This can be seen from the power series

$$\left(\frac{it}{2} \right)^{1-\nu} J_\nu'(it) = \frac{\frac{1}{2}\nu}{\Gamma(\nu + 1)} + \sum_{s=1}^{\infty} \frac{(s + \frac{1}{2}\nu)}{s! \Gamma(s + \nu + 1)} \left(\frac{t}{2} \right)^{2s},$$

where t is real. The function defined by the right-hand side of this equation is an even function, negative at $t = 0$, and monotonically increases to infinity as $t \to +\infty$. Therefore this function vanishes for two real values of t, completing the proof. $\qquad\square$

9.3 Integral representations

An integral representation of J_ν can be obtained by using the gauge transformation $u(x) = x^\nu e^{ix} v(x)$ introduced in Section 3.7 to reduce (9.0.1) to the canonical form (3.7.3):

$$x v'' + (2\nu + 1 + 2ix) v' + (2\nu + 1) i v = 0. \tag{9.3.1}$$

The change of variables $y = -2ix$ gives

$$y w'' + (2\nu + 1 - y) w' - \left(\nu + \frac{1}{2}\right) w = 0.$$

For ν not an integer, any solution of this last equation that is regular at the origin is a multiple of the Kummer function with indices $\nu + \frac{1}{2}$, $2\nu + 1$. It follows that J_ν is a multiple of the solution

$$x^\nu e^{ix} M\left(\nu + \frac{1}{2}, 2\nu + 1; -2ix\right).$$

Comparing behavior as $x \to 0$ gives the identity

$$J_\nu(x) = \frac{1}{\Gamma(\nu+1)} \left(\frac{x}{2}\right)^\nu e^{ix} M\left(\nu + \frac{1}{2}, 2\nu + 1; -2ix\right). \tag{9.3.2}$$

In view of the integral representation (8.1.3) of the Kummer function when $\operatorname{Re}(\nu + \frac{1}{2}) > 0$, we may deduce an integral representation for J_ν. Making use of the duplication identity (2.3.1) in the form

$$\frac{\Gamma(2\nu+1)}{\Gamma(\nu+\frac{1}{2})} = \frac{2^{2\nu}}{\sqrt{\pi}} \Gamma(\nu+1),$$

we obtain the identity

$$J_\nu(x) = \frac{(2x)^\nu}{\sqrt{\pi}\, \Gamma(\nu+\frac{1}{2})} e^{ix} \int_0^1 e^{-2ixs} s^{\nu-\frac{1}{2}} (1-s)^{\nu-\frac{1}{2}}\, ds, \tag{9.3.3}$$

when $\operatorname{Re}(\nu + \frac{1}{2}) > 0$.

The change of variables $t = 1 - 2s$ in (9.3.3) leads to the *Poisson representation* [331]:

$$J_\nu(x) = \frac{1}{\sqrt{\pi}\, \Gamma(\nu+\frac{1}{2})} \left(\frac{x}{2}\right)^\nu \int_{-1}^1 \cos xt \,(1-t^2)^{\nu-\frac{1}{2}}\, dt. \tag{9.3.4}$$

The expansion (9.1.2) can be recovered from (9.3.4) by using the series expansion of $\cos xt$:

$$\int_0^1 \cos xt \,(1-t^2)^{\nu-\frac{1}{2}}\, dt = \sum_{m=0}^\infty (-1)^m \frac{\Gamma(m+\frac{1}{2})\Gamma(\nu+\frac{1}{2})x^{2m}}{\Gamma(m+\nu+1)(2m)!},$$

where we used the identity

$$\int_{-1}^{1} t^{2m}(1-t^2)^{\nu-\frac{1}{2}} \, dt = \int_{0}^{1} s^{m-\frac{1}{2}}(1-s)^{\nu-\frac{1}{2}} \, ds = B\left(m+\frac{1}{2}, \nu+\frac{1}{2}\right).$$

Since $(2m)! = 2^{2m}m!(\frac{1}{2})_m$, we obtain (9.1.2).

A second approach to the Bessel functions J_ν is closely associated with a second integral representation. We noted in Section 3.6 that functions of the form $\exp(i\mathbf{k}\cdot\mathbf{x})$ are solutions of the Helmholtz equation $\Delta u + |\mathbf{k}|^2 u = 0$. In particular, $\exp(ix_2)$ is a solution of $\Delta u + u = 0$. In cylindrical coordinates, $x_2 = r\sin\theta$ and the Helmholtz equation is

$$\left\{\frac{\partial^2}{\partial r^2} + \frac{1}{r}\frac{\partial}{\partial r} + \frac{1}{r^2}\frac{\partial^2}{\partial\theta^2} + 1\right\}\left[e^{ir\sin\theta}\right] = 0. \tag{9.3.5}$$

Consider the Fourier expansion

$$e^{ir\sin\theta} = \sum_{n=-\infty}^{\infty} j_n(r)\, e^{in\theta}, \tag{9.3.6}$$

where the coefficients are given by

$$j_n(r) = \frac{1}{2\pi}\int_{0}^{2\pi} e^{ir\sin\theta} e^{-in\theta} \, d\theta. \tag{9.3.7}$$

Since

$$e^{-in\theta} = \frac{i}{n}\frac{d}{d\theta}\left[e^{-in\theta}\right],$$

repeated integrations by parts show that for every integer $k \geq 0$,

$$|j_n(r)| \leq C_k \frac{(1+r)^k}{|n|^k}, \quad n \neq 0,$$

and similar estimates hold for derivatives of j_n. It follows that the expansion (9.3.6) may be differentiated term by term. By the uniqueness of Fourier coefficients, it follows that (9.3.5) implies Bessel's equation

$$r^2 j_n''(r) + r j_n'(r) + (r^2 - n^2)j_n(r) = 0, \quad n = 0, \pm 1, \pm 2, \ldots$$

We shall show that the functions j_n are precisely the Bessel functions J_n by computing the series expansion. The first step is to take $t = e^{i\theta}$ in the integral representation (9.3.7), which becomes

$$j_n(r) = \frac{1}{2\pi i}\int_C \exp\left(\frac{1}{2}r\left[t-\frac{1}{t}\right]\right) t^{-n-1} \, dt, \tag{9.3.8}$$

where C is the unit circle $\{|t| = 1\}$. Since the integrand is holomorphic in the plane minus the origin, we may replace the circle with a more convenient contour, one that begins and ends at $-\infty$ and encircles the origin in the positive (counter-clockwise) direction. With this choice of contour we may define j_ν for all ν:

$$j_\nu(r) = \frac{1}{2\pi i} \int_C \exp\left(\frac{1}{2}r\left[t - \frac{1}{t}\right]\right) t^{-\nu-1} dt, \qquad (9.3.9)$$

where we take the curve to lie in the complement of the ray $(-\infty, 0]$ and take the argument of t in the interval $(-\pi, \pi)$. Taking $s = rt/2$ as the new variable of integration, the curve can be taken to be the same as before, and (9.3.9) is

$$j_\nu(r) = \frac{1}{2\pi i} \left(\frac{r}{2}\right)^\nu \int_C \exp\left(s - \frac{r^2}{4s}\right) s^{-\nu-1} ds$$

$$= \frac{1}{2\pi i} \left(\frac{r}{2}\right)^\nu \sum_{m=0}^{\infty} \frac{(-1)^m}{m!} \frac{r^{2m}}{2^{2m}} \int_C e^s s^{-m-\nu-1} ds. \qquad (9.3.10)$$

According to Hankel's integral formula (2.2.8), the contour integral in the sum in (9.3.10) is $2\pi i/\Gamma(\nu + m + 1)$. Comparison with (9.1.2) shows that $j_\nu = J_\nu$. The integral representation (9.3.9) is due to Schlömilch [357].

Summarizing, we have an integral formula due to Bessel [43],

$$J_n(x) = \frac{1}{2\pi} \int_0^{2\pi} e^{ix\sin\theta - in\theta} d\theta \qquad (9.3.11)$$

for integer n, and

$$J_\nu(x) = \frac{1}{2\pi i} \int_C \exp\left(\frac{1}{2}x\left[t - \frac{1}{t}\right]\right) t^{-\nu-1} dt$$

for arbitrary real ν. This last integral can also be put in the form known as the *Sommerfeld representation* [373],

$$J_\nu(x) = \frac{1}{2\pi} \int_C e^{ix\sin\theta - i\nu\theta} d\theta, \qquad (9.3.12)$$

by taking $t = e^{i\theta}$ and taking the path of integration in (9.3.12) to be the boundary of the strip defined by the inequalities

$$-\pi < \mathrm{Re}\, z < \pi, \quad \mathrm{Im}\, z > 0.$$

9.4 Hankel functions

Our starting point here is (9.3.1), which is in the canonical form discussed in Section 5.9:

$$pv'' + qv' + \lambda_\mu v = w^{-1}(pwv')' + \lambda_\mu v = 0,$$

where

$$\lambda_\mu = -q'\mu - \frac{1}{2}\mu(\mu - 1)p''.$$

As noted in that section, under certain conditions on the curve C, the integral

$$\int_C \frac{w(t)}{w(x)} \frac{p^\mu(t)\,dt}{(x-t)^{\mu+1}}$$

is a solution of the equation. In (9.3.1), $p(x) = x$, $q(x) = 2\nu + 1 + 2ix$, and $\lambda_\mu = (2\nu + 1)i$, so $\mu = -\nu - \frac{1}{2}$. Since $(pw)' = qw$, we take $w(x) = x^{2\nu}e^{2ix}$. Thus the proposed solutions $u(x) = x^\nu e^{ix}v(x)$ of the original equation (9.0.1) have the form

$$x^{-\nu}e^{-ix}\int_C e^{2it}t^{\nu-\frac{1}{2}}(x-t)^{\nu-\frac{1}{2}}\,dt. \tag{9.4.1}$$

If we take C to be the interval $[0,x]$, then the conditions of Theorem 5.9.2 are satisfied at a finite endpoint 0 or x so long as $\operatorname{Re}\nu > \frac{3}{2}$. Moreover, the integral converges so long as $\operatorname{Re}(\nu + \frac{1}{2}) > 0$, and by analytic continuation it continues to define a solution of (9.0.1).

Up to the multiplicative constant

$$c_\nu = \frac{2^\nu}{\sqrt{\pi}\,\Gamma(\nu + \frac{1}{2})}, \tag{9.4.2}$$

(9.4.1) becomes (9.3.3) after the change of variables $t = x - sx$ in the integral. There are two natural alternative choices for a path of integration: the positive imaginary axis $\{is; s > 0\}$, and the ray $\{x + is; s \geq 0\}$. This leads to the *Bessel functions of the third kind*, or *Hankel functions*, defined for $\operatorname{Re}(\nu + \frac{1}{2}) > 0$ by the Poisson representations

$$H_\nu^{(1)}(x) = -2ic_\nu x^{-\nu}e^{-ix}\int_0^\infty e^{2i(x+is)}[(x+is)(-is)]^{\nu-\frac{1}{2}}\,ds$$

$$= 2c_\nu x^{-\nu}e^{i(x-\frac{1}{4}\pi-\frac{1}{2}\nu\pi)}\int_0^\infty e^{-2s}[s(x+is)]^{\nu-\frac{1}{2}}\,ds; \tag{9.4.3}$$

$$H_\nu^{(2)}(x) = 2ic_\nu x^{-\nu}e^{-ix}\int_0^\infty e^{-2s}[is(x-is)]^{\nu-\frac{1}{2}}\,ds$$

$$= 2c_\nu x^{-\nu}e^{-i(x-\frac{1}{4}\pi-\frac{1}{2}\nu\pi)}\int_0^\infty e^{-2s}[s(x-is)]^{\nu-\frac{1}{2}}\,ds. \tag{9.4.4}$$

The reason for the choice of constants is that it leads to the identity

$$J_\nu(x) = \frac{1}{2}\left[H_\nu^{(1)}(x) + H_\nu^{(2)}(x)\right], \quad \operatorname{Re}\left(\nu + \frac{1}{2}\right) > 0. \tag{9.4.5}$$

Indeed, $2J_\nu(x) - H_\nu^{(1)}(x) - H_\nu^{(2)}(x)$ is an integral around a contour from $i\infty$ to $x + i\infty$ of a function that is holomorphic in the half-strip $\{0 < \operatorname{Re}z < x, \operatorname{Im}z > 0\}$

and vanishes exponentially as $\operatorname{Im} s \to +\infty$, so by Cauchy's theorem the result is zero.

It is easy to determine the asymptotics of the Hankel functions as $x \to +\infty$, and this can be used to determine the asymptotics of J_ν. As $x \to +\infty$,

$$\int_0^\infty e^{-2s}[s(x \pm is)]^{\nu-\frac{1}{2}}\, ds \sim \int_0^\infty e^{-2s}(sx)^{\nu-\frac{1}{2}}\, ds$$

$$= \frac{x^{\nu-\frac{1}{2}}}{2^{\nu+\frac{1}{2}}} \int_0^\infty e^{-t} t^{\nu-\frac{1}{2}}\, dt = \frac{x^{\nu-\frac{1}{2}}\,\Gamma(\nu+\frac{1}{2})}{2^{\nu+\frac{1}{2}}}.$$

Therefore as $|x| \to \infty$,

$$H_\nu^{(1)}(x) \sim \frac{\sqrt{2}}{\sqrt{\pi x}} e^{i(x-\frac{1}{4}\pi-\frac{1}{2}\nu\pi)};$$

$$\tag{9.4.6}$$

$$H_\nu^{(2)}(x) \sim \frac{\sqrt{2}}{\sqrt{\pi x}} e^{-i(x-\frac{1}{4}\pi-\frac{1}{2}\nu\pi)}.$$

Writing

$$(x-t)^{\nu-\frac{1}{2}} = x^{\nu-\frac{1}{2}}\left(1-\frac{t}{x}\right)^{\nu-\frac{1}{2}}$$

$$\sim x^{\nu-\frac{1}{2}}\left[1-\left(\nu-\frac{1}{2}\right)\frac{t}{x}+\cdots\right],$$

we may extend the asymptotics to full asymptotic series:

$$H_\nu^{(1)}(x) \sim \frac{\sqrt{2}}{\sqrt{\pi x}} e^{i(x-\frac{1}{4}\pi-\frac{1}{2}\nu\pi)}$$

$$\times \sum_{m=0}^\infty \frac{(-i)^m(-\nu+\frac{1}{2})_m(\nu+\frac{1}{2})_m}{2^m m!} x^{-m};$$

$$\tag{9.4.7}$$

$$H_\nu^{(2)}(x) \sim \frac{\sqrt{2}}{\sqrt{\pi x}} e^{-i(x-\frac{1}{4}\pi-\frac{1}{2}\nu\pi)}$$

$$\times \sum_{m=0}^\infty \frac{i^m(-\nu+\frac{1}{2})_m(\nu+\frac{1}{2})_m}{2^m m!} x^{-m}.$$

$$\tag{9.4.8}$$

The verification is left as an exercise.

It follows from (9.4.6) that the Hankel functions are independent and that

$$J_\nu(x) \sim \frac{\sqrt{2}}{\sqrt{\pi x}}\cos\left(x-\frac{1}{4}\pi-\frac{1}{2}\nu\pi\right), \qquad \operatorname{Re}(2\nu+1) > 0. \tag{9.4.9}$$

This result is due to Poisson [331] ($\nu = 0$), Hansen [178] ($\nu = 1$), Jacobi [199] (ν an integer), and Hankel [177]. The recurrence identity (9.1.5) implies that

$-J_{\nu-2}$ has the same principal asymptotic behavior as J_ν, so (9.4.9) extends to all complex ν. This allows us to compute the asymptotics of Y_ν using (9.1.11):

$$Y_\nu(x) \sim \frac{\sqrt{2}}{\sqrt{\pi x}\sin\nu\pi}$$

$$\times \left[\cos\nu\pi \, \cos\left(x - \frac{1}{4}\pi - \frac{1}{2}\nu\pi\right) - \cos\left(x - \frac{1}{4}\pi + \frac{1}{2}\nu\pi\right)\right].$$

The trigonometric identity

$$\cos 2b \cos(a-b) - \cos(a+b) = \sin 2b \sin(a-b),$$

with $a = x - \frac{1}{4}\pi$ and $b = \frac{1}{2}\nu\pi$, gives

$$Y_\nu(x) \sim \frac{\sqrt{2}}{\sqrt{\pi x}} \sin\left(x - \frac{1}{4}\pi - \frac{1}{2}\nu\pi\right). \tag{9.4.10}$$

For $\mathrm{Re}(\nu + \frac{1}{2}) > 0$, Y_ν is a linear combination of the Hankel functions. It follows from (9.4.6) and (9.4.10) that

$$Y_\nu(x) = \frac{1}{2i}\left[H_\nu^{(1)}(x) - H_\nu^{(2)}(x)\right]. \tag{9.4.11}$$

Conversely, (9.4.5) and (9.4.11) imply

$$\begin{aligned}
H_\nu^{(1)}(x) &= J_\nu(x) + iY_\nu(x); \\
H_\nu^{(2)}(x) &= J_\nu(x) - iY_\nu(x)
\end{aligned} \tag{9.4.12}$$

for $\mathrm{Re}\left(\nu + \frac{1}{2}\right) > 0$. We may use (9.4.12) to *define* the Hankel functions for all values of ν. Then the identities (9.4.5) and (9.4.11) are valid for all complex ν, as are the asymptotics (9.4.6) and (by analytic continuation) the asymptotic series (9.4.7) and (9.4.8). In particular,

$$\begin{aligned}
H_{\frac{1}{2}}^{(1)}(x) &= -i\frac{\sqrt{2}}{\sqrt{\pi x}}e^{ix}; & H_{-\frac{1}{2}}^{(1)}(x) &= \frac{\sqrt{2}}{\sqrt{\pi x}}e^{ix}; \\
H_{\frac{1}{2}}^{(2)}(x) &= i\frac{\sqrt{2}}{\sqrt{\pi x}}e^{-ix}; & H_{-\frac{1}{2}}^{(2)}(x) &= \frac{\sqrt{2}}{\sqrt{\pi x}}e^{-ix}.
\end{aligned} \tag{9.4.13}$$

It follows from (9.4.12) that the Wronskian is

$$W(H_\nu^{(1)}, H_\nu^{(2)})(x) = -2iW(J_\nu, Y_\nu)(x) = \frac{4}{\pi ix}. \tag{9.4.14}$$

It also follows from (9.4.12) that the Hankel functions satisfy the analogues of (9.1.5), (9.1.6), (9.1.14), and (9.1.15):

$$H_{\nu-1}^{(1)}(x) + H_{\nu+1}^{(1)}(x) = \frac{2\nu}{x}H_\nu^{(1)}(x); \tag{9.4.15}$$

$$H_{\nu-1}^{(2)}(x) - H_{\nu+1}^{(2)}(x) = 2\left[H_\nu^{(2)}\right]'(x). \tag{9.4.16}$$

For a given index ν, the functions J_ν, $J_{-\nu}$, Y_ν, $Y_{-\nu}$, $H_\nu^{(1)}$, $H_\nu^{(2)}$, $H_{-\nu}^{(1)}$, and $H_{-\nu}^{(2)}$ are all solutions of (9.0.1), so any choice of three of these functions satisfies a linear relation. The relations not already given above can be obtained easily from the asymptotics (9.4.6), (9.4.9), and (9.4.10). In particular,

$$H_{-\nu}^{(1)}(x) = e^{i\pi\nu} H_\nu^{(1)}(x);$$

$$H_{-\nu}^{(2)}(x) = e^{-i\pi\nu} H_\nu^{(2)}(x). \tag{9.4.17}$$

9.5 Modified Bessel functions

Replacing x by ix in Bessel's equation (9.0.1) yields

$$x^2 u''(x) + x u'(x) - (x^2 + \nu^2) u(x) = 0. \tag{9.5.1}$$

Solutions of this equation are known as *modified Bessel functions*. The most obvious way to obtain solutions is to evaluate the Bessel and Hankel functions on the positive imaginary axis. It is natural to choose one solution by modifying $J_\nu(ix)$ so that it takes real values. The result is

$$I_\nu(x) = \sum_{m=0}^{\infty} \frac{1}{\Gamma(\nu+1+m)m!} \left(\frac{x}{2}\right)^{\nu+2m} = e^{-\frac{1}{2}i\nu\pi} J_\nu(ix). \tag{9.5.2}$$

It follows from (9.4.6) that $H_\nu^{(1)}(ix)$ decays exponentially as $x \to +\infty$, while $H_\nu^{(2)}(ix)$ grows exponentially. Therefore it is natural to obtain a second solution by modifying $H_\nu^{(1)}(ix)$:

$$K_\nu(x) = \frac{\pi}{2} e^{\frac{1}{2}i(\nu+1)\pi} H_\nu^{(1)}(ix) = \frac{\pi}{2} \frac{I_{-\nu}(x) - I_\nu(x)}{\sin \pi \nu}. \tag{9.5.3}$$

The Poisson integral representations (9.3.4) and (9.4.3) lead to

$$I_\nu(x) = \frac{1}{\sqrt{\pi}\,\Gamma(\nu+\frac{1}{2})} \left(\frac{x}{2}\right)^\nu \int_{-1}^{1} \cosh(xt)(1-t^2)^{\nu-\frac{1}{2}}\, dt; \tag{9.5.4}$$

$$K_\nu(x) = \frac{\sqrt{\pi}}{\sqrt{2x}\,\Gamma(\nu+\frac{1}{2})} e^{-x} \int_0^{\infty} e^{-t}\left(t + \frac{t^2}{2x}\right)^{\nu-\frac{1}{2}}\, dt. \tag{9.5.5}$$

A consequence is that I_ν and K_ν are positive, $0 < x < \infty$.

The derivative formula

$$\frac{d}{dx}\left[x^{\nu}I_{\nu}(x)\right] = x^{\nu}I_{\nu-1}(x)$$

follows from (9.1.3) or directly from the expansion (9.5.2), and leads to the relations

$$I_{\nu-1}(x) - I_{\nu+1}(x) = \frac{2\nu}{x}I_{\nu}(x);$$

$$I_{\nu-1}(x) + I_{\nu+1}(x) = 2I_{\nu}'(x).$$

These imply the corresponding relations for K_{ν}:

$$K_{\nu-1}(x) - K_{\nu+1}(x) = -\frac{2\nu}{x}K_{\nu}(x);$$

$$K_{\nu-1}(x) + K_{\nu+1}(x) = -2K_{\nu}'(x).$$

The asymptotic relation (9.4.9) implies

$$I_{\nu}(x) \sim \frac{e^{x}}{\sqrt{2\pi x}};$$

$$K_{\nu}(x) \sim \frac{\sqrt{\pi}e^{-x}}{\sqrt{2x}}. \tag{9.5.6}$$

Full asymptotic expansions may be obtained from (9.4.7) and (9.4.8). The principal terms in the derivatives come from differentiating (9.5.6), so the Wronskian is

$$W(K_{\nu}, I_{\nu})(x) = \frac{1}{x}.$$

9.6 Addition theorems

In Section 9.3 we established the identity

$$e^{ix\sin\theta} = \sum_{n=-\infty}^{\infty} J_n(x)e^{in\theta}. \tag{9.6.1}$$

Taking $t = e^{i\theta}$, we may write this in the form of a generating function for the Bessel functions of integral order:

$$G(x,t) = \sum_{n=-\infty}^{\infty} J_n(x)t^n = e^{\frac{1}{2}x(t-1/t)}, \quad |t| = 1. \tag{9.6.2}$$

Moreover,

$$\sum_{n=-\infty}^{\infty} J_n(x+y)\,e^{in\theta} = e^{i(x+y)\sin\theta} = \sum_{n=-\infty}^{\infty} J_n(x)\,e^{in\theta} \sum_{n=-\infty}^{\infty} J_n(y)\,e^{in\theta}.$$

Equating coefficients of $e^{in\theta}$ gives the *addition formula*

$$J_n(x+y) = \sum_{m=-\infty}^{\infty} J_m(x)J_{n-m}(y).$$

This is a special case of a more general addition formula. Consider a plane triangle whose vertices, in polar coordinates, are the origin and the points (r_1,θ_1) and (r_2,θ_2). Let r be the length of the third side. To fix ideas, suppose that $0 < \theta_1 < \theta_2 < \frac{1}{2}\pi$. Then the triangle lies in the first quadrant and the angle θ opposite the side of length r is $\theta_2 - \theta_1$. Projecting onto the vertical axis gives the identity

$$r\sin(\theta_2+\varphi) = r_2\sin\theta_2 - r_1\sin\theta_1 = r_2\sin\theta_2 - r_1\sin(\theta_2-\theta),$$

where $\theta = \theta_2 - \theta_1$ is the angle opposite the side with length r and φ is the angle opposite the side with length r_1. By analytic continuation, this identity carries over to general values of θ_2, with

$$r^2 = r_1^2 + r_2^2 - 2r_1r_2\cos\theta, \quad \theta = \theta_2 - \theta_1.$$

According to (9.3.12),

$$\begin{aligned}
J_\nu(r)\,e^{i\nu\varphi} &= \frac{1}{2\pi}\int_C e^{ir\sin(\theta_2+\varphi)-i\nu\theta_2}\,d\theta_2 \\
&= \frac{1}{2\pi}\int_C e^{ir_2\sin\theta_2-ir_1\sin(\theta_2-\theta)-i\nu\theta_2}\,d\theta_2 \\
&= \frac{1}{2\pi}\int_C e^{ir_1\sin(\theta-\theta_2)}\,e^{ir_2\sin\theta_2-i\nu\theta_2}\,d\theta_2.
\end{aligned}$$

By (9.6.1),

$$e^{ir_1\sin(\theta-\theta_2)} = \sum_{n=-\infty}^{\infty} J_n(r_1)\,e^{in(\theta-\theta_2)}.$$

Inserting this into the preceding integral and interchanging integration and summation gives

$$J_\nu(r)\,e^{i\nu\varphi} = \sum_{n=-\infty}^{\infty} J_n(r_1)\,e^{in\theta} \cdot \frac{1}{2\pi}\int_C e^{ir_2\sin\theta_2-i(\nu+n)\theta_2}\,d\theta_2.$$

Using (9.3.12) again, we have *Graf's addition formula* [171]:

$$J_\nu(r)e^{i\nu\varphi} = \sum_{n=-\infty}^{\infty} J_n(r_1)J_{\nu+n}(r_2)e^{in\theta}. \tag{9.6.3}$$

This generalizes an addition formula of Neumann; see [425], Section 11.1. A deeper result is *Gegenbauer's addition formula* [158]:

$$\frac{J_\nu(r)}{r^\nu} = 2^\nu \, \Gamma(\nu) \sum_{n=0}^{\infty} (\nu+n)\frac{J_{\nu+n}(r_1)J_{\nu+n}(r_2)}{r_1^\nu \, r_2^\nu} \, C_n^\nu(\cos\theta), \tag{9.6.4}$$

where C_n^ν are the Gegenbauer polynomials, expressed in terms of Jacobi polynomials as

$$C_n^\nu(x) = \frac{(2\nu)_n}{(\nu+\frac{1}{2})_n} \, P_n^{(\nu-\frac{1}{2},\nu-\frac{1}{2})}(x).$$

For a simple derivation of (9.6.4) from Graf's formula when ν is an integer, see [10]. For a proof in the general case see [305].

9.7 Fourier transform and Hankel transform

If $f(x) = f(x_1,x_2)$ is absolutely integrable, i.e.

$$\int_{-\infty}^{\infty}\int_{-\infty}^{\infty} |f(x_1,x_2)| \, dx_1 \, dx_2 \, < \, \infty,$$

then the *Fourier transform* of f is the function

$$\widehat{f}(\xi_1,\xi_2) = \frac{1}{2\pi} \int_{-\infty}^{\infty}\int_{-\infty}^{\infty} e^{-i(x_1\xi_1+x_2\xi_2)} f(x_1,x_2) \, dx_1 \, dx_2.$$

If f is continuous and $\widehat{f}$ is also absolutely integrable, then

$$f(x_1,x_2) = \frac{1}{2\pi} \int_{-\infty}^{\infty}\int_{-\infty}^{\infty} e^{i(x_1\xi_1+x_2\xi_2)} \widehat{f}(\xi_1,\xi_2) \, d\xi_1 \, d\xi_2;$$

see Appendix B.

If f is an absolutely integrable function on the half-line $(0,\infty)$, its nth *Hankel transform* is

$$g(y) = \int_0^{\infty} J_n(xy)f(x)x\,dx. \tag{9.7.1}$$

This Fourier inversion formula above can be used to show that the Hankel transform is its own inverse: if f is an absolutely integrable function on the half-line $[0,\infty)$ and its nth Hankel transform g is also integrable, then f is the *n*th Hankel transform of g:

$$f(x) = \int_0^\infty J_n(xy)g(y)y\,dy. \tag{9.7.2}$$

To prove (9.7.2) given (9.7.1), we first write the two-variable Fourier transform in polar coordinates (x,θ) and (y,φ):

$$\widehat{F}(y,\varphi) = \frac{1}{2\pi}\int_0^\infty\int_0^{2\pi} e^{-ixy\cos(\theta-\varphi)}F(x,\theta)\,d\theta\,x\,dx; \tag{9.7.3}$$

$$F(x,\theta) = \frac{1}{2\pi}\int_0^\infty\int_0^{2\pi} e^{ixy\cos(\theta-\varphi)}\widehat{F}(y,\varphi)\,d\varphi\,y\,dy. \tag{9.7.4}$$

Take $F(x,\theta) = f(x)e^{-in\theta}$. Then the integration with respect to θ in (9.7.3) gives

$$\int_0^{2\pi} e^{-ixy\cos(\theta-\varphi)-in\theta}\,d\theta = e^{-in\varphi}\int_0^{2\pi} e^{-ixy\cos\theta-in\theta}\,d\theta, \tag{9.7.5}$$

where we used periodicity to keep the limits of integration unchanged when changing the variable of integration. Since $-\cos\theta = \sin\left(\theta - \frac{1}{2}\pi\right)$, we may change variables once again and conclude that (9.7.5) is

$$(-i)^n e^{-in\varphi}\int_0^{2\pi} e^{ixy\sin\theta-in\theta}\,d\theta = (-i)^n e^{-in\varphi}2\pi\,J_n(xy) \tag{9.7.6}$$

by (9.3.11). Therefore the Fourier transform of $f(x)e^{-in\theta}$ is

$$(-i)^n e^{-in\varphi}g(y), \quad g(y) = \int_0^\infty J_n(xy)f(x)x\,dx.$$

In (9.7.4), therefore, the integral with respect to φ is

$$\int_0^{2\pi} e^{ixy\cos(\theta-\varphi)-in\varphi}\,d\varphi = e^{-in\theta}\int_0^{2\pi} e^{ixy\cos\varphi-in\varphi}\,d\varphi. \tag{9.7.7}$$

Proceeding as above and using the identity $\cos\varphi = \sin\left(\varphi + \frac{1}{2}\pi\right)$, we find that (9.7.7) is

$$i^n e^{-in\theta}2\pi\,J_n(xy).$$

Therefore the right-hand side of (9.7.4) is

$$e^{-in\theta}\int_0^\infty J_n(xy)g(y)y\,dy.$$

This proves (9.7.2), given (9.7.1).

9.8 Integrals of Bessel functions

We have shown that $J_\nu(x) \sim c_\nu x^\nu$ as $x \to 0+$ and $J_\nu(x) = O\left(x^{-\frac{1}{2}}\right)$ as $x \to \infty$. It follows that if f is any continuous function such that

$$\int_0^1 |f(x)| x^\nu \, dx + \int_1^\infty |f(x)| \frac{dx}{\sqrt{x}} < \infty, \qquad (9.8.1)$$

then the product fJ_ν is absolutely integrable on $(0, \infty)$. For $\nu > -1$, it is not difficult to show that the integral can be obtained as a series by using the power series expansion (9.1.2) and integrating term by term:

$$\int_0^\infty f(x) J_\nu(x) \, dx = \sum_{m=0}^\infty \frac{(-1)^m}{\Gamma(\nu+m+1) m!} \int_0^\infty f(x) \left(\frac{x}{2}\right)^{\nu+2m} dx, \qquad (9.8.2)$$

so long as the last series is absolutely convergent. As an example, let

$$f(x) = x^{a-1} e^{-sx^2}, \quad \mathrm{Re}(a+\nu) > 0, \quad \mathrm{Re}\, s > 0.$$

Then

$$\int_0^\infty f(x) \left(\frac{x}{2}\right)^{\nu+2m} dx = 2^{-\nu-2m-1} \int_0^\infty e^{-sy} y^{\frac{1}{2}(\nu+a)+m} \frac{dy}{y}$$

$$= \frac{\Gamma\left(\frac{1}{2}\nu + \frac{1}{2}a + m\right)}{2^{\nu+2m+1} s^{\frac{1}{2}(\nu+a)+m}},$$

so

$$\int_0^\infty x^{a-1} e^{-sx^2} J_\nu(x) \, dx = \frac{1}{2^{\nu+1} s^{\frac{1}{2}(\nu+a)}} \sum_{m=0}^\infty \frac{\Gamma\left(\frac{1}{2}\nu + \frac{1}{2}a + m\right)}{\Gamma(\nu+m+1) m!} \left(-\frac{1}{4s}\right)^m$$

$$= \frac{\Gamma\left(\frac{1}{2}\nu + \frac{1}{2}a\right)}{\Gamma(\nu+1) 2^{\nu+1} s^{\frac{1}{2}(\nu+a)}} \sum_{m=0}^\infty \frac{\left(\frac{1}{2}\nu + \frac{1}{2}a\right)_m}{(\nu+1)_m m!} \left(-\frac{1}{4s}\right)^m.$$

The last sum is the Kummer function M with indices $\frac{1}{2}(\nu+a), \nu+1$ evaluated at $-1/4s$, so for $\mathrm{Re}(\nu+a) > 0$ and $\mathrm{Re}\, s > 0$,

$$\int_0^\infty x^{a-1} e^{-sx^2} J_\nu(x) \, dx$$

$$= \frac{\Gamma\left(\frac{1}{2}\nu + \frac{1}{2}a\right)}{\Gamma(\nu+1) 2^{\nu+1} s^{\frac{1}{2}(\nu+a)}} M\left(\frac{1}{2}\nu + \frac{1}{2}a, \nu+1; -\frac{1}{4s}\right). \qquad (9.8.3)$$

In particular, if $a = \nu + 2$ so that $\frac{1}{2}(\nu+a) = \nu+1$, this simplifies to

$$\int_0^\infty x^{\nu+1} e^{-sx^2} J_\nu(x) \, dx = \frac{e^{-1/4s}}{(2s)^{\nu+1}}.$$

As a second example, let

$$f(x) = x^{a-1}e^{-sx}, \quad a+v > 0, \quad s > 0,$$

to compute the Laplace transform of $x^{a-1}J_v(x)$. Then

$$\int_0^\infty f(x)\left(\frac{x}{2}\right)^{v+2m} dx = \frac{1}{2^{v+2m}}\int_0^\infty e^{-sx}x^{v+a+2m-1}\,dx$$

$$= \frac{\Gamma(v+a+2m)}{2^{v+2m}\,s^{v+a+2m}}.$$

Now

$$\Gamma(v+a+2m) = \Gamma(v+a)(v+a)_{2m}$$

$$= \Gamma(v+a)2^{2m}\left(\frac{1}{2}v+\frac{1}{2}a\right)_m\left(\frac{1}{2}v+\frac{1}{2}a+\frac{1}{2}\right)_m.$$

Therefore

$$\int_0^\infty x^{a-1}e^{-sx}J_v(x)\,dx$$

$$= \frac{\Gamma(v+a)}{\Gamma(v+1)2^v s^{v+a}}\sum_{m=0}^\infty \frac{\left(\frac{1}{2}v+\frac{1}{2}a\right)_m\left(\frac{1}{2}v+\frac{1}{2}a+\frac{1}{2}\right)_m}{(v+1)_m\,m!}\left(-\frac{1}{s^2}\right)^m.$$

The last sum converges for $s > 1$ to the hypergeometric function with indices $\frac{1}{2}(v+a)$, $\frac{1}{2}(v+a+1)$, $v+1$ evaluated at $-1/s^2$. By analytic continuation, the following identity holds for all s, v, and a with $\operatorname{Re} s > 0$, $\operatorname{Re}(v+a) > 0$:

$$\int_0^\infty x^{a-1}e^{-sx}J_v(x)\,dx$$

$$= \frac{\Gamma(v+a)}{\Gamma(v+1)2^v s^{v+a}}F\left(\frac{1}{2}v+\frac{1}{2}a,\frac{1}{2}v+\frac{1}{2}a+\frac{1}{2},v+1;-\frac{1}{s^2}\right).$$

Corresponding to the various cases (10.6.2), (10.6.5), and (10.6.6) in Section 10.6, we obtain

$$\int_0^\infty e^{-xs}x^v\,J_v(x)\,dx = \frac{\Gamma(2v+1)}{\Gamma(v+1)2^v}\cdot\frac{1}{(1+s^2)^{v+\frac{1}{2}}}; \qquad (9.8.4)$$

$$\int_0^\infty e^{-xs}x^{v+1}J_v(x)\,dx = \frac{\Gamma(2v+2)}{\Gamma(v+1)2^v}\cdot\frac{s}{(1+s^2)^{v+\frac{3}{2}}}; \qquad (9.8.5)$$

$$\int_0^\infty e^{-xs}x^{-1}J_v(x)\,dx = \frac{1}{v(s+\sqrt{1+s^2})^v}; \qquad (9.8.6)$$

$$\int_0^\infty e^{-xs}J_v(x)\,dx = \frac{1}{\sqrt{1+s^2}(s+\sqrt{1+s^2})^v}. \qquad (9.8.7)$$

9.9 Airy functions

If v is a solution of Bessel's equation (9.0.1) and u is defined by

$$u(x) = x^a\, v(bx^c),$$

where a, b, and c are constants, then u is a solution of *Lommel's equation*

$$x^2 u''(x) + (1 - 2a)xu'(x) + \left[b^2 c^2 x^{2c} + a^2 - c^2 v^2\right] u(x) = 0. \qquad (9.9.1)$$

If v is, instead, a solution of the modified Bessel equation (9.5.1), then u is a solution of the corresponding modified Lommel equation

$$x^2 u''(x) + (1 - 2a)xu'(x) - [b^2 c^2 x^{2c} - a^2 + c^2 v^2]u(x) = 0. \qquad (9.9.2)$$

The particular case of (9.9.2) with $v^2 = \frac{1}{9}$, $a = \frac{1}{2}$, $b = \frac{2}{3}$, $c = \frac{3}{2}$ gives the *Airy equation*

$$u''(x) - xu(x) = 0. \qquad (9.9.3)$$

The calculation is reversible: any solution of (9.9.3) for $x > 0$ has the form

$$u(x) = x^{\frac{1}{2}}\, v\left(\frac{2}{3}x^{\frac{3}{2}}\right),$$

where v is a solution of the modified Bessel equation. The standard choices are the *Airy functions*

$$\mathrm{Ai}(x) = \frac{\sqrt{x}}{\pi\sqrt{3}}\, K_{\frac{1}{3}}\left(\frac{2}{3}x^{\frac{3}{2}}\right)$$

$$= \frac{\sqrt{x}}{3}\left[I_{-\frac{1}{3}}\left(\frac{2}{3}x^{\frac{3}{2}}\right) - I_{\frac{1}{3}}\left(\frac{2}{3}x^{\frac{3}{2}}\right)\right], \qquad (9.9.4)$$

and

$$\mathrm{Bi}(x) = \frac{\sqrt{x}}{\sqrt{3}}\left[I_{-\frac{1}{3}}\left(\frac{2}{3}x^{\frac{3}{2}}\right) + I_{\frac{1}{3}}\left(\frac{2}{3}x^{\frac{3}{2}}\right)\right]. \qquad (9.9.5)$$

It follows from (9.5.2) that the series expansions of the Airy functions are

$$\mathrm{Ai}(x) = \sum_{n=0}^{\infty}\left[\frac{x^{3n}}{3^{2n+\frac{2}{3}}\Gamma(n+\frac{2}{3})n!} - \frac{x^{3n+1}}{3^{2n+\frac{4}{3}}\Gamma(n+\frac{4}{3})n!}\right];$$

$$\qquad (9.9.6)$$

$$\mathrm{Bi}(x) = \sqrt{3}\sum_{n=0}^{\infty}\left[\frac{x^{3n}}{3^{2n+\frac{2}{3}}\Gamma(n+\frac{2}{3})n!} + \frac{x^{3n+1}}{3^{2n+\frac{4}{3}}\Gamma(n+\frac{4}{3})n!}\right],$$

which show that these are entire functions of x. It follows from these expansions that the initial conditions are

$$\mathrm{Ai}(0) = \frac{1}{3^{\frac{2}{3}}\Gamma\left(\frac{2}{3}\right)}, \qquad \mathrm{Ai}'(0) = -\frac{1}{3^{\frac{4}{3}}\Gamma\left(\frac{4}{3}\right)};$$

$$\mathrm{Bi}(0) = \frac{1}{3^{\frac{1}{6}}\Gamma\left(\frac{2}{3}\right)}, \qquad \mathrm{Bi}'(0) = \frac{1}{3^{\frac{5}{6}}\Gamma\left(\frac{4}{3}\right)}.$$

The Wronskian is constant, so the constant is

$$W(\mathrm{Ai},\mathrm{Bi})(x) = \frac{2}{3^{\frac{3}{2}}\Gamma\left(\frac{2}{3}\right)\Gamma\left(\frac{4}{3}\right)}.$$

The asymptotics of the Airy functions as $x \to +\infty$ follow from the asymptotics of the modified Bessel functions. The leading terms are

$$\mathrm{Ai}(x) \sim \frac{e^{-\frac{2}{3}x^{\frac{3}{2}}}}{2\sqrt{\pi}\,x^{\frac{1}{4}}}; \qquad \mathrm{Bi}(x) \sim \frac{e^{\frac{2}{3}x^{\frac{3}{2}}}}{\sqrt{\pi}\,x^{\frac{1}{4}}}. \tag{9.9.7}$$

The principal terms in the asymptotics of the derivatives are obtained by differentiating (9.9.7). This gives a second determination of the Wronskian:

$$W(\mathrm{Ai},\mathrm{Bi})(x) = \frac{1}{\pi}.$$

The asymptotics of the Airy functions for $x \to -\infty$ can be obtained from the asymptotics of the Bessel functions J_ν. Replacing x by $-x$ in the series expansions shows that

$$\mathrm{Ai}(-x) = \frac{\sqrt{x}}{3}\left[J_{-\frac{1}{3}}\left(\frac{2}{3}x^{\frac{3}{2}}\right) + J_{\frac{1}{3}}\left(\frac{2}{3}x^{\frac{3}{2}}\right)\right];$$

$$\mathrm{Bi}(-x) = \frac{\sqrt{x}}{\sqrt{3}}\left[J_{-\frac{1}{3}}\left(\frac{2}{3}x^{\frac{3}{2}}\right) - J_{\frac{1}{3}}\left(\frac{2}{3}x^{\frac{3}{2}}\right)\right].$$

It follows from (9.4.9) that as $x \to +\infty$,

$$\mathrm{Ai}(-x) \sim \frac{\cos\left(\frac{2}{3}x^{\frac{3}{2}} - \frac{1}{4}\pi\right)}{\sqrt{\pi}\,x^{\frac{1}{4}}}; \quad \mathrm{Bi}(-x) \sim -\frac{\sin\left(\frac{2}{3}x^{\frac{3}{2}} - \frac{1}{4}\pi\right)}{\sqrt{\pi}\,x^{\frac{1}{4}}}.$$

The original function that arose in Airy's research on optics was defined by the integral

$$\frac{1}{\pi}\int_0^\infty \cos\left(\frac{1}{3}t^3 + xt\right)dt. \tag{9.9.8}$$

To see that this integral is equal to the function (9.9.6), we first make a change of variable $t = \tau/i$, so that the integral becomes

$$\frac{1}{\pi i} \int_0^{i\infty} \cosh\left(\frac{1}{3}\tau^3 - x\tau\right) d\tau = \frac{1}{2\pi i} \int_{-i\infty}^{i\infty} \exp\left(\frac{1}{3}\tau^3 - x\tau\right) d\tau. \qquad (9.9.9)$$

By Cauchy's theorem, the vertical line of integration can be deformed into a contour C which begins at infinity in the sector $-\pi/2 < \arg\tau < -\pi/6$ and ends at infinity in the sector $\pi/6 < \arg\tau < \pi/2$. So far we have restricted x to be real, but the function

$$\mathrm{Ai}(z) = \frac{1}{2\pi i} \int_C \exp\left(\frac{1}{3}\tau^3 - z\tau\right) d\tau \qquad (9.9.10)$$

is an entire function in z, since the integrand vanishes rapidly at the endpoints of C. Furthermore, by differentiation under the integral sign, one can readily see that this integral satisfies the Airy equation (9.9.3). From (9.9.10), it can be verified that the integral in (9.9.8) has the Maclaurin expansion given in (9.9.6).

Let $\omega = \exp(2\pi i/3)$. In addition to $\mathrm{Ai}(z)$, the functions $\mathrm{Ai}(\omega z)$ and $\mathrm{Ai}(\omega^2 z)$ are also solutions of (9.9.3). With the aid of Cauchy's theorem, one can use (9.9.10) to show that these three solutions are connected by the relation

$$\mathrm{Ai}(z) + \omega\mathrm{Ai}(wz) + \omega^2\mathrm{Ai}(\omega^2 z) = 0. \qquad (9.9.11)$$

Returning to (9.9.9), we restrict x to be positive:

$$\mathrm{Ai}(x) = \frac{1}{2\pi i} \int_{-i\infty}^{i\infty} \exp\left(\frac{1}{3}\tau^3 - x\tau\right) d\tau.$$

By deforming the imaginary axis into the parallel vertical line $\mathrm{Re}\,\tau = \sqrt{x}$, one can show that

$$\mathrm{Ai}(x) = \frac{1}{2\pi} e^{-\frac{2}{3}x^{3/2}} \int_{-\infty}^{\infty} e^{-\sqrt{x}\rho^2 - \frac{1}{3}i\rho^3} d\rho$$

$$= \frac{1}{\pi} e^{-\frac{2}{3}x^{3/2}} \int_0^{\infty} e^{-\sqrt{x}\rho^2} \cos\left(\frac{1}{3}\rho^3\right) d\rho. \qquad (9.9.12)$$

By analytic continuation, this holds for $|\arg x| < \pi$. Let us now replace x by z, and make the change of variable $\rho^2 = u$, so that (9.9.12) becomes

$$\mathrm{Ai}(z) = \frac{1}{2\pi} e^{-\frac{2}{3}z^{3/2}} \int_0^{\infty} e^{-\sqrt{z}u} \cos\left(\frac{1}{3}u^{\frac{3}{2}}\right) \frac{du}{\sqrt{u}}, \qquad (9.9.13)$$

valid for $|\arg z| < \pi$. By expanding the cosine function into a Maclaurin series and integrating term by term, we obtain a sharper version of (9.9.7):

the asymptotic expansion

$$\text{Ai}(z) \sim \frac{1}{2\pi z^{\frac{1}{4}}} e^{-\frac{2}{3}z^{\frac{3}{2}}} \sum_{n=0}^{\infty} \frac{\Gamma\left(3n+\frac{1}{2}\right)(-1)^n}{3^{2n}(2n)!} \frac{1}{z^{\frac{3}{2}n}}$$

as $z \to \infty$ in $|\arg z| < \pi$.

9.10 Exercises

1. Show that

$$\lim_{a \to +\infty} M\left(a, v+1; -\frac{x}{a}\right) = \Gamma(v+1)x^{-\frac{1}{2}v} J_v\left(2\sqrt{x}\right).$$

2. Show that the Fourier transform of the restriction to the interval $-1 < x < 1$ of the Legendre polynomial P_n,

$$\frac{1}{\sqrt{2\pi}} \int_{-1}^{1} e^{-ix\xi} P_n(x)\,dx,$$

   is

$$\frac{(-i)^n}{\sqrt{\xi}} J_{n+\frac{1}{2}}(\xi).$$

   Hint: use the Rodrigues formula.

3. Use Exercise 2 and the Fourier inversion formula to show that

$$\frac{1}{\sqrt{2\pi}} \int_{-\infty}^{\infty} e^{ix\xi} J_{n+\frac{1}{2}}(\xi)\frac{d\xi}{\sqrt{\xi}} = \begin{cases} i^n P_n(x), & |x| < 1, \\ 0, & |x| > 1. \end{cases}$$

4. Use Exercise 2 to show that the expansion of the plane wave $e^{i\kappa x}$ as a sum of Legendre polynomials is

$$e^{i\kappa x} = \sum_{n=0}^{\infty} i^n \left(n+\frac{1}{2}\right) \frac{\sqrt{2\pi}}{\sqrt{\kappa}} J_{n+\frac{1}{2}}(\kappa) P_n(x).$$

5. Use Exercise 4 and the orthogonality properties of the Legendre polynomials to prove the integral formula

$$J_{n+\frac{1}{2}}(\kappa) = (-i)^n \frac{\sqrt{\kappa}}{\sqrt{2\pi}} \int_{-1}^{1} e^{i\kappa x} P_n(x)\,dx.$$

6. Show that the expansion of the plane wave $e^{i\kappa x}$ in terms of Gegenbauer polynomials is

$$e^{i\kappa x} = \Gamma(\lambda) \sum_{n=0}^{\infty} i^n (n+\lambda) \left(\frac{\kappa}{2}\right)^{-\lambda} J_{\lambda+n}(\kappa) C_n^{\lambda}(x).$$

7. Use the orthogonality property of the Gegenbauer polynomials to derive an integral formula for $J_{\lambda+n}(\kappa)$ involving $C_n^\lambda(x)$.

8. Prove (9.1.16).

9. Verify the following relations:

$$J_{\nu+1}(x)J_{-\nu}(x)+J_\nu(x)J_{-(\nu+1)}(x) = -\frac{2\sin\nu\pi}{\pi x};$$

$$J_{\nu+1}(x)Y_\nu(x)-J_\nu(x)Y_{\nu+1}(x) = \frac{2}{\pi x};$$

$$I_\nu(x)K_{\nu+1}(x)+I_{\nu+1}(x)K_\nu(x) = \frac{1}{x}.$$

10. Prove that the positive zeros of any two linearly independent real cylinder functions of the same order are interlaced.

11. Let $y_{\nu,n}$ denote the nth positive zero of $Y_\nu(x)$. Prove that when $\nu > -\frac{1}{2}$,

$$y_{\nu,1} < j_{\nu,1} < y_{\nu,2} < j_{\nu,2} < \cdots$$

12. Assume that for fixed n, we know that $j_{\nu,n}$ is a differentiable function of ν in $(-1,\infty)$.

    (a) Differentiate the equation $J_\nu(j_{\nu,n}) = 0$ to get

    $$J_\nu'(j_{\nu,n})\frac{dj_{\nu,n}}{d\nu} + \left[\frac{\partial J_\nu(x)}{\partial\nu}\right]_{x=j_{\nu,n}} = 0.$$

    (b) Verify by differentiation

    $$\int_0^x \frac{J_\mu(y)J_\nu(y)}{y}\,dy = \frac{x\{J_\mu'(x)J_\nu(x) - J_\mu(x)J_\nu'(x)\}}{\mu^2 - \nu^2}, \qquad \mu^2 \neq \nu^2.$$

    (c) Letting $\mu \to \nu$, show that for $\nu > 0$,

    $$\int_0^{j_{\nu,n}} \frac{J_\nu^2(x)}{x}\,dx = -\frac{j_{\nu,n}}{2\nu}J_\nu'(j_{\nu,n})\left[\frac{\partial J_\nu(x)}{\partial\nu}\right]_{x=j_{\nu,n}}.$$

    (d) Establish the representation

    $$\frac{dj_{\nu,n}}{d\nu} = \frac{2\nu}{j_{\nu,n}\{J_\nu'(j_{\nu,n})\}^2}\int_0^{j_{\nu,n}}\frac{J_\nu^2(x)}{x}\,dx, \qquad \nu > 0,$$

    which shows that when $\nu > 0, j_{\nu,n}$ is an increasing function of ν.

13. Prove the first statement of Theorem 9.2.2 by adapting the method of Proposition 3.5.1.

14. Prove the second statement of Theorem 9.2.2.

15. Show that the sequence (9.2.6) has a limit and determine it, assuming that u has the form (9.2.1).

16. Given an index $\nu \geq -\frac{1}{2}$ and a constant $\lambda > 0$, define $f_\lambda(x) = J_\nu(\lambda x)$, $x > 0$.

(a) Show that $x^2 f_\lambda'' + x f_\lambda' + (\lambda^2 x^2 - \nu^2) f_\lambda = 0$.

(b) Let $W(f_\lambda, f_\mu)$ be the Wronskian. Show that

$$[x W(x)]' = (\lambda^2 - \mu^2) x f_\lambda(x) f_\mu(x),$$

and deduce that if $J_\nu(\lambda) = 0 = J_\nu(\mu)$ and $\lambda \neq \mu$, then

$$\int_0^1 x J_\nu(\lambda x) J_\nu(\mu x) \, dx = 0.$$

(c) Suppose that $J_\nu(\lambda) = 0$. Show that

$$\int_0^1 x J_\nu(\lambda x)^2 \, dx = \lim_{\mu \to \lambda} \int_0^1 x J_\nu(\lambda x) J_\nu(\mu x) \, dx = \frac{1}{2} J_\nu'(\lambda)^2.$$

(d) Use (9.1.3) to show that $J_\nu(\lambda) = 0$ implies $J_\nu'(\lambda) = -J_{\nu+1}(\lambda)$.

(e) Suppose that

$$f(x) = \sum_{k=0}^\infty a_k J_\nu(\lambda_k x), \quad 0 < x < 1, \tag{9.10.1}$$

where the $\{\lambda_k\}$ are the positive zeros of J_ν numbered in increasing order. Assume that the series converges uniformly. Show that

$$a_k = \frac{2}{J_{\nu+1}(\lambda_k)^2} \int_0^1 x f(x) J_\nu(\lambda_k x) \, dx. \tag{9.10.2}$$

The expansion (9.10.1), (9.10.2) is called the *Fourier–Bessel expansion* of the function f. In particular, the Fourier–Bessel expansion of a function f converges to $f(x)$, $0 < x < 1$, if f is differentiable for $0 < x < 1$ and

$$\int_0^1 x^{\frac{1}{2}} |f(x)| \, dx < \infty;$$

see Watson [425], Chapter 18.

17. Show that

$$|J_n(x)| \le 1, \qquad x \ge 0, \; n = 0, 1, 2, \ldots$$

18. Show that

$$W\left(H_\nu^{(1)}, H_\nu^{(2)}\right)(x) = -\frac{4i}{\pi x}; \qquad W\left(I_\nu, I_{-\nu}\right)(x) = -\frac{2 \sin \nu \pi}{\pi x}.$$

19. Deduce from the generating function for the Bessel functions of integer order that

$$\sin x = 2 \sum_{n=0}^\infty (-1)^n J_{2n+1}(x);$$

$$\cos x = J_0(z) + 2\sum_{n=1}^{\infty}(-1)^n J_{2n}(x);$$

$$x\cos x = 2\sum_{n=1}^{\infty}(-1)^{n+1}(2n-1)^2 J_{2n-1}(x).$$

20. Deduce from (9.5.5) that

$$K_\nu(x) = \frac{\sqrt{\pi}\left(\frac{1}{2}x\right)^\nu}{\Gamma\left(\nu+\frac{1}{2}\right)}\int_1^{\infty} e^{-xt}(t^2-1)^{\nu-\frac{1}{2}}dt, \quad \text{Re } \nu > -\frac{1}{2}.$$

Use this formula and the beta function integral to show that

$$\int_0^{\infty} t^{\mu-1}K_\nu(t)dt = 2^{\mu-2}\Gamma\left(\frac{\mu+\nu}{2}\right)\Gamma\left(\frac{\mu-\nu}{2}\right), \quad \text{Re } \mu > |\text{Re } \nu|.$$

21. Define functions θ_n, y_n by

$$\theta_n(x) = \sqrt{\frac{2}{\pi}}x^{n+\frac{1}{2}}e^x K_{n+\frac{1}{2}}(x);$$

$$y_n(x) = x^n\theta_n(1/x).$$

(a) Show that these functions are polynomials. The y_n are known as the *Bessel polynomials* and the θ_n are known as the *reverse Bessel polynomials*; see Krall and Frink [230].
(b) Show that

$$x^2 y_n''(x) + (2x+1)y_n'(x) = n(n+1)y_n(x).$$

22. Show that

$$e^{x\cos t} = \sum_{n=-\infty}^{\infty}(\cos nt)I_n(x) = I_0(x) + 2\sum_{n=1}^{\infty}(\cos nt)I_n(x).$$

23. Show that

$$\int_0^x \cos(x-t)J_0(t)dt = xJ_0(x).$$

24. Deduce from (9.1.6) that

$$2^m\frac{d^m}{dx^m}J_n(x) = \sum_{k=0}^{m}(-1)^{m-k}\binom{m}{k}J_{n+m-2k}(x).$$

25. Show that for $a > 0$, $b > 0$,

$$\int_0^{\infty} e^{-ax}J_0(bx)dx = \frac{1}{\sqrt{a^2+b^2}}.$$

26. Show that for $a > 0$, $b > 0$, Re $\nu > -1$,

$$\int_0^\infty e^{-a^2 x^2} J_\nu(bx) x^{\nu+1} dx = \frac{b^\nu}{(2a^2)^{\nu+1}} e^{-b^2/4a^2}.$$

27. Using Exercise 23, show that for $x > 0$ and $|\text{Re } \nu| < \frac{1}{2}$,

$$J_\nu(x) = \frac{2\left(\frac{1}{2}x\right)^{-\nu}}{\sqrt{\pi}\,\Gamma\left(\frac{1}{2}-\nu\right)} \int_1^\infty \frac{\sin xt}{(t^2-1)^{\nu+\frac{1}{2}}} dt;$$

$$Y_\nu(x) = -\frac{2\left(\frac{1}{2}x\right)^{-\nu}}{\sqrt{\pi}\,\Gamma\left(\frac{1}{2}-\nu\right)} \int_1^\infty \frac{\cos xt}{(t^2-1)^{\nu+\frac{1}{2}}} dt.$$

28. Derive from (9.1.3) the recurrence relation

$$\int_0^x t^\mu J_\nu(t) dt = x^\mu J_{\nu+1}(x) - (\mu-\nu-1) \int_0^x t^{\mu-1} J_{\nu+1}(t) dt$$

for Re $(\mu+\nu) > -1$.

29. Show that

$$\int_0^x J_\nu(t) dt = 2 \sum_{n=0}^\infty J_{\nu+2n+1}(x), \qquad \text{Re } \nu > -1.$$

30. Verify *Sonine's first finite integral formula* [374]: for Re $\mu > -1$ and Re $\nu > -1$,

$$\int_0^{\frac{\pi}{2}} J_\mu(x\sin\theta) \sin^{\mu+1}\theta \cos^{2\nu+1}\theta\, d\theta = \frac{2^\nu \Gamma(\nu+1)}{x^{\nu+1}} J_{\mu+\nu+1}(x).$$

31. Verify the identities (9.8.4)–(9.8.7).
32. Show that $w(x) = [\text{Ai}(x)]^2$ satisfies the third-order equation

$$w^{(3)} - 4xw' - 2w = 0.$$

33. Show that the solutions of the differential equation

$$x^4 w^{(4)} + 2x^3 w^{(3)} - (1+2\nu^2)(x^2 w'' - xw') + (\nu^4 - 4\nu^2 + x^4)w = 0$$

are the *Kelvin functions* $\text{bei}_\nu(x)$, $\text{ber}_{-\nu}(x)$, and $\text{bei}_{-\nu}(x)$, defined by

$$\text{ber}_\nu(x) \pm i\text{bei}_\nu(x) = J_\nu(xe^{\pm 3\pi i/4}) = e^{\pm \nu\pi i/2} I_\nu(xe^{\pm \pi i/4}).$$

These functions were introduced by Kelvin [215] for $\nu = 0$ and by Russell [348] and Whitehead [433] for general ν and other types (not *the* Russell and Whitehead, however!)

34. Show that

$$\int_0^\infty \text{Ai}(t) t^{\alpha-1} dt = \frac{\Gamma(\alpha)}{3^{(\alpha+2)/3}\Gamma\left(\frac{1}{3}\alpha + \frac{2}{3}\right)}.$$

35. Prove that for $0 < t < \infty$,

$$0 \leq \mathrm{Ai}(t) \leq \frac{1}{2\sqrt{\pi}}\, t^{-1/4} \exp\left(-\frac{2}{3}t^{3/2}\right).$$

36. Use Gauss's formula (2.3.6) to show that the two determinations of the Wronskian of the Airy functions Ai, Bi are the same.

9.11 Remarks

Bessel's equation is closely related to Riccati's equation [341], a case of which was investigated by Johann and Daniel Bernoulli starting in 1694 [41]; see Exercises 23 and 24 of Chapter 3. It also arose in investigations of the oscillations of a heavy chain (D. Bernoulli, 1734 [38]), vibrations of a circular membrane (Euler, 1759 [121]), and heat conduction in a cylinder (Fourier, 1822 [141]) or sphere (Poisson, 1823 [331]). Daniel Bernoulli gave a power series solution which is J_0. The functions J_n for integer n occur in Euler [121]; he found the series expansions and looked for a second solution, finding Y_0 but not Y_n. The J_n also appear as coefficients in studies of planetary motion (Lagrange [238] and Laplace [242]). The early history is discussed in some detail in Dutka [111].

Bessel's 1824 investigation of the J_n [43] and Schlömilch's memoir [357] left Bessel's name attached to both the functions of integer order and the functions J_ν for arbitrary ν, which were introduced by Lommel [263] in 1868.

Up to factors, the Bessel function of the second kind Y_n was introduced by Hankel [177], Weber [428], and Schläfli [354]. Neumann [298] introduced a different version. The functions $H_\nu^{(i)}$ were introduced by Nielsen [302] and named in honor of Hankel. Up to factors, the functions I_ν and K_ν were introduced by Bassett [31]. The function K_ν also appears in [266] and is sometimes called *Macdonald's function*.

Airy's integral was introduced and studied by Airy in 1838 [6]. The current notation $\mathrm{Ai}(x)$, $\mathrm{Bi}(x)$ is due to Jeffreys [204].

The theory, the history, and the extensive literature on cylinder functions through the early twentieth century are surveyed in Watson's classic treatise [425]. Other references, with an emphasis on applications, are Korenev [227] and Petiau [322].

10

Hypergeometric functions

Hypergeometric functions were introduced briefly in Chapters 1 and 3. In this chapter, we discuss at length the solutions of the hypergeometric equation

$$x(1-x)u''(x) + [c - (a+b+1)x]\, u'(x) - ab\, u(x) = 0. \qquad (10.0.1)$$

There are two classic transformations (Pfaff and Euler) from one solution to another.

Normalizing at the singular points $x = 0$, $x = 1$, and $x = \infty$ gives three natural pairs of solutions of (10.0.1). Any three solutions must satisfy a linear relation. In particular, for most values of the parameters (a, b, c), each solution of one normalized pair is a linear combination of the two solutions in each of the other two pairs. We find a fundamental set of such relations.

When the parameter c is an integer, the standard solutions coincide ($c = 1$), or one of them is not defined. A second solution is found by a limiting process.

Three hypergeometric functions whose respective parameters (a, b, c) differ by integers satisfy a linear relation. A basis for such relations, due to Gauss, is derived.

When the three parameters (a, b, c) satisfy certain relations, a quadratic transformation of the independent variable converts a hypergeometric function to the product of a hypergeometric function and a power of a rational function. The basic such quadratic transformations are thus found.

Hypergeometric functions can be transformed into other hypergeometric functions by certain combinations of multiplication, differentiation, and integration. As a consequence, we obtain some useful evaluations in closed form and some useful integral representations. Jacobi polynomials, rescaled to the interval $[0, 1]$, are multiples of hypergeometric functions. This leads to some additional explicit evaluations, recalled from Chapter 5.

Let us note in passing the relation to the "confluent hypergeometric" case of Chapter 8. The terminology comes from the idea of replacing x by x/b in (10.0.1), so that the singularities of the equation are at 0, b, and ∞, and then letting $b \rightarrow +\infty$ so that the latter two singularities flow together. The formal limit of the series in (10.1.2) is the series for the Kummer function $M(a,c;x)$.

10.1 Solutions of the hypergeometric equation

The operator associated with (10.0.1) is the hypergeometric operator

$$L_{abc} = x(1-x)\frac{d^2}{dx^2} + [c - (a+b+1)x]\frac{d}{dx} - ab. \qquad (10.1.1)$$

Any solution of $L_{abc}F = 0$ in a region in the complex plane extends analytically to any simply connected plane region that does not contain the singular points $x = 0$, $x = 1$. The series

$$F(a,b,c;x) = \sum_{n=0}^{\infty} \frac{(a)_n (b)_n}{(c)_n n!} x^n \qquad (10.1.2)$$

gives, for $|x| < 1$, the solution that is regular at the origin and satisfies $F(0) = 1$. It has a single-valued analytic continuation to the complement of the real ray $\{x \geq 1\}$. In the various formulas that follow, we choose principal branches on the complement of this ray. Note the identity

$$\frac{d}{dx}[F(a,b,c;x)] = \frac{ab}{c}F(a+1,b+1,c+1;x). \qquad (10.1.3)$$

See Exercise 2.

With $D = D_x = x(d/dx)$, the hypergeometric operator (10.1.1) is

$$L_{abc} = x^{-1}D(D+c-1) - (D+a)(D+b). \qquad (10.1.4)$$

Recall (8.1.5): for any constant b,

$$x^{-b}D\{x^b u(x)\} = (D+b)u(x). \qquad (10.1.5)$$

It follows that conjugating by x^{1-c} converts the operator (10.1.1) to the operator

$$x^{c-1}L_{abc}x^{1-c}$$
$$= x^{-1}D(D+1-c) - (D+a+1-c)(D+b+1-c)$$
$$= L_{a+1-c,b+1-c,2-c}.$$

Therefore a second solution of the equation $L_{abc}F = 0$ is provided through the gauge transformation $u(x) = x^{1-c}v(x)$:

$$x^{1-c}F(a+1-c,b+1-c,2-c;x), \qquad (10.1.6)$$

provided that c is not a positive integer. (This is one of many provisos that exclude certain integer values of combinations of the indices a,b,c. The exceptional cases will be discussed separately.)

The hypergeometric operators L_{abc} can be characterized as the linear second-order differential operators that have exactly three singular points on the Riemann sphere, $0, 1, \infty$, each of them regular. (For the concepts of regular and irregular singular points, see Coddington and Levinson [84], Hille [185], or Ince [193]; for a look at irregular singular points, see the exercises for Chapter 13.) We mention this because it explains an important invariance property of the set of hypergeometric operators $\{L_{abc}\}$: this set is invariant under changes of coordinates on the Riemann sphere $\mathbf{C} \cup \{\infty\}$ by linear fractional transformations that map the set of singular points $\{0, 1, \infty\}$ to itself. This provides a way of producing solutions that have specified behavior at $x = 1$ or at $x = \infty$. These transformations are generated by the transformation $y = 1 - x$ that interchanges 0 and 1 and fixes ∞ and the transformation $y = 1/x$ that interchanges 0 and ∞ and fixes 1.

Consider the first of these transformations. Let $u(x) = v(1 - x)$. Then (10.0.1) is equivalent to

$$y(1 - y)v''(y) + \left[c' - (a + b + 1)y \right] v'(x) - ab\, v(y) = 0,$$
$$c' = a + b + 1 - c.$$

Therefore there is a solution that is regular at $y = 0$ $(x = 1)$ and another that behaves like $y^{1-c'} = (1 - x)^{1-c'}$ at $y = 0$. These solutions are multiples of the two solutions

$$F(a, b, a + b + 1 - c; 1 - x),$$

$$(1 - x)^{c-a-b} F(c - a, c - b, 1 + c - a - b; 1 - x),$$

(10.1.7)

respectively, provided that $c - a - b$ is not an integer.

The inversion $y = 1/x$ takes D_x to $-D_y$, so

$$L_{abc} = yD_y(D_y + 1 - c) - (D_y - a)(D_y - b);$$

$$(-x)^a L_{abc}(-x)^{-a} = -\left[D_y(D_y - b + a) - y(D_y + a)(D_y + 1 - c + a) \right],$$

where we have made use of (10.1.5). It follows from this and from interchanging the roles of a and b that there are solutions that behave like

$(-x)^{-a}$ and $(-x)^{-b}$ at ∞. They are multiples of the two solutions

$$(-x)^{-a}F\left(a, 1-c+a, a-b+1; \frac{1}{x}\right),$$

$$(10.1.8)$$

$$(-x)^{-b}F\left(1-c+b, b, b-a+1; \frac{1}{x}\right),$$

respectively, provided that $a - b$ is not an integer.

These results can be used to generate identities for hypergeometric functions. For example, composing $x \to 1 - x$ with inversion and composing the result with $x \to 1 - x$ gives the map $y = x(x-1)^{-1}$ that fixes the origin and interchanges 1 and ∞. This leads to a different expression for the solution that is regular at the origin, given as a function of $x(1-x)^{-1}$; this is *Pfaff's identity* [324]:

$$F(a,b,c;x) = (1-x)^{-b}F\left(c-a, b, c; \frac{x}{x-1}\right). \qquad (10.1.9)$$

(The special case of this with $b = 1$ was known to Stirling [386].) There is, of course, a companion identity with a and b interchanged:

$$F(a,b,c;x) = (1-x)^{-a}F\left(a, c-b, c; \frac{x}{x-1}\right). \qquad (10.1.10)$$

If we iterate (10.1.9), fixing the index $c - a$ on the right, we obtain an identity due to Euler [125]:

$$F(a,b,c;x) = (1-x)^{c-a-b}F(c-a, c-b, c; x). \qquad (10.1.11)$$

Another derivation of Pfaff's identity is given in the next section.

Kummer [235] listed 24 solutions of the hypergeometric equation. Four of them occur in (10.1.9), (10.1.10), and (10.1.11). The remaining 20 are generated in the same way, starting with the five solutions (10.1.6), (10.1.7), and (10.1.8).

If we replace a by $a+\nu$ and b by $-\nu$, the operator (10.1.1) has the form

$$x(1-x)\frac{d^2}{dx^2} + [c-(a+1)x]\frac{d}{dx} + \nu(a+\nu),$$

so that $\lambda(\nu) = \nu(a+\nu)$ appears as a natural parameter associated with the fixed operator

$$x(1-x)\frac{d^2}{dx^2} + [c-(a+1)x]\frac{d}{dx}.$$

The following asymptotic result due to Darboux [93] will be proved in Chapter 13:

$$
\begin{aligned}
&F\left(a+\nu,-\nu,c;\sin^2\left(\frac{1}{2}\theta\right)\right)\\
&=\frac{\Gamma(c)\cos\left(\nu\theta+\frac{1}{2}a\theta-\frac{1}{2}c\pi+\frac{1}{4}\pi\right)+O(\nu^{-1})}{\sqrt{\pi}\left(\nu\sin\frac{1}{2}\theta\right)^{c-\frac{1}{2}}\left(\cos\frac{1}{2}\theta\right)^{\frac{1}{2}+(a-c)}}
\end{aligned}
\tag{10.1.12}
$$

as $\nu \to +\infty$, for $0 < \theta < \pi$.

10.2 Linear relations of solutions

In the previous section, we identified six solutions of the hypergeometric equation $L_{abc}F = 0$ that have specified behavior at the singular points $\{0, 1, \infty\}$: as $x \to 0$,

$$
F(a,b,c;x) \sim 1,
\tag{10.2.1}
$$

$$
x^{1-c}F(a+1-c,b+1-c,2-c;x) \sim x^{1-c};
$$

as $x \to 1$,

$$
F(a,b,a+b+1-c;1-x) \sim 1,
\tag{10.2.2}
$$

$$
(1-x)^{c-a-b}F(c-a,c-b,1+c-a-b;1-x) \sim (1-x)^{c-a-b};
$$

as $x \to -\infty$,

$$
(-x)^{-a}F(a,1+a-c,1+a-b;1/x) \sim (-x)^{-a},
\tag{10.2.3}
$$

$$
(-x)^{-b}F(b,1+b-c,1+b-a;1/x) \sim (-x)^{-b}.
$$

The hypergeometric operator has degree 2, so any three solutions must be linearly related. Our principal tool for computing coefficients in these relations is an integral representation due to Euler [123], which is obtained in the same way as the integral representation (8.1.3) for Kummer's confluent hypergeometric function.

Proposition 10.2.1 (Euler's integral representation) *Suppose* $\operatorname{Re} c > \operatorname{Re} a > 0$. *Then*

$$
F(a,b,c;x) = \frac{1}{B(a,c-a)}\int_0^1 s^{a-1}(1-s)^{c-a-1}(1-sx)^{-b}\,ds. \tag{10.2.4}
$$

Proof Since

$$
\frac{(a)_n}{(c)_n} = \frac{\Gamma(a+n)\Gamma(c)}{\Gamma(a)\Gamma(c+n)} = \frac{B(a+n,c-a)}{B(a,c-a)},
$$

the integral representation (2.1.7) for the beta function implies

$$F(a,b,c;x) = \frac{1}{B(a,c-a)} \int_0^1 s^{a-1}(1-s)^{c-a-1} \left[\sum_{n=0}^{\infty} \frac{(b)_n}{n!} (sx)^n \right] ds.$$

Summing the series in brackets gives (10.2.4).   □

For another integral representation due to Barnes, see (12.3.17).

The identity (10.2.4) provides an explicit analytic continuation for x in the complement of $[1,\infty)$ when $\operatorname{Re} c > \operatorname{Re} a > 0$.

The change of variables $t = 1 - s$ in (10.2.4) converts the integral to

$$(1-x)^{-b} \int_0^1 t^{c-a-1}(1-t)^{a-1} \left(1 - \frac{tx}{x-1} \right)^{-b} dt.$$

In view of (10.2.4) and the values at $x = 0$, we obtain Pfaff's identity (10.1.9) in the case $\operatorname{Re} c > \operatorname{Re} a > 0$. Analytic continuation in the parameters gives (10.1.9) for all values.

Let us return to the six solutions (10.2.1), (10.2.2), and (10.2.3). In principle, the first is a linear combination of the third and fourth:

$$F(a,b,c;x) = C_1(a,b,c) F(a,b,a+b+1-c;1-x)$$
$$+ C_2(a,b,c)(1-x)^{c-a-b}$$
$$\times F(c-a,c-b,1+c-a-b;1-x). \quad (10.2.5)$$

The coefficients C_1 and C_2 are analytic functions of the parameters, so it is enough to compute these coefficients under special assumptions. Assuming that $c - a - b > 0$, the fourth solution vanishes at $x = 1$, so the coefficient of the third solution is $F(a,b,c;1)$. This value can be obtained immediately from (10.2.4) if we also assume that $c > a > 0$ or $c > b > 0$:

$$C_1(a,b,c) = F(a,b,c;1) = \frac{\Gamma(c)\Gamma(c-a-b)}{\Gamma(c-a)\Gamma(c-b)}.$$

This extends by analytic continuation to the full case, giving *Gauss's summation formula* [150]:

$$\sum_{n=0}^{\infty} \frac{(a)_n (b)_n}{(c)_n n!} = \frac{\Gamma(c)\Gamma(c-a-b)}{\Gamma(c-a)\Gamma(c-b)}, \quad \operatorname{Re}(c-a-b) > 0. \quad (10.2.6)$$

If b is a negative integer, the sum is finite, and (10.2.6) reduces to a combinatorial identity usually attributed to Vandermonde in 1772 [410], but known to Chu [80] in 1303; see Lecture 7 of [18]:

$$F(a,-n,c;1) = \sum_{k=0}^{n} (-1)^k \binom{n}{k} \frac{(a)_k}{(c)_k} = \frac{(c-a)_n}{(c)_n}, \quad (10.2.7)$$

valid for $c \neq 0, -1, -2, \ldots$

We may use Euler's identity (10.1.11) to rewrite (10.2.5) as

$$
F(c-a, c-b, c; x)
$$
$$
= C_1(a,b,c)(1-x)^{a+b-c} F(a,b,a+b+1-c;1-x)
$$
$$
+ C_2(a,b,c) F(c-a, c-b, 1+c-a-b; 1-x).
$$

Assuming that $a+b > c$, we evaluate C_2 by computing $F(c-a, c-b, c; 1)$. By (10.2.6) (with a change of indices), the result is

$$
C_2(a,b,c) = F(c-a, c-b, c; 1) = \frac{\Gamma(c)\Gamma(a+b-c)}{\Gamma(a)\Gamma(b)}.
$$

Therefore

$$
F(a,b,c;x) = \frac{\Gamma(c)\Gamma(c-a-b)}{\Gamma(c-a)\Gamma(c-b)} F(a,b,a+b+1-c;1-x)
$$
$$
+ \frac{\Gamma(c)\Gamma(a+b-c)}{\Gamma(a)\Gamma(b)} (1-x)^{c-a-b}
$$
$$
\times F(c-a, c-b, 1+c-a-b; 1-x). \tag{10.2.8}
$$

This identity can be inverted by replacing x by $1-x$ and c by $a+b+1-c$, in order to express the first of the solutions normalized at $x = 1$ as a linear combination of the solutions normalized at $x = 0$:

$$
F(a,b,a+b+1-c;1-x)
$$
$$
= \frac{\Gamma(a+b+1-c)\Gamma(1-c)}{\Gamma(a+1-c)\Gamma(b+1-c)} F(a,b,c;x)
$$
$$
+ \frac{\Gamma(a+b+1-c)\Gamma(c-1)}{\Gamma(a)\Gamma(b)} x^{1-c}
$$
$$
\times F(a+1-c, b+1-c, 2-c; x). \tag{10.2.9}
$$

This identity and a change of indices allows one to obtain the second of the solutions normalized at $x = 1$ as a linear combination of the solutions normalized at $x = 0$.

Similarly, the first solution in (10.2.1) is a linear combination of the two in (10.2.3), and it is enough to obtain one coefficient, under the assumption that $c > a > b$. Under this assumption, take $x \to -\infty$ in (10.2.4) to obtain

$$
F(a,b,c;x) \sim \frac{(-x)^{-b}}{B(a,c-a)} \int_0^1 s^{a-b}(1-s)^{c-a} \frac{ds}{s(1-s)}
$$
$$
= \frac{(-x)^{-b}}{B(a,c-a)} B(a-b, c-a)
$$
$$
= (-x)^{-b} \frac{\Gamma(c)\Gamma(a-b)}{\Gamma(c-b)\Gamma(a)}.
$$

By symmetry and analytic continuation, we obtain

$$F(a,b,c;x) = \frac{\Gamma(c)\Gamma(b-a)}{\Gamma(c-a)\Gamma(b)}(-x)^{-a}F\left(a,1+a-c,1+a-b;\frac{1}{x}\right)$$
$$+ \frac{\Gamma(c)\Gamma(a-b)}{\Gamma(c-b)\Gamma(a)}(-x)^{-b}F\left(b,1+b-c,1+b-a;\frac{1}{x}\right). \quad (10.2.10)$$

This can be inverted to give

$$(-x)^{-a}F\left(a,1+a-c,1+a-b;\frac{1}{x}\right)$$
$$= \frac{\Gamma(1+a-b)\Gamma(1-c)}{\Gamma(a+1-c)\Gamma(1-b)}F(a,b,c;x)$$
$$+ \frac{\Gamma(1+a-b)\Gamma(c-1)}{\Gamma(c-b)\Gamma(a)}x^{1-c}$$
$$\times F(a+1-c,b+1-c,2-c;x). \quad (10.2.11)$$

The identities (10.2.8), (10.2.9), (10.2.10), and (10.2.11) are valid when all coefficients are well-defined, no third index is a nonpositive integer, and all arguments x, $1-x$, $1/x$ are in the complement of the ray $[1,\infty)$. Additional identities can be obtained by applying Pfaff's identity (10.1.9) or the Euler identity (10.1.11) to some or all of the terms.

10.3 Solutions when c is an integer

As is the case for the confluent hypergeometric functions, when c is an integer $\neq 1$, one of the two solutions (10.1.2) and (10.1.6) of the hypergeometric equation (10.0.1) is not defined, while if $c = 1$ these two solutions coincide. We can find a second solution by adapting the procedure used in Section 8.3. (Nørlund [306] contains a very comprehensive discussion.)

Assume first that neither a nor b is an integer. Assuming that $c \neq 0, -1, -2, \ldots$, let

$$N(a,b,c;x) \equiv \frac{\Gamma(a)\Gamma(b)}{\Gamma(c)}F(a,b,c;x) = \sum_{n=0}^{\infty} \frac{\Gamma(a+n)\Gamma(b+n)}{\Gamma(c+n)n!}x^n.$$

The series expansion is well-defined for all values of c. Note that if $c = -k$ is a nonpositive integer, then the first $k+1$ terms of the series vanish. In particular, if $c = m$ is a positive integer,

$$N(a+1-m, b+1-m, 2-m; x) \tag{10.3.1}$$

$$= \sum_{n=m-1}^{\infty} \frac{\Gamma(a+1-m+n)\Gamma(b+1-m+n)}{\Gamma(2-m+n)n!} x^n$$

$$= x^{m-1} \sum_{k=0}^{\infty} \frac{\Gamma(a+k)\Gamma(b+k)}{\Gamma(m+k)k!} x^k = x^{m-1} N(a, b, m; x).$$

We define a solution of (10.0.1) by analogy with the Kummer function of the second kind:

$$U(a, b, c; x) = \frac{\Gamma(1-c)}{\Gamma(a+1-c)\Gamma(b+1-c)} F(a, b, c; x)$$

$$+ \frac{\Gamma(c-1)}{\Gamma(a)\Gamma(b)} x^{1-c} F(a+1-c, b+1-c, 2-c; x)$$

$$= \frac{\pi}{\sin \pi c \, \Gamma(a)\Gamma(a+1-c)\Gamma(b)\Gamma(b+1-c)}$$

$$\times \left[N(a, b, c; x) - x^{1-c} N(a+1-c, b+1-c, 2-c; x) \right]. \tag{10.3.2}$$

In view of (10.3.1), the difference in brackets has limit 0 as $c \to m$, m a positive integer. Therefore, by l'Hôpital's rule,

$$U(a, b, m; x) = \frac{(-1)^m}{\Gamma(a)\Gamma(a+1-m)\Gamma(b)\Gamma(b+1-m)}$$

$$\times \frac{\partial}{\partial c} \left[N(a, b, c; x) - x^{1-c} N(a+1-c, b+1-c, 2-c; x) \right] \Big|_{c=m}.$$

For noninteger values of a and b and positive integer values of m, calculating the derivative shows that

$$U(a, b, m; x) = \frac{(-1)^m}{\Gamma(a+1-m)\Gamma(b+1-m)(m-1)!}$$

$$\times \left\{ F(a, b, m; x) \log x + \sum_{n=0}^{\infty} \frac{(a)_n (b)_n}{(m)_n n!} \right.$$

$$\times \left[\psi(a+n) + \psi(b+n) - \psi(n+1) - \psi(m+n) \right] x^n \Big\}$$

$$+ \frac{(m-2)!}{\Gamma(a)\Gamma(b)} x^{1-m} \sum_{n=0}^{m-2} \frac{(a+1-m)_n (b+1-m)_n}{(2-m)_n n!} x^n, \tag{10.3.3}$$

where $\psi(b) = \Gamma'(b)/\Gamma(b)$, and the last sum is taken to be zero if $m = 1$.

The function in (10.3.3) is well-defined for all values of a and b. By a continuity argument, it is a solution of (10.0.1) for all values of $a,b,c = m$ and all values of $x \notin (-\infty, 0]$. If neither a nor b is an integer less than m, then $U(a,b,m;x)$ has a logarithmic singularity at $x = 0$ and is therefore independent of the solution $F(a,b,m;x)$. If neither a nor b is a nonpositive integer and one or both is an integer less than m, then the coefficient of the term in brackets vanishes and $U(a,b,m;x)$ is the finite sum, which is a rational function that is again independent of $F(a,b,m;x)$.

If a and/or b is a nonpositive integer, then $U(a,b,m;x) \equiv 0$. To obtain a solution in this case we start with noninteger a and b and multiply (10.3.3) by $\Gamma(a)$ and/or $\Gamma(b)$. The limiting value of the resulting function as a and/or b approaches a nonpositive integer is a well-defined solution of (10.0.1) that has a logarithmic singularity at $x = 0$.

We have found a second solution of (10.0.1) when c is a positive integer. When c is a negative integer, we may take advantage of the identity

$$U(a,b,c;x) = x^{1-c} U(a+1-c,b+1-c,2-c;x). \tag{10.3.4}$$

See Exercise 18.

The various identities of Section 10.2 can be extended to exceptional cases by using the solution U. As an example, to obtain the analogue of (10.2.8) when $1 + a + b - c$ is a nonpositive integer, we may start with the general case and express the right-hand side of (10.2.8) using the solutions

$$(1-x)^{c-a-b} F(c-a,c-b,1+c-a-b;1-x), \qquad U(a,b,1+a+b-c;1-x).$$

The result in this case is

$$F(a,b,c;x) = \Gamma(c) U(a,b,1+a+b-c;1-x).$$

10.4 Contiguous functions

As in the case of confluent hypergeometric functions, two hypergeometric functions are said to be *contiguous* if two of the indices of one function are the same as those of the other, and if the third indices differ by 1. Gauss [150] showed that there is a linear relationship between a hypergeometric function and any two of its contiguous functions, with coefficients that are linear functions of the indices a,b,c and the variable x. By iteration, it follows that if the respective indices of three hypergeometric functions differ by integers, then they satisfy a linear relationship, with coefficients that are rational functions of the indices a,b,c and the variable x.

It is convenient to again use a shorthand notation: fixing indices a,b,c, let F be the function $F(a,b,c;x)$ and denote the six contiguous functions by $F(a\pm)$,

$F(b\pm)$, $F(c\pm)$, where

$$F(a\pm) = F(a\pm 1, b, c; x),$$

and so on. Since there are 15 pairs of these six functions, there are 15 contiguous relations. Because of the symmetry between a and b, however, there are nine distinct relations: we do not need the five that involve b but not a, and the relation that involves $F(a-)$ and $F(b+)$ follows from the one that involves $F(a+)$ and $F(b-)$.

These relations can be derived in a way similar to that used for Kummer functions in Section 8.5. The coefficient of x^n in the expansion of F is

$$\varepsilon_n = \frac{(a)_n (b)_n}{(c)_n \, n!}.$$

The coefficients of x^n in the expansions of $F(a+)$ and $F(c-)$ are

$$\frac{(a+1)_n}{(a)_n} \varepsilon_n = \frac{a+n}{a} \varepsilon_n, \qquad \frac{(c)_n}{(c-1)_n} \varepsilon_n = \frac{c-1+n}{c-1} \varepsilon_n,$$

respectively. Since $D[x^n] = nx^n$, the corresponding coefficient for DF is $n\varepsilon_n$. It follows that

$$DF = a\,[F(a+) - F] = b\,[F(b+) - F] = (c-1)\,[F(c-) - F].$$

These identities give

$$(a-b)F = aF(a+) - bF(b+); \tag{10.4.1}$$

$$(a-c+1)F = aF(a+) - (c-1)F(c-). \tag{10.4.2}$$

The coefficient of x^n in the expansion of F' is

$$\frac{(n+1)(a)_{n+1}(b)_{n+1}}{(c)_{n+1}(n+1)!} = \frac{(a+n)(b+n)}{(c+n)} \varepsilon_n. \tag{10.4.3}$$

Now

$$\frac{(a+n)(b+n)}{c+n} = n + (a+b-c) + \frac{(c-a)(c-b)}{c+n},$$

while the coefficient of x^n in the expansion of $F(c+)$ is

$$\frac{(c)_n}{(c+1)_n} \varepsilon_n = \frac{c}{c+n} \varepsilon_n.$$

Therefore (10.4.3) implies that

$$F' = DF + (a+b-c)F + \frac{(c-a)(c-b)}{c} F(c+).$$

Multiplying by x gives

$$(1-x)DF = x\left[(a+b-c)F + \frac{(c-a)(c-b)}{c} F(c+)\right].$$

Since

$$(1-x)DF = (1-x)a\,[F(a+)-F],$$

it follows that

$$[a+(b-c)x]\,F = a(1-x)F(a+) - \frac{(c-a)(c-b)x}{c}F(c+). \qquad (10.4.4)$$

The procedure that led to (10.4.4) can be applied to $F(a-)$ to yield another such relation. In fact the coefficient of x^n in $F'(a-)$ is

$$\frac{(a-1)(b+n)}{c+n}\,\varepsilon_n = \left[(a-1) - \frac{(a-1)(c-b)}{c+n}\right]\varepsilon_n.$$

Multiplying by x gives

$$DF(a-) = (a-1)xF - \frac{(a-1)(c-b)x}{c}F(c+).$$

Replacing a by $a-1$ in a previous identity gives

$$DF(a-) = (a-1)[F - F(a-)].$$

Therefore

$$(1-x)F = F(a-) - \frac{(c-b)x}{c}F(c+). \qquad (10.4.5)$$

Identities (10.4.1)–(10.4.5) can be used to generate the remaining five identities. Eliminating $F(c+)$ from (10.4.4) and (10.4.5) gives

$$[2a-c+(b-a)x]\,F = a(1-x)F(a+) - (c-a)F(a-). \qquad (10.4.6)$$

Eliminating $F(a+)$ from (10.4.2) and (10.4.4) gives

$$[(c-1)+(a+b+1-2c)x]\,F$$
$$= (c-1)(1-x)F(c-) - \frac{(c-a)(c-b)x}{c}F(c+). \qquad (10.4.7)$$

Eliminating $F(c+)$ from (10.4.5) and (10.4.7) gives

$$[1-a+(c-b-1)x]\,F$$
$$= (c-a)F(a-) - (c-1)(1-x)F(c-). \qquad (10.4.8)$$

Eliminating $F(c+)$ from (10.4.4) and (10.4.5), with a replaced by b in (10.4.5), gives

$$(a+b-c)F = a(1-x)F(a+) - (c-b)F(b-). \qquad (10.4.9)$$

Eliminating $F(a+)$ from (10.4.6) and (10.4.9) gives

$$(b-a)(1-x)F = (c-a)F(a-) - (c-b)F(b-). \qquad (10.4.10)$$

10.5 Quadratic transformations

Suppose that φ is a quadratic transformation of the Riemann sphere, i.e. a two-to-one rational map. Under what circumstances is the function $F(a,b,c;\varphi(x))$ a hypergeometric function

$$F(a,b,c;\varphi(x)) = F(a',b',c';x)? \qquad (10.5.1)$$

Assume first that φ is a polynomial of degree 2. An equation of the form (10.5.1) implies that φ takes the singular points $\{0,1\}$ of the equation satisfied by the right side of (10.5.1) to the singular points $\{0,1\}$ of the equation satisfied by the left side. Furthermore, by comparison of these two equations, it is readily seen that the unique double point of φ, the zero of φ', must go to 0 or 1. The right side is holomorphic at $x = 0$, so necessarily $\varphi(0) = 0$. Finally, the origin must be a simple zero of φ. The unique such polynomial φ is $4x(1-x)$. Considering asymptotics at infinity for the two sides of (10.5.1), we must have $\{a',b'\} = \{2a,2b\}$. Considering behavior at 0 and at 1, we must have $1-c = 1-c'$ and $1-c = c'-a'-b'$ so, up to interchanging a' and b',

$$a' = 2a, \quad b' = 2b, \quad c = c' = a+b+\frac{1}{2}.$$

A comparison of the two differential equations and of the behavior at $x = 0$ shows that these necessary conditions are also sufficient:

$$F\left(a,b,a+b+\frac{1}{2};4x(1-x)\right) = F\left(2a,2b,a+b+\frac{1}{2};x\right). \qquad (10.5.2)$$

This can also be written in the inverse form

$$F\left(a,b,a+b+\frac{1}{2};x\right) = F\left(2a,2b,a+b+\frac{1}{2};\frac{1}{2}-\frac{1}{2}\sqrt{1-x}\right). \qquad (10.5.3)$$

(See also Exercise 13.)

The general quadratic two-to-one rational map of the sphere has the form $\varphi(x) = p(x)/q(x)$, where p and q are polynomials of degree ≤ 2 with no common factor, and at least one has degree 2. The requirement for an identity of the form (10.5.1) is that φ map the set $\{0,1,\infty\}$, together with any double points, into the set $\{0,1,\infty\}$, and that the origin be a simple zero. This last requirement can be dropped if we look for a more general form

$$F(a,b,c;\varphi(x)) = (1-\alpha x)^{\beta} F(a',b',c';x).$$

Indeed, if $\alpha\beta = a'b'/c'$, then the right side will have a double zero at the origin, i.e. the derivative will vanish at $x = 0$. A candidate for φ here is $x^2/(2-x)^2$, which takes both 1 and ∞ to 1. In this case we would expect to take $\alpha = \frac{1}{2}$ to compensate for the singularity of the left side at $x = 2$. Since

$$1 - \frac{x^2}{(2-x)^2} = \frac{4(1-x)}{(2-x)^2} \sim 4(1-x), \quad x \to 1;$$

$$\sim -\frac{4}{x}, \quad x \to \infty,$$

comparison of the behavior of the two sides as $x \to 1$ gives the condition $c - a - b = c' - a' - b'$, while comparison of the two sides as $x \to \infty$ gives $\{0, c - a - b\} = \{\beta - a', \beta - b'\}$. Up to interchanging a' and b', these conditions together with $\frac{1}{2}\beta = a'b'/c'$ imply that $\beta = b'$, $c' = 2a'$. Comparison of the behavior as $x \to 2$ shows that, up to interchanging a and b, we should have $2a = \beta$, $2b = \beta + 1$. Then $c = a' + \frac{1}{2}$. Our proposed identity is therefore

$$F\left(a, a + \frac{1}{2}, c; \frac{x^2}{(2-x)^2}\right) = \left(1 - \frac{x}{2}\right)^{2a} F\left(c - \frac{1}{2}, 2a, 2c - 1; x\right). \quad (10.5.4)$$

It can be shown that both sides satisfy the same modification of the hypergeometric equation. The inverse form is

$$F\left(a, a + \frac{1}{2}, c; x\right) = (1 + \sqrt{x})^{-2a} F\left(c - \frac{1}{2}, 2a, 2c - 1; \frac{2\sqrt{x}}{1 + \sqrt{x}}\right). \quad (10.5.5)$$

Starting with (10.5.2), inverting it, applying Pfaff's identity (10.1.9) to the right side, and repeating this process yields the following sequence of identities. (At each step we reset the indices a, b, c.)

$$F\left(a, b, \frac{1}{2}(a + b + 1); x\right)$$

$$= F\left(\frac{1}{2}a, \frac{1}{2}b, \frac{1}{2}(a + b + 1); 4x(1 - x)\right); \quad (10.5.6)$$

$$F\left(a, b, a + b + \frac{1}{2}; x\right)$$

$$= F\left(2a, 2b, a + b + \frac{1}{2}; \frac{1}{2} - \frac{1}{2}\sqrt{1 - x}\right) \quad (10.5.7)$$

$$= \left(\frac{1}{2} + \frac{1}{2}\sqrt{1 - x}\right)^{-2a} F\left(2a, a - b + \frac{1}{2}, a + b + \frac{1}{2}; \frac{\sqrt{1-x} - 1}{\sqrt{1-x} + 1}\right); \quad (10.5.8)$$

$$F(a, b, a - b + 1; x)$$

$$= (1 - x)^{-a} F\left(\frac{1}{2}a, \frac{1}{2}a - b + \frac{1}{2}, a - b + 1; -\frac{4x}{(1-x)^2}\right) \quad (10.5.9)$$

$$= (1 + x)^{-a} F\left(\frac{1}{2}a, \frac{1}{2}a + \frac{1}{2}, a - b + 1; \frac{4x}{(1+x)^2}\right); \quad (10.5.10)$$

$$F\left(a, a+\frac{1}{2}, c; x\right)$$

$$= \left(\frac{1}{2}+\frac{1}{2}\sqrt{1-x}\right)^{-2a} F\left(2a, 2a-c+1, c; \frac{1-\sqrt{1-x}}{1+\sqrt{1-x}}\right) \qquad (10.5.11)$$

$$= (1-x)^{-a} F\left(2a, 2c-2a-1, c; \frac{\sqrt{1-x}-1}{2\sqrt{1-x}}\right); \qquad (10.5.12)$$

$$F\left(a, b, \frac{1}{2}(a+b+1), x\right)$$

$$= (1-2x)^{-2a} F\left(\frac{1}{2}a, \frac{1}{2}a+\frac{1}{2}, \frac{1}{2}(a+b+1); -\frac{4x(1-x)}{(1-2x)^2}\right). \qquad (10.5.13)$$

Applying (10.1.9) to the right side of (10.5.13) returns us to (10.5.6).

Similarly, starting with (10.5.4), inverting it, applying (10.1.9) to the right side, and repeating this process yields the following sequence of identities. (Again we reset the indices a, b, c.)

$$F(a, b, 2b; x)$$

$$= \left(1-\frac{1}{2}x\right)^{-a} F\left(\frac{1}{2}a, \frac{1}{2}a+\frac{1}{2}, b+\frac{1}{2}; \frac{x^2}{(2-x)^2}\right); \qquad (10.5.14)$$

$$F\left(a, a+\frac{1}{2}, c; x\right)$$

$$= (1+\sqrt{x})^{-2a} F\left(2a, c-\frac{1}{2}, 2c-1; \frac{2\sqrt{x}}{1+\sqrt{x}}\right) \qquad (10.5.15)$$

$$= (1-\sqrt{x})^{-2a} F\left(2a, c-\frac{1}{2}, 2c-1; -\frac{2\sqrt{x}}{1-\sqrt{x}}\right); \qquad (10.5.16)$$

$$F(a, b, 2b; x)$$

$$= (1-x)^{-\frac{1}{2}a} F\left(\frac{1}{2}a, b-\frac{1}{2}a, b+\frac{1}{2}; \frac{x^2}{4x-4}\right). \qquad (10.5.17)$$

Applying (10.1.9) to the right side of (10.5.17) returns us to (10.5.14).

One more collection of identities can be generated by starting with (10.5.17) and following it with (10.5.8), then proceeding to invert and apply (10.1.9) on the right:

$$F(a, b, 2b; x) = \left(\frac{1}{2}+\frac{1}{2}\sqrt{1-x}\right)^{-2a}$$

$$\times F\left(a, a-b+\frac{1}{2}, b+\frac{1}{2}; \left[\frac{1-\sqrt{1-x}}{1+\sqrt{1-x}}\right]^2\right); \qquad (10.5.18)$$

$$F(a,b,a-b+1;x) = (1+\sqrt{x})^{-2a}$$

$$\times F\left(a,a-b+\frac{1}{2},2a-2b+1;\frac{4\sqrt{x}}{(1+\sqrt{x})^2}\right) \quad (10.5.19)$$

$$= (1-\sqrt{x})^{-2a}$$

$$\times F\left(a,a-b+\frac{1}{2},2a-2b+1;-\frac{4\sqrt{x}}{(1-\sqrt{x})^2}\right); \quad (10.5.20)$$

$$F(a,b,2b,x) = (1-x)^{-\frac{1}{2}a}$$

$$\times F\left(a,2b-a,b+\frac{1}{2};-\frac{(1-\sqrt{1-x})^2}{4\sqrt{1-x}}\right). \quad (10.5.21)$$

Applying (10.1.9) to the right side of (10.5.21) returns us to (10.5.18).

Additional identities can be generated from (10.5.6)–(10.5.21) by applying (10.1.9) (in the first or second index) and (10.1.11) to one or both sides, or by a change of variables. We mention in particular the identity obtained by applying (10.1.11) to the left-hand side of (10.5.2),

$$(1-2x)F\left(a+\frac{1}{2},b+\frac{1}{2},a+b+\frac{1}{2};4x(1-x)\right)$$

$$= F\left(2a,2b,a+b+\frac{1}{2};x\right), \quad (10.5.22)$$

and an identity obtained by a change of variable in (10.5.18),

$$F\left(a,b,2b;\frac{4x}{(1+x)^2}\right)$$

$$= (1+x)^{2a}F\left(a,a-b+\frac{1}{2},b+\frac{1}{2};x^2\right). \quad (10.5.23)$$

Each of the identities (15.3.15)–(15.3.32) in [4] can be obtained in this way. The basic identities are due to Kummer [235]; a complete list can be found in Goursat [168].

10.6 Integral transformations and special values

A given hypergeometric function may be transformed into another by operations that involve multiplication and differentiation or integration. Two examples are

$$F(a,b,c;x) = x^{1+k-a} \frac{d^k}{dx^k} \left[\frac{x^{a-1}}{(a-k)_k} F(a-k,b,c;x) \right]$$

$$= x^{1-c} \frac{d^k}{dx^k} \left[\frac{x^{c+k-1}}{(c)_k} F(a,b,c+k;x) \right] \tag{10.6.1}$$

for $k = 0,1,2,\ldots$ The proofs are left as Exercise 21.

Formulas like this are primarily of interest when the hypergeometric function on the right-hand side can be expressed in closed form, such as

$$F(a,b,a;x) = (1-x)^{-b}. \tag{10.6.2}$$

This example, together with (10.6.1), shows that $F(a,b,c;x)$ can be written in closed form whenever $c - a$ or $c - b$ is a nonnegative integer.

The identities (10.6.1) allow us to decrease the indices a,b by integers or to increase c by an integer. The following integral transform allows us, under certain conditions, to increase an upper index or to decrease the lower index by integer or fractional amounts.

Given complex constants α and β with positive real parts, we define an integral transform $E_{\alpha,\beta}$, the *Euler transform* that acts on functions that are defined on an interval containing the origin:

$$E_{\alpha,\beta} f(x) = \frac{\Gamma(\alpha+\beta)}{\Gamma(\alpha)\Gamma(\beta)} \int_0^1 s^{\alpha-1}(1-s)^{\beta-1} f(sx)\,ds. \tag{10.6.3}$$

Then

$$E_{\alpha,\beta}[x^n] = \frac{\Gamma(\alpha+\beta)}{\Gamma(\alpha)\Gamma(\beta)} B(\alpha+n,\beta) x^n = \frac{(\alpha)_n}{(\alpha+\beta)_n} x^n.$$

Taking $\alpha = c$, or $\alpha + \beta = a$, respectively, we obtain

$$E_{c,\beta} F(a,b,c;\cdot) = F(a,b,c+\beta;\cdot)$$

and

$$E_{a,a-\alpha} F(a,b,c;\cdot) = F(\alpha,b,c;\cdot).$$

Reversing the point of view, we may write a given hypergeometric function as an integral. For $\operatorname{Re} a > 0$ and $\operatorname{Re}\varepsilon > 0$ or for $\operatorname{Re} c > \operatorname{Re}\varepsilon > 0$, respectively,

$$F(a,b,c;x) = E_{a,\varepsilon}[F(a+\varepsilon,b,c;\cdot)](x)$$

$$= \frac{\Gamma(a+\varepsilon)}{\Gamma(a)\Gamma(\varepsilon)} \int_0^1 s^{a-1}(1-s)^{\varepsilon-1} F(a+\varepsilon,b,c;sx)\,ds;$$

$$F(a,b,c;x) = E_{c-\varepsilon,\varepsilon}[F(a,b,c-\varepsilon;\cdot)](x)$$

$$= \frac{\Gamma(c)}{\Gamma(c-\varepsilon)\Gamma(\varepsilon)} \int_0^1 s^{c-\varepsilon-1}(1-s)^{\varepsilon-1} F(a,b,c-\varepsilon;sx)\,ds.$$

$$\tag{10.6.4}$$

Therefore if $\operatorname{Re} c > \operatorname{Re} a > 0$, we may take $\varepsilon = c - a$ and use (10.6.2) and (10.6.4) to recover the integral form (10.2.4)

$$F(a,b,c;x) = E_{a,c-a}[(1-x)^{-b}].$$

The collection of well-understood hypergeometric functions can be enlarged by combining (10.6.2) with the various quadratic transformations in the previous section, choosing values of the indices so that one of the first two is equal to the third and (10.6.2) is applicable. In many cases the result is a simple algebraic identity; for example, taking $c = a + \frac{1}{2}$ in (10.5.5) yields

$$(1-x)^{-a} = (1+\sqrt{x})^{-2a}\left(\frac{1-\sqrt{x}}{1+\sqrt{x}}\right)^{-a}.$$

In addition, however, one can obtain some less obvious identities.

Taking $b = a + \frac{1}{2}$ in (10.5.3) gives

$$F\left(a, a+\frac{1}{2}, 2a+1; x\right) = \left(\frac{1+\sqrt{1-x}}{2}\right)^{-2a}. \qquad (10.6.5)$$

Taking $c = 2a$ in (10.5.11) gives

$$F\left(a, a+\frac{1}{2}, 2a; x\right) = \frac{1}{\sqrt{1-x}}\left(\frac{1+\sqrt{1-x}}{2}\right)^{1-2a}. \qquad (10.6.6)$$

Another category of special values occurs when one of the first two indices (a,b) is a nonpositive integer, say $a = -n$, so that $F(a,b,c;x)$ is a polynomial of degree n. As we noted in Chapter 5, this polynomial is a rescaling of a Jacobi polynomial, provided $c > 0$ and $b - c - n + 1 > 0$. The identity (5.5.13) is equivalent to

$$F(a+n, -n, c; x) = \frac{n!}{(c)_n} P_n^{(c-1, a-c)}(1-2x). \qquad (10.6.7)$$

The arguments that led to the formulas (5.6.14) and (5.6.19) for the Chebyshev polynomials, as well as to (5.6.24) and (5.6.25), are valid for general values of n. It follows from this fact and from (10.6.7) (carried over to general values of n) that

$$F\left(v, -v, \frac{1}{2}; \frac{1}{2}(1-\cos\theta)\right) = \cos v\theta;$$

$$F\left(v+2, -v, \frac{3}{2}; \frac{1}{2}(1-\cos\theta)\right) = \frac{\sin(v+1)\theta}{(v+1)\sin\theta};$$

$$F\left(v+1,-v,\frac{1}{2};\frac{1}{2}(1-\cos\theta)\right) = \frac{\cos\left(v+\frac{1}{2}\right)\theta}{\cos\frac{1}{2}\theta};$$

$$F\left(v+1,-v,\frac{3}{2};\frac{1}{2}(1-\cos\theta)\right) = \frac{\sin\left(v+\frac{1}{2}\right)\theta}{(2v+1)\sin\frac{1}{2}\theta}.$$

These can be rewritten, setting $x = \frac{1}{2}(1-\cos\theta)$ to obtain

$$F\left(v,-v,\frac{1}{2};x\right) = \mathrm{Re}\,\{[1-2x+i\sqrt{4x(1-x)}]^v\}; \tag{10.6.8}$$

$$F\left(v+2,-v,\frac{3}{2};x\right) = \frac{\mathrm{Im}\,\{[1-2x+i\sqrt{4x(1-x)}]^{v+1}\}}{(v+1)\sqrt{4x(1-x)}}; \tag{10.6.9}$$

$$F\left(v+1,-v,\frac{1}{2};x\right) = \frac{\mathrm{Re}\,\{[1-2x+i\sqrt{4x(1-x)}]^{v+\frac{1}{2}}\}}{\sqrt{1-x}}; \tag{10.6.10}$$

$$F\left(v+1,-v,\frac{3}{2};x\right) = \frac{\mathrm{Im}\,\{[1-2x+i\sqrt{4x(1-x)}]^{v+\frac{1}{2}}\}}{(2v+1)\sqrt{x}}. \tag{10.6.11}$$

Integrating the series representations of $(1+t)^{-1}$, $(1+t^2)^{-1}$, $(1-t^2)^{-\frac{1}{2}}$, $(1-t^2)^{-1}$, and $(1+t^2)^{-\frac{1}{2}}$ from $t=0$ to $t=\pm x$ or $t=\pm x^2$ gives the identities

$$\frac{\log(1+x)}{x} = F(1,1,2;-x); \tag{10.6.12}$$

$$\frac{\tan^{-1}x}{x} = F\left(\frac{1}{2},1,\frac{3}{2};-x^2\right); \tag{10.6.13}$$

$$\frac{\sin^{-1}x}{x} = F\left(\frac{1}{2},\frac{1}{2},\frac{3}{2};x^2\right); \tag{10.6.14}$$

$$\frac{\tanh^{-1}x}{x} = F\left(\frac{1}{2},1,\frac{3}{2};x^2\right); \tag{10.6.15}$$

$$\frac{\sinh^{-1}x}{x} = F\left(\frac{1}{2},\frac{1}{2},\frac{3}{2};-x^2\right). \tag{10.6.16}$$

Manipulation of the coefficients in the series expansions leads to the identities

$$\frac{1}{2}\left[(1+x)^{-a}+(1-x)^{-a}\right] = F\left(\frac{1}{2}a,\frac{1}{2}a+\frac{1}{2},\frac{1}{2};x^2\right); \tag{10.6.17}$$

$$\frac{1}{2a}\left[(1+x)^{-a}-(1-x)^{-a}\right] = -xF\left(\frac{1}{2}a+\frac{1}{2},\frac{1}{2}a+1,\frac{3}{2};x^2\right). \tag{10.6.18}$$

An integral transform that is more specialized in its application is

$$E_{\alpha,\beta}^{(2)} f(x) = \frac{\Gamma(2\alpha + 2\beta)}{\Gamma(2\alpha)\Gamma(2\beta)} \int_0^1 s^{2\alpha-1}(1-s)^{2\beta-1} f(s^2 x)\, ds. \tag{10.6.19}$$

Then

$$E_{\alpha,\beta}^{(2)}\left[x^n\right] = \frac{(2\alpha)_{2n}}{(2\alpha + 2\beta)_{2n}} x^n = \frac{(\alpha)_n (\alpha + \frac{1}{2})_n}{(\alpha + \beta)_n (\alpha + \beta + \frac{1}{2})_n} x^n.$$

It follows that

$$F\left(a, a + \frac{1}{2}, c; x\right) = E_{a,\frac{1}{4}}^{(2)}\left[F\left(a + \frac{1}{4}, a + \frac{3}{4}, c; x\right)\right]. \tag{10.6.20}$$

For use in Chapter 11, we give two examples of these considerations. The first uses (10.6.3) and (10.6.10):

$$F(\nu + 1, -\nu, 1; x) = E_{\frac{1}{2},\frac{1}{2}}\left[F\left(\nu + 1, -\nu, \frac{1}{2}; x\right)\right]$$

$$= \frac{1}{\pi} \int_0^1 \frac{\operatorname{Re}\{[1 - 2sx + i\sqrt{4sx(1-sx)}]^{\nu+\frac{1}{2}}\}}{\sqrt{1-sx}} \frac{ds}{\sqrt{s(1-s)}}. \tag{10.6.21}$$

The second uses (10.6.20) and (10.6.6): for $\operatorname{Re}\nu > -1$,

$$F\left(\frac{1}{2}\nu + \frac{1}{2}, \frac{1}{2}\nu + 1, \nu + \frac{3}{2}; x\right)$$

$$= E_{\frac{1}{2}(\nu+1),\frac{1}{4}}^{(2)}\left[F\left(\frac{1}{2}\nu + \frac{3}{4}, \frac{1}{2}\nu + \frac{5}{4}, \nu + \frac{3}{2}; x\right)\right]$$

$$= \frac{\Gamma\left(\nu + \frac{3}{2}\right)}{\sqrt{\pi}\,\Gamma(\nu + 1)} \int_0^1 \left(\frac{1 + \sqrt{1 - s^2 x}}{2}\right)^{-\nu - \frac{1}{2}} \frac{s^\nu\, ds}{\sqrt{1 - s^2 x}\sqrt{1 - s}}. \tag{10.6.22}$$

10.7 Exercises

1. Show that

$$\lim_{b \to +\infty} F\left(a, b, c; \frac{x}{b}\right) = M(a, c; x), \quad |x| < 1.$$

This can be interpreted as sending the singular point at $x = b$ to ∞, so as to coincide with the singular point at ∞, thus explaining the terminology "confluent hypergeometric."

2. Verify the identity (10.1.3).

3. Show that for $|t| < 1$ and $|(1 - x)t| < 1$,

$$\sum_{n=0}^{\infty} \frac{(c)_n}{n!} F(a, -n, c; x) t^n = (1 - t)^{a-c} [1 - (1 - x)t]^{-a}.$$

4. Show that for $\operatorname{Re} a > 0$ and $\operatorname{Re} b > 0$,

$$\int_0^x t^{a-1} (1 - t)^{b-1} \, dt = \frac{x^a}{a} F(a, 1 - b, a + 1; x).$$

The integral is called the *incomplete beta function*, denoted $B_x(a, b)$. The identity is due to Gauss [150].

5. Use the reflection formula (2.2.7) and the integral formula (10.2.4) to verify the integral formula for $\operatorname{Re} c > \operatorname{Re} a > 0$:

$$F(a, b, c; x) = e^{-i\pi a} \frac{\Gamma(1 - a)\Gamma(c)}{\Gamma(c - a)} \frac{1}{2\pi i} \int_C \frac{t^{a-1}(1 + t)^{b-c}}{(1 + t - xt)^b} \, dt,$$

where the curve C runs from $+\infty$ to 0 along the upper edge of the cut on $[0, \infty)$ and returns to $+\infty$ along the lower edge.

6. Verify the integral formula, for $\operatorname{Re} a > 0$ and arbitrary complex b, c:

$$F(a, b, c; x)$$
$$= \frac{\Gamma(c)\Gamma(1 + a - c)}{\Gamma(a)} \frac{1}{2\pi i} \int_C s^{a-1}(s - 1)^{c-a-1}(1 - xs)^{-b} \, ds,$$

where C is a counter-clockwise loop that passes through the origin and encloses the point $s = 1$. Hint: assume first that $\operatorname{Re} c > \operatorname{Re} a > 0$ and change the contour to run along the interval $[0, 1]$ and back.

7. Derive Kummer's identity (8.1.10) from Pfaff's transformation (10.1.9).

8. Show that

$$\int_0^{\pi/2} \frac{d\varphi}{\sqrt{1 - k^2 \sin^2 \varphi}} = \frac{\pi}{2} F\left(\frac{1}{2}, \frac{1}{2}, 1; k^2\right), \quad |k| < 1;$$

$$\int_0^{\pi/2} \sqrt{1 - k^2 \sin^2 \varphi} \, d\varphi = \frac{\pi}{2} F\left(\frac{1}{2}, -\frac{1}{2}, 1; k^2\right), \quad |k| < 1.$$

The functions $K = K(k)$ and $E = E(k)$ that are defined by these integrals are called the complete elliptic integral of the first and second kind, respectively; see Chapter 14. Hint: evaluate

$$\int_0^{\pi/2} \sin^{2n} \varphi \, d\varphi.$$

9. Let u_1 and u_2 be two solutions of the hypergeometric equation (10.0.1). Show that the Wronskian has the form

$$W(u_1, u_2)(x) = A x^{-c}(1-x)^{c-a-b-1}$$

for some constant A.

10. Denote the following six solutions of the hypergeometric equation by

$$F_1(x) = F(a,b,c;x);$$
$$F_2(x) = x^{1-c} F(a+1-c, b+1-c, 2-c; x);$$
$$F_3(x) = F(a,b, a+b+1-c; 1-x);$$
$$F_4(x) = (1-x)^{c-a-b} F(c-a, c-b, 1+c-a-b; 1-x);$$
$$F_5(x) = (-x)^{-a} F(a, 1-c+a, a-b+1; 1/x);$$
$$F_6(x) = (-x)^{-b} F(1-c+b, b, b-a+1; 1/x).$$

(a) Compute the Wronskians

$$W(F_1, F_2)(x), \quad W(F_3, F_4)(x), \quad W(F_5, F_6)(x).$$

(b) Compute the Wronskians

$$W(F_1, F_3)(x), \quad W(F_1, F_4)(x), \quad W(F_1, F_5)(x), \quad W(F_1, F_6)(x).$$

Hint: use (a) and the relations in Section 10.2.

11. Exercise 10 lists six of Kummer's 24 solutions of the hypergeometric equation. Use (10.1.9)–(10.1.11) to find the remaining 18 solutions.

12. Verify the contiguous relations (10.4.6)–(10.4.10).

13. Suppose that $Q(x)$ is a quadratic polynomial and $u(x) = F(a', b', c'; Q(x))$ satisfies the hypergeometric equation (10.0.1) with indices a, b, c.
   (a) Let $y = Q(x)$. Show that $x(1-x)[Q'(x)]^2 = Ay(1-y)$ for some constant A.
   (b) Show that $A = 4$ and $Q(x) = 4x(1-x)$.
   (c) Show that $c = \frac{1}{2}(a+b+1)$ and $a' = \frac{1}{2}a$, $b' = \frac{1}{2}b$, $c' = c$.

14. Show that

$$F(a, 1-a, c; x) = (1-x)^{c-1} F\left(\frac{c-a}{2}, \frac{c+a-1}{2}, c; 4x(1-x)\right).$$

15. Show that

$$F\left(a, 1-a, c; \frac{1}{2}\right) = 2^{1-c} \frac{\Gamma(c)\Gamma(\frac{1}{2})}{\Gamma\left(\frac{1}{2}[c+a]\right)\Gamma\left(\frac{1}{2}[c-a+1]\right)}.$$

16. Use (10.5.23) (with a and b interchanged) to derive Kummer's quadratic transformation (8.1.11).

17. Show that

$$F\left(a,b,\frac{a+b+1}{2};\frac{1}{2}\right) = \frac{\Gamma\left(\frac{1}{2}[a+b+1]\right)\Gamma\left(\frac{1}{2}\right)}{\Gamma\left(\frac{1}{2}[a+1]\right)\Gamma\left(\frac{1}{2}[b+1]\right)}.$$

18. Verify the identity (10.3.4).
19. Suppose $c > 0$, $c' > 0$. Let $w(x) = x^{c-1}(1-x)^{c'-1}$, $0 < x < 1$.
    (a) Show that the hypergeometric operator

    $$L = x(1-x)\frac{d^2}{dx^2} + \left[c - (c+c')x\right]\frac{d}{dx}$$

    is symmetric in the Hilbert space $L^2([0,1],w(x)dx)$.
    (b) Given $\lambda > 0$ and $f \in L^2_w$, the equation $Lu + \lambda u = f$ has a unique
    solution $u \in L^2_w$, expressible in the form

    $$u(x) = \int_0^1 G_\lambda(x,y)f(y)dy.$$

    (Note that if we set $a = c + c' - 1$ and $v > \max\{a,0\}$ is chosen so that
    $\lambda = v(v-a)$, then $L + \lambda$ is the hypergeometric operator with indices
    $(a-v,v,c)$.) Compute the Green's function G_λ. Hint: see Section 3.3.
    The appropriate boundary conditions here are regularity at $x = 0$ and
    at $x = 1$.
20. Let c, c', $w(x)$, L, and a be as in Exercise 19. Let $F_v(x) = F(a+v,-v,$
    $c;x)$, so that F_0, F_1, F_2, ..., are orthogonal polynomials for the weight w.
    What is wrong with the following argument? Suppose that
    $(a+v)v \neq (a+n)n$, $n = 0,1,2,...$ Then

    $$-\lambda_v(F_v,F_n)_w = (LF_v,F_n)_w = (F_v,LF_n)_w = -\lambda_n(F_v,F_n)_w.$$

    Therefore $(F_v,F_n)_w = 0$ for all $n = 0,1,2,...$ By Theorem 4.1.5, the
    orthogonalized polynomials

    $$P_n(x) = ||F_n||_w^{-1}F_n(x)$$

    are complete in L^2_w. Therefore $F_v = 0$.
21. Verify the identities (10.6.1). Show that similar identities hold for the
    Kummer functions.
22. Change a to $2a$ and b to $a+1$ in (10.5.10), and show that the resulting
    identity is a special case of (10.6.1).
23. Change a to $2a+1$ and b to a in (10.5.14), and show that the resulting
    identity is a special case of (10.6.1).
24. Verify the identities (10.6.12)–(10.6.16).
25. Verify the identities (10.6.17) and (10.6.18).

26. Show that for $c > 2$ and $x > 0$,

$$F(1,1,c;-x) = \frac{\Gamma(c)}{\Gamma(c-2)} \int_0^1 (1-s)^{c-3} \frac{\log(1+sx)}{x} \, ds.$$

27. Show that

$$\log \frac{1+x}{1-x} = 2xF\left(\frac{1}{2}, 1, \frac{3}{2}; x^2\right).$$

28. Show that

$$\cos ax = F\left(\frac{1}{2}a, -\frac{1}{2}a, \frac{1}{2}; \sin^2 x\right);$$

$$\sin ax = a \sin x F\left(\frac{1}{2} + \frac{1}{2}a, \frac{1}{2} - \frac{1}{2}a, \frac{3}{2}; \sin^2 x\right).$$

10.8 Remarks

The hypergeometric equation was studied by Euler and Gauss, among others. According to Askey in the addendum to Szegő's article [393], both the hypergeometric equation and its series solution were probably first written down by Euler in a manuscript dated 1778 and published in 1794 [125]. The six solutions in Section 10.1 and the relations among them were obtained by Kummer [235] in 1836; see the discussion by Prosser [335]. Another method for obtaining such relations was introduced by Barnes [30]; see Section 12.3 below. The contiguous relations were obtained by Gauss in his 1812 paper on hypergeometric series [150]. They include as special cases many of the recurrence relations given elsewhere.

Riemann [343] studied hypergeometric functions from a function-theoretic point of view. Riemann's study of the conformal mappings defined by the quotient of two solutions of the same equation was carried out systematically by Schwarz [359]. The theory and history, pre-Riemann and post-Riemann, are treated in Klein's classic work [219], which has an extensive bibliography. The history, and its roots in the work of Wallis, Newton, and Stirling, is also discussed in Dutka [109].

Hypergeometric functions are discussed in almost every text on special functions; see the relatively recent books by Andrews, Askey, and Roy [10] and Seaborn [360]. In particular, [10] contains a different treatment of quadratic transformations and more information about general hypergeometric series. Various algorithms have been developed to establish identities involving hypergeometric series; see Koepf [223], Petkovšek, Wilf, and Zeilberger [323], and Wilf and Zeilberger [437].

Separation of variables in various singular or degenerate partial differential equations leads to solutions that are expressible in terms of hypergeometric functions. These include the Euler–Poisson–Darboux equation (see Darboux [94], and Copson [87]), the Tricomi equation (see Delache and Leray [102]), certain subelliptic operators (see Beals, Gaveau, and Greiner [33]), and certain singular or degenerate hyperbolic operators (see Beals and Kannai [34]).

A modified version of the hypergeometric series is called "basic hypergeometric series." The basic series that corresponds to the series for the Gauss hypergeometric function $F = {}_2F_1$ is

$$
{}_2\Phi_1(a,b,c;q;x) = \sum_{n=0}^{\infty} \frac{(a;q)_n (b;q)_n}{(c;q)_n (q;q)_n} x^n,
$$

in which the shifted factorials $(a)_n$, and so on, are replaced by the products

$$
(a;q)_n = (1-a)(1-aq)\cdots(1-aq^{n-1}),
$$

and so on. Identities involving these expressions also go back to Euler and Gauss. This "q-analysis" has applications in number theory and in combinatorial analysis; see Andrews, Askey, and Roy [10], Bailey [28], Fine [135], Gasper and Rahman [149], and Slater [369].

11

Spherical functions

Spherical functions are solutions of the equation

$$(1 - x^2)u''(x) - 2xu'(x) + \left[v(v+1) - \frac{m^2}{1 - x^2} \right] u(x) = 0, \qquad (11.0.1)$$

which arises from separating variables in Laplace's equation $\Delta u = 0$ in spherical coordinates. *Surface harmonics* are the restrictions to the unit sphere of harmonic functions (solutions of Laplace's equation) in three variables. For surface harmonics, m and v are nonnegative integers. The case $m = 0$ is Legendre's equation

$$(1 - x^2)u''(x) - 2xu'(x) + v(v+1)u(x) = 0, \qquad (11.0.2)$$

with v a nonnegative integer. The solutions to (11.0.1) that satisfy the associated boundary conditions are Legendre polynomials and certain multiples of their derivatives. These functions are the building blocks for all surface harmonics. They satisfy a number of important identities.

For general values of the parameter v, Legendre's equation (11.0.2) has linearly independent solutions $P_v(z)$, holomorphic for z in the complement of $(-\infty, -1]$, and $Q_v(z)$, holomorphic in the complement of $(-\infty, 1]$. These Legendre functions satisfy a number of identities and have several representations as integrals.

For most values of the parameter v, the four functions

$$P_v(z), \quad P_v(-z), \quad Q_v(z), \quad Q_{-v-1}(z)$$

are distinct solutions of (11.0.2), so there are linear relations connecting any three. Integer and half-integer values of v are exceptional cases.

The solutions of the spherical harmonic equation (11.0.1) with $m = 1, 2, \ldots$, are known as *associated Legendre functions*. They are closely related to derivatives $P_v^{(m)}$ and $Q_v^{(m)}$ of the Legendre functions. The associated Legendre functions also satisfy a number of important identities and have representations as integrals.

11.1 Harmonic polynomials and surface harmonics

A polynomial $P(x,y,z)$ in three variables is said to be *homogeneous* of degree n if it is a linear combination of monomials of degree n:

$$P(x,y,z) = \sum_{j+k+l=n} a_{jkl} x^j y^k z^l.$$

By definition, the zero polynomial is homogeneous of all degrees. A homogeneous polynomial P of degree n is said to be *harmonic* if it satisfies Laplace's equation

$$\Delta P = P_{xx} + P_{yy} + P_{zz} = 0.$$

The homogeneous polynomials of degree n and the harmonic polynomials of degree n are vector spaces over real or complex numbers. (Note that our convention here is that a harmonic polynomial is, by definition, homogeneous of some degree.)

Proposition 11.1.1 *The space of homogeneous polynomials of degree n has dimension $(n+2)(n+1)/2$. The space of harmonic polynomials of degree n has dimension $2n+1$.*

Proof The monomials of degree n are a basis for the homogeneous polynomials of degree n. Such a monomial has the form

$$x^{n-j-k} y^j z^k,$$

and is uniquely determined by the pair $(j, j+k+1)$, which corresponds to a choice of two distinct elements from the set $\{0, 1, \ldots, n+1\}$. This proves the first statement. To prove the second, we note that Δ maps homogeneous polynomials of degree n to homogeneous polynomials of degree $n-2$. It is enough to show that this map is surjective (onto), since then its kernel – the space of harmonic polynomials of degree n – has dimension

$$\frac{(n+2)(n+1)}{2} - \frac{n(n-1)}{2} = 2n+1.$$

To prove surjectivity, we note that $\Delta x^n = n(n-1)x^{n-2}$ and, in general,

$$\Delta \left(x^{n-j-k} y^j z^k \right) = (n-j-k)(n-j-k-1)x^{n-2-j-k} y^j z^k + R_{njk},$$

where R_{njk} has total degree $j+k-2$ in y and z. It follows by recursion on j that monomials $x^{n-2-j} y^j$ are in the range of Δ, and then by recursion on k that monomials $x^{n-2-j-k} y^j z^k$ are in the range of Δ. □

In spherical coordinates (3.6.5), monomials take the form

$$x^j y^k z^l = r^{j+k+l} \cos^j \varphi \, \sin^k \varphi \, \sin^{j+k} \theta \, \cos^l \theta.$$

In particular, a harmonic polynomial of degree n has the form

$$P(r,\theta,\varphi) = r^n Y(\theta,\varphi),$$

where Y is a trigonometric polynomial in θ and φ, of degree at most n in each, and

$$\frac{1}{\sin^2\theta} Y_{\varphi\varphi} + \frac{1}{\sin\theta} [\sin\theta\, Y_\theta]_\theta + n(n+1) Y = 0; \qquad (11.1.1)$$

see (3.6.8). The function Y can be regarded as the restriction of the harmonic polynomial P to the unit sphere $\{x^2 + y^2 + z^2 = 1\}$; it is called a *surface harmonic* of degree n.

As in Section 3.6, we may seek to solve (11.1.1) by separating variables: if $Y(\theta,\varphi) = \Theta(\theta)\Phi(\varphi)$, then wherever the product $\Theta\Phi \neq 0$ we must have

$$\frac{\Phi''}{\Phi} + \left\{ \sin\theta \frac{[\sin\theta\, \Theta']'}{\Theta} + n(n+1)\sin^2\theta \right\} = 0.$$

The first summand is a function of φ alone and the second is a function of θ alone, so each summand is constant. Since Y is a trigonometric polynomial in φ of degree at most n, it follows that $\Phi'' = -m^2\Phi$ for some integer $m = -n,$ $-n+1,\ldots,n-1,n$. This gives us $2n+1$ linearly independent choices for Φ:

$$\cos m\varphi, \quad m = 0,1,\ldots,n; \qquad \sin m\varphi, \quad m = 1,2,\ldots,n,$$

or the complex version

$$e^{im\varphi}, \quad m = 0,\pm 1,\ldots,\pm n.$$

Let us take $\Phi(x) = e^{imx}$. Then Θ is a solution of the equation

$$\frac{1}{\sin\theta} \frac{d}{d\theta} \left[\sin\theta \frac{d\Theta}{d\theta} \right] + \left[n(n+1) - \frac{m^2}{\sin^2\theta} \right] \Theta = 0.$$

As noted in Section 3.6, the change of variables $x = \cos\theta$ converts the preceding equation to the spherical harmonic equation

$$(1-x^2)u''(x) - 2xu'(x) + \left[n(n+1) - \frac{m^2}{1-x^2} \right] u(x) = 0, \qquad (11.1.2)$$

$0 < x < 1$. Suppose for the moment that $m \geq 0$. As noted in Section 3.7, the gauge transformation $u(x) = (1-x^2)^{\frac{1}{2}m}v(x)$ reduces (11.1.2) to the equation

$$(1-x^2)v'' - 2(m+1)xv' + (n-m)(n+m+1)v = 0. \qquad (11.1.3)$$

This has as a solution the Jacobi polynomial $P_{n-m}^{(m,m)}$, so (11.1.2) has a solution

$$(1-x^2)^{\frac{1}{2}m} P_{n-m}^{(m,m)}(x).$$

In view of (5.2.10), $P_{n-m}^{(m,m)}$ is a multiple of the mth derivative $P_n^{(m)}$ of the Legendre polynomial of degree n:

$$P_{n-m}^{(m,m)} = \frac{2^m n!}{(n+m)!} P_n^{(m)}. \tag{11.1.4}$$

Therefore Θ_{nm} is a multiple of the associated Legendre function

$$P_n^m(x) = (1-x^2)^{\frac{1}{2}m} P_n^{(m)}(x). \tag{11.1.5}$$

Formula (11.1.5) may be used to obtain an integral formula for P_n^m. The Rodrigues formula (5.1.12) for P_n and the Cauchy integral representation for the derivative give

$$P_n^m(x) = (-1)^n \frac{1}{2^n n!} (1-x^2)^{\frac{1}{2}m} \frac{(n+m)!}{2\pi i} \int_C \frac{(1-s^2)^n \, ds}{(s-x)^{n+m+1}},$$

where C is a contour enclosing x. Let us take C to be the circle centered at x with radius $\rho = \sqrt{1-x^2}$:

$$s(\alpha) = x + \rho e^{i\alpha} = x + \sqrt{1-x^2}\, e^{i\alpha}.$$

Then

$$1 - s(\alpha)^2 = -2\rho e^{i\alpha}(x + i\rho \sin \alpha),$$

so for $m \geq 0$,

$$P_n^m(\cos\theta) = \frac{(m+n)!}{n!} \frac{1}{2\pi} \int_0^{2\pi} e^{-im\alpha}(\cos\theta + i\sin\theta \sin\alpha)^n \, d\alpha. \tag{11.1.6}$$

In defining Θ_{nm} and Y_{nm}, we normalize so that the solution has L^2 norm 1 on the interval $[-1,1]$, or equivalently, the corresponding multiple of $P_{n-m}^{(m,m)}$ has L^2 norm 1 with respect to the weight $(1-x^2)^m$. Using (5.1.15) and the Rodrigues formula, we obtain the solution

$$\Theta_{nm}(x) = \frac{1}{2^m n!} \left[\left(n + \frac{1}{2}\right)(n-m)!(n+m)! \right]^{\frac{1}{2}}$$

$$\times (1-x^2)^{\frac{1}{2}m} P_{n-m}^{(m,m)}(x)$$

$$= (-1)^{n-m} \frac{1}{2^n n!} \left[\frac{2n+1}{2} \frac{(n+m)!}{(n-m)!} \right]^{\frac{1}{2}}$$

$$\times (1-x^2)^{-\frac{1}{2}m} \frac{d^{n-m}}{dx^{n-m}} \left[(1-x^2)^n \right]. \tag{11.1.7}$$

Combining (11.1.7), (11.1.4), and the Rodrigues formula for P_n, we obtain

$$\Theta_{nm}(x) = \left[\frac{2n+1}{2} \frac{(n-m)!}{(n+m)!} \right]^{\frac{1}{2}}$$

$$\times (1-x^2)^{\frac{1}{2}m} P_n^{(m)}(x)$$

$$= (-1)^n \frac{1}{2^n n!} \left[\frac{2n+1}{2} \frac{(n-m)!}{(n+m)!} \right]^{\frac{1}{2}}$$

$$\times (1-x^2)^{\frac{1}{2}m} \frac{d^{n+m}}{dx^{n+m}} \left[(1-x^2)^n \right]. \tag{11.1.8}$$

The expressions (11.1.7) and (11.1.8) were obtained under the assumption that $m \geq 0$. Comparing the last part of each, it is natural to define Θ_{nm} for $m = -1, -2, \ldots$, by

$$\Theta_{n,-m}(x) = (-1)^m \Theta_{nm}(x), \qquad m = 1, 2, \ldots, n. \tag{11.1.9}$$

Then the last parts of (11.1.7) and (11.1.8) are valid for all integers $m = -n, -n+1, \ldots, n$.

The previous considerations lead to the surface harmonics

$$Y_{nm}(\theta, \varphi) = \frac{1}{\sqrt{2\pi}} e^{im\varphi} \Theta_{nm}(\cos\theta), \qquad -n \leq m \leq n. \tag{11.1.10}$$

The factor $1/\sqrt{2\pi}$ is introduced so that the $\{Y_{nm}\}$ are orthonormal with respect to the normalized measure $\sin\theta \, d\theta \, d\varphi$ on the sphere.

Combining (11.1.6) with (11.1.10) and (11.1.8) gives Laplace's integral [242]:

$$Y_{nm}(\theta, \varphi) = \frac{A_{nm}}{2\pi} \int_0^{2\pi} e^{im(\varphi-\alpha)} (\cos\theta + i\sin\theta \sin\alpha)^n \, d\alpha, \tag{11.1.11}$$

$$A_{nm} = \frac{1}{n!} \left[\frac{2n+1}{4\pi} (n-m)!(n+m)! \right]^{\frac{1}{2}}. \tag{11.1.12}$$

In view of (11.1.8), this is valid for all $m = -n, \ldots, n$. This leads to an integral formula for the corresponding harmonic polynomial:

$$r^n Y_{nm}(\theta, \varphi) = \frac{A_{nm}}{2\pi} \int_0^{2\pi} e^{-im\alpha} [r\cos\theta + ir\sin\theta \sin(\alpha+\varphi)]^n \, d\alpha.$$

We have proved the following.

Theorem 11.1.2 *The functions Y_{nm} of (11.1.10) are a basis for the surface harmonics of degree n. They are orthonormal with respect to the normalized surface measure $\sin\theta \, d\theta \, d\varphi$. The corresponding harmonic polynomials of* *degree n have the form*

$$r^n Y_{nm}(\theta,\varphi) = \frac{A_{nm}}{2\pi} \int_0^{2\pi} e^{-im\alpha}(z + ix\sin\alpha + iy\cos\alpha)^n \, d\alpha, \qquad (11.1.13)$$

where A_{nm} is given by (11.1.12).

Remarks. 1. If we allow both indices n,m to vary, the functions Y_{nm} are still orthonormal. This is clear when the second indices differ, since the factors involving φ are orthogonal. When the second indices are the same, but the first indices differ, the second factors involve polynomials $P_k^{(m,m)}$ with different indices k, and again are orthogonal.

2. A second solution of the equation $r^2 R'' + 2rR' - n(n+1)R = 0$ is $R(r) = r^{-n-1}$. Thus to each surface harmonic Y_{nm} there corresponds a function

$$r^{-n-1} Y_{nm}(\theta,\varphi),$$

which is harmonic away from the origin, and, in particular, in the exterior of the sphere.

Suppose that O is a rotation about the origin in $\mathbf{R}^3$ (an orthogonal transformation with determinant 1). The Laplacian Δ is invariant with respect to O:

$$\Delta[f(Op)] = [\Delta f](Op), \qquad p \in \mathbf{R}^3.$$

If $P(p)$ is a homogeneous polynomial of degree n, then so is $Q(p) = P(Op)$. Therefore the space of harmonic polynomials of degree n is invariant under rotations. The surface measure $\sin\theta \, d\theta \, d\varphi$ is also invariant under rotations. It follows that any rotation takes the orthonormal basis $\{Y_{nm}\}$ to an orthonormal basis; thus it induces a unitary transformation in the (complex) space of spherical harmonics of degree n. This implies that the sum

$$F_n(\theta,\varphi;\theta',\varphi') = \sum_{m=-n}^{n} Y_{nm}(\theta,\varphi) \overline{Y}_{nm}(\theta',\varphi'),$$

where $\overline{Y}_{nm}$ is the complex conjugate, is left unchanged if both points (θ,φ) and (θ',φ') are subjected to the same rotation O. It follows that this sum is a function of the inner product between the points with spherical coordinates $(1,\theta,\varphi)$ and $(1,\theta',\varphi')$:

$$(\cos\varphi\sin\theta, \sin\varphi\sin\theta, \cos\theta) \cdot (\cos\varphi'\sin\theta', \sin\varphi'\sin\theta', \cos\theta')$$
$$= \cos(\varphi - \varphi')\sin\theta\sin\theta' + \cos\theta\cos\theta'.$$

If we take $(\theta',\varphi') = (0,0)$, which corresponds to the point with Cartesian coordinates $(0,0,1)$, then the inner product is $\cos\theta$. With this choice, the function is independent of rotations around the z-axis, which implies that it is a multiple of Y_{n0}:

$$c_n Y_{n0}(\theta, \varphi) = \sum_{m=-n}^{n} Y_{nm}(\theta, \varphi) \overline{Y}_{nm}(0,0). \qquad (11.1.14)$$

The constant c_n may be determined by multiplying both sides by $\overline{Y}_{n0}(\theta, \varphi)$ and integrating over the sphere. This computation, together with (11.1.10), (11.1.7), and (5.5.4), yields

$$c_n = \overline{Y}_{n0}(0,0) = \frac{1}{\sqrt{2\pi}} \Theta_{n0}(1) = \frac{\sqrt{2n+1}}{\sqrt{4\pi}}. \qquad (11.1.15)$$

The identities (11.1.8) and (11.1.9), together with (11.1.10), imply that

$$\frac{\sqrt{2n+1}}{\sqrt{4\pi}} Y_{n0}(\theta, \varphi) = \frac{2n+1}{4\pi} P_n(\cos\theta);$$

$$Y_{n0}(\theta, \varphi) \overline{Y}_{n0}(\theta', \varphi') = \frac{2n+1}{4\pi} P_n(\cos\theta) P_n(\cos\theta');$$

$$Y_{n,\pm m}(\theta, \varphi) \overline{Y}_{n,\pm m}(\theta', \varphi')$$
$$= \frac{2n+1}{4\pi} \frac{(n-m)!}{(n+m)!} e^{\pm im(\varphi-\varphi')} P_n^m(\cos\theta) P_n^m(\cos\theta'), \qquad m = 1, 2, \ldots, n.$$

This shows that the left side and each summand on the right side of (11.1.14) are functions of the difference $\varphi - \varphi'$. We take $\varphi' = 0$ and obtain Legendre's *addition formula* [251]:

$$P_n(\cos\varphi \sin\theta \sin\theta' + \cos\theta \cos\theta')$$
$$= P_n(\cos\theta) P_n(\cos\theta')$$
$$+ 2 \sum_{m=1}^{n} \frac{(n-m)!}{(n+m)!} \cos(m\varphi) P_n^m(\cos\theta) P_n^m(\cos\theta'). \qquad (11.1.16)$$

We have chosen the surface harmonics Y_{nm} to be orthonormal in $L^2(\Sigma)$, the space of functions on the sphere whose squares are integrable with respect to the measure $\sin\theta \, d\theta \, d\varphi$.

Theorem 11.1.3 *The surface harmonics $\{Y_{nm}\}$ are a complete orthonormal set in $L^2(\Sigma)$.*

Proof The restrictions to the sphere Σ of polynomials are dense in the space $L^2(\Sigma)$. Therefore it is sufficient to show that any such restriction can be written as a sum of harmonic polynomials. It is enough to consider homogeneous polynomials. For this, see Exercise 2. $\qquad \square$

11.2 Legendre functions

A Legendre function is a solution of the Legendre differential equation

$$(1 - z^2)u''(z) - 2zu'(z) + v(v + 1)u(z) = 0, \tag{11.2.1}$$

where the parameter v is not necessarily a nonnegative integer. When v is a nonnegative integer, one solution is the Legendre polynomial P_v. We know from Section 5.2 that this solution has the integral representation

$$P_v(z) = \frac{1}{2^{v+1}\pi i} \int_C \frac{(t^2 - 1)^v}{(t - z)^{v+1}} \, dt, \tag{11.2.2}$$

where C is a contour that encloses z. As noted in Section 5.9, the function P_v defined by (11.2.2) for general values of v is still a solution of (5.6.1) if C is a closed curve lying in a region where $[(t^2 - 1)/(t - z)]^v$ is holomorphic. We use (11.2.2) to define a solution that is holomorphic in the complement of the interval $(-\infty, -1]$, by choosing C in this complement in such a way as to enclose both $t = z$ and $t = 1$. The resulting solution satisfies $P_v(1) = 1$, and is called the *Legendre function of the first kind*.

As noted in Sections 3.4, 5.5, and 11.1, the change of variables $y = \frac{1}{2}(1 - z)$ converts (11.2.1) to the hypergeometric equation

$$y(1 - y)v''(y) + (1 - 2y)v'(y) + v(v + 1)v(y) = 0.$$

It follows that

$$P_v(z) = F\left(v + 1, -v, 1; \frac{1 - z}{2}\right). \tag{11.2.3}$$

Since the hypergeometric function is unchanged if the first two indices are interchanged, it follows that

$$P_{-v-1}(z) = P_v(z). \tag{11.2.4}$$

Suppose that $z > 1$. Then we may take the curve C in (11.2.2) to be the circle of radius $R = \sqrt{z^2 - 1}$ centered at z. Let

$$t = t(\varphi) = z + \sqrt{z^2 - 1}\, e^{i\varphi}, \quad -\pi \le \varphi \le \pi.$$

Then (11.2.2) becomes the general form of Laplace's integral:

$$P_v(z) = \frac{1}{\pi} \int_0^\pi \left[z + \sqrt{z^2 - 1}\, \cos\varphi\right]^v d\varphi. \tag{11.2.5}$$

Combining this with (11.2.4) gives

$$P_v(z) = \frac{1}{\pi} \int_0^\pi \frac{d\varphi}{\left[z + \sqrt{z^2 - 1}\, \cos\varphi\right]^{v+1}}. \tag{11.2.6}$$

Each of these formulas extends analytically to z in the complement of $(-\infty, 1]$.

The change of variables

$$e^\alpha = z + \sqrt{z^2 - 1} \cos\varphi$$

leads to the Dirichlet–Mehler integral representation [105, 279]:

$$P_\nu(\cosh\theta) = \frac{1}{\pi} \int_{-\theta}^{\theta} \frac{e^{-(\nu+\frac{1}{2})\alpha}\, d\alpha}{\sqrt{2\cosh\theta - 2\cosh\alpha}}. \tag{11.2.7}$$

This representation shows that

$$P_\nu(x) > 0 \quad \text{for } 1 < x < \infty, \ \nu \text{ real}. \tag{11.2.8}$$

Note that if u is a solution of (11.2.1), then so is $v(x) = u(-x)$. The third index in the hypergeometric function in (11.2.3) is the sum of the first two, so this is one of the exceptional cases discussed in Section 10.3; for noninteger ν the hypergeometric function in (11.2.3) has a logarithmic singularity at $z = 0$. For noninteger ν, $P_\nu(x)$ and $P_\nu(-x)$ are independent solutions of (11.2.1). However, for integer ν, $P_\nu = P_{-\nu-1}$ is a Legendre polynomial and $P_\nu(-x) = (-1)^\nu P_\nu(x)$.

To find a second solution of (11.2.1), we proceed as suggested in Section 5.9 by adapting the formula (11.2.2) to a different contour. The *Legendre function of the second kind* is defined to be

$$Q_\nu(z) = \frac{1}{2^{\nu+1}} \int_{-1}^{1} \frac{(1-s^2)^\nu\, ds}{(z-s)^{\nu+1}}, \quad \nu \neq -1, -2, \ldots \tag{11.2.9}$$

Here we take the principal branch of the power $w^{-\nu}$ on the complement of the negative real axis, so that Q_ν is holomorphic in the complement of the interval $(-\infty, 1]$.

The function Q_ν is a multiple of a hypergeometric function, but with argument z^{-2}. To verify this, suppose that $|z| > 1$ and $\mathrm{Re}(\nu+1) > 0$. Then

$$Q_\nu(z) = \frac{1}{(2z)^{\nu+1}} \int_{-1}^{1} \frac{(1-s^2)^\nu\, ds}{(1-s/z)^{\nu+1}}$$

$$= \frac{1}{(2z)^{\nu+1}} \sum_{k=0}^{\infty} \left[\frac{(\nu+1)_k}{k!} \int_{-1}^{1} (1-s^2)^\nu s^k\, ds \right] z^{-k}.$$

The integral vanishes for odd k. For $k = 2n$,

$$\int_{-1}^{1} (1-s^2)^\nu s^{2n}\, ds = \int_{0}^{1} (1-t)^\nu t^{n-\frac{1}{2}}\, dt$$

$$= B\left(\nu+1, n+\frac{1}{2}\right) = \frac{\Gamma(\nu+1)\Gamma(n+\frac{1}{2})}{\Gamma(\nu+n+\frac{3}{2})}$$

$$= \frac{\Gamma(\nu+1)\Gamma(\frac{1}{2})(\frac{1}{2})_n}{\Gamma(\nu+\frac{3}{2})(\nu+\frac{3}{2})_n}.$$

Also,

$$\frac{(\nu+1)_{2n}}{(2n)!} = \frac{2^{2n} \left(\frac{1}{2}\nu+\frac{1}{2}\right)_n \left(\frac{1}{2}\nu+1\right)_n}{2^{2n} \left(\frac{1}{2}\right)_n n!}.$$

Therefore, for $|z| > 1$ and $\mathrm{Re}\,(\nu+1) > 0$,

$$\begin{aligned}
Q_\nu(z) &= \frac{\Gamma(\nu+1)\Gamma(\frac{1}{2})}{\Gamma(\nu+\frac{3}{2})(2z)^{\nu+1}} \sum_{n=0}^{\infty} \frac{(\frac{1}{2}\nu+\frac{1}{2})_n (\frac{1}{2}\nu+1)_n}{(\nu+\frac{3}{2})_n n!} z^{-2n} \\
&= \frac{\Gamma(\nu+1)\sqrt{\pi}}{\Gamma(\nu+\frac{3}{2})(2z)^{\nu+1}} F\left(\frac{1}{2}\nu+\frac{1}{2},\frac{1}{2}\nu+1,\nu+\frac{3}{2};\frac{1}{z^2}\right). \quad (11.2.10)
\end{aligned}$$

This formula allows us to define Q_ν for all values of ν except $\nu = -1,-2,\ldots$ The apparent difficulty for $\nu + \frac{3}{2}$ a nonpositive integer is overcome by noting that the coefficient of $z^{-\nu-1-2n}$ has $\Gamma(\nu+\frac{3}{2}+n)$ rather than $(\nu+\frac{3}{2})_n$ in the denominator. A consequence is that when $\nu = -m-\frac{1}{2}$ for m a positive integer, these coefficients vanish for $n < m$.

We take the principal branch of the power z^ν. It follows that

$$Q_\nu(-z) = -e^{\nu\pi i} Q_\nu(z), \qquad \mathrm{Im}\,z > 0. \qquad (11.2.11)$$

Suppose $z > 1$. The change of variables

$$s = z - \sqrt{z^2 - 1}\, e^\theta$$

in the integral (11.2.9) gives the integral representation

$$Q_\nu(z) = \int_0^\alpha \left[z - \sqrt{z^2-1}\,\cosh\theta\right]^\nu d\theta, \qquad (11.2.12)$$

where $\coth\alpha = z$. This extends by analytic continuation to all z in the complement of $(-\infty, 1]$.

A further change of variables

$$[z-\sqrt{z^2-1}\,\cosh\theta]^{-1} = z+\sqrt{z^2-1}\,\cosh\varphi$$

leads to a form similar to (11.2.6), due to Heine [180]:

$$Q_\nu(z) = \int_0^\infty \frac{d\varphi}{\left[z+\sqrt{z^2-1}\,\cosh\varphi\right]^{\nu+1}}, \qquad \mathrm{Re}\,\nu > 0. \qquad (11.2.13)$$

Let $z = \cosh\theta$. The change of variables

$$e^\alpha = \cosh\theta + \sinh\theta\,\cosh\varphi$$

leads to a form similar to (11.2.7):

$$Q_\nu(\cosh\theta) = \int_\theta^\infty \frac{e^{-(\nu+\frac{1}{2})\alpha}\,d\alpha}{\sqrt{2\cosh\alpha - 2\cosh\theta}}. \tag{11.2.14}$$

This representation shows that

$$Q_\nu(x) > 0 \quad \text{for } 1 < x < \infty, \ \nu \text{ real, } \nu \neq -1, -2, \ldots \tag{11.2.15}$$

11.3 Relations among the Legendre functions

Although $P_\nu = P_{-\nu-1}$, the same is not generally true for Q_ν. To sort out the remaining relationships among the solutions

$$P_\nu(z), \quad P_\nu(-z), \quad Q_\nu(z), \quad Q_{-\nu-1}(z)$$

of the Legendre equation (11.2.1), we start by computing the Wronskian of Q_ν and $Q_{-\nu-1}$. If u_1 and u_2 are any two solutions, the Wronskian $W = u_1 u_2' - u_2 u_1'$ satisfies

$$(1 - z^2)\,W'(z) = 2z\,W(z).$$

Therefore wherever $W(z) \neq 0$ it satisfies $W'(z)/W(z) = -(1 - z^2)'/(1 - z^2)$:

$$W(u_1, u_2)(z) = \frac{C}{1 - z^2}, \quad C = \text{constant}.$$

To compute the constant, it is enough to compute asymptotics of u_j and u_j', say as $z = x \to +\infty$. It follows from (11.2.10) or directly from (11.2.9) that as $x \to +\infty$,

$$Q_\nu(x) \sim c_\nu x^{-\nu-1}, \qquad Q_\nu'(x) \sim -c_\nu(\nu+1)x^{-\nu-2};$$

$$c_\nu = \frac{\Gamma(\nu+1)\sqrt{\pi}}{\Gamma(\nu+\frac{3}{2})2^{\nu+1}},$$

so long as ν is not a negative integer. Therefore the Wronskian is asymptotically

$$c_\nu c_{-\nu-1} \begin{vmatrix} x^{-\nu-1} & x^\nu \\ -(\nu+1)x^{-\nu-2} & \nu x^{\nu-1} \end{vmatrix} = \frac{(2\nu+1)c_\nu c_{-\nu-1}}{x^2}.$$

The reflection formula (2.2.7) gives

$$c_\nu c_{-\nu-1} = \frac{\Gamma(\nu+1)\Gamma(-\nu)\pi}{(\nu+\frac{1}{2})\Gamma(\nu+\frac{1}{2})\Gamma(-\nu+\frac{1}{2})2}$$

$$= \frac{\pi\,\sin(\nu\pi+\frac{1}{2}\pi)}{(2\nu+1)\sin(-\nu\pi)} = -\frac{\pi\,\cos\nu\pi}{(2\nu+1)\sin\nu\pi}.$$

It follows that

$$W(Q_\nu, Q_{-\nu-1})(z) = \frac{\pi \cot \nu \pi}{1 - z^2}. \tag{11.3.1}$$

Recall that Q_ν is not defined when ν is a negative integer, so $Q_{-\nu-1}$ is not defined when ν is a nonnegative integer. Therefore the right side of (11.3.1) is well-defined for all admissible values, and these two solutions of the Legendre equation are independent if and only if $\nu + \frac{1}{2}$ is not an integer.

Assuming that ν is neither an integer nor a half-integer, every solution of the Legendre equation (11.2.1) is a linear combination of Q_ν and $Q_{-\nu-1}$. The coefficients for

$$P_\nu = A_\nu Q_\nu + B_\nu Q_{-\nu-1}$$

are analytic functions of ν, so it is enough to consider the case $-\frac{1}{2} < \nu < 0$. Then the integral form (10.2.4) gives

$$P_\nu(x) = \frac{1}{\Gamma(\nu+1)\Gamma(-\nu)} \int_0^1 s^\nu (1-s)^{-\nu-1} \left(1 - s\frac{1-x}{2}\right)^\nu ds. \tag{11.3.2}$$

It follows that as $x \to +\infty$,

$$P_\nu(x) \sim \frac{B(2\nu+1, -\nu)}{2^\nu \Gamma(\nu+1)\Gamma(-\nu)} x^\nu$$

$$= \frac{\Gamma(2\nu+1)}{2^\nu \Gamma(\nu+1)^2} x^\nu = \frac{\Gamma(\nu+\frac{1}{2})}{\Gamma(\nu+1)\sqrt{\pi}} (2x)^\nu,$$

where we have used Legendre's duplication formula (2.3.1). In the range $-\frac{1}{2} < \nu < 0$, $Q_{-\nu-1} \sim c_{-\nu-1} x^\nu$ and Q_ν decays more rapidly, so the coefficient B_ν is the ratio

$$\frac{\Gamma(\nu+\frac{1}{2})2^\nu}{\Gamma(\nu+1)\sqrt{\pi}} \left[\frac{\Gamma(-\nu)\sqrt{\pi}}{\Gamma(-\nu+\frac{1}{2})2^{-\nu}}\right]^{-1}$$

$$= \frac{\Gamma(\nu+\frac{1}{2})\Gamma(-\nu+\frac{1}{2})}{\Gamma(\nu+1)\Gamma(-\nu)\pi} = -\frac{\sin \nu \pi}{\sin(\nu \pi + \frac{1}{2}\pi)\pi} = -\frac{\tan \nu \pi}{\pi}.$$

Since $P_{-\nu-1} = P_\nu$, the coefficient $B_\nu = A_{-\nu-1} = -A_\nu$. Thus

$$P_\nu = \frac{\tan \nu \pi}{\pi} \left[Q_\nu - Q_{-\nu-1}\right]. \tag{11.3.3}$$

Equations (11.3.3) and (11.3.1) allow the computation of the Wronskian

$$W(Q_\nu, P_\nu)(z) = -\frac{\tan \nu \pi}{\pi} W(Q_\nu, Q_{-\nu-1})(z) = -\frac{1}{1-z^2}. \tag{11.3.4}$$

The identity (11.3.3) can also be derived by a complex variable argument. Given $0 < \theta < 2\pi$, the function $f(\alpha) = \cos \theta - \cos \alpha$ has no zeros in the strip

$0 < \operatorname{Im}\alpha < 2\pi$ and is periodic with period 2π, continuous up to the boundary except at $\alpha = \pm\theta$ and $\alpha = \pm\theta + 2\pi$. Assuming that $-1 < \nu < 0$,

$$\frac{e^{i(\nu+\frac{1}{2})\alpha}}{\sqrt{2\cos\alpha - 2\cos\theta}}$$

is integrable over the boundary and has integral zero over the oriented boundary. Keeping track of the argument on the various portions of the boundary, the result is again (11.3.3).

Multiplying (11.3.3) by $\cos\nu\pi$ and taking the limit as ν approaches a half-integer $n - \frac{1}{2}$, we obtain

$$Q_{n-\frac{1}{2}} = Q_{-n-\frac{1}{2}}, \qquad n = 0, \pm 1, \pm 2, \ldots \tag{11.3.5}$$

Recall that P_ν is holomorphic for $z \notin (-\infty, -1]$, and therefore continuous at the interval $(-1, 1)$. We show next that Q_ν has finite limits $Q_\nu(x \pm i0)$ from the upper and lower half-planes for $x \in (-1, 1)$ and that these limits are linear combinations of $P_\nu(x)$ and $P_\nu(-x)$. It follows from (11.3.4) that each limit is independent of P_ν. We define Q_ν on the interval to be the average:

$$Q_\nu(x) = \frac{1}{2}[Q_\nu(x+i0) + Q_\nu(x-i0)], \qquad -1 < x < 1. \tag{11.3.6}$$

To compute the average and the jump, note that (11.3.3) and (11.2.11) imply that

$$P_\nu(-z) = -\frac{\tan\nu\pi}{\pi}\left[e^{\nu\pi i}Q_\nu(z) + e^{-\nu\pi i}Q_{-\nu-1}(z)\right], \qquad \operatorname{Im}z > 0.$$

Eliminating $Q_{-\nu-1}$ between this equation and (11.3.3) gives

$$Q_\nu(z) = \frac{\pi}{2\sin\nu\pi}\left[e^{-\nu\pi i}P_\nu(z) - P_\nu(-z)\right], \qquad \operatorname{Im}z > 0. \tag{11.3.7}$$

Similarly,

$$Q_\nu(z) = \frac{\pi}{2\sin\nu\pi}\left[e^{\nu\pi i}P_\nu(z) - P_\nu(-z)\right], \qquad \operatorname{Im}z < 0. \tag{11.3.8}$$

It follows from the previous two equations that the average Q_ν is

$$Q_\nu(x) = \frac{\pi}{2}\cot\nu\pi\, P_\nu(x) - \frac{\pi}{2\sin\nu\pi}P_\nu(-x), \qquad -1 < x < 1, \tag{11.3.9}$$

while the jump is

$$Q_\nu(x+i0) - Q_\nu(x-i0) = -i\pi\, P_\nu(x), \qquad -1 < x < 1. \tag{11.3.10}$$

In terms of hypergeometric functions, for $-1 < x < 1$,

$$Q_\nu(x) = \frac{\pi}{2} \cot \nu\pi \, F\left(\nu+1, -\nu, 1; \frac{1-x}{2}\right)$$
$$- \frac{\pi}{2\sin \nu\pi} F\left(\nu+1, -\nu, 1; \frac{1+x}{2}\right). \qquad (11.3.11)$$

The recurrence and derivative identities satisfied by the Legendre polynomials (5.6.3), (5.6.4), and (5.6.5) carry over to the Legendre functions of the first and second kind. They can be derived from the integral formulas, as in Section 5.9, or checked directly from the series expansions (see the next section). As an alternative, the identities

$$(\nu+1)P_{\nu+1}(x) - (2\nu+1)xP_\nu(x) + \nu P_{\nu-1}(x) = 0,$$
$$(11.3.12)$$
$$(\nu+1)Q_{\nu+1}(x) - (2\nu+1)xQ_\nu(x) + \nu Q_{\nu-1}(x) = 0,$$

and

$$P'_{\nu+1}(x) - P'_{\nu-1}(x) - (2\nu+1)P_\nu(x) = 0,$$
$$(11.3.13)$$
$$Q'_{\nu+1}(x) - Q'_{\nu-1}(x) - (2\nu+1)Q_\nu(x) = 0,$$

follow easily from the integral representations (11.2.7) and (11.2.14). To derive (11.3.12) for P_ν from (11.2.7), we write

$$-\frac{d}{d\alpha}\left\{e^{-(\nu+\frac{1}{2})\alpha}\sqrt{2\cosh\theta - 2\cosh\alpha}\right\}$$
$$= e^{-(\nu+\frac{1}{2})\alpha}\frac{(\nu+\frac{1}{2})[2\cosh\theta - e^\alpha - e^{-\alpha}] + \frac{1}{2}(e^\alpha - e^{-\alpha})}{\sqrt{2\cosh\theta - 2\cosh\alpha}}$$

as a linear combination of the integrands of

$$P_{\nu+1}(\cosh\theta), \quad \cosh\theta\, P_\nu(\cosh\theta), \quad P_{\nu-1}(\cosh\theta)$$

and integrate. The proof for Q_ν is essentially the same, using (11.2.14).

To derive (11.3.13) from the integral representation, note that the integrand of

$$P_{\nu+1}(\cosh\theta) - P_{\nu-1}(\cosh\theta)$$

is $e^{-(\nu+\frac{1}{2})\alpha}$, multiplied by the derivative with respect to α of

$$2\sqrt{2\cosh\theta - 2\cosh\alpha}.$$

Integrating by parts,

$$P_{v+1}(\cosh\theta) - P_{v-1}(\cosh\theta)$$

$$= \frac{2v+1}{\pi} \int_{-\theta}^{\theta} e^{-(v+\frac{1}{2})\alpha} \sqrt{2\cosh\theta - 2\cosh\alpha}\, d\alpha.$$

Differentiation with respect to θ gives (11.3.13) for P_v. The proof for Q_v is essentially the same, using (11.2.14).

Differentiating (11.3.12) and combining the result with (11.3.13) gives

$$P'_{v+1}(x) - xP'_v(x) = (v+1)P_v(x);$$

$$\tag{11.3.14}$$

$$Q'_{v+1}(x) - xQ'_v(x) = (v+1)Q_v(x).$$

Subtracting (11.3.14) from (11.3.13) gives

$$xP'_v(x) - P'_{v-1}(x) = v P_v(x);$$

$$\tag{11.3.15}$$

$$xQ'_v(x) - Q'_{v-1}(x) = v Q_v(x).$$

Multiplying (11.3.15) by x and subtracting it from the version of (11.3.14) with v replaced by $v-1$ gives

$$(1-x^2)P'_v(x) = -vxP_v(x) + v P_{v-1}(x);$$

$$\tag{11.3.16}$$

$$(1-x^2)Q'_v(x) = -vxQ_v(x) + v Q_{v-1}(x).$$

11.4 Series expansions and asymptotics

Expansions of the Legendre functions $P_v(z)$ and $Q_v(z)$ as $z \to \infty$ in the complement of $(-\infty, 1]$ follow from the representation (11.2.10) of Q_v as a hypergeometric function, together with the representation (11.3.3) of P_v as a multiple of $Q_v - Q_{-v-1}$, for v not a half-integer.

To find expansions for P_v and Q_v on the interval $(-1, 1)$ at $x = 0$, we begin by noting that taking $y = z^2$ converts the Legendre equation (11.2.1) to the hypergeometric equation

$$y(1-y)v''(y) + \left(\frac{1}{2} - \frac{3}{2}y\right) v'(y) + \frac{1}{4}(v+1)v\, v(y) = 0.$$

It follows that P_v is a linear combination of two solutions

$$P_v(x) = A_v F\left(\frac{v+1}{2}, -\frac{v}{2}, \frac{1}{2}; x^2\right) + B_v x F\left(\frac{1}{2}v+1, -\frac{v-1}{2}, \frac{3}{2}; x^2\right).$$

Then

$$P_v(0) = A_v, \qquad P'_v(0) = B_v.$$

To determine $P_\nu(0)$ and $P'_\nu(0)$, we assume $-1 < \nu < 0$ and use (11.3.2) to find that

$$P_\nu(0) = \frac{1}{\Gamma(\nu+1)\Gamma(-\nu)} \int_0^1 s^\nu (1-s)^{-\nu-1} \left(1 - \frac{1}{2}s\right)^\nu ds;$$

$$P'_\nu(0) = \frac{\nu}{2\,\Gamma(\nu+1)\Gamma(-\nu)} \int_0^1 s^{\nu+1}(1-s)^{-\nu-1} \left(1 - \frac{1}{2}s\right)^{\nu-1} ds.$$

To evaluate the integrals we first let $s = 1 - t$, so that the first integral becomes

$$2^{-\nu} \int_0^1 (1 - t^2)^\nu \, t^{-\nu-1} \, dt.$$

This suggests letting $u = t^2$, so that the integral is

$$2^{-1-\nu} \int_0^1 (1-u)^\nu u^{-\frac{1}{2}\nu-1} du = 2^{-1-\nu} \, B\left(\nu+1, -\frac{1}{2}\nu\right).$$

Use of the reflection formula (2.2.7) and the duplication formula (2.3.1) leads to the evaluation

$$P_\nu(0) = \frac{\Gamma(\frac{1}{2}\nu+\frac{1}{2})}{\Gamma(\frac{1}{2}\nu+1)\sqrt{\pi}} \cos\left(\frac{\nu\pi}{2}\right).$$

The same procedure applied to the integral in the expression for $P'_\nu(0)$ leads to

$$2^{-\nu} \int_0^1 (1-u)^{\nu-1}(1 - 2\sqrt{u} + u)u^{-\frac{1}{2}\nu-1} du$$

$$= 2^{-\nu} \left[B\left(\nu, -\frac{1}{2}\nu\right) - 2B\left(\nu, -\frac{1}{2}\nu+\frac{1}{2}\right) + B\left(\nu, -\frac{1}{2}\nu+1\right) \right].$$

The first and third summands cancel. Since $\nu\,\Gamma(\nu) = \Gamma(\nu + 1)$, use of the reflection formula and the duplication formula leads to the evaluation

$$P'_\nu(0) = \frac{2\Gamma(\frac{1}{2}\nu+1)}{\Gamma(\frac{1}{2}\nu+\frac{1}{2})\sqrt{\pi}} \sin\left(\frac{\nu\pi}{2}\right). \tag{11.4.1}$$

The next result expresses P_ν as the sum of an even function and an odd function of x:

$$P_\nu(x)$$
$$= \frac{\Gamma(\frac{1}{2}\nu+\frac{1}{2})}{\Gamma(\frac{1}{2}\nu+1)\sqrt{\pi}} \cos\left(\frac{1}{2}\nu\pi\right) F\left(\frac{1}{2}\nu+\frac{1}{2}, -\frac{1}{2}\nu, \frac{1}{2}; x^2\right)$$

$$+ \frac{2\Gamma(\frac{1}{2}\nu+1)}{\Gamma(\frac{1}{2}\nu+\frac{1}{2})\sqrt{\pi}} \sin\left(\frac{1}{2}\nu\pi\right) xF\left(\frac{1}{2}\nu+1, \frac{1}{2}-\frac{1}{2}\nu, \frac{3}{2}; x^2\right). \tag{11.4.2}$$

Thus P_ν is an even function precisely when ν is an even integer, and an odd function when ν is an odd integer: the case of Legendre polynomials.

The identities (11.3.7) and (11.3.8), together with (11.4.2), give the corresponding expression for Q_ν in general:

$$Q_\nu(x) = e^{\mp\frac{1}{2}\nu\pi i} \left[\frac{\Gamma(\frac{\nu}{2}+1)\sqrt{\pi}}{\Gamma(\frac{1}{2}\nu+\frac{1}{2})} xF\left(\frac{1}{2}\nu+1, \frac{1}{2}-\frac{1}{2}\nu, \frac{3}{2};x^2\right) \right.$$
$$\left. \mp i\frac{\Gamma(\frac{1}{2}\nu+\frac{1}{2})\sqrt{\pi}}{2\Gamma(\frac{1}{2}\nu+1)} F\left(\frac{1}{2}\nu+\frac{1}{2}, -\frac{1}{2}\nu, \frac{1}{2};x^2\right) \right], \qquad (11.4.3)$$

$\pm\operatorname{Im}x > 0$. This is mainly of interest on the interval $(-1,1)$, where (11.3.6) gives a formula dual to (11.4.2) as a sum of odd and even parts:

$$Q_\nu(x) = \frac{\Gamma(\frac{1}{2}\nu+1)\sqrt{\pi}}{\Gamma(\frac{1}{2}\nu+\frac{1}{2})} \cos\left(\frac{1}{2}\nu\pi\right) xF\left(\frac{1}{2}\nu+1, \frac{1}{2}-\frac{1}{2}\nu, \frac{3}{2};x^2\right)$$
$$- \frac{\Gamma(\frac{1}{2}\nu+\frac{1}{2})\sqrt{\pi}}{2\Gamma(\frac{1}{2}\nu+1)} \sin\left(\frac{1}{2}\nu\pi\right) F\left(\frac{1}{2}\nu+\frac{1}{2}, -\frac{1}{2}\nu, \frac{1}{2};x^2\right).$$

$$(11.4.4)$$

The following asymptotic results of Laplace [242], Darboux [93], and Heine [180] will be proved in Chapter 13:

$$P_\nu(\cosh\theta) = \frac{e^{(\nu+\frac{1}{2})\theta}}{\sqrt{2\nu\pi\sinh\theta}} \left[1 + O(|\nu|^{-1})\right];$$

$$(11.4.5)$$

$$Q_\nu(\cosh\theta) = \frac{\sqrt{\pi}\,e^{-(\nu+\frac{1}{2})\theta}}{\sqrt{2\nu\sinh\theta}} \left[1 + O(|\nu|^{-1})\right],$$

$|\arg\nu| \le \frac{1}{2}\pi - \delta$, as $|\nu| \to \infty$, uniformly for $0 < \delta \le \theta \le \delta^{-1}$.

Asymptotics on the interval $-1 < x < 1$ can be computed from (11.2.3) and (11.3.9) using (10.1.12):

$$P_\nu(\cos\theta) = \frac{\sqrt{2}\cos\left(\nu\theta + \frac{1}{2}\theta - \frac{1}{4}\pi\right) + O\left(|\nu|^{-1}\right)}{\sqrt{\nu\pi\sin\theta}}$$

$$(11.4.6)$$

$$Q_\nu(\cos\theta) = -\frac{\sqrt{\pi}\sin\left(\nu\theta + \frac{1}{2}\theta - \frac{1}{4}\pi\right) + O\left(|\nu|^{-1}\right)}{\sqrt{2\nu\sin\theta}}$$

as $|\nu| \to \infty$, uniformly for $0 < \delta \le \theta \le \pi - \delta$.

11.5 Associated Legendre functions

The *associated Legendre functions* are the solutions of the spherical harmonic equation (11.1.2),

$$(1-z^2)u''(z) - 2zu'(z) + \left[v(v+1) - \frac{m^2}{1-z^2}\right]u(z) = 0, \qquad (11.5.1)$$

$m = 1, 2, \ldots$ As in Section 11.1, the gauge transformation $u(z) = (1-z^2)^{\frac{1}{2}m}v(z)$ reduces this to

$$(1-z^2)v''(z) - 2(m+1)zv'(z) + (v-m)(v+m+1)v(z) = 0. \qquad (11.5.2)$$

Repeated differentiation shows that the mth derivative of a solution of the Legendre equation (11.2.1) is a solution of (11.5.2). Therefore the functions

$$P_v^m(z) = (z^2-1)^{\frac{1}{2}m}\frac{d^m}{dz^m}[P_v(z)],$$

$$\qquad (11.5.3)$$

$$Q_v^m(z) = (z^2-1)^{\frac{1}{2}m}\frac{d^m}{dz^m}[Q_v(z)]$$

are solutions of (11.5.1). These are conveniently normalized for $|z| > 1$. Various normalizations are used for associated Legendre functions on the interval $(-1,1)$, including

$$P_v^m(x) = (-1)^m(1-x^2)^{\frac{1}{2}m}\frac{d^m}{dx^m}[P_v(x)],$$

$$\qquad (11.5.4)$$

$$Q_v^m(x) = (-1)^m(1-x^2)^{\frac{1}{2}m}\frac{d^m}{dx^m}[Q_v(x)].$$

In this section, z will always denote a complex number in the complement of $(-\infty, 1]$ and x will denote a real number in the interval $(-1,1)$. It follows from (11.2.3), (10.1.3), and the definitions (11.5.3) and (11.5.4) that

$$P_v^m(z) = (-1)^m\frac{(v+1)_m(-v)_m}{2^m m!}(z^2-1)^{\frac{1}{2}m}$$

$$\times F\left(v+1+m, m-v, m+1; \frac{1}{2}(1-z)\right)$$

$$= \frac{\Gamma(v+m+1)}{2^m\Gamma(v-m+1)m!}(z^2-1)^{\frac{1}{2}m}$$

$$\times F\left(v+1+m, m-v, m+1; \frac{1}{2}(1-z)\right);$$

$$P_\nu^m(x) = (-1)^m \frac{\Gamma(\nu+m+1)}{2^m \Gamma(\nu-m+1)m!} (1-x^2)^{\frac{1}{2}m}$$
$$\times F\left(\nu+1+m, m-\nu, m+1; \frac{1}{2}(1-x)\right).$$

To obtain a representation of Q_ν^m as a multiple of a hypergeometric function, we differentiate the series representation (11.2.10) m times. Since

$$\frac{d^m}{dz^m}\left[z^{-\nu-1-2n}\right] = (-1)^m (\nu+1+2n)_m z^{-\nu-1-2n-m}$$

$$= (-1)^m \frac{(\nu+1)_m(\nu+1+m)_{2n}}{(\nu+1)_{2n}} z^{-\nu-1-2n-m}$$

$$= (-1)^m \frac{(\nu+1)_m(\frac{1}{2}\nu+\frac{1}{2}m+\frac{1}{2})_n(\frac{1}{2}\nu+\frac{1}{2}m+1)_n}{(\frac{1}{2}\nu+\frac{1}{2})_n(\frac{1}{2}\nu+1)_n} z^{-\nu-1-2n-m},$$

it follows from the series expansion in (11.2.10) that

$$\frac{d^m}{dz^m}[Q_\nu(z)] = (-1)^m \frac{\Gamma(\nu+1+m)\sqrt{\pi}}{\Gamma(\nu+\frac{3}{2})2^{\nu+1}z^{\nu+1+m}}$$

$$\times F\left(\frac{1}{2}\nu+\frac{1}{2}m+\frac{1}{2}, \frac{1}{2}\nu+\frac{1}{2}m+1, \nu+\frac{3}{2}; \frac{1}{z^2}\right).$$

Therefore

$$Q_\nu^m(z) = (-1)^m (z^2-1)^{\frac{1}{2}m} \frac{\Gamma(\nu+1+m)\sqrt{\pi}}{\Gamma(\nu+\frac{3}{2})2^{\nu+1}z^{\nu+1+m}}$$

$$\times F\left(\frac{1}{2}\nu+\frac{1}{2}m+\frac{1}{2}, \frac{1}{2}\nu+\frac{1}{2}m+1, \nu+\frac{3}{2}; \frac{1}{z^2}\right). \qquad (11.5.5)$$

To obtain integral representations of P_ν^m and Q_ν^m, we begin with the representations (11.2.6) and (11.2.13). Set

$$A(z,\varphi) = z + \sqrt{z^2-1}\cos\varphi,$$

so that (11.2.6) is

$$P_\nu(z) = \frac{1}{\pi}\int_0^\pi \frac{d\varphi}{A(z,\varphi)^{\nu+1}}.$$

The identity

$$\frac{\partial}{\partial z}\left[\frac{\sin^{2k}\varphi}{A(z,\varphi)^r}\right] = \frac{r(r-2k-1)}{2k+1}\frac{\sin^{2k+2}\varphi}{A(z,\varphi)^{r+1}}$$

$$-\frac{r}{2k+1}\frac{\partial}{\partial\varphi}\left[\frac{\sin^{2k+1}\varphi}{A(z,\varphi)^r\sqrt{z^2-1}}\right] \qquad (11.5.6)$$

implies that

$$\frac{d}{dz}\int_0^\pi \frac{\sin^{2k}\varphi\,d\varphi}{A(z,\varphi)^r} = \frac{r(r-2k-1)}{2k+1}\int_0^\pi \frac{\sin^{2k+2}\varphi\,d\varphi}{A(z,\varphi)^{r+1}}.$$

Applying this identity m times starting with $k=0$, $r=v+1$ shows that (11.2.6) and (11.5.3) imply that

$$P_v^m(z) = (z^2-1)^{\frac{1}{2}m}\frac{(-m+v+1)_{2m}}{\pi\,2^m(\frac{1}{2})_m}$$

$$\times \int_0^\pi \frac{\sin^{2m}\varphi\,d\varphi}{[z+\sqrt{z^2-1}\cos\varphi]^{v+1+m}}. \tag{11.5.7}$$

Similarly, let

$$B(z,\varphi) = z+\sqrt{z^2-1}\,\cosh\varphi.$$

Then

$$\frac{\partial}{\partial z}\left[\frac{\sinh^{2k}\varphi}{B(z,\varphi)^r}\right] = -\frac{r(r-2k-1)}{2k+1}\frac{\sinh^{2k+2}\varphi}{B(z,\varphi)^{r+1}}$$

$$-\frac{r}{2k+1}\frac{\partial}{\partial\varphi}\left[\frac{\sinh^{2k+1}\varphi}{B(z,\varphi)^r\sqrt{z^2-1}}\right]. \tag{11.5.8}$$

Together with (11.2.13) and (11.5.3), this implies that

$$Q_v^m(z) = (-1)^m(z^2-1)^{\frac{1}{2}m}\frac{(-m+v+1)_{2m}}{2^m(\frac{1}{2})_m}$$

$$\times \int_0^\infty \frac{\sinh^{2m}\varphi\,d\varphi}{[z+\sqrt{z^2-1}\cosh\varphi]^{v+1+m}}. \tag{11.5.9}$$

These integral representations will be used in Chapter 13 to obtain the asymptotics on the interval $-1 < x < \infty$:

$$P_v^m(\cosh\theta) = \frac{e^{(v+\frac{1}{2})\theta}}{\sqrt{2\pi\sinh\theta}}(m+1+v)^{m-\frac{1}{2}}$$

$$\times\left[1+O\left(\{m+1+v\}^{-\frac{1}{2}}\right)\right];$$

$$Q_v^m(\cosh\theta) = (-1)^m\frac{e^{-(v+\frac{1}{2})\theta}\sqrt{\pi}}{\sqrt{2\sinh\theta}}(m+1+v)^{m-\frac{1}{2}}$$

$$\times\left[1+O\left(\{m+1+v\}^{-\frac{1}{2}}\right)\right] \tag{11.5.10}$$

as $v\to\infty$, uniformly on intervals $0 < \delta \le \theta \le \delta^{-1}$.

11.6 Relations among associated functions

As noted above, the mth derivatives $P_\nu^{(m)}$ and $Q_\nu^{(m)}$ are solutions of (11.5.2). Putting these derivatives into the equation and using (11.5.3), we obtain the recurrence relations

$$P_\nu^{m+2}(z) + \frac{2(m+1)z}{\sqrt{z^2-1}} P_\nu^{m+1}(z) - (\nu-m)(\nu+m+1) P_\nu^m(z) = 0;$$

$$Q_\nu^{m+2}(z) + \frac{2(m+1)z}{\sqrt{z^2-1}} Q_\nu^{m+1}(z) - (\nu-m)(\nu+m+1) Q_\nu^m(z) = 0$$

(11.6.1)

for $z \notin (-\infty, 1]$. Similarly,

$$P_\nu^{m+2}(x) + \frac{2(m+1)x}{\sqrt{1-x^2}} P_\nu^{m+1}(x) + (\nu-m)(\nu+m+1) P_\nu^m(x) = 0;$$

$$Q_\nu^{m+2}(x) + \frac{2(m+1)x}{\sqrt{1-x^2}} Q_\nu^{m+1}(x) + (\nu-m)(\nu+m+1) Q_\nu^m(x) = 0$$

(11.6.2)

for $-1 < x < 1$. Now $P_\nu^0 = P_\nu$, and it follows from (11.3.16) and (11.5.3) that

$$P_\nu^1(z) = \frac{\nu z}{\sqrt{z^2-1}} P_\nu(z) - \frac{\nu}{\sqrt{z^2-1}} P_{\nu-1}(z),$$

so P_ν^m and Q_ν^m can be computed recursively from the case $m = 0$.

Other relations can be obtained from general results for Jacobi polynomials. As an alternative, we may use the results above for the case $m = 0$. Differentiating (11.3.12) and (11.3.13) m times and multipling by $(z^2 - 1)^{\frac{1}{2}m}$ gives analogous identities involving the P_ν^m, Q_ν^m. Proceeding from these identities as in the derivation of (11.3.14)–(11.3.16) gives identities that generalize (11.3.14)–(11.3.16).

In addition, a pure recurrence relation can be obtained. Differentiate (11.3.13) $(m-1)$ times, differentiate (11.3.12) m times, use the latter to eliminate the term $P_\nu^{(m-1)}$ from the former, and multiply the result by $(z^2-1)^{\frac{1}{2}m}$ to obtain the recurrence relations

$$(\nu+1-m) P_{\nu+1}^m(z) = (2\nu+1)z P_\nu^m(z) - (\nu+m) P_{\nu-1}^m(z);$$

$$(\nu+1-m) Q_{\nu+1}^m(z) = (2\nu+1)z Q_\nu^m(z) - (\nu+m) Q_{\nu-1}^m(z).$$

(11.6.3)

Differentiating the relations (11.2.4), (11.2.11), (11.3.3), (11.3.5), (11.3.7), and (11.3.8), and multiplying by $(z^2 - 1)^{\frac{1}{2}m}$, gives the corresponding relations for the various solutions of (11.1.2):

$$P_\nu^m(z) = P_{-\nu-1}^m(z);$$

$$Q_{n-\frac{1}{2}}^m(z) = Q_{-n-\frac{1}{2}}^m(z);$$

$$Q_\nu^m(-z) = -e^{\pm\nu\pi i}Q_\nu^m(z), \quad \pm\mathrm{Im}\,z > 0;$$

$$P_\nu^m(z) = \frac{\tan\nu\pi}{\pi}\left[Q_\nu^m(z) - Q_{-\nu-1}^m(z)\right];$$

$$Q_\nu^m(z) = \frac{\pi}{2\sin\nu\pi}\left[e^{\mp\nu\pi i}P_\nu^m(z) - P_\nu^m(-z)\right], \quad \pm\mathrm{Im}\,z > 0.$$

In view of these relations, to compute Wronskians we only need to compute $W(P_\nu^m, Q_\nu^m)$. Differentiating (11.5.3) gives

$$\left[P_\nu^m\right]'(z) = (z^2-1)^{-\frac{1}{2}}P_\nu^{m+1}(z) + mz(z^2-1)^{-1}P_\nu^m(z);$$

$$\left[Q_\nu^m\right]'(z) = (z^2-1)^{-\frac{1}{2}}Q_\nu^{m+1}(z) + mz(z^2-1)^{-1}Q_\nu^m(z).$$

It follows that

$$W(P_\nu^m, Q_\nu^m)(z) = (z^2-1)^{-\frac{1}{2}}\left[P_\nu^m(z)Q_\nu^{m+1}(z) - P_\nu^{m+1}(z)Q_\nu^m(z)\right].$$

The recurrence relation (11.6.1) implies that

$$P_\nu^m Q_\nu^{m+1} - P_\nu^{m+1}Q_\nu^m = (\nu+m)(m-\nu-1)\left[P_\nu^{m-1}Q_\nu^m - P_\nu^m Q_\nu^{m-1}\right],$$

so that the computation leads to

$$P_\nu(z)Q_\nu^1(z) - P_\nu^1(z)Q_\nu(z) = (z^2-1)^{\frac{1}{2}}\left[P_\nu(z)Q_\nu'(z) - P_\nu'(z)Q_\nu(z)\right]$$

$$= (z^2-1)^{\frac{1}{2}}W(P_\nu, Q_\nu)(z) = \frac{(z^2-1)^{\frac{1}{2}}}{1-z^2}.$$

Combining these results,

$$W(P_\nu^m, Q_\nu^m)(z) = \frac{\Gamma(\nu+1+m)\,\Gamma(-\nu+m)}{\Gamma(\nu+1)\,\Gamma(-\nu)}\,\frac{1}{1-z^2}$$

$$= \frac{(\nu+1)_m(-\nu)_m}{1-z^2}.$$

11.7 Exercises

1. Suppose that $f(\mathbf{x})$ and $g(\mathbf{x})$ are smooth functions in $\mathbf{R}^n$. Show that

$$\Delta(fg) = (\Delta f)g + 2\sum_{j=1}^{n} \frac{\partial f}{\partial x_j}\frac{\partial g}{\partial x_j} + f\,\Delta g.$$

Show that if $f(\lambda\mathbf{x}) = \lambda^m f(\mathbf{x})$ for all $\lambda > 0$, then

$$\sum_{j=1}^{n} x_j \frac{\partial f}{\partial x_j} = mf.$$

2. Suppose that $p(x,y,z)$ is a homogeneous polynomial of degree n. Use Exercise 1 to show that there are harmonic polynomials p_{n-2j} of degree $n-2j$, $0 \le j \le n/2$, such that

$$p = p_n + r^2 p_{n-2} + r^4 p_{n-4} + \cdots$$

3. Suppose that p is a harmonic polynomial of degree m in three variables and $r^2 = x^2 + y^2 + z^2$. Show that

$$\Delta(r^k p) = k(k+2m+1)r^{k-2}p.$$

   Use this fact to show that the harmonic polynomials in Exercise 2 can be identified in sequence, starting with p_0 if n is even, or p_1 if n is odd. This procedure is due to Gauss [154].

4. Use the Taylor expansion of $f(s) = 1/r(x,y,z-s)$ to show that

$$P_n\left(\frac{z}{r}\right) = \frac{(-1)^n r^{n+1}}{n!}\frac{\partial^n}{\partial z^n}\left[\frac{1}{r}\right].$$

5. Verify that the change of variables after (11.2.4) converts (11.2.2) to (11.2.5).

6. Show that the change of variables after (11.2.6) gives the identity

$$(z^2 - 1)\sin^2\varphi = e^{\alpha}(2z - 2\cosh\alpha)$$

   and show that (11.2.6) becomes (11.2.7).

7. Use (11.2.2) to prove that for $|z| < 1$,

$$P_\nu(z) = \frac{1}{2\pi}\int_0^{2\pi}\left(z + i\sqrt{1-z^2}\sin\varphi\right)^\nu d\varphi.$$

8. Show that for $|z| < 1$,

$$P_\nu(z) = \frac{1}{2\pi}\int_0^{2\pi}\frac{1}{\left(z + i\sqrt{1-z^2}\sin\varphi\right)^{\nu+1}}\,d\varphi.$$

9. Use the identity

$$\frac{1}{a^{\nu+1}} = \frac{1}{\Gamma(\nu+1)} \int_0^\infty e^{-as} s^\nu \, ds, \quad \mathrm{Re}\,\alpha > 0,$$

to show that

$$P_\nu(\cos\theta) = \frac{1}{\Gamma(\nu+1)} \int_0^\infty e^{-\cos\theta s} J_0(\sin\theta s) s^\nu \, ds,$$

where J_0 is the Bessel function of order 0.

10. Show that

$$P_\nu(\cosh\theta) = \frac{1}{\Gamma(\nu+1)} \int_0^\infty e^{-\cosh\theta s} I_0(\sinh\theta s) s^\nu \, ds,$$

where I_0 is one of the modified Bessel functions.

11. Prove the Rodrigues-type identity

$$Q_n(z) = \frac{(-1)^n}{2^{n+1} n!} \frac{d^n}{dz^n} \int_{-1}^1 \frac{(1-s^2)^n}{z-s} \, ds, \quad z \neq (-\infty, 1].$$

12. Prove the generating function identity

$$\sum_{n=0}^\infty Q_n(z) t^n = \frac{1}{2R} \log \frac{z-t+R}{z-t-R}, \quad R = \sqrt{1+t^2-2tz}.$$

13. Show that

$$Q_n\left(\frac{z}{r}\right) = \frac{(-1)^n r^{n+1}}{n!} \frac{\partial^n}{\partial z^n} \left[\frac{1}{2r} \log \frac{r+z}{r-z}\right],$$

where $r^2 = x^2 + y^2 + z^2$.

14. Show that the change of variables after (11.2.11) converts (11.2.9) to (11.2.12).

15. Show that the change of variables after (11.2.12) leads to the identities

$$R^2 \sinh^2\varphi = s^2 - 2sz + 1 = s^2 R^2 \sinh^2\theta,$$

where

$$R = \sqrt{z^2 - 1}, \quad s = z + R\cosh\varphi = [z - R\cosh\theta]^{-1}.$$

16. Use Exercise 15 to show that (11.2.12) implies (11.2.13).

17. Show that the change of variables after (11.2.13) leads to the identity

$$e^\alpha(2\cosh\alpha - 2\cosh\theta) = \sinh^2\theta \sinh^2\varphi$$

and show that (11.2.13) implies (11.2.14).

18. Show that for $x < -1$,

$$P_\nu(x+i0) - P_\nu(x-i0) = 2i \sin\nu\pi \, P_\nu(-x);$$

$$Q_\nu(x+i0) - Q_\nu(x-i0) = 2i \sin\nu\pi \, Q_\nu(-x).$$

19. Use the method of proof of (11.4.2) to prove

$$P_\nu(z) = F\left(\frac{1}{2}\nu + \frac{1}{2}, -\frac{1}{2}\nu, 1; 1 - z^2\right), \quad |1 - z^2| < 1.$$

20. Prove (11.4.6). Hint: $2\cos A \cos B = \cos(A+B) + \cos(A-B)$, and $\cos(A+B) - \cos(A-B) = -2\sin A \sin B$.
21. Use (10.2.10) and (11.4.6) to give a different proof of (11.3.3).
22. Prove by induction the Jacobi lemma

$$\frac{d^{m-1}}{d\mu^{m-1}}\left[\sin^{2m-1}\varphi\right] = (-1)^{m-1}\frac{(2m)!}{m2^m m!}\sin(m\varphi),$$

where $\mu = \cos\varphi$, and deduce that

$$\cos(m\varphi) = (-1)^{m-1}\frac{2^m m!}{(2m)!}\frac{d^m}{d\mu^m}\left[\sin^{2m-1}\varphi\right]\frac{d\mu}{d\varphi}.$$

23. Use Exercise 22 to give another derivation of the integral representation (11.5.7).
24. Use (11.2.2) and (11.5.3) to derive the identities

$$P_\nu^m(z) = (z^2 - 1)^{\frac{1}{2}m}\frac{(\nu+1)_m}{2^{\nu+1}\pi i}\int_C \frac{(t^2 - 1)^\nu}{(t-z)^{\nu+1+m}}\,dt$$

and

$$P_\nu^m(z) = \frac{(\nu+1)_m}{\pi}\int_0^\pi \left[z + \sqrt{z^2 - 1}\cos\varphi\right]^\nu \cos(m\varphi)\,d\varphi$$

$$= \frac{(-\nu)_m}{\pi}\int_0^\pi \frac{\cos(m\varphi)\,d\varphi}{\left[z + \sqrt{z^2 - 1}\cos\varphi\right]^{\nu+1}}.$$

25. Use (11.2.9) and (11.5.3) to derive the identity

$$Q_\nu^m(z) = (-1)^m(z^2 - 1)^{\frac{1}{2}m}\frac{(\nu+1)_m}{2^{\nu+1}}\int_{-1}^1 \frac{(1-t^2)^\nu}{(z-t)^{\nu+1+m}}\,dt$$

$$= (-1)^m(\nu+1)_m$$

$$\times \int_0^\alpha \left[z - \sqrt{z^2 - 1}\cosh\varphi\right]^\nu \cosh(m\varphi)\,d\varphi, \quad \coth\alpha = z.$$

26. Verify (11.5.6) and (11.5.8).

11.8 Remarks

The surface harmonics Y_{n0} occur in Laplace's work on celestial mechanics [242]. In his study of potential theory for celestial bodies, he introduced

Laplace's equation and found solutions by separating variables in spherical coordinates. As noted earlier, Legendre polynomials were studied by Legendre in 1784 [250, 251]. See the end of Chapter 5 for remarks on the early history of Legendre polynomials. The functions P_ν were defined for general ν by Schläfli [355]. Associated Legendre functions for nonnegative integer values of ν were introduced by Ferrers [134], and for general values of ν by Hobson [187].

Laplace [242] gave the principal term in the asymptotics (11.4.5) of P_n for positive integer n, and Heine [180] investigated asymptotics for both P_n and Q_n for integer n. Darboux [93] proved (11.4.5) and (11.4.6) for integer n. Hobson [187] proved the general form of (11.4.5), (11.4.6), and (11.5.10).

The term "spherical harmonics" is often used to include all the topics covered in this chapter. Two classical treatises are Ferrers [134] and Heine [180]. More recent references include Hobson [188], MacRobert [268], Robin [344], and Sternberg and Smith [379]. Müller [295] treats the subject in $\mathbf{R}^n$ and in $\mathbf{C}^n$.

12

Generalized hypergeometric functions; G-functions

The series expansions of the confluent hypergeometric function $M(a,c;x)$ and the hypergeometric function $F(a,b,c;x)$ are particular cases of a family of expansions with quotients of more or fewer factors of the form $(a)_n$, $(c)_n$. Such expansions are referred to as generalized hypergeometric series, or simply as hypergeometric series.

Like M and F, the corresponding functions are solutions of ordinary differential equations of special form. These equations can be characterized by a special property, the same property as discussed in Section 1.1.

Seeking a solution to a generalized hypergeometric equation in the form of an integral leads naturally to a class of solutions, the G-functions, introduced by Meijer. This family of functions is closed under several natural operations and transformations, and plays an important role in a number of applications, including the evaluation of integrals involving special functions. The integral form specializes to the Barnes integral representations of the confluent hypergeometric functions and hypergeometric functions, and the Mellin–Barnes integral representation of generalized hypergeometric functions.

12.1 Generalized hypergeometric series

We have seen that the solutions of the confluent hypergeometric equation and the hypergeometric equation

$$x u''(x) + (c - x) u'(x) - a u(x) = 0; \qquad (12.1.1)$$

$$x(1 - x) u''(x) + (c - [a + b + 1]x) u'(x) - ab u(x) = 0, \qquad (12.1.2)$$

with $u(0) = 1$, have the series representations

$$u(x) = \sum_{n=0}^{\infty} \frac{(a)_n}{(c)_n n!} x^n; \tag{12.1.3}$$

$$u(x) = \sum_{n=0}^{\infty} \frac{(a)_n (b)_n}{(c)_n n!} x^n. \tag{12.1.4}$$

Similarly, the third equation derived in Section 1.1,

$$x u''(x) + c u'(x) + u(x) = 0, \tag{12.1.5}$$

has a solution with the series representation

$$u(x) = \sum_{n=0}^{\infty} \frac{1}{(c)_n n!} (-x)^n.$$

(Recall that the shifted factorials are defined by

$$(a)_n = a(a+1)(a+2)\cdots(a+n-1) = \frac{\Gamma(a+n)}{\Gamma(a)}.)$$

These examples suggest the following generalization.

A *(generalized) hypergeometric series* is a power series of the form

$$\sum_{n=0}^{\infty} \frac{(a_1)_n (a_2)_n \cdots (a_p)_n}{(c_1)_n (c_2)_n \cdots (c_q)_n n!} x^n. \tag{12.1.6}$$

It is assumed in (12.1.6) that no c_j is a nonpositive integer. If any a_j is a nonpositive integer, then (12.1.6) is a polynomial; we exclude this case in the following general remarks.

The radius of convergence of the series (12.1.6) is zero if $p > q+1$, is 1 if $p = q+1$, and is infinite if $p \le q$ (ratio test). Therefore it is usually assumed that $p \le q+1$. The function defined by (12.1.6) is denoted by

$$_pF_q \begin{pmatrix} a_1, \ldots, a_p \\ c_1, \ldots, c_q \end{pmatrix}; x \end{pmatrix}. \tag{12.1.7}$$

In this notation the Kummer function and the hypergeometric function are

$$M(a,c;x) = {}_1F_1 \begin{pmatrix} a \\ c \end{pmatrix}; x \end{pmatrix};$$

$$F(a,b,c;x) = {}_2F_1 \begin{pmatrix} a,b \\ c \end{pmatrix}; x \end{pmatrix}.$$

The integral formulas (8.1.3), for the Kummer function, and (10.2.4), for the hypergeometric function, can be extended to the general case using the same argument.

Proposition 12.1.1 *If* $\operatorname{Re} c_1 > \operatorname{Re} a_1 > 0$, *then*

$$\frac{\Gamma(a_1)\Gamma(c_1 - a_1)}{\Gamma(c_1)} {}_pF_q \left(\begin{matrix} a_1, \ldots, a_p \\ c_1, \ldots, c_q \end{matrix} ; x \right)$$

$$= \int_0^1 s^{a_1 - 1} (1 - s)^{c_1 - a_1 - 1} {}_{p-1}F_{q-1} \left(\begin{matrix} a_2, \ldots, a_p \\ c_2, \ldots, c_q \end{matrix} ; sx \right) ds. \quad (12.1.8)$$

If $\operatorname{Re} c_2 > \operatorname{Re} a_2 > 0$, etc., this can be iterated.

These series satisfy a number of identities, including contiguous relations generalizing those for the confluent hypergeometric and hypergeometric functions, and evaluations at $x = 1$ like Gauss's summation formula (10.2.6). Perhaps the earliest such identity was found in 1828 by Clausen [83], who posed the question that, in modern notation, is as follows: for what values of the parameters is it true that

$$\left[{}_2F_1 \left(\begin{matrix} a, b \\ c \end{matrix} ; x \right) \right]^2 = {}_3F_2 \left(\begin{matrix} a', b', c' \\ d', e' \end{matrix} ; x \right) ?$$

(For the question in its original form, see the title of [83].) The answer is

$$c = a + b + \frac{1}{2}, \quad a' = 2a, \quad b' = 2b,$$

$$c' = a + b, \quad d' = c, \quad e' = 2a + 2b.$$

For other identities, see the references in the remarks at the end of this chapter.

12.1.1 Examples of generalized hypergeometric functions

An empty product (no factors) is always taken to be 1, so

$$ {}_0F_0 \left(\begin{matrix} - \\ - \end{matrix} ; x \right) = \sum_{n=0}^{\infty} \frac{x^n}{n!} = e^x.$$

The binomial expansion gives

$$ {}_1F_0 \left(\begin{matrix} a \\ - \end{matrix} ; x \right) = \sum_{n=0}^{\infty} \frac{(a)_n}{n!} x^n = \frac{1}{(1-x)^a}.$$

Since

$$\frac{1}{(2n)!} = \frac{1}{4^n (\frac{1}{2})_n n!} \quad \text{and} \quad \frac{1}{(2n+1)!} = \frac{1}{4^n (\frac{3}{2})_n n!},$$

it follows that

$$_0F_1\left(\begin{matrix}-\\\frac{1}{2}\end{matrix};-\frac{x^2}{4}\right)=\cos x;\qquad _0F_1\left(\begin{matrix}-\\\frac{1}{2}\end{matrix};\frac{x^2}{4}\right)=\cosh x;$$

$$_0F_1\left(\begin{matrix}-\\\frac{3}{2}\end{matrix};-\frac{x^2}{4}\right)=\frac{\sin x}{x};\qquad _0F_1\left(\begin{matrix}-\\\frac{3}{2}\end{matrix};\frac{x^2}{4}\right)=\frac{\sinh x}{x}.$$

It is clear from the series representation (9.1.2) that the Bessel function J_ν is

$$J_\nu(x)=\frac{(x/2)^\nu}{\Gamma(\nu+1)}\cdot {_0F_1}\left(\begin{matrix}-\\\nu+1\end{matrix};-\frac{x^2}{4}\right).$$

12.2 The generalized hypergeometric equation

In Section 1.1, we asked what second-order differential equations are *recursive*, in the sense that the sequence of equations for the coefficients of a formal power series solution at the origin determine these coefficients by a particularly simple recursion. Here we consider the question for equations of arbitrary order. The answer is, essentially, the equations that determine the generalized hypergeometric functions.

Again we use the operator $D=xd/dx$, which is diagonalized by powers of x:

$$D[x^s]=sx^s. \tag{12.2.1}$$

The general homogeneous linear ordinary differential equation of order M,

$$a_M(x)\frac{d^M}{dx^M}u(x)+a_{M-1}(x)\frac{d^{M-1}}{dx^{M-1}}u(x)+\cdots+a_0(x)u(x)=0,$$

can, after multiplication by x^M, be rewritten in the form

$$b_M(x)D^Mu(x)+b_{M-1}(x)D^{M-1}u(x)+\cdots+b_0(x)u(x)=0. \tag{12.2.2}$$

As in Section 1.1, suppose that the coefficients $b_j(x)$ are analytic near $x=0$, and consider the standard power series method: determine the coefficients of a formal power series solution $\sum_{n=0}^{\infty}u_nx^n$ with constant term $u_0=1$ by expanding (12.2.2) and collecting the coefficients of like powers of x. The general form of the question discussed in Section 1.1 is as follows.

When do the linear equations for the coefficients $\{u_n\}$ in the series expansion of (12.2.2) reduce to a two-term recursion of the form $c_nu_n=d_nu_{n-1}$?

Suppose that the coefficient b_k in (12.2.2) has the power series expansion $b_k(x)=\sum b_{km}x^m$. Dividing by some power of x, if necessary, we may assume that some constant term b_{k0} is not zero. Define the polynomials

$$P_m(t) = \sum_{k=0}^{M} b_{km} t^k.$$

Then the differential operator in (12.2.2) has a formal expansion

$$P_0(D) + x P_1(D) + x^2 P_2(D) + \cdots + x^n P_n(D) + \cdots$$

Our assumption implies that P_0 is not identically zero.

In the formal power series expansion of (12.2.2), the coefficient of x^n is

$$u_n P_0(n) + u_{n-1} P_1(n-1) + u_{n-2} P_2(n-2) + \cdots + u_0 P_n(0). \tag{12.2.3}$$

Suppose this reduces to a two-term recursion as above. Then for each $m \geq 2$ and every $n \geq m$, $P_m(n-m) = 0$. Thus each polynomial P_m, $m \geq 2$, is the zero polynomial, and the operator is

$$P_0(D) + x P_1(D).$$

Moreover, (12.2.3) implies that $P_0(0) = 0$. We may divide by the leading coefficient of P_0, so that the equation has the form

$$Q(D) u(x) - \alpha x P(D) u(x) = 0,$$

where Q and P are monic polynomials (leading coefficient 1), and $Q(0) = 0$. In view of (12.2.3), the recursion for the coefficients of a formal series solution is

$$Q(n) u_n = \alpha P(n-1) u_{n-1}. \tag{12.2.4}$$

We write the degree of Q as $q+1$ and the degree of P as p, so that $M = \max\{p, q+1\}$.

If $\alpha = 0$, (12.2.4) trivializes, e.g. if the roots r_j of Q are distinct, any solution is a linear combination of powers x^{r_j}. If $\alpha \neq 0$, we may take advantage of the scale invariance of the operator D to take αx as the independent variable and reduce to the case $\alpha = 1$.

Thus, excluding trivial cases, and up to normalization, the answer to the question above is that (12.2.2) must be a *generalized hypergeometric equation*: factoring the polynomials gives the form

$$\left[D \prod_{j=1}^{q} (D + c_j - 1) - x \prod_{j=1}^{p} (D + a_j) \right] u(x) = 0. \tag{12.2.5}$$

The associated recursion

$$u_0 = 1, \qquad u_n = \frac{P(n-1)}{Q(n)} u_{n-1} = \frac{\prod_{j=1}^{p} (a_j + n - 1)}{n \prod_{j=1}^{q} (c_j + n - 1)} u_{n-1} \tag{12.2.6}$$

gives us the expansion (12.1.6).

Assume that $p \le q+1$, so the order of the differential operator in (12.2.5) is $q+1$. The case $p=1$, $q=1$ is the confluent hypergeometric equation (12.1.1), the case $p=2$, $q=1$ is the standard hypergeometric equation (12.1.2), and the remaining second-order equation, $p=0$, $q=1$ is (12.1.5).

It is convenient to remove the asymmetry in the first product in (12.2.5). By a *generalized hypergeometric operator*, we mean an operator of the form

$$L = \prod_{j=1}^{q+1}(D+c_j-1) - x\prod_{j=1}^{p}(D+a_j). \tag{12.2.7}$$

This reduces to the operator in (12.2.5) when $c_{q+1}=1$. As in the case of the hypergeometric equation, we use the following identities for $D=xd/dx=D_x$:

$$\begin{aligned}
D(x^a f) &= x^a(D+a)f; \\
D_y &= D_x && \text{if } y=-x; \\
D_y &= -D_x && \text{if } y=1/x,
\end{aligned} \tag{12.2.8}$$

to identify solutions of the corresponding homogeneous equation $Lu=0$.

Proposition 12.2.1 *If $p \le q+1$ and no two c_j differ by an integer, then the equation*

$$\left[\prod_{j=1}^{q+1}(D+c_j-1) - x\prod_{j=1}^{p}(D+a_j)\right]u(x) = 0 \tag{12.2.9}$$

has a basis of $q+1$ solutions

$$u_j(x) = x^{1-c_j}\,{}_pF_q\left(\begin{matrix}a_1+1-c_j,\ldots,a_p+1-c_j \\ c_1+1-c_j,\ldots,c_{q+1}+1-c_j\end{matrix};x\right), \tag{12.2.10}$$

$j=1,\ldots,q+1$, where c_j+1-c_j is omitted from the lower line.

Proof The first of the identities (12.2.8) implies that the u_j are solutions of (12.2.9). They are clearly linearly independent and the equation has degree $q+1$, so they are a basis; see Theorem 3.2.4. □

Proposition 12.2.2 *If $p \ge q+1$ and no two a_j differ by an integer, then (12.2.9) has a basis of p solutions*

$$v_j(x) = \left[(-1)^{p-q}x\right]^{-a_j}$$

$$\times {}_{q+1}F_{p-1}\left(\begin{matrix}1-c_1+a_j,\ldots,1-c_{q+1}+a_j \\ 1-a_1+a_j,\ldots,1-a_p+a_j\end{matrix};\frac{(-1)^{q+1-p}}{x}\right), \tag{12.2.11}$$

$j=1,\ldots,p$, where $1-a_j+a_j$ is omitted from the bottom line.

Proof It follows from the identities (12.2.8) that under the change of variables $x \to 1/x$, followed by multiplication by $(-1)^{p-1}x$, the operator L of (12.2.7) goes into

$$L' = \prod_{j=1}^{p}(D - a_j) - (-1)^{q+1-p}x\prod_{j=1}^{q+1}(D + 1 - c_j). \qquad (12.2.12)$$

This explains the choice of sign of the argument in (12.2.11). Proposition 12.2.1 implies that the $\{v_j\}$ are a basis of solutions. (The choice of the opposite sign for the powers of $\pm x$ is dictated by the use in the case $p = q + 1$ below, a generalization of (10.2.10).) □

The two preceding results imply that if $p = q + 1$, a generalized hypergeometric function $_pF_q$ is a linear combination of solutions that have very specific behaviors as $x \to \infty$. The coefficients can be determined by an extension of the method used in Section 10.2 for the hypergeometric function; see Exercise 2.

Theorem 12.2.3 *If $p = q + 1$ and no two a_j differ by an integer, then*

$$_pF_q\left(\begin{matrix}a_1,\ldots,a_p\\c_1,\ldots,c_q\end{matrix}; x\right) = \sum_{j=1}^{p}A_j\,v_j(x), \qquad (12.2.13)$$

where the v_j are the functions (12.2.11) with $c_p = 1$ and the coefficients are

$$A_j = \prod_{k=1,k\neq j}^{q+1}\frac{\Gamma(a_k - a_j)}{\Gamma(a_k)}\prod_{k=1}^{q}\frac{\Gamma(c_k)}{\Gamma(c_k - a_j)}.$$

When $p < q + 1$, the asymptotics as $x \to +\infty$ are very different.

Theorem 12.2.4 *Suppose that $p < q + 1$ and that no two a_j differ by an integer. Then $_pF_q$ has an asymptotic expansion*

$$F(x) \sim \sum_{n=0}^{\infty}a_n x^{\nu - n/k}\exp\left(kx^{1/k}\right) \quad \text{as } x \to +\infty, \qquad (12.2.14)$$

where $k = q + 1 - p$ and

$$\nu = \frac{1}{k}\left[\sum_{j=1}^{p}a_j - \sum_{j=1}^{q}c_j + \frac{q-p}{2}\right].$$

See [313], p. 411. A proof of this result is outlined in the exercises.

12.3 Meijer *G*-functions

These functions arise naturally if we look for solutions of the generalized hypergeometric equation (12.2.5) in a certain integral form.

To conform to standard notation, we modify the operator (12.2.7) by replacing the indices $\{a_j\}$ and $\{c_k\}$ with $\{1 - a_j\}$ and $\{1 - b_k\}$, respectively. The range of the index k is taken to be $1 \le k \le q$ rather than $1 \le k \le q+1$. The associated generalized hypergeometric equation is

$$\left[\prod_{j=1}^{q}(D - b_j) - x \prod_{j=1}^{p}(D + 1 - a_j) \right] u(x) = 0. \tag{12.3.1}$$

We also consider the version obtained by the change of variables $x \to -x$:

$$\left[\prod_{j=1}^{q}(D - b_j) + x \prod_{j=1}^{p}(D + 1 - a_j) \right] u(x) = 0. \tag{12.3.2}$$

The change of variables $x \to (-1)^{p-q}/x$ converts each of the equations (12.3.1) and (12.3.2) to an equation of the same type with p and q interchanged, so *we usually assume that $p \le q$*.

The operator D is diagonalized by powers of x: $D[x^a] = ax^a$, while multiplication by x simply shifts the exponent by 1. Therefore it is natural to look for a solution of (12.3.2) or (12.3.1) as a sum of exponentials, or as a "continuous sum" of exponentials:

$$u(x) = \frac{1}{2\pi i} \int_C \Phi(s) x^s \, ds. \tag{12.3.3}$$

Here C is a curve in the complex plane, and $1/2\pi i$ is the usual normalizing factor for such integrals.

To begin, we proceed in a formal way, with calculations to be justified later. Putting (12.3.3) into (12.3.1) leads to

$$0 = \frac{1}{2\pi i} \int_C \Phi(s) \left[\prod_{j=1}^{q}(s - b_j)x^s - \prod_{j=1}^{p}(s + 1 - a_j)x^{s+1} \right] ds$$

$$= \frac{1}{2\pi i} \int_C \Phi(s) \prod_{j=1}^{q}(s - b_j)x^s \, ds$$

$$- \frac{1}{2\pi i} \int_{C+1} \Phi(s - 1) \prod_{j=1}^{p}(s - a_j)x^s \, ds.$$

Here $C + 1$ denotes the original contour C shifted one unit to the right. If the contour has the property that $C+1$ can be deformed to C without crossing any singularities of the integrand, then we are led to the continuous version of the recursion (12.2.6):

$$\Phi(s)\prod_{j=1}^{q}(s-b_j) = \Phi(s-1)\prod_{j=1}^{p}(s-a_j). \tag{12.3.4}$$

If we had started with (12.3.2) instead of (12.3.1), the recursion would be

$$\Phi(s)\prod_{j=1}^{q}(s-b_j) = -\Phi(s-1)\prod_{j=1}^{p}(s-a_j). \tag{12.3.5}$$

To motivate the remainder of the discussion, we look at the simplest nontrivial example: the case $q=1, p=0$. The recursion (12.3.4) reduces to

$$\frac{\Phi(s)}{\Phi(s-1)} = \frac{1}{s-b} = \frac{\Gamma(s-b)}{\Gamma(s+1-b)}. \tag{12.3.6}$$

Therefore we could take $\Phi(s) = 1/\Gamma(s+1-b)$. However, under reasonable constraints on the contour, the resulting function

$$u(x) = \frac{1}{2\pi i}\int_C \frac{x^s\,ds}{\Gamma(s+1-b)}$$

is identically zero, by Cauchy's theorem. This leads us to look for a second solution of (12.3.6). The quotient φ of any two solutions is characterized by periodicity: $\varphi(s-1) = \varphi(s)$. Euler's reflection formula,

$$\Gamma(z)\Gamma(1-z) = \frac{\pi}{\sin \pi z}, \tag{12.3.7}$$

gives us a periodic function in the context of gamma functions:

$$\varphi(s) = \Gamma(s+1-b)\Gamma(b-s).$$

This satisfies $\varphi(s-1) = -\varphi(s)$, so we would expect the resulting function

$$u(x) = \frac{1}{2\pi i}\int_C \Gamma(b-s)x^s\,ds \tag{12.3.8}$$

to satisfy the $p=0, q=1$ version of (12.3.2). Indeed, let us take for C a loop from $+\infty$ to $+\infty$, oriented so that the poles of the integrand lie to the right of the curve. The residue of $\Gamma(b-s)$ at $s=b+n$ is $(-1)^{n+1}/n!$, and the curve is oriented so that the residue calculus picks up the negatives of the residues. Therefore the function (12.3.8) has the series expansion

$$u(x) = x^b\sum_{n=0}^{\infty}\frac{(-x)^n}{n!} = x^b\,{}_0F_0\left(\begin{matrix}-\\-\end{matrix}\middle|-x\right).$$

Remark. This choice of negative orientation of the curve relative to sequences of poles tending to the right, in order to undo the effect of the negative sign of s in terms of the form $\Gamma(b-s)$, will be used in the general case.

Let us now return to the general recursion (12.3.4). The solution of this recursion that corresponds to our first solution in the special case is

$$\Phi(s) = \frac{\prod_{j=1}^{p} \Gamma(s+1-a_j)}{\prod_{j=1}^{q} \Gamma(s+1-b_j)}. \tag{12.3.9}$$

The corresponding solution of (12.3.1), in Meijer's notation, is the G-function

$$G_{p,q}^{0,p}\left(x; \begin{matrix} a_1,\ldots,a_p \\ b_1,\ldots,b_q \end{matrix}\right) = \frac{1}{2\pi i} \int_C \frac{\prod_{j=1}^{p} \Gamma(1-a_j+s)}{\prod_{j=1}^{q} \Gamma(1-b_j+s)} x^s \, ds. \tag{12.3.10}$$

(The contour C must be specified; we return to this below.)

We may obtain other solutions of (12.3.4) or (12.3.5) by multiplying (12.3.9) by terms of the form

$$\Gamma(s+1-b_j)\Gamma(b_j-s) = \frac{\pi}{\sin\pi(b_j-s)}$$

and/or of the form

$$\frac{1}{\Gamma(s+1-a_j)\Gamma(a_j-s)} = \frac{\sin\pi(a_j-s)}{\pi}.$$

Using an *even* number of such factors will lead (at least formally) to a solution of (12.3.1), while an *odd* number of factors will lead to a solution of (12.3.2).

The general Meijer G-function [282] associated with (12.3.2) and (12.3.1) is denoted

$$G_{p,q}^{m,n}\left(x; \begin{matrix} a_1',\ldots.a_p' \\ b_1',\ldots,b_q' \end{matrix}\right), \qquad 0 \le m \le q, \ 0 \le n \le p. \tag{12.3.11}$$

The indices $\{a_j'\}$ and $\{b_j'\}$ are permutations of the original indices $\{a_j\}$ and $\{b_j\}$, respectively. The upper indices m, n indicate that the *first* m factors in the *denominator* of the integrand have been changed to factors $\Gamma(b_j-s)$ in the numerator, and the *last* $p-n$ factors in the *numerator* have been changed to factors $\Gamma(a_j-s)$ in the denominator. The result is $m+n$ factors in the numerator and $p+q-(n+m)$ factors in the denominator. Formally, at least, this will be a solution of (12.3.1) if $m+n+p$ is even, and a solution of (12.3.2) if $m+n+p$ is odd.

Thus

$$G_{p,q}^{m,n}\left(x;\begin{array}{c}a_1,\ldots,a_p\\b_1,\ldots,b_q\end{array}\right)$$

$$=\frac{1}{2\pi i}\int_C\frac{\prod_{j=1}^m\Gamma(b_j-s)\prod_{k=1}^n\Gamma(s+1-a_k)}{\prod_{j=m+1}^q\Gamma(s+1-b_j)\prod_{k=n+1}^p\Gamma(a_k-s)}x^s\,ds.\quad(12.3.12)$$

The following conditions guarantee that *each* of the integrands in (12.3.11) will have only simple poles:

- $i\neq j$ implies b_i-b_j is not an integer;

- $k\neq l$ implies a_k-a_l is not an integer; $\quad\quad\quad\quad(12.3.13)$

- for all $i,k,\quad a_k-b_i$ is not a positive integer.

For the particular form (12.3.12), we need these conditions only for those indices b_i, b_j, a_k, a_l that correspond to the factors in the *numerator* of the integrand, i.e. for $i,j\le m$ and $k,l\le n$.

Taking into account the invariance of the integrand under permutations of the original indices a_j and b_j, (12.3.11) (with argument $-x$ if $m+n+p$ is odd) accounts, formally, for 2^{p+q} solutions of (12.3.1) – one for each subset of the $p+q$ indices. However if $p+q>1$, these solutions are not distinct. For example, if $p<q$, then the 2^p solutions with $m=0$ vanish identically, like our first solution in the case $p=0$, $q=1$. If $p=q$, only the solution with $m=n=0$ vanishes identically. (The contour C is specified below.)

We turn now to the problem of choosing contours of integration. A general principle is that *the contour in a given case* (12.3.11) *must separate the poles due to the factors* $\Gamma(s+1-a_j)$ *that remain in the numerator from the poles due to the factors* $\Gamma(b_k-s)$ *that have been raised to the numerator*. The third of the conditions (12.3.13), for $i\le n$ and $k\le m$, guarantees that this is possible.

We shall see in Section 12.4 that the following choices can be made for the contour C.

- If $p<q$, or if $p=q$ and $|x|<1$, we take C to begin and end at $s=+\infty$, oriented so that the sequences of poles tending to $+\infty$ lie to the *right* of C, as in the example above.
- If $p>q$, or if $p=q$ and $|x|>1$, we take C to begin and end at $s=-\infty$, oriented so that the sequences of poles tending to $-\infty$ lie to the *left* of C.
- Under certain conditions, the contour chosen before, beginning and ending at $\pm\infty$, can be deformed so as to begin at $-i\infty$ and end at $+i\infty$; see Theorem 12.4.2.

Remark. Up to this point, we have taken for granted that the functions (12.3.11) (with argument $(-1)^{m+n+p}x$) are the solutions of (12.3.1) that we used to derive the forms (12.3.11). The reasoning that led to the recursion (12.3.4) can be reversed, so long as the contours C and $C+1$ can be deformed into each other without crossing poles of the integrand. A condition that guarantees that C can be chosen so as to have this property is that no a_j and b_k lie on the same horizontal line: $\operatorname{Im} a_j \neq \operatorname{Im} b_k, j \leq n, k \leq m$. The general case follows from this case by analytic continuation.

12.3.1 Examples

1. The Kummer function M: in the notation of this section, the indices for (12.1.1) are $a_1 = 1 - a$ and $b_1 = 0, b_2 = 1 - c$. Thus

$$G_{1,2}^{1,1}\left(-x; \begin{matrix} 1-a \\ 0, 1-c \end{matrix}\right)$$

is a solution of the equation. The integral representation shows that this solution is regular at the origin, so it must be a multiple of $M(a, c; x)$. In fact, the integrand has poles precisely at the nonnegative integers, leading to the power series expansion. The first residue is $\Gamma(a)/\Gamma(c)$, so

$$M(a, c; x) = \frac{\Gamma(c)}{\Gamma(a)} G_{1,2}^{1,1}\left(-x; \begin{matrix} 1-a \\ 0, 1-c \end{matrix}\right)$$

$$= \frac{1}{2\pi i} \int_C \frac{\Gamma(c)\Gamma(a+s)\Gamma(-s)}{\Gamma(a)\Gamma(c+s)} (-x)^s \, ds, \quad (12.3.14)$$

where the curve C begins and ends at $+\infty$, encloses the nonnegative integers in a negative direction, and separates them from the poles of $\Gamma(a+s)$. This is the *Barnes integral* [30] for the Kummer function. The identity (12.3.14) is valid for $|\arg(-x)| < \pi$.

2. A second solution of (12.1.1) is given by

$$G_{1,2}^{2,1}\left(x; \begin{matrix} 1-a \\ 0, 1-c, \end{matrix}\right), \qquad (12.3.15)$$

assuming that neither a nor $a - c$ is a negative integer. The path of integration can be taken to begin at $-i\infty$ and end at $+i\infty$, and can be deformed to one that consists of a small circle around $s = -a$ and a curve that lies in a region $\operatorname{Re} s \leq b < -\operatorname{Re} a$. As $x \to +\infty$, the leading term comes from the residue at $s = -a$, and is $O(x^{-a})$. If $\operatorname{Re} a > 0$, we can conclude that (12.3.15) is a multiple of the Kummer function of the second kind U.

Computing the residue, we obtain

$$U(a,c;x) = \frac{1}{\Gamma(a)\Gamma(1-c+a)} \, G_{1,2}^{2,1}\left(x; \begin{matrix} 1-a \\ 0, 1-c \end{matrix}\right)$$

$$= \frac{1}{2\pi i} \int_C \frac{\Gamma(a+s)\Gamma(1-c-s)\Gamma(-s)}{\Gamma(a)\Gamma(1-c+a)} x^s \, ds, \quad (12.3.16)$$

the Barnes integral representation of the Kummer function of the second kind. This is valid for $|\arg x| < \pi$. Using this result and the identity (8.2.5), analytic continuation allows us to remove the restriction $\operatorname{Re} a > 0$ and verify (8.2.2) for arbitrary a.

3. The hypergeometric function F: in the notation of this section, the indices for (12.1.2) are $a_1 = 1-a$, $a_2 = 1-b$, $b_1 = 0$, $b_2 = 1-c$. Thus

$$G_{2,2}^{1,2}\left(-x; \begin{matrix} 1-a, 1-b \\ 0, 1-c \end{matrix}\right)$$

is a solution of the equation. As in Example 1, we conclude that

$$F(a,b,c;x) = \frac{\Gamma(c)}{\Gamma(a)\Gamma(b)} \, G_{2,2}^{1,2}\left(-x; \begin{matrix} 1-a, 1-b \\ 0, 1-c \end{matrix}\right) \quad (12.3.17)$$

$$= \frac{1}{2\pi i} \int_C \frac{\Gamma(c)\Gamma(a+s)\Gamma(b+s)\Gamma(-s)}{\Gamma(a)\Gamma(b)\Gamma(c+s)} (-x)^s \, ds,$$

where again C encloses the nonnegative integers in a negative sense and separates them from the other poles of the integrand. This is the Barnes integral for the hypergeometric function, valid for $|x| < 1$ and, by continuation, valid for $|\arg(-x)| < \pi$.

4. With the same kind of change of indices, we are led to conclude that if $p \le q+1$,

$$_pF_q\left(\begin{matrix} a_1,\ldots,a_p \\ b_1,\ldots,b_q \end{matrix}; x\right)$$

$$= \frac{\prod_{j=1}^{q}\Gamma(b_j)}{\prod_{j=1}^{p}\Gamma(a_j)} \, G_{p,q+1}^{1,p}\left(-x; \begin{matrix} 1-a_1,\ldots,1-a_p \\ 0, 1-b_1,\ldots,1-b_q \end{matrix}\right). \quad (12.3.18)$$

Writing out the integral form of the right-hand side gives the Mellin–Barnes integral representation of $_pF_q$.

12.3.2 Invariance properties

The family $\mathcal{G}$ of functions of the form (12.3.12) has important invariance properties. On the one hand, if G belongs to $\mathcal{G}$ and c is a constant, $c \neq 0, 1$, then in general cG does not belong to $\mathcal{G}$. If G_1 and G_2 belong to $\mathcal{G}$, then in general $G_1 + G_2$ does not belong to $\mathcal{G}$. In other words, unlike many families of functions that are considered in analysis, $\mathcal{G}$ is far from being a vector space. On the other hand, $\mathcal{G}$ is invariant under the change of variables $x \to 1/x$, differentiation, multiplication by powers of x, and (under some conditions on the parameters) integration.

Proposition 12.3.1 *Suppose that the function $G = G(x)$ belongs to $\mathcal{G}$. Then*

$$G(1/x), \quad G'(x), \quad \text{and} \quad x^\nu G(x) \quad \text{belong to } \mathcal{G}. \tag{12.3.19}$$

Moreover, if $p \leq q$, G has the form (12.3.12) and $\mathrm{Re}\, b_k > -1$, $k = 1,\ldots,m$, then

$$\int_0^x G(y)\,dy \quad \text{belongs to } \mathcal{G}. \tag{12.3.20}$$

The proof is left as an exercise.

The family $\mathcal{G}$ is also invariant under the Laplace transform $\mathcal{L}$, introduced in Exercise 23 of Chapter 5, a suitably normalized version of the Euler transform $E_{\alpha,\beta}$, introduced in Section 10.6, and the Riemann–Liouville fractional integral transform I_α: for $\mathrm{Re}\,\alpha > 0$, $\mathrm{Re}\,\beta > 0$,

$$\mathcal{L}f(x) = \int_0^\infty e^{-xy} f(y)\,dy;$$

$$E_{\alpha,\beta}f(x) = \frac{\Gamma(\alpha + \beta)}{\Gamma(\alpha)\Gamma(\beta)} \int_0^1 s^{\alpha-1}(1-s)^{\beta-1} f(sx)\,ds;$$

$$I_\alpha f(x) = \frac{1}{\Gamma(\alpha)} \int_0^x (x-y)^{\alpha-1} f(y)\,dy.$$

Proposition 12.3.2 *Suppose that the function G belongs to $\mathcal{G}$. Then*

$$\mathcal{L}G \text{ belongs to } \mathcal{G}; \tag{12.3.21}$$

$$\frac{\Gamma(\alpha)}{\Gamma(\alpha + \beta)} \cdot E_{\alpha,\beta} G \text{ belongs to } \mathcal{G}; \tag{12.3.22}$$

$$I_\alpha G \text{ belongs to } \mathcal{G}. \tag{12.3.23}$$

The proofs are left as exercises.

An important, and subtler, form of invariance is connected with the compositions

$$g_1 * g_2(x) = \int_0^\infty g_1\left(\frac{x}{y}\right) g_2(y) \frac{dy}{y} \tag{12.3.24}$$

(multiplicative convolution) and twisted convolution

$$g_1 \# g_2(x) = \int_0^\infty g_1(xy) g_2(y) \frac{dy}{y}. \qquad (12.3.25)$$

We assume here that g_1 and g_2 are absolutely integrable with respect to the measure dx/x:

$$\int_0^\infty |g_j(x)| \frac{dx}{x} < \infty, \quad j = 1, 2. \qquad (12.3.26)$$

Theorem 12.3.3 *Suppose that the functions G_1 and G_2 belong to $\mathcal{G}$ and are absolutely integrable with respect to dx/x. Then the functions $G_1 * G_2$ and $G_1 \# G_2$ belong to $\mathcal{G}$.*

The proof is postponed to Section 12.6, but here are some preliminary remarks that are easily checked. First, let $\widetilde{g}(x) = g(1/x)$. If g satisfies the integrability condition (12.3.26), then so does $\tilde{g}$. Moreover,

$$g_1 \# g_2 = g_1 * \widetilde{g}_2.$$

Since $G \in \mathcal{G}$ implies $\widetilde{G} \in \mathcal{G}$, the two assertions in Theorem 12.3.3 are equivalent. A useful observation is that if both G_1 and G_2 satisfy (12.3.26), then so does the composed function $G_1 * G_2$. As we shall see in Appendix B,

$$\mathcal{M}[G_1 * G_2] = \mathcal{M}G_1 \cdot \mathcal{M}G_2, \qquad (12.3.27)$$

where $\mathcal{M}$ denotes the Mellin transform, defined in Section 12.6. The proof of Theorem 12.3.3 amounts to showing that (12.3.27) can be inverted:

$$G_1 * G_2 = \mathcal{M}^{-1}[\mathcal{M}G_1 \cdot \mathcal{M}G_2].$$

Theorem 12.3.3 is at the heart of many tables of integrals, such as Gradshteyn and Ryzhik [170], Prudnikov, Brychkov, and Marichev [336], and Brychkov [56]. For a discussion of this point, see Wolfram [442].

12.4 Choices of contour of integration

Consider the kernel associated with the G-function in (12.3.12):

$$\Phi(s) = \frac{\prod_{j=1}^m \Gamma(b_j - s) \prod_{j=1}^n \Gamma(1 - a_j + s)}{\prod_{j=m+1}^q \Gamma(1 - b_j + s) \prod_{j=n+1}^p \Gamma(a_j - s)}. \qquad (12.4.1)$$

To find a contour of integration, we need to estimate $|\Phi(s)|$.

As in Section 12.3, we may convert from the case $p > q$ to the case $p < q$ by changing indices and taking $1/x$ as the variable. For G-functions, changing x

to $1/x$ amounts to changing s to $-s$:

$$\frac{1}{2\pi i} \int_C x^s \, \Phi(-s) \, ds = G_{q,p}^{n,m}\left(x; \begin{array}{c} 1-b_1,\ldots,1-b_q \\ 1-a_1,\ldots,1-a_p \end{array}\right). \tag{12.4.2}$$

We assume here that $p \le q$. We want to find uniform estimates on $|\Phi(s)|$ in a region of the form

$$\Omega = \{s \,|\, \mathrm{Re}\, s \ge -N, \, |\mathrm{Im}\, s| \ge K\}, \tag{12.4.3}$$

where

$$|\mathrm{Im}\, a_j| < K, \quad 1 \le j \le p, \quad |\mathrm{Im}\, b_k| < K, \quad 1 \le k \le q, \tag{12.4.4}$$

and $N \in \mathbf{R}$ is arbitrary (but fixed). The condition on K implies that all zeros and poles of Φ are excluded from Ω.

Lemma 12.4.1 *Suppose $p \le q$. Then for $s \in \Omega$, $|\Phi(s)|$ is bounded above and below by positive multiples of*

$$\left(\frac{|s|}{e}\right)^{(p-q)\mathrm{Re}\, s} |s|^{B-A-(q-p)/2} e^{\{-|\mathrm{Im}\, s|[l\pi + (q-p)(\pi/2 - |\theta|)]\}}, \tag{12.4.5}$$

where $l = (m+n) - (p+q)/2$, $\theta = \arg s$, $|\theta| < \pi$, and

$$B = \sum_{j=1}^{q} \mathrm{Re}\, b_j, \quad A = \sum_{j=1}^{p} \mathrm{Re}\, a_j. \tag{12.4.6}$$

Proof As a first step, we use the reflection formula to convert each factor $\Gamma(b_j - s)$ in the numerator and each factor $\Gamma(a_j - s)$ in the denominator:

$$\Phi(s) = \frac{\pi^m \prod_{j=n+1}^{p} \sin \pi(a_j - s)}{\pi^{p-n} \prod_{j=1}^{m} \sin \pi(b_j - s)} \cdot \frac{\prod_{j=1}^{p} \Gamma(1 - a_j + s)}{\prod_{j=1}^{q} \Gamma(1 - b_j + s)}. \tag{12.4.7}$$

Since

$$|\sin(x + iy)|^2 = |\sin x \cosh y + i \cos x \sinh y|^2 = \sin^2 x + \sinh^2 y,$$

it follows that the quotients

$$\frac{|\sin \pi(a_j - s)|}{|\sinh(\pi \, \mathrm{Im}\, s)|}, \quad \frac{|\sinh(\pi \, \mathrm{Im}\, s)|}{|\sin \pi(b_j - s)|}$$

are bounded above and below, uniformly, for $s \in \Omega$. Moreover

$$\frac{\sinh(\pi \,|\mathrm{Im}\, s|)}{\exp(\pi \,|\mathrm{Im}\, s|)}$$

is bounded above and below, uniformly, for $s \in \Omega$. Therefore we may replace $|\Phi(s)|$ by

$$\exp\{\pi(p-n-m)|\operatorname{Im} s|\} \cdot \left| \frac{\prod_{j=1}^{p} \Gamma(1-a_j+s)}{\prod_{j=1}^{q} \Gamma(1-b_j+s)} \right|. \tag{12.4.8}$$

By Corollary 2.5.4, $|\Gamma(c+s)/\Gamma(s)|$ is bounded above and below, uniformly in Ω, by positive multiples of $|s|^{\operatorname{Re} c}$. Therefore we may replace (12.4.8) by

$$|s|^{B-A+p-q} \exp\{\pi(p-m-n)|\operatorname{Im} s|\} |\Gamma(s)|^{p-q}. \tag{12.4.9}$$

According to (2.5.6),

$$\Gamma(s) = \left(\frac{s}{e}\right)^s \sqrt{\frac{2\pi}{s}} \left[1 + O(|s|^{-1})\right] \tag{12.4.10}$$

as $s \to \infty$, uniformly for $s \in \Omega$. Now

$$|s^s| = |\exp(s \log s)| = \exp\{\operatorname{Re}(s \log s)\} = \exp(\operatorname{Re} s \log |s| - \theta \operatorname{Im} s). \tag{12.4.11}$$

Also, $\theta \operatorname{Im} s = |\theta| |\operatorname{Im} s|$. Combining this last remark with (12.4.9), (12.4.10), and (12.4.11), and noting that $p - n - m = (p-q)/2 - l$, we find that $|\Phi(s)|$ is asymptotically bounded above and below by multiples of (12.4.5). Since Φ has no zeros in Ω, such bounds hold everywhere in Ω. $\qquad\square$

Remark. For $p \geq q$, a similar estimate for Φ holds in the left half-plane. In fact, changing s to $-s$ in (12.4.1) amounts to interchanging p with q and m with n, and writing a_j for $1 - b_j$ and b_k for $1 - a_k$.

Theorem 12.4.2 *Suppose that Φ is given by (12.4.1) and that $p \leq q$. Let C be a curve that passes through no poles of Φ, comes from $+\infty$ along the line $\operatorname{Im} s = -K$, and returns to $+\infty$ along the line $\operatorname{Im} s = K$, where K satisfies the condition (12.4.4). Then the integral*

$$\frac{1}{2\pi i} \int_C x^s \Phi(s) \, ds \tag{12.4.12}$$

converges absolutely for any fixed x if $p < q$, and for $|x| < 1$ if $p = q$.

The curve C can be deformed, preserving absolute integrability, to a curve that begins at $-i\infty$ and ends at $+i\infty$ along a line $\operatorname{Re} s = -N$, under any of the following conditions:

(a) $p < q$, $m + n > (p+q)/2$, $x > 0$; $\qquad$ (12.4.13)

(b) $p = q$, $m + n > (p+q)/2$, $0 < x < 1$; $\qquad$ (12.4.14)

(c) $p = q$, $m + n = (p+q)/2$, $B - A < -1$, $0 < x < 1$, (12.4.15)

where again $A = \sum_{j=1}^{p} \operatorname{Re} a_j$ and $B = \sum_{j=1}^{q} \operatorname{Re} b_j$.

Proof The estimate (12.4.5) shows that there are constants a and b such that for $|\operatorname{Im} s| = K$, $s \in \Omega$, and any fixed x,

$$|x^s \Phi(s)| \le a |x|^{\operatorname{Re} s} |s|^{b-(q-p)\operatorname{Re} s} e^{(q-p)\operatorname{Re} s}.$$

If $p < q$, this integrand decays at a greater than exponential rate as $s \to \infty$ with $|\operatorname{Im} s| = K$. If $p = q$, it decays at an exponential rate if $0 < x < 1$.

Suppose now that $m + n \ge (p+q)/2$, so l in (12.4.5) is nonnegative. Given an angle α, $0 \le \alpha \le \pi/2$, let $C_\alpha^\pm$ be the ray

$$s = -N \pm iK + r e^{\pm i\alpha}, \qquad 0 < r < \infty.$$

On this ray, for $x > 0$, (12.4.5) gives the estimate

$$|x^s \Phi(s)| \le a x^{\operatorname{Re} s} |s|^{b+(p-q)\operatorname{Re} s} e^{-|\operatorname{Im} s|[l\pi + c(s)] - (p-q)\operatorname{Re} s}, \tag{12.4.16}$$

where a and b are fixed and

$$c(s) = (p - q)[|\arg s| - \pi/2] = (p - q)[\alpha - \pi/2] + O(|s|^{-1}).$$

If $l > 0$, and either $p < q$ or $0 < x < 1$, we have exponential decay along each ray, uniformly with respect to the angle α.

Suppose now that $p = q$, $l = 0$, and $0 < x < 1$. The estimate (12.4.16) becomes

$$|x^s \Phi(s)| \le a x^{\operatorname{Re} s} |s|^{B-A}.$$

If $B - A < -1$, $x^s \Phi(s)$ is integrable on the rays $C_\alpha^\pm$, uniformly with respect to α. Therefore the original curve, which may be assumed to include the rays $C_0^\pm$, can be deformed to begin and end with the vertical rays $C_{\pi/2}^\pm$ under any of the conditions (a), (b), or (c). $\qquad \square$

12.5 Expansions and asymptotics

Suppose $p \le q$. As a solution of (12.3.1), the G-function (12.3.11) has an expansion as in Proposition 12.2.1 with the appropriate change of indices. Let

$$u_j(x) = x^{b_j} {}_pF_{q-1}\left(\begin{matrix} 1 - a_1 + b_j, \ldots, 1 - a_p + b_j \\ 1 - b_1 + b_j, \ldots, 1 - b_q + b_j \end{matrix} ; x \right), \tag{12.5.1}$$

$j = 1, \ldots, q$, with $1 - b_j + b_j$ omitted.

Theorem 12.5.1 *Suppose that $p \le q$ and that the conditions (12.3.13) are satisfied for $i, j \le m$ and $k, l \le n$. Then the G-function*

$$G(x) = G_{p,q}^{m,n}\left((-1)^{m+n+p} x; \begin{matrix} a_1, \ldots, a_p \\ b_1, \ldots, b_q \end{matrix} \right) \tag{12.5.2}$$

has an expansion

$$G(x) = \sum_{j=1}^{m} (-1)^{(m+n+p)b_j} A_j u_j(x), \qquad (12.5.3)$$

where

$$A_j = \frac{\prod_{i \neq j} \Gamma(b_i - b_j) \prod_{k=1}^{n} \Gamma(b_j + 1 - a_k)}{\prod_{i=m+1}^{q} \Gamma(b_j + 1 - b_i) \prod_{k=n+1}^{p} \Gamma(a_k - b_j)}. \qquad (12.5.4)$$

Proof As noted above, Proposition 12.2.1 guarantees that there is an expansion in the u_j, $j = 1, \ldots, q$. Writing the integral for (12.5.2) as a sum of residues shows that it is a sum of m series of the form

$$\sum_{v=0}^{\infty} \alpha_{jv} x^{b_j + v}, \qquad j = 1, \ldots, m.$$

Therefore only the first m of the u_j occur in the expansion. The coefficient of u_j in (12.5.3) is α_{j0}, which is the residue of the integrand at $s = b_j$. This residue, in turn, is (12.5.4). $\qquad \square$

The expansion (12.5.3) gives the detailed behavior of the function G as $x \to 0$. If $p \geq q$, and especially if $p > q$, the change of variables $x \to 1/x$ gives the behavior of G as $x \to \infty$:

$$G_{p,q}^{m,n}\left(x; \begin{matrix} a_1, \ldots, a_p \\ b_1, \ldots, b_q \end{matrix}\right) = G_{q,p}^{n,m}\left(\frac{1}{x}; \begin{matrix} 1 - b_1, \ldots, 1 - b_q \\ 1 - a_1, \ldots, 1 - a_p \end{matrix}\right). \qquad (12.5.5)$$

The functions

$$v_j(x) = x^{1-a_j} {}_qF_{p-1}\left(\begin{matrix} 1 + b_1 - a_j, \ldots, 1 + b_q - a_j \\ 1 + a_1 - a_j, \ldots, 1 + a_p - a_j \end{matrix}; x\right), \qquad (12.5.6)$$

$j = 1, \ldots, p$, with $1 + a_j - a_j$ omitted, correspond to the functions (12.5.1) used in the expansion (12.5.3) in the case $p \leq q$. In view of Proposition 12.2.2, the functions $v_j((-1)^{p-q}/x)$ are solutions of (12.3.1). Therefore Theorem 12.5.1 implies the following.

Theorem 12.5.2 *Suppose that $p \geq q$ and that the conditions (12.3.13) are satisfied for $i, j \leq m$ and $k, l \leq n$. Then the G-function*

$$G(x) = G_{p,q}^{m,n}\left((-1)^{m+n+p} x; \begin{matrix} a_1, \ldots, a_p \\ b_1, \ldots, b_q \end{matrix}\right) \qquad (12.5.7)$$

has an expansion

$$G(x) = \sum_{j=1}^{n} (-1)^{(m+n+q)(1-a_j)} B_j v_j\left(\frac{(-1)^{p-q}}{x}\right), \qquad (12.5.8)$$

where

$$B_j = \frac{\prod_{j \neq k} \Gamma(a_j - a_k) \prod_{i=1}^{m} \Gamma(1 - a_j + b_i)}{\prod_{k=n+1}^{p} \Gamma(1 - a_j + a_k) \prod_{i=m+1}^{q} \Gamma(a_j - b_i)}. \tag{12.5.9}$$

The expansions (12.5.3), for $p \leq q$, and (12.5.8), for $q \leq p$, are valid asymptotically whenever $m + n > (p + q)/2$. We consider any formal $_pF_q$ as defining an asymptotic series

$$_pF_q \left(\begin{matrix} a_1, \ldots, a_p \\ b_1, \ldots, b_q \end{matrix}; x \right) = \sum_{k=0}^{N} \frac{(a_1)_k \cdots (a_p)_k}{(b_1)_k \cdots (b_q)_k} \frac{x^k}{k!} + O(x^{N+1}).$$

With A_j, u_j and B_j, v_j given by (12.5.4), (12.5.1), (12.5.9), and (12.5.6), define the asymptotic series by

$$G_+(x) = \sum_{j=1}^{m} (-1)^{(m+n+p)b_j} A_j u_j(x);$$

$$G_-(x) = \sum_{j=1}^{n} (-1)^{(m+n+q)(1-a_j)} B_j v_j \left(\frac{(-1)^{p-q}}{x} \right). \tag{12.5.10}$$

Theorem 12.5.3 *Suppose that the conditions* (12.3.13) *are satisfied for* $i, j \leq m$ *and* $k, l \leq n$. *Let* G *be the G-function* (12.5.7). *Suppose that* $m + n > (p + q)/2$.
(a) If $p > q$, *then* G *has the asymptotic expansion* G_+ *as* $x \to 0+$.
(b) If $p < q$, *then* G *has the asymptotic expansion* G_- *as* $x \to +\infty$.

Proof Suppose that $p < q$. It follows from Theorem 12.4.2 that the original curve beginning and ending at ∞ can be deformed to one that coincides, for large $|s|$, with any given vertical line $\operatorname{Re} s = -N$. We take N so that no poles of Φ lie on the vertical line $\operatorname{Re} s = -N$. For $N \gg 0$, the original curve can be deformed to consist of this line, together with small circles around the poles of Φ at the points $s = a_j - 1$, $a_j - 2, \ldots$, $1 \leq j \leq n$ that lie to the right of the line. Therefore $G(x)$ consists of the sum of the residues at these points together with an integral that is $O(x^{-N})$ as $x \to +\infty$. Computing the residues shows that their sum is precisely the part of the expansion of G_- that consists of terms $x^{a_j - 1 - k}$, $j \leq n$, with $\operatorname{Re} a_j - 1 - k > -N$. (We leave details of this calculation as an exercise.)

The same argument applies in case (a), this time moving the original curve that begins and ends at $-\infty$ to one that consists in part of a vertical line $\operatorname{Re} s = N \gg 0$, using the dual version of Theorem 12.4.2. $\qquad \square$

For a slightly different presentation of the above result as $x \to +\infty$, see Lin and Wong [259].

12.6 The Mellin transform and *G*-functions

For a function g defined for $0 < x < \infty$, the Mellin transform $\mathcal{M}g(z)$ is defined by

$$\mathcal{M}g(z) = \int_0^\infty x^z g(x) \frac{dx}{x} \tag{12.6.1}$$

for any complex number z for which the integral is absolutely convergent:

$$\int_0^\infty |x^z g(x)| \frac{dx}{x} < \infty. \tag{12.6.2}$$

If this condition is satisfied for a particular value $z = z_0$, then it is satisfied for every z on the vertical line $\operatorname{Re} z = \operatorname{Re} z_0$.

Suppose that the condition (12.6.2) holds for a given real value $z = c$, so that $\mathcal{M}g(c + it)$ is defined for all real t. The integrability condition implies that the $\mathcal{M}g(c + it)$ is a bounded and continuous function of t. Suppose also that it is absolutely integrable:

$$\int_{-\infty}^\infty |\mathcal{M}g(c + it)| \, dt < \infty.$$

Then g can be recovered by the inverse formula

$$g(x) = \frac{1}{2\pi i} \int_{c-i\infty}^{c+i\infty} x^{-s} \mathcal{M}g(s) \, ds; \tag{12.6.3}$$

see Appendix B.

Inspection of (12.6.3) shows that under some circumstances we may expect the Mellin transform of a *G*-function

$$G(x) = \frac{1}{2\pi i} \int_C x^s \Phi(s) \, ds$$

to be $\Phi(-s)$. Before identifying such circumstances, let us note that the left inverse

$$\mathcal{M}^{-1}\varphi(x) = \frac{1}{2\pi i} \int_{c-i\infty}^{c+i\infty} x^{-s} \varphi(s) \, ds \tag{12.6.4}$$

is also a right inverse $\mathcal{M}\mathcal{M}^{-1}\varphi = \varphi$; see Appendix B. This, together with the remarks at the end of Section 12.3.2, proves Theorem 12.3.3.

Our goal now is to determine the conditions under which a *G*-function is an inverse Mellin transform.

Theorem 12.6.1 *Let G be the G-function*

$$G(x) = G_{p,q}^{m,n}\left(x; \begin{matrix} a_1, \ldots, a_p \\ b_1, \ldots, b_q \end{matrix}\right) = \frac{1}{2\pi i} \int_C x^s \Phi(s) \, ds, \tag{12.6.5}$$

where

$$\Phi(s) = \frac{\prod_{j=1}^{m} \Gamma(b_j - s) \prod_{k=1}^{n} \Gamma(s + 1 - a_k)}{\prod_{j=m+1}^{q} \Gamma(s + 1 - b_j) \prod_{k=n+1}^{p} \Gamma(a_k - s)}. \tag{12.6.6}$$

Suppose that

$$\text{Re}\, a_k - 1 \; < \; -c \; < \; \text{Re}\, b_j, \qquad 1 \leq k \leq n, \; 1 \leq j \leq m. \tag{12.6.7}$$

Let $B = \sum_{j=1}^{q} \text{Re}\, b_j$, $A = \sum_{k=1}^{p} \text{Re}\, a_k$. *If* $p < q$ *and* $m + n > (p+q)/2$, *then* G
and Φ *are related by the Mellin transform*

$$\mathcal{M}G(s) = \Phi(-s), \qquad G(1/x) = \mathcal{M}^{-1}\Phi(x). \tag{12.6.8}$$

Proof The line $C' = \{\text{Re}\, s = -c\}$ separates the sequences of poles

$$a_j - 1, \; a_j - 2, \; a_j - 3, \ldots, \quad j \leq n,$$

from the sequences of poles

$$b_j, \; b_j + 1, \; b_j + 2, \ldots, \quad j \leq m.$$

Therefore C' is a candidate for a curve that can be used to define G as an integral transform of Φ. According to Theorem 12.4.2, the conditions above imply that the original defining curve for G can be deformed to the line C'. Moreover, either condition guarantees absolute integrability of $\Phi(s)x^s$ along this line for $x > 0$.

Theorem 12.5.1 shows that as $x \to 0+$,

$$G(x) = O(x^b), \qquad b = \min\{\text{Re}\, b_j \,; \, 1 \leq j \leq m\}.$$

Theorem 12.5.3 shows that as $x \to +\infty$,

$$G(x) = O(x^{a-1}), \qquad a = \max\{\text{Re}\, a_j \,; \, 1 \leq j \leq n\}.$$

Therefore (12.6.7) is the necessary and sufficient condition that $x^c G(x)$ be absolutely integrable on the half-line with respect to the measure dx/x. We showed earlier that these integrability conditions imply that $\mathcal{M}^{-1}\Phi$ is well-defined using the contour $\text{Re}\, s = -c$, and that (12.6.8) holds. $\qquad\square$

12.7 Exercises

1. Iterate the integral representation (12.1.8) to write $_pF_q$ as a $(p-1)$-fold integral, under some restrictions on the parameters.
2. Use Exercise 1 to prove Theorem 12.2.3.
3. Use the residue calculus to compute each term of the series expansion in (12.3.14) and verify that it coincides with (12.1.3).

4. Use the residue calculus to obtain the complete asymptotic expansion (8.2.3) for the Kummer function of the second kind from (12.3.16).

5. Use the residue calculus to compute each term of the series expansion in (12.3.17) and verify that it coincides with (12.1.4).

6. In this and the following family of exercises, we assume that no two of the upper indices of the functions F of type ${}_pF_q$ differ by an integer. Let $\mathcal{F}_k$, $k = 1, 2, 3, \ldots$, be the family of functions F of type ${}_pF_q$ with $q + 1 = p + k$ that have the following property: F has an asymptotic expansion as a formal series of the form

$$\sum_{n=0}^{\infty} a_n x^{\nu - n/k} \exp\left(kx^{1/k}\right) \quad \text{as } x \to +\infty,$$

and the expansion can be differentiated term by term. Show that if F belongs to $\mathcal{F}_k$ and $\operatorname{Re}\alpha > 0$, $\operatorname{Re}\beta > 0$, then the Euler transform $E_{\alpha,\beta}F$ belongs to $\mathcal{F}_k$. (See the discussion of the asymptotics of the Kummer function M in Section 8.1.)

7. Suppose one of the lower indices of the function F of type ${}_pF_q$ is c. Show that $x^{1-c}D[x^{c-1}F](x)/(c-1)$ has the same form but with the index c replaced by $c - 1$.

8. Let

$$F(x) = {}_0F_{k-1}\left(\begin{matrix} - \\ \frac{1}{k}, \frac{2}{k}, \ldots, \frac{k-1}{k} \end{matrix}; x\right).$$

Use the identity

$$\left(\frac{1}{k}\right)_n \left(\frac{2}{k}\right)_n \cdots \left(\frac{k-1}{k}\right)_n n! = \frac{1}{k^{nk}}(nk)!$$

to show that, with $\omega = \exp(2\pi i/k)$,

$$F(x) = \frac{1}{k}\sum_{j=0}^{k-1} \exp\left(\omega^j k x^{1/k}\right).$$

Conclude that F belongs to $\mathcal{F}_k$, with

$$F(x) \sim \frac{\exp\left(kx^{1/k}\right)}{k} \quad \text{as } x \to +\infty.$$

9. Use Exercises 6, 7, and 8 to prove that every F of type ${}_0F_{k-1}$ belongs to $\mathcal{F}_k$.

10. Show that if $\operatorname{Re}a > 0$, then every F of type ${}_1F_k$ with upper index a belongs to $\mathcal{F}_k$. (Hint: if any of the lower indices equals a, then F coincides with a function of type ${}_0F_{k-1}$. Use Exercise 7 to modify the lower index.)

11. Suppose that $F(x)$ is of type ${}_pF_q$ with indices $a_1, \ldots, a_p, c_1, \ldots, c_q$. Let F_- denote the function with a_1 replaced by $a_1 - 1$ and the remaining indices

unchanged. Let F_+ denote the function with a_1 unchanged but with all other indices increased by 1. Prove the contiguous relation

$$F_-(x) = F(x) - x \frac{\prod_{j=2}^{p} a_j}{\prod_{k=1}^{q} c_k} F_+(x).$$

12. (a) Use Exercise 11 to remove the restriction $\operatorname{Re} a > 0$ in Exercise 10 and show that every F of type $_1F_k$ belongs to $\mathcal{F}_k$.
    (b) Show that every F of type $_pF_q$ with $q+1 = p+k$ belongs to $\mathcal{F}_k$.
    (c) Complete the proof of Theorem 12.2.4 by using the differential equation to show that for the formal series described in Exercise 6, the index v is given by

$$v = \frac{1}{k} \left[\sum_{j=1}^{p} a_j - \sum_{j=1}^{q} c_j + \frac{q-p}{2} \right].$$

13. Use (12.3.18) to write each of the examples in Section 12.1.1 as an integral.
14. Prove (12.3.21).
15. Prove (12.3.22).
16. Prove (12.3.23).
17. Prove (12.3.27).
18. Use Theorem 12.5.2 to give another proof of (10.2.10).
19. Use Theorem 12.5.2 to give another proof of Proposition 12.2.3.
20. Provide the missing details of Theorem 12.5.3 by computing the residues and comparing with the expansions of G_+, G_-.
21. Apply $\mathcal{M}$ to prove the following relation between the Laplace and Mellin transforms

$$\mathcal{L}f = \left[\mathcal{M}^{-1}\varphi \right](s), \quad \text{where} \quad \varphi(s) = \Gamma(s) \left[\mathcal{M}f \right](1-s).$$

22. Use a change of variables $x = e^y$ to show that the Mellin transform and Mellin inversion formula in the case $c = 0$ are the same as the Fourier transform and Fourier inversion formula; see Appendix B.

12.8 Remarks

Generalizations of the hypergeometric function $_2F_1$ go back at least to Clausen's $_3F_2$ in 1828 [83]. The $_3F_2$ case was investigated by Thomae in 1870 [398]. The general case $_pF_q$ was treated by Goursat in 1882 [169]. Nørlund [307] has an extensive discussion of the case $p = q + 1$. For many special identities see Bailey [28], Erdélyi, Magnus, Oberhettinger, and Tricomi [116], Slater [369], Rainville [339], Wimp [440], Luke [264], and Dwork [112].

Algorithms have been developed to establish identities involving hypergeometric series; see Koepf [223], Petkovšek, Wilf, and Zeilberger [323], Wilf and Zeilberger [437], and Larsen [244].

The idea of using an integral transform whose kernel is a quotient of products of gamma functions is due to Pincherle [326, 327]. This idea was subsequently exploited by Barnes for the classical hypergeometric function [30] and by Mellin [286] in general. The history is discussed by Mainardi and Pagnini [271], who show how Pincherle's idea leads naturally to the G-function, as in Section 12.3. A generalization in a different direction (more singular points) was suggested by Pochhammer [330] in 1870. The presentation here originated with an expository note by Beals and Szmigielski [36]. As in the hypergeometric case $_2F_1$, there is a q-version of the generalized hypergeometric case; see the remarks at the end of Chapter 10 and the references there.

The G-functions were first defined by Meijer [281] as linear combinations of generalized hypergeometric functions – the expansion (12.5.3). The integral characterization (12.3.12) was given by Meijer in 1941 [282]. In a long series of papers, he investigated the relations among, and the asymptotics of, these functions [283]. See also Erdélyi, Magnus, Oberhettinger and Tricomi [116]. The last-cited reference has a list of 75 identities involving G-functions. For applications to physical and statistical problems, see Luke [264] and Mathai and Saxena [273]. For a discussion of the use of G-functions in computing tables of integrals, see [442].

Generalizations of hypergeometric functions to two or more variables have been proposed by Appell [13], Horn [191], Lauricella [246], Appell and Kampé de Fériet [14], and Burchnall and Chaundy [58, 59, 68]; see Horn [192] and the extensive discussion in Erdélyi, Magnus, Oberhettinger, and Tricomi [116]. More recent versions have been proposed by Gelfand, Kapranov, and Zelevinsky [163] and Gelfand, Graev, and Retakh [162]; see Dwork [112], Mathai and Saxena [273], and Varchenko [412]. For an expository account, see Beukers [44].

13

Asymptotics

In this chapter, we prove various asymptotic results for special functions and classical orthogonal polynomials that have been stated without proof in previous chapters.

The method of proof used in the first three sections is to reduce the second-order differential equation to the point where it takes one of the following two forms:

$$u''(x) + \lambda^2 u(x) = f(x)u(x),$$

a perturbation of the wave equation, or

$$v''(\lambda x) + \frac{1}{\lambda x} v'(\lambda x) + \left(1 - \frac{v^2}{(\lambda x)^2}\right) v(\lambda x) = g(\lambda x)v(\lambda x),$$

a perturbation of Bessel's equation. In each case, λ is a large parameter and we are interested in the asymptotic behavior of solutions as $\lambda \to +\infty$.

Taking into account initial conditions, these equations can be converted to integral equations of the form

$$u(x) = u_0(x) + \frac{1}{\lambda} \int_0^x \sin(\lambda x - \lambda y) f(y) u(y) \, dy$$

or

$$v(\lambda x) = v_0(\lambda x) + \frac{1}{\lambda} \int_0^{\lambda x} G_v(\lambda x, \lambda y) g(\lambda y) v(\lambda y) \, d(\lambda y),$$

where u_0 and v_0 are solutions of the unperturbed equations

$$u_0'' + \lambda^2 u_0 = 0; \qquad x^2 v_0'' + x v_0' + (x^2 - v^2) v_0 = 0,$$

and G_v is a Green's function for Bessel's equation. The asymptotic results follow easily.

Asymptotics of the Legendre and associated Legendre functions P_v^m, Q_v^m on the interval $1 < x < \infty$ are obtained from integral representations of these functions, by concentrating on the portion of the contour of integration that contributes most strongly to the value.

Other asymptotic results can be obtained from integral formulas by means of elaborations of this method. One of these is the "method of steepest descents," while another is the "method of stationary phase." These two methods are illustrated in the final section with alternative derivations of the asymptotics of the Laguerre polynomials and the asymptotics of Bessel functions of the first kind, J_ν.

13.1 Hermite and parabolic cylinder functions

The equations (8.6.1) and (8.6.2),

$$u''(x) \mp \frac{x^2}{4} u(x) + \left(\nu + \frac{1}{2} \right) u(x) = 0, \tag{13.1.1}$$

can be viewed as perturbations of the equation

$$u''(x) + \left(\nu + \frac{1}{2} \right) u(x) = 0. \tag{13.1.2}$$

The solutions of the corresponding inhomogeneous equation

$$v''(x) + \left(\nu + \frac{1}{2} \right) v(x) = f(x)$$

have the form

$$v(x) = u_0(x) + \int_0^x \frac{\sin \lambda(x-y)}{\lambda} f(y) \, dy, \quad \lambda = \sqrt{\nu + \frac{1}{2}},$$

where u_0 is a solution of (13.1.2). Note that $v(0) = u_0(0)$, $v'(0) = u_0'(0)$. Thus solving (13.1.1) is equivalent to solving the integral equation

$$u(x) = u_0(x) \pm \int_0^x \frac{\sin \lambda(x-y)}{4\lambda} y^2 u(y) \, dy, \tag{13.1.3}$$

where u_0 is a solution of (13.1.2). For any given choice of u_0, the solution of (13.1.3) can be obtained by the method of successive approximations, i.e. as the limit

$$u(x) = \lim_{m \to \infty} u_m(x),$$

where $u_{-1}(x) \equiv 0$ and

$$u_m(x) = u_0(x) \pm \int_0^x \frac{\sin \lambda(x-y)}{4\lambda} y^2 u_{m-1}(y) \, dy, \quad m = 0, 1, 2, \dots.$$

Since $|u_0(x)| \leq A$ for some constant A, it follows by induction that

$$|u_m(x) - u_{m-1}(x)| \leq \frac{A}{m!} \frac{x^{3m}}{(12\lambda)^m}. \tag{13.1.4}$$

Therefore the sequence $\{u_m\}$ converges uniformly on bounded intervals, and

$$u(x) = u_0(x) + O(\lambda^{-1}),$$

uniformly on bounded intervals.

To apply this to the parabolic cylinder functions D_ν given by (8.6.6), we take

$$u_0(x) = D_\nu(0) \cos \lambda x + D'_\nu(0) \frac{\sin \lambda x}{\lambda}.$$

It follows from (8.6.10) and (2.2.7) that

$$D_\nu(0) = \frac{2^{\frac{1}{2}\nu} \sqrt{\pi}}{\Gamma(\frac{1}{2} - \frac{1}{2}\nu)} = \frac{2^{\frac{1}{2}\nu}}{\sqrt{\pi}} \Gamma\left(\frac{1}{2}\nu + \frac{1}{2}\right) \cos \frac{1}{2}\nu\pi;$$

$$D'_\nu(0) = -\frac{2^{\frac{1}{2}\nu + \frac{1}{2}} \sqrt{\pi}}{\Gamma(-\frac{1}{2}\nu)} = \frac{2^{\frac{1}{2}\nu + \frac{1}{2}}}{\sqrt{\pi}} \Gamma\left(\frac{1}{2}\nu + 1\right) \sin \frac{1}{2}\nu\pi.$$

Formula (2.1.9) implies that

$$\Gamma\left(\frac{1}{2}\nu + 1\right) \sim \left(\frac{1}{2}\nu\right)^{\frac{1}{2}} \Gamma\left(\frac{1}{2}\nu + \frac{1}{2}\right)$$

as $\nu \to +\infty$, so we obtain

$$u_0 \sim \frac{2^{\frac{1}{2}\nu}}{\sqrt{\pi}} \Gamma\left(\frac{1}{2}\nu + \frac{1}{2}\right) \left[\cos \frac{1}{2}\nu\pi \cos \lambda x + \sin \frac{1}{2}\nu\pi \sin \lambda x\right]$$

$$= \frac{2^{\frac{1}{2}\nu}}{\sqrt{\pi}} \Gamma\left(\frac{1}{2}\nu + \frac{1}{2}\right) \cos\left(\sqrt{\nu + \frac{1}{2}} x - \frac{1}{2}\nu\pi\right),$$

which gives (8.6.19).

The asymptotics of the Hermite polynomials can be obtained in a similar way. Recall that H_n is the solution of

$$H_n''(x) - 2x H_n'(x) + 2n H_n(x) = 0,$$

with

$$H_{2m}(0) = (-1)^m \frac{(2m)!}{m!}, \quad H'_{2m}(0) = 0$$

and

$$H_{2m+1}(0) = 0, \quad H'_{2m+1}(0) = (-1)^m \frac{2(2m+1)!}{m!}.$$

The gauge transformation $H_n(x) = e^{\frac{1}{2}x^2} h_n(x)$ gives the equation

$$h_n''(x) - x^2 h_n(x) + (2n+1) h_n(x) = 0,$$

with conditions

$$h_n(0) = H_n(0), \quad h_n'(0) = H_n'(0).$$

Stirling's approximation (2.5.1) implies that

$$\frac{(2m)!}{m!} \sim 2^{m+\frac{1}{2}} \left(\frac{2m}{e}\right)^m \sim 2^m \frac{[(2m)!]^{\frac{1}{2}}}{(m\pi)^{\frac{1}{4}}};$$

$$\frac{2(2m+1)!}{\sqrt{4m+3}\,m!} \sim 2^{m+1} \left(\frac{2m+1}{e}\right)^{m+\frac{1}{2}} \sim 2^{m+\frac{1}{2}} \frac{[(2m+1)!]^{\frac{1}{2}}}{[(m+\frac{1}{2})\pi]^{\frac{1}{4}}}$$

as $n \to \infty$. It follows that for even $n = 2m$,

$$H_n(x) = (-1)^m 2^m \frac{[(2m)!]^{\frac{1}{2}}}{(m\pi)^{\frac{1}{4}}} e^{\frac{1}{2}x^2} \left[\cos\left(\sqrt{2n+1}\,x\right) + O(n^{-\frac{1}{2}})\right]$$

as $n \to \infty$, while for odd $n = 2m + 1$,

$$H_n(x) = (-1)^m 2^{m+\frac{1}{2}} \frac{[(2m+1)!]^{\frac{1}{2}}}{[(m+\frac{1}{2})\pi]^{\frac{1}{4}}} e^{\frac{1}{2}x^2} \left[\sin\left(\sqrt{2n+1}\,x\right) + O(n^{-\frac{1}{2}})\right]$$

as $n \to \infty$. These two results can be combined as

$$H_n(x) = 2^{\frac{1}{2}n} \frac{2^{\frac{1}{4}}(n!)^{\frac{1}{2}}}{(n\pi)^{\frac{1}{4}}} e^{\frac{1}{2}x^2} \left[\cos\left(\sqrt{2n+1}\,x - \frac{1}{2}n\pi\right) + O(n^{-\frac{1}{2}})\right],$$

which is (5.3.20).

13.2 Confluent hypergeometric functions

Starting with the confluent hypergeometric equation

$$xu''(x) + (c-x)u'(x) - au(x) = 0, \tag{13.2.1}$$

we use the Liouville transformation. This begins with the change of variables $y = 2\sqrt{x}$ that leads to

$$v''(y) + \left[\frac{2c-1}{y} - \frac{y}{2}\right] v'(y) - av(y) = 0,$$

followed by the gauge transformation

$$v(y) = y^{\frac{1}{2}-c} e^{\frac{1}{8}y^2} w(y)$$

to remove the first-order term. This leads to the equation

$$w''(y) - \left[\frac{(c-1)^2 - \frac{1}{4}}{y^2} + \frac{y^2}{16} - \frac{c}{2} + a\right]w(y) = 0. \tag{13.2.2}$$

We are considering this equation on the half-line $y > 0$. If we could identify $u(y_0)$ and $u'(y_0)$ at some point, we could use the method of the previous section: solve (13.2.2) as a perturbation of $w'' - aw = 0$. The singularity at $y = 0$ prevents the use of $y_0 = 0$ and it is difficult to determine values at other points, such as $y_0 = 1$. Instead, we shall introduce a first-order term in such a way that (13.2.2) can be converted to a perturbation of Bessel's equation. The gauge transformation

$$w(y) = y^{\frac{1}{2}} W(y)$$

converts (13.2.2) to

$$W''(y) + \frac{1}{y} W'(y) - \left[\frac{(c-1)^2}{y^2} - \frac{c}{2} + a + \frac{y^2}{16}\right]W(y) = 0.$$

Let us assume that $c - 2a$ is positive. The final step in getting to a perturbation of Bessel's equation is to take $y = (\frac{1}{2}c - a)^{-\frac{1}{2}}z$, so that the previous equation becomes

$$V''(z) + \frac{1}{z}V'(z) + \left[1 - \frac{(c-1)^2}{z^2}\right]V(z) = \frac{z^2}{(2c-4a)^2}V(z). \tag{13.2.3}$$

This is indeed a perturbation of Bessel's equation with index $\nu = c - 1$. Tracing this argument back, a solution of the confluent hypergeometric equation (13.2.1) has the form

$$y^{1-c}e^{\frac{1}{8}y^2}V(\beta y), \quad \beta = \sqrt{\frac{1}{2}c - a},$$

where V is a solution of (13.2.3). Since $y = 2\sqrt{x}$, this has the form

$$x^{\frac{1}{2}(1-c)}e^{\frac{1}{2}x}V\left(\sqrt{(2c-4a)x}\right). \tag{13.2.4}$$

The Kummer function $u(x) = M(a, c; x)$ is the solution of (13.2.1) with $u(0) = 1$. The corresponding function $V(z)$ must look like a multiple of $z^{c-1} = z^\nu$ as $z \to 0$. Let us assume for the moment that $c \geq 1$. To obtain a solution of (13.2.3) with this behavior, we begin with the Bessel function J_ν and convert (13.2.3) to the integral equation

$$V(z) = J_\nu(z) + \gamma \int_0^z G_\nu(z,\zeta)\zeta^2 V(\zeta)d\zeta, \quad \gamma = \frac{1}{(2c-4a)^2}, \quad (13.2.5)$$

where $G_\nu(z,\zeta)$ is the Green's function

$$G_\nu(z,\zeta) = \frac{Y_\nu(z)J_\nu(\zeta) - J_\nu(z)Y_\nu(\zeta)}{W(J_\nu(\zeta), Y_\nu(\zeta))}.$$

The solution of (13.2.5) can be obtained by the method of successive approximations

$$V(z) = \lim_{m\to\infty} V_m(z),$$

where $V_{-1}(z) \equiv 0$ and

$$V_m(z) = J_\nu(z) + \gamma \int_0^z G_\nu(z,\zeta)\zeta^2 V_{m-1}(\zeta)d\zeta, \quad m = 0,1,2,\ldots$$

It follows from the results in Sections 9.1 and 9.4 that there are constants $A = A_\nu$ and $B = B_\nu$ such that

$$|J_\nu(z)| \le Az^\nu, \quad |G_\nu(z,\zeta)| \le B\left(\frac{z}{\zeta}\right)^\nu.$$

It follows from these inequalities and by induction that

$$|V_m(z) - V_{m-1}(z)| \le Az^\nu \frac{(\gamma Bz^3)^m}{3^m m!}. \quad (13.2.6)$$

Therefore the sequence $\{V_m\}$ converges uniformly on bounded intervals, and

$$V(z) = J_\nu(z) + O\left([c-2a]^{-2}\right),$$

uniformly on bounded intervals. Up to a constant factor, the Kummer function $M(a,c;x)$ is given by (13.2.4). As $x \to 0$, the function defined by (13.2.4) has the limiting value

$$\left[\frac{\sqrt{2c-4a}}{2}\right]^{c-1} \frac{1}{\Gamma(\nu+1)}.$$

Therefore, for $c \ge 1$ and $c - 2a > 0$,

$$M(a,c;x) = \Gamma(\nu+1)\left[\frac{1}{2}cx - ax\right]^{\frac{1}{2}(1-c)} e^{\frac{1}{2}x} V\left(\sqrt{2cx-4ax}\right) \quad (13.2.7)$$

$$= \Gamma(c)\left[\frac{1}{2}cx - ax\right]^{\frac{1}{2}-\frac{1}{2}c} e^{\frac{1}{2}x}\left[J_{c-1}\left(\sqrt{2cx-4ax}\right) + O\left([c-2a]^{-2}\right)\right].$$

According to (9.4.9),

$$J_{c-1}(z) = \frac{\sqrt{2}}{\sqrt{\pi z}} \left[\cos\left(z - \frac{1}{2}c\pi + \frac{1}{4}\pi\right) + O(z^{-1}) \right]$$

as $z \to +\infty$. Combining this with (13.2.7) gives

$$M(a,c;x) = \frac{\Gamma(c)}{\sqrt{\pi}} \left(\frac{1}{2}cx - ax\right)^{\frac{1}{4}-\frac{1}{2}c} e^{\frac{1}{2}x} \tag{13.2.8}$$
$$\times \left[\cos\left(\sqrt{2cx - 4ax} - \frac{1}{2}c\pi + \frac{1}{4}\pi\right) + O\left(|a|^{-\frac{1}{2}}\right) \right]$$

as $a \to -\infty$, which is the Erdélyi–Schmidt–Fejér result (8.1.12). The convergence is uniform for x in any interval $0 < \delta \le x \le \delta^{-1}$.

We have proved (13.2.8) under the assumption $c \ge 1$. Suppose that $0 < c < 1$. The contiguous relation (8.5.6) with c replaced by $c+1$ is

$$M(a,c;x) = \frac{c+x}{c} M(a,c+1;x) + \frac{a-c-1}{c(c+1)} x M(a,c+2;x).$$

The asymptotics of the two terms on the right are given by (13.2.8). They imply that

$$M(a,c;x) \sim \frac{ax}{c(c+1)} M(a,c+2;x)$$
$$\sim \frac{\Gamma(c)}{\sqrt{\pi}} \frac{-ax}{(\frac{1}{2}cx - ax + x)^{\frac{3}{4}+\frac{1}{2}c}} e^{\frac{1}{2}x}$$
$$\times \cos\left(\sqrt{(2c - 4a + 4)x} - \frac{1}{2}c\pi + \frac{1}{4}\pi\right)$$

as $a \to -\infty$. Since $-ax$, $\frac{1}{2}cx - ax + x$, and $\frac{1}{2}cx - ax$ agree up to $O\left([-a]^{-1}\right)$ for $x > 0$, it follows that (13.2.8) is true for $c > 0$. This argument can be iterated to show that (13.2.8) is valid for all real indices $c \ne 0, -1, -2, \ldots$.

The asymptotics of the function $U(a,c;x)$ can be obtained from (13.2.8) and the identity (8.2.5). We use the reflection formula (2.2.7) to rewrite (8.2.5):

$$U(a,c;x) = \frac{\Gamma(1-c)\Gamma(c-a)\sin(c-a)\pi}{\pi} M(a,c;x)$$
$$+ \frac{\Gamma(c-1)\Gamma(1-a)\sin a\pi}{\pi} x^{1-c} M(a+1-c,2-c;x).$$

Therefore the asymptotics are

$$U(a,c;x)$$
$$\sim \frac{\Gamma(1-c)\Gamma(c)\Gamma(c-a)}{\pi^{\frac{3}{2}}}(-ax)^{\frac{1}{4}-\frac{1}{2}c}e^{\frac{1}{2}x}$$
$$\times \sin(c\pi - a\pi)\cos\left(y - \frac{1}{2}c\pi + \frac{1}{4}\pi\right)$$
$$+ \frac{\Gamma(c-1)\Gamma(2-c)\Gamma(1-a)}{\pi^{\frac{3}{2}}}e^{\frac{1}{2}x}x^{1-c}(-ax)^{\frac{1}{2}c-\frac{3}{4}}$$
$$\times \sin a\pi \cos\left(y + \frac{1}{2}c\pi - \frac{3}{4}\pi\right)$$
$$= \frac{e^{\frac{1}{2}x}x^{\frac{1}{4}-\frac{1}{2}c}}{\sqrt{\pi}\,\sin c\pi}\left[(-a)^{\frac{1}{4}-\frac{1}{2}c}\Gamma(c-a)\sin(c\pi - a\pi)\cos\left(y - \frac{1}{2}c\pi + \frac{1}{4}\pi\right)\right.$$
$$\left. -(-a)^{\frac{1}{2}c-\frac{3}{4}}\Gamma(1-a)\sin a\pi \cos\left(y + \frac{1}{2}c\pi - \frac{3}{4}\pi\right)\right],$$

where $y = \sqrt{(2c-4a)x}$; we used (2.2.7) again. It follows from (2.1.9) that

$$(-a)^{\frac{1}{4}-\frac{1}{2}c}\Gamma(c-a) \sim \Gamma\left(\frac{1}{2}c-a+\frac{1}{4}\right) \sim (-a)^{\frac{1}{2}c-\frac{3}{4}}\Gamma(1-a).$$

Let $z = y - \frac{1}{2}c\pi + \frac{1}{4}\pi$. Then

$$\sin(c\pi - a\pi)\cos\left(y - \frac{1}{2}c\pi + \frac{1}{4}\pi\right) - \sin a\pi \cos\left(y + \frac{1}{2}c\pi - \frac{3}{4}\pi\right)$$
$$= \sin(c\pi - a\pi)\cos z + \sin a\pi \cos(z + c\pi)$$
$$= \sin c\pi \cos(z + a\pi) = \sin c\pi \cos\left(y - \frac{1}{2}c\pi + a\pi + \frac{1}{4}\pi\right).$$

Therefore

$$U(a,c;x) \sim \frac{\Gamma(\frac{1}{2}c-a+\frac{1}{4})}{\sqrt{\pi}}x^{\frac{1}{4}-\frac{1}{2}c}e^{\frac{1}{2}x}$$
$$\times \cos\left(\sqrt{2cx-4ax} - \frac{1}{2}c\pi + a\pi + \frac{1}{4}\pi\right),$$

which gives (8.2.10). As noted in (5.4.10), the Laguerre polynomial

$$L_n^{(a)}(x) = \frac{(\alpha+1)_n}{n!}M(-n,\alpha+1;x).$$

It follows from this, (2.1.10), and (13.2.8) that as $n \to \infty$,

$$L_n^{(a)}(x) = \frac{e^{\frac{1}{2}x}n^{\frac{1}{2}a-\frac{1}{4}}}{\sqrt{\pi}\,x^{\frac{1}{2}a+\frac{1}{4}}}\left[\cos\left(2\sqrt{nx} - \frac{1}{2}a\pi - \frac{1}{4}\pi\right) + O(n^{-\frac{1}{2}})\right],$$

which is Fejér's result (5.4.12).

13.3 Hypergeometric functions and Jacobi polynomials

Consider the eigenvalue problem

$$x(1-x)u''(x)+[c-(a+1)x]u'(x)+\lambda u(x)=0. \tag{13.3.1}$$

As we have noted several times, the change of variables $y=1-2x$ converts the hypergeometric equation (13.3.1) to the equation

$$(1-y^2)v''(y)+[a+1-2c-(a+1)y]v'(y)+\lambda v(y)=0.$$

Setting $\alpha=c-1$, $\beta=a-c$, this equation is

$$(1-y^2)v''(y)+[\beta-\alpha-(\alpha+\beta+2)y]v'(y)+\lambda v(y)=0. \tag{13.3.2}$$

We noted in Section 5.5 that the change of variables $\theta=\cos^{-1}y$ followed by the gauge transformation

$$v(y)=\left(\sin\frac{1}{2}\theta\right)^{-\alpha-\frac{1}{2}}\left(\cos\frac{1}{2}\theta\right)^{-\beta-\frac{1}{2}}w(\theta)$$

converts (13.3.2) to

$$w''(\theta)+\frac{1-4\alpha^2}{16\sin^2\frac{1}{2}\theta}w(\theta)+\left[\frac{1-4\beta^2}{16\cos^2\frac{1}{2}\theta}+\lambda_1\right]w(\theta)=0, \tag{13.3.3}$$

where $\lambda_1=\lambda+\frac{1}{4}(\alpha+\beta+1)^2$. Now

$$\frac{1}{\sin^2\frac{1}{2}\theta}=\frac{4}{\theta^2}+\frac{1}{3}+O(\theta^2);\qquad \frac{1}{\cos^2\frac{1}{2}\theta}=1+O(\theta^2),$$

so (13.3.3) can be written as

$$w''(\theta)+\frac{1-4\alpha^2}{4\theta^2}w(\theta)+\left[\frac{1-4\alpha^2}{48}+\frac{1-4\beta^2}{16}+\lambda_1\right]w(\theta)=r(\theta)w(\theta),$$

where $|r(\theta)|\leq A\theta^2$. As in the previous section, we can convert this to a perturbation of Bessel's equation. First, taking $w(\theta)=\theta^{\frac{1}{2}}W(\theta)$ leads to

$$W''(\theta)+\frac{1}{\theta}W'(\theta)+\left[\mu^2-\frac{\alpha^2}{\theta^2}\right]W(\theta)=r(\theta)W(\theta), \tag{13.3.4}$$

where

$$\mu^2=\frac{1-4\alpha^2}{48}+\frac{1-4\beta^2}{16}+\lambda+\frac{(\alpha+\beta+1)^2}{4}.$$

We are interested in the limit $\lambda\to+\infty$, so suppose that μ is positive. The change of variables $z=\mu\theta$ leads to the equation

$$V''(z)+\frac{1}{z}V'(z)+\left[1-\frac{(c-1)^2}{z^2}\right]V(z)=\frac{R(z)V(z)}{\mu^4}, \tag{13.3.5}$$

with $|R(z)| \le Az^2$. This equation is very close to (13.2.3) and can be analyzed in exactly the same way.

So far, we have shown that a solution of (13.3.1) has the form

$$u(x) = v(1 - 2x) = \left(\sin\frac{1}{2}\theta\right)^{-a-\frac{1}{2}} \left(\cos\frac{1}{2}\theta\right)^{-\beta-\frac{1}{2}} w(\theta)$$

$$= \theta^{\frac{1}{2}} \left(\sin\frac{1}{2}\theta\right)^{-a-\frac{1}{2}} \left(\cos\frac{1}{2}\theta\right)^{-\beta-\frac{1}{2}} W(\theta)$$

$$= \theta^{\frac{1}{2}} \left(\sin\frac{1}{2}\theta\right)^{-a-\frac{1}{2}} \left(\cos\frac{1}{2}\theta\right)^{-\beta-\frac{1}{2}} V(\mu\theta),$$

where V is a solution of the perturbed Bessel equation (13.3.5). Note that

$$x = \frac{1-y}{2} = \frac{1-\cos\theta}{2} = \sin^2\frac{1}{2}\theta, \qquad (13.3.6)$$

so that as $x \to 0+, \theta \sim 2\sqrt{x}$ and

$$u(x) \sim \sqrt{2}x^{-\frac{1}{2}a} V(\mu\theta) = \sqrt{2}x^{\frac{1}{2}(1-c)} V(\mu\theta).$$

We suppose first that $c - 1 > 0$, and we want to choose V so that $u(x)$ is a hypergeometric function, the solution of (13.3.1) that is characterized by $u(0) = 1$. Since

$$J_{c-1}(\mu\theta) \sim \left(\frac{\mu\theta}{2}\right)^{c-1} \frac{1}{\Gamma(c)} \sim \frac{\mu^{c-1}}{\Gamma(c)} x^{\frac{1}{2}(c-1)}$$

as $x \to 0$, we obtain $V = \lim_{m\to\infty} V_m$ as in the previous section, with

$$V_0(z) = \frac{\Gamma(c)}{\sqrt{2}} \mu^{1-c} J_{c-1}(z),$$

which has the asymptotics

$$V_0(z) \sim \frac{\Gamma(c)}{\sqrt{\pi z}} \mu^{1-c} \cos\left(z - \frac{1}{2}c\pi + \frac{1}{4}\pi\right)$$

as $z \to +\infty$.

Taking $\lambda = v(a+v)$, the hypergeometric solution to (13.3.1) is the function with indices $a + v, -v, c$. As $v \to +\infty$, we may replace $\mu = \sqrt{\lambda}$ in the asymptotics by v, or by the more precise $v + \frac{1}{2}a$. Putting all this together, we have, for $0 < \theta < \pi$,

$$F\left(a+v, -v, c; \sin^2\left(\frac{1}{2}\theta\right)\right)$$

$$= \frac{\Gamma(c)\cos\left(v\theta + \frac{1}{2}a\theta - \frac{1}{2}c\pi + \frac{1}{4}\pi\right) + O(v^{-1})}{\sqrt{\pi}\,(v\sin\frac{1}{2}\theta)^{c-\frac{1}{2}}(\cos\frac{1}{2}\theta)^{\frac{1}{2}+(a-c)}}$$

as $\nu \to +\infty$. This is Darboux's result (10.1.12).

We have proved this asymptotic result under the assumption $c > 1$. The contiguous relation (10.4.7) shows that the asymptotics with index $c - 1$ can be computed from those with indices c and $c + 1$ in the range $c > 1$, and that the result extends to $c > 0$. By induction, it extends to all real values of c, $c \neq 0, -1, -2, \ldots$

As a corollary, we obtain Darboux's asymptotic result (5.5.12) for Jacobi polynomials. In fact,

$$P_n^{(\alpha,\beta)}(\cos\theta) = \frac{(\alpha + 1)_n}{n!} F\left(\alpha + n, -n, c; \sin^2\left(\frac{1}{2}\theta\right)\right),$$

with $c = \alpha + 1$, $a = \alpha + \beta + 1$. Since

$$\frac{(c)_n}{n!} = \frac{\Gamma(c + n)}{\Gamma(c)\Gamma(n + 1)} \sim \frac{n^{c-1}}{\Gamma(c)}$$

as $n \to \infty$, by (2.1.9), it follows that

$$P_n^{(\alpha,\beta)}(\cos\theta) = \frac{\cos\left(n\theta + \frac{1}{2}[\alpha + \beta + 1]\theta - \frac{1}{2}\alpha\pi - \frac{1}{4}\pi\right) + O(n^{-1})}{\sqrt{n\pi}\,(\sin\frac{1}{2}\theta)^{\alpha+\frac{1}{2}}(\cos\frac{1}{2}\theta)^{\beta+\frac{1}{2}}}.$$

13.4 Legendre functions

As noted in Section 11.4, the asymptotics in ν for the Legendre functions P_ν and Q_ν on the interval $-1 < x < 1$ follow from the asymptotics of the hypergeometric function. To determine the asymptotics on the interval $1 < x < \infty$ for the Legendre functions and associated Legendre functions, we make use of the integral representations in Section 11.5, written in the form

$$P_\nu^m(\cosh\theta) = (\sinh\theta)^m \frac{(-m + \nu + 1)_{2m}}{2^m(\frac{1}{2})_m \pi}$$

$$\times \int_0^\pi \frac{(\sin\alpha)^{2m}\,d\alpha}{(\cosh\theta - \sinh\theta \cos\alpha)^{m+1+\nu}}; \quad (13.4.1)$$

$$Q_\nu^m(\cosh\theta) = (-1)^m(\sinh\theta)^m \frac{(-m + \nu + 1)_{2m}}{2^m(\frac{1}{2})_m}$$

$$\times \int_0^\infty \frac{(\sinh\alpha)^{2m}\,d\alpha}{(\cosh\theta + \sinh\theta \cosh\alpha)^{m+1+\nu}}. \quad (13.4.2)$$

These representations, and the asymptotic formulas below, apply in particular to the Legendre functions $P_\nu = P_\nu^0$ and $Q_\nu = Q_\nu^0$.

As $\nu \to \infty$, the principal contribution to each integral comes where the denominator is smallest, which is near $\alpha = 0$. We use Laplace's method: make a change of variables so that the denominator takes a form for which the asymptotics are easily computed.

For the integral in (13.4.1), we take

$$s(\alpha) = \log(A - B\cos\alpha), \quad A = e^\theta \cosh\theta, \quad B = e^\theta \sinh\theta,$$

so

$$\cosh\theta - \sinh\theta \cos\alpha = e^{-\theta} e^{s(\alpha)}.$$

Then s is strictly increasing for $0 \le \alpha \le \pi$, with $s(0) = 0$, $s(\pi) = 2\theta$. For $\alpha \approx 0$,

$$A - B\cos\alpha = 1 + \frac{1}{2}B\alpha^2 + O(\alpha^4), \quad s(\alpha) = \frac{1}{2}B\alpha^2 + O(\alpha^4),$$

so

$$\frac{d\alpha}{ds} = \frac{1}{\sqrt{2Bs}}[1 + O(s)], \quad \sin\alpha = \frac{\sqrt{2s}}{\sqrt{B}}[1 + O(s)].$$

Therefore the integral in (13.4.1) is

$$\frac{2^{m-\frac{1}{2}} e^{(m+1+\nu)\theta}}{B^{m+\frac{1}{2}}} \int_0^{2\theta} e^{-(m+1+\nu)s} s^{m-\frac{1}{2}} [1 + O(s)] \, ds$$

$$= \frac{2^{m-\frac{1}{2}} e^{(m+1+\nu)\theta}}{B^{m+\frac{1}{2}}(m+1+\nu)^{m+\frac{1}{2}}}$$

$$\times \int_0^{(m+1+\nu)2\theta} e^{-t} t^{m-\frac{1}{2}} \left[1 + O(t[m+1+\nu]^{-1})\right] dt.$$

Up to an error that is exponentially small in ν, we may replace the preceding integral by an integral over the line and write it as the sum of two parts. The first part is $\Gamma\left(m + \frac{1}{2}\right)$, and the second part is dominated by

$$\frac{\Gamma(m+1+\frac{1}{2})}{m+1+\nu} = \frac{m+\frac{1}{2}}{m+1+\nu} \Gamma\left(m + \frac{1}{2}\right).$$

Therefore the integral in (13.4.1) is

$$\frac{2^{m-\frac{1}{2}} e^{(m+1+\nu)\theta} \Gamma(m+\frac{1}{2})}{B^{m+\frac{1}{2}}(m+1+\nu)^{m+\frac{1}{2}}} \left[1 + O\left([m+1+\nu]^{-1}\right)\right] \quad (13.4.3)$$

as $\nu \to \infty$.

To complete the calculation we note that as $\nu \to \infty$, (2.1.9) implies that

$$(-m+\nu+1)_{2m} = \frac{\Gamma(m+\nu+1)}{\Gamma(-m+\nu+1)} = (m+\nu+1)^{2m} \left[1 + O\left([m+1+\nu]^{-1}\right)\right],$$

while

$$\frac{\Gamma(m+\frac{1}{2})}{(\frac{1}{2})_m} = \Gamma\left(\frac{1}{2}\right) = \sqrt{\pi}.$$

Combining these with (13.4.1) and (13.4.3) gives the Hobson–Darboux–Laplace result

$$P_v^m(\cosh\theta) = \frac{e^{(v+\frac{1}{2})\theta}}{\sqrt{2\pi\sinh\theta}}(m+1+v)^{m-\frac{1}{2}}\left[1+O\left([m+1+v]^{-1}\right)\right]. \quad (13.4.4)$$

The same idea is used to compute the asymptotics of Q_v^m: the integral in (13.4.2) is rewritten using the variable

$$s(\alpha) = \log(A+B\cosh\alpha), \quad A = e^\theta\cosh\theta, \ B = e^\theta\sinh\theta.$$

The calculation is essentially the same and gives Hobson's result

$$Q_v^m(\cosh\theta) = (-1)^m\frac{e^{-(v+\frac{1}{2})\theta}\sqrt{\pi}}{\sqrt{2\sinh\theta}}(m+1+v)^{m-\frac{1}{2}}$$

$$\times\left[1+O\left([m+1+v]^{-1}\right)\right]. \quad (13.4.5)$$

13.5 Steepest descents and stationary phase

In this section, we briefly discuss two methods of deriving asymptotics from integral representations. Both methods address integrals that are in, or can be put into, the form

$$I(\lambda) = \int_C e^{\lambda\varphi(t)}f(t)\,dt, \quad (13.5.1)$$

and one is interested in the behavior as the parameter $\lambda \to +\infty$.

Suppose that the functions φ and f in (13.5.1) are holomorphic in some domain that contains the contour C, except perhaps the endpoints. If φ is real on the contour, then clearly the main contribution to $I(\lambda)$ for large λ will occur near points where φ has a maximum value; any such point t_0 that is not an endpoint of C will be a critical value $\varphi'(t_0) = 0$. The idea is to deform the contour C, if possible, so that φ is real along C and attains a global maximum. If this is not possible, we look for a contour such that $\operatorname{Re}\varphi$ attains a global maximum and φ is real near any points where the maximum is attained. Thus in a neighborhood of such a point the curve will follow the paths of steepest descent, the paths along which the real part decreases most rapidly. This accounts for the terminology *method of steepest descents*. One can then use a method such as that used in Section 13.4 to get the asymptotic result.

As an illustration, we consider the Laguerre polynomials. The generating function is given by (5.4.6):

$$\sum_{n=0}^{\infty}L_n^{(\alpha)}(x)z^n = (1-z)^{-\alpha-1}\exp\left(-\frac{xz}{1-z}\right), \quad |z| < 1.$$

By Cauchy's theorem,

$$L_n^{(\alpha)}(x) = \frac{1}{2\pi i} \int_{C_0} (1-z)^{-\alpha-1} \exp\left(-\frac{xz}{1-z}\right) \frac{dz}{z^{n+1}}, \tag{13.5.2}$$

where the path of integration encloses $z = 0$ but not $z = 1$. The change of variable $z = \exp(t/\sqrt{n})$ converts (13.5.2) to

$$L_n^{(\alpha)}(x) = \frac{1}{2\pi \sqrt{ni}} \int_C \exp\left\{\sqrt{n}\left(\frac{x}{t} - t\right)\right\} f\left(\frac{t}{\sqrt{n}}\right) dt, \tag{13.5.3}$$

where the amplitude function

$$f(s) = (1 - e^s)^{-\alpha-1} \exp\left\{-x\left(\frac{e^s}{1 - e^s} + \frac{1}{s}\right)\right\}$$

is holomorphic in the strip $\{|\operatorname{Im} s| < 2\pi\}$ cut along $[0, \infty)$. The contour C starts at $+\infty$, follows the lower edge of the cut, encircles the origin in the clockwise direction, and returns to $+\infty$ along the upper edge of the cut.

The real part of the phase function $\varphi(t) = -t + x/t$ has the same sign as $\operatorname{Re} t(x - |t|^2)$, and its critical points for $x > 0$ are $t = \pm i\sqrt{x}$. The second derivative at the point $\pm i\sqrt{x}$ is $\pm 2i/\sqrt{x}$. It follows that there is a change of variables $u = u(t)$ near $\pm i\sqrt{x}$ such that

$$t = \sqrt{x}\left(\pm i \pm e^{\pm i\pi/4} u\right) + O(u^2), \quad \frac{x}{t} - t = \sqrt{x}(\mp 2i - u^2).$$

Therefore each of these points is a *saddle point* for the real part of φ: at $t = i\sqrt{x}$, the real part decreases in the southwest and northeast directions and increases in the southeast and northwest directions, while the opposite is true at $t = -i\sqrt{x}$. The path C can be chosen so that it lies in the region where $\operatorname{Re} \varphi(t) < 0$, except at the points $\pm i\sqrt{x}$, while near these points it coincides with the paths of steepest descent.

These considerations imply that, apart from quantities that are exponentially small in $\sqrt{n}$ as $n \to \infty$, the integral in (13.5.3) can be replaced by the sum of two integrals $I_\pm$, along short segments through the points $t = \pm i\sqrt{x}$ coinciding with the paths of steepest descent. Moreover, up to terms of lower order in $\sqrt{n}$, we may replace $f(t/\sqrt{n})$ by its value at $t = \pm i\sqrt{x}$. Taking $u(t)$ as above, and taking into account the orientation of C, the integral I_+ is then

$$\int_{-\delta}^{\delta} f\left(\frac{i\sqrt{x}}{\sqrt{n}}\right) \exp\{-\sqrt{nx}(2i + u^2)\} \sqrt{x}\, e^{i\pi/4} du$$

$$= f\left(\frac{i\sqrt{x}}{\sqrt{n}}\right) e^{i(-2\sqrt{nx} + \frac{1}{4}\pi)} \sqrt{x} \int_{-\delta}^{\delta} \exp(-\sqrt{nx}\, u^2) du,$$

while I_- is the negative of the complex conjugate of I_+.

Up to an error that is exponentially small in $\sqrt{n}$, we may replace the last integral over $[-\delta, \delta]$ with the integral over the line and use

$$\int_{-\infty}^{\infty} e^{-\sqrt{nx}u^2} \, du = \frac{1}{(nx)^{\frac{1}{4}}} \int_{-\infty}^{\infty} e^{-s^2} \, ds = \frac{\sqrt{\pi}}{(nx)^{\frac{1}{4}}}.$$

As $n \to \infty$,

$$f\left(\frac{i\sqrt{x}}{\sqrt{n}}\right) \approx e^{\frac{1}{2}(\alpha+1)\pi i} x^{-\frac{1}{2}(\alpha+1)} n^{\frac{1}{2}(\alpha+1)} e^{\frac{1}{2}x}.$$

Collecting these results, we find again Fejér's result

$$L_n^{(\alpha)}(x) = \frac{1}{\sqrt{\pi}} x^{-\frac{1}{2}\alpha-\frac{1}{4}} n^{\frac{1}{2}\alpha-\frac{1}{4}} e^{\frac{1}{2}x} \cos\left(2\sqrt{nx} - \frac{1}{2}\alpha\pi - \frac{1}{4}\pi\right)$$
$$+ O(n^{\frac{1}{2}\alpha-\frac{3}{4}}).$$

Suppose now that the function in the exponent of (13.5.1) is purely imaginary. We change the notation and write $\exp(i\lambda\varphi)$. We assume that the functions f and φ have some degree of smoothness but are not necessarily analytic. The idea here is that wherever $\varphi' \neq 0$, the exponential factor oscillates rapidly as $\lambda \to \infty$, and therefore there is cancellation in the integral. If φ' does not vanish at any point of C, we may integrate by parts using the identity

$$e^{i\lambda\varphi(t)} = \frac{1}{i\lambda\varphi'(t)} \frac{d}{dt}\left[e^{i\lambda\varphi(t)}\right]$$

to introduce a factor $1/\lambda$. On the other hand, if, say, $\varphi'(t_0) = 0$ and $f(t_0)\varphi''(t_0) \neq 0$, then it turns out that the part of the integral near t_0 contributes an amount that is $O(\lambda^{-\frac{1}{2}})$. Thus, once again, the main contributions to the asymptotics come from points where $\varphi' = 0$, i.e. points of stationary phase, hence the terminology *method of stationary phase*.

As an example, we consider the asymptotics of the Bessel function $J_\nu(x)$ as $x \to +\infty$. In Chapter 9, the asymptotics were obtained by expressing J_ν in terms of Hankel functions. Here we give a direct argument by the method of stationary phase applied to the Sommerfeld–Bessel integral (9.3.12):

$$J_\nu(x) = \frac{1}{2\pi} \int_C e^{ix\sin\theta - i\nu\theta} \, d\theta,$$

where the path of integration consists of the boundary of the strip $\{|\mathrm{Re}\,\theta| < \pi, \mathrm{Im}\,\theta > 0\}$. The critical points of the phase function $\sin\theta$ on this path are at $\theta = \pm\pi/2$. As noted above, on any straight part of the path of integration that does not contain a critical point, we may integrate by parts to get an estimate that is $O(x^{-1})$. Thus we are led to consider integration over two small intervals, each containing one of the two points $\pm\pi/2$. We may change the variables $u = u(\theta)$ in these intervals in such a way that

$$\theta = \pm\frac{\pi}{2} + u + O(u^2), \quad \sin\theta = \pm\left(1 - \frac{u^2}{2}\right).$$

Up to terms of lower order, we may replace $\exp(-iv\theta)$ by $\exp(\mp iv\pi/2)$. Thus the integral over the interval that contains $\theta = \pi/2$ becomes

$$\frac{e^{ix - \frac{1}{2}iv\pi}}{2\pi} \int_{-\delta}^{\delta} e^{-\frac{1}{2}ixu^2} \, du,$$

while the integral over the interval that contains $\theta = -\pi/2$ becomes the complex conjugate. Up to quantities of order $1/\sqrt{x}$, we may replace the last integral by the integral over the line, interpreted as

$$\int_{-\infty}^{\infty} e^{-\frac{1}{2}ixu^2} \, du = \lim_{\varepsilon \to 0+} \int_{-\infty}^{\infty} e^{-\frac{1}{2}x(i+\varepsilon)u^2} \, du. \tag{13.5.4}$$

Recall that for $a > 0$,

$$\int_{-\infty}^{\infty} e^{-au^2} \, du = a^{-1/2} \int_{-\infty}^{\infty} e^{-t^2} \, dt = \frac{\sqrt{\pi}}{\sqrt{a}}.$$

This formula remains valid, by analytic continuation, for complex a with $\mathrm{Re}\, a > 0$, so (13.5.4) gives

$$\int_{-\infty}^{\infty} e^{-\frac{1}{2}ixu^2} \, du = \lim_{\varepsilon \downarrow 0} \frac{\sqrt{2\pi}}{\sqrt{x(i+\varepsilon)}} = \frac{\sqrt{2\pi}}{\sqrt{x}} e^{-\frac{1}{4}\pi i}.$$

Combining these results, we obtain the Jacobi–Hankel result

$$J_v(x) = \frac{\sqrt{2}}{\sqrt{\pi x}} \cos\left(x - \frac{1}{2}v\pi - \frac{1}{4}\right) + O(x^{-1})$$

as $x \to +\infty$, which is (9.4.9).

13.6 Exercises

1. Verify (13.1.4).
2. Verify (13.2.6).
3. Consider the second-order differential equation

$$w''(z) + f(z)\, w'(z) + g(z)\, w(z) = 0,$$

where $f(z)$ and $g(z)$ have convergent power series expansions

$$f(z) = \sum_{n=0}^{\infty} \frac{f_n}{z^n}, \qquad g(z) = \sum_{n=0}^{\infty} \frac{g_n}{z^n}.$$

Assume that not all coefficients f_0, g_0, and g_1 are zero, i.e. $z = \infty$ is an *irregular singular point*. Show that this equation has a formal series solution of the form

$$w(z) = e^{\lambda z} z^{\mu} \sum_{n=0}^{\infty} \frac{a_n}{z^n},$$

where

$$\lambda^2 + f_0 \lambda + g_0 = 0, \qquad (f_0 + 2\lambda)\mu = -(f_1 \lambda + g_1),$$

and the coefficients a_n are constants. Derive the equation that determines the coefficients a_n recursively.

4. The quadratic equation of λ in Exercise 3 is called the *characteristic equation*, and it has two *characteristic roots* λ_1, λ_2. Solutions with the kinds of expansions given in Exercise 3 are called *normal solutions*. Suppose that $f_0^2 = 4g_0$, so $\lambda_1 = \lambda_2$. Show that the *Fabry transformation* [127]

$$w(z) = e^{-f_0 z/2} W, \qquad t = z^{1/2}$$

takes the differential equation in Exercise 3 into the new equation

$$W''(t) + F(t) W'(t) + G(t) W(t) = 0,$$

where

$$F(t) = \frac{2f_1 - 1}{t} + \frac{2f_2}{t^3} + \cdots, \qquad G(t) = (4g_1 - 2f_0 f_1) + \frac{4g_2 - 2f_0 f_2}{t^2} + \cdots.$$

If $4g_1 = 2f_0 f_1$, then infinity is a regular singular point of the new equation, so it has a convergent power series solution. If $4g_1 \neq 2f_0 f_1$, then the new equation has distinct characteristic roots. By Exercise 3, we can write down formal series solutions in the form given there, with z replaced by t, or equivalently, $z^{1/2}$. Series solutions of this kind are called *subnormal solutions*.

5. Let λ_1 and λ_2 be the two characteristic values in Exercise 3, and suppose that $\operatorname{Re} \lambda_1 \geq \operatorname{Re} \lambda_2$. Consider first $j = 1$ and, for convenience, drop the subscript and write

$$w(z) = L_n(z) + \varepsilon_n(z), \qquad L_n(z) = e^{\lambda z} z^{\mu} \sum_{m=0}^{n-1} \frac{a_m}{z^m}.$$

Use the recursive formulas in Exercise 3 to show that

$$L_n''(z) + f(z) L_n'(z) + g(z) L_n(z) = e^{\lambda z} z^{\mu} R_n(z),$$

where $R_n(z) = O(z^{-n-1})$ as $z \to \infty$. Show that the error term $\varepsilon_n(z)$ satisfies the integral equation

$$\varepsilon_n(z) = \int_z^{e^{-i\omega}\infty} K(z,t)\{e^{\lambda t}t^\mu R_n(t) + [f(t) - f_0]\varepsilon_n'(t) + [g(t) - g_0]\varepsilon_n(t)\}dt,$$

where

$$K(z,t) = \frac{e^{\lambda_1(z-t)} - e^{\lambda_2(z-t)}}{\lambda_1 - \lambda_2}, \qquad \omega = \arg(\lambda_2 - \lambda_1).$$

Recall that we have used λ for λ_1. Use the method of successive approximation to prove that for sufficiently large n,

$$\varepsilon_n(z) = O(e^{\lambda_1 z}z^{\mu_1 - n})$$

as $z \to \infty$ in the sector $|\arg(\lambda_2 z - \lambda_1 z)| \leq \pi$.

6. In Exercise 5, we established that for all sufficiently large n, the differential equation in Exercise 3 has a solution $w_{n,1}(z)$ given by

$$w_{n,1}(z) = e^{\lambda_1 z}z^{\mu_1}\left[\sum_{m=0}^{n-1} \frac{a_{m,1}}{z^m} + O\left(\frac{1}{z^n}\right)\right]$$

as $z \to \infty$, $|\arg(\lambda_2 z - \lambda_1 z)| \leq \pi$. By relabeling, we get another solution

$$w_{n,2}(z) = e^{\lambda_2 z}z^{\mu_2}\left[\sum_{m=0}^{n-1} \frac{a_{m,2}}{z^m} + O\left(\frac{1}{z^n}\right)\right]$$

as $z \to \infty$, $|\arg(\lambda_2 z - \lambda_1 z)| \leq \pi$. Show that $w_{n,1}(z)$ and $w_{n,2}(z)$ are independent of n.

7. Show that the equation

$$\frac{d^2w}{dz^2} = \left(1 + \frac{4}{z}\right)^{\frac{1}{2}} w$$

has two linearly independent solutions given by

$$w_1(z) \sim \frac{1}{z}e^{-z}\left(1 - \frac{2}{z} + \frac{5}{z^2} - \frac{44}{3z^3} + \cdots\right)$$

as $z \to \infty$, $|\arg z| \leq \pi$, and

$$w_2(z) \sim ze^z\left(1 + \frac{1}{z} - \frac{1}{2z^2} + \frac{2}{3z^3} + \cdots\right)$$

as $z \to \infty$, $|\arg(-z)| \leq \pi$.

8. Find a change of variable $z \to \xi$ that transforms Airy's equation $w'' - zw = 0$ into the equation

$$w''(\xi) + \frac{1}{3\xi}w'(\xi) - w(\xi) = 0.$$

Show that the new equation has two linearly independent asymptotic solutions

$$w_1(\xi) \sim e^{-\xi}\xi^{-1/6}\left(1 - \frac{5}{2^3 \cdot 3^2}\frac{1}{\xi} + \frac{5\cdot 7\cdot 11}{2^7 \cdot 3^4}\frac{1}{\xi^2} + \cdots\right);$$

$$w_2(\xi) \sim e^{\xi}\xi^{-1/6}\left(1 + \frac{5}{2^3 \cdot 3^2}\frac{1}{\xi} + \frac{5\cdot 7\cdot 11}{2^7 \cdot 3^4}\frac{1}{\xi^2} + \cdots\right),$$

the first valid in the sector $|\arg\xi| \le \pi$, and the second in the sector $|\arg(-\xi)| \le \pi$.

9. Show that the modified Bessel equation (9.5.1) has an irregular singular point at ∞, and that for x real, two linearly independent asymptotic solutions are

$$w_1(x) \sim x^{-1/2}e^x\left[1 - \frac{(4v^2-1^2)}{1!8x} + \frac{(4v^2-1^2)(4v^2-3^2)}{2!(8x)^2} - \cdots\right],$$

$$w_2(x) \sim x^{-1/2}e^{-x}\left[1 + \frac{(4v^2-1^2)}{1!8x} + \frac{(4v^2-1^2)(4v^2-3^2)}{2!(8x)^2} + \cdots\right]$$

as $x \to \infty$. The modified Bessel function $I_v(x)$ grows exponentially as $x \to \infty$, and its asymptotic expansion is given by $w_1(x)$ multiplied by the constant $(2\pi)^{-1/2}$. The function $K_v(x)$ decays exponentially as $x \to \infty$, and its asymptotic expansion is given by $w_2(x)$ multiplied by $(\pi/2)^{1/2}$; see (7.5.6).

10. Show that the equation

$$xy''(x) - (x+1)y(x) = 0, \qquad x > 0,$$

has solutions of the forms

$$y_1(x) \sim x^{\frac{1}{2}}e^x\sum_{n=0}^{\infty}\frac{a_n}{x^n}, \qquad y_2(x) \sim x^{-\frac{1}{2}}e^{-x}\sum_{n=0}^{\infty}\frac{b_n}{x^n}.$$

Furthermore, the two series involved are both divergent for all values of x.

11. Show that the equation

$$y''(x) - x^4 y(x) = 0$$

has two linearly independent formal solutions given by

$$y_1(x) \sim e^{\frac{1}{3}x^3}\sum_{n=0}^{\infty}\frac{(3n)!}{18^n(n!)^2}x^{-3n-1}, \qquad y_2(x) \sim e^{-\frac{1}{3}x^3}\sum_{n=0}^{\infty}\frac{(-1)^n(3n)!}{18^n(n!)^2}x^{-3n-1}$$

as $x \to \infty$; see de Bruijn [96].

12. Consider the equation

$$y''(x) + \left[\lambda^2 a(x) + b(x)\right] y(x) = 0.$$

Assume that $a(x)$ is positive and twice continuously differentiable in a finite or infinite interval (a_1, a_2) and that $b(x)$ is continuous. Show that the Liouville transformation

$$\xi = \int a^{1/2}(x) dx, \qquad w = a^{1/4}(x) y(x)$$

takes this equation into

$$\frac{d^2 w}{d\xi^2} + \left[\lambda^2 + \psi(\xi)\right] w = 0,$$

where

$$\psi(\xi) = \frac{5}{16} \frac{a'^2(x)}{a^3(x)} - \frac{1}{4} \frac{a''(x)}{a^2(x)} + \frac{b(x)}{a(x)}.$$

Discarding ψ in the transformed equation gives two linearly independent solutions $e^{\pm i\lambda\xi}$. In terms of the original variables, one obtains formally the WKB approximation

$$y(x) \sim A a^{-1/4}(x) \exp\left\{ i\lambda \int a^{1/2}(x) dx \right\}$$

$$+ B a^{-1/4} \exp\left\{ -i\lambda \int a^{1/2}(x) dx \right\}$$

as $\lambda \to \infty$, where A and B are arbitrary constants. (See the remarks at the end of this chapter concerning the history and terminology of the WKB approximation.) This formula sometimes also holds as $x \to \infty$ with λ fixed. For a discussion of this double asymptotic feature of the WKB approximation, see Chapter 6, p. 203 of Olver [311]. Use this idea to show that the equation in Exercise 3 has solutions

$$w(z) \sim C \exp(\lambda z + \mu \log z),$$

where C is a constant and

$$\lambda = \frac{-f_0 \pm (f_0^2 - 4g_0)^{1/2}}{2}, \qquad \mu = -\frac{f_1 \lambda + g_1}{f_0 + 2\lambda}.$$

This is the leading term of the formal series solution given in Exercise 3.

13. In the transformed equation in Exercise 12, substitute $w(\xi) = e^{i\lambda\xi}[1 + h(\xi)]$. Show that

$$h''(\xi) + i2\lambda h'(\xi) = -\psi(\xi)[1 + h(\xi)].$$

Convert this to the integral equation

$$h(\xi) = -\frac{1}{i2\lambda} \int_\alpha^\xi \{1 - e^{i2\lambda(v-\xi)}\} \psi(v)[1 + h(v)]dv,$$

where α is the value of $x = c$, $c = a_1$ or $c = a_2$. Assume that α is finite. Verify that any solution of this integral equation is also a solution of the second-order differential equation satisfied by $h(\xi)$. Use the method of successive approximation to show that

$$|h(\xi)| \leq \exp\left\{\frac{1}{\lambda}\Psi(\xi)\right\} - 1,$$

where

$$\Psi(\xi) = \int_\alpha^\xi |\psi(v)|dv.$$

14. Introduce the *control function*

$$F(x) = \int \left\{\frac{1}{a^{1/4}}\frac{d^2}{dx^2}\left[\frac{1}{a^{1/4}}\right] + \frac{b}{a^{1/2}}\right\}dx$$

and the notation

$$\mathcal{V}_{c,x}(F) = \int_c^x |F'(t)|dt$$

for the total variation of F over the interval (c,x). Verify that $\Psi(\xi) = \mathcal{V}_{c,x}(F)$, where Ψ is defined in Exercise 13. By summarizing the results in Exercises 12 and 13, show that the equation

$$y''(x) + \{\lambda^2 a(x) + b(x)\}y(x) = 0$$

has two linearly independent solutions

$$y_1(x) = a^{-1/4}(x)\exp\left\{i\lambda \int a^{1/2}(x)dx\right\}[1 + \varepsilon_1(\lambda,x)],$$

$$y_2(x) = a^{-1/4}(x)\exp\left\{-i\lambda \int a^{1/2}(x)dxt\right\}[1 + \varepsilon_2(\lambda,x)],$$

where

$$|\varepsilon_j(\lambda,x)| \leq \exp\left\{\frac{1}{\lambda}\mathcal{V}_{c,x}(F)\right\} - 1, \quad j = 1,2.$$

For fixed x and large λ, the right-hand side of the last equation is $O(\lambda^{-1})$. Hence a general solution has the asymptotic behavior given by the WKB approximation.

15. By following the argument outlined in Exercises 12–14, prove that the equation

$$y''(x) - \{\lambda^2 a(x) + b(x)\}y(x) = 0$$

has two linearly independent solutions

$$y_1(x) = a^{-1/4}(x) \exp\left\{\lambda \int a^{1/2}(x)dx\right\}[1+\varepsilon_1(\lambda,x)],$$

$$y_2(x) = a^{-1/4}(x) \exp\left\{-\lambda \int a^{1/2}(x)dx\right\}[1+\varepsilon_2(\lambda,x)],$$

where

$$|\varepsilon_j(\lambda,x)| \le \exp\left\{\frac{1}{2\lambda}\mathcal{V}_{c,x}(F)\right\} - 1, \quad j = 1,2.$$

If $\mathcal{V}_{a_1,a_2}(F) < \infty$, then these inequalities imply that

$$y_j(x) = a^{-1/4}(x) \exp\left\{(-1)^{j-1}\lambda \int a^{1/2}(x)dx\right\}[1+O(\lambda^{-1})],$$

$j = 1,2$, uniformly for $x \in (a_1,a_2)$.

16. Let $N = 2n+1$ and $x = N^{1/2}\zeta$. From the generating function of the Hermite polynomial $H_n(x)$ given in (5.3.6), derive the integral representation

$$H_n(x) = \frac{(-1)^n n!}{N^{n/2}} \frac{1}{2\pi i} \int_C g(t)e^{-Nf(t,\zeta)}dt,$$

where

$$g(t) = t^{-1/2}, \qquad f(t,\zeta) = 2\zeta t + t^2 + \frac{1}{2}\log t,$$

and C is the steepest descent path passing through the two saddle points

$$t_\pm = \frac{-\zeta \pm \sqrt{\zeta^2-1}}{2}.$$

Note that for fixed $x \in (0,\infty)$, $\zeta \to 0^+$ as $n \to \infty$, and hence, $t_\pm$ are two well-separated complex numbers both approaching the imaginary axis. Show that

$$\frac{(-1)^n n!}{2\pi i N^{n/2}} g(t_+)e^{-Nf(t_+,\zeta)}\left(\frac{-2\pi}{Nf''(t_+,\zeta)}\right)^{1/2}$$

$$\sim e^{x^2/2 - i\sqrt{N}x}2^{(n-1)/2}\left(\frac{n}{e}\right)^{n/2}e^{n\pi i/2} \qquad \text{as } n \to \infty.$$

Use this to prove that

$$H_n(x) \sim 2^{(n+1)/2}\left(\frac{n}{e}\right)^{n/2}e^{x^2/2}\cos\left(\sqrt{2n+1}x - \frac{1}{2}n\pi\right)$$

as $n \to \infty$, thus establishing (5.3.20).

17. Returning to (13.5.3), we deform the contour C into an infinite loop which consists of (i) a circle centered at the origin with radius $\sqrt{x}$ and (ii) two straight lines along the upper and lower edges of the positive real axis from

$\sqrt{x}$ to $+\infty$. (a) Show that the contribution from the two straight lines is of the magnitude $\varepsilon_n(x) = O(n^{\frac{1}{2}|a|+1} e^{-\sqrt{nx}})$. On the circle we introduce the parameterization $t = \sqrt{x} e^{i\theta}$, $\theta \in (0, 2\pi)$. (b) Show that

$$L_n^{(\alpha)}(x) = \int_0^{2\pi} \psi(\theta) e^{i\lambda\phi(\theta)} \, d\theta + \varepsilon_n(x),$$

where $\lambda = 2\sqrt{nx}$, $\phi(\theta) = -\sin\theta$,

$$\psi(\theta) = -\frac{\sqrt{x}}{2\pi\sqrt{n}} e^{i\theta} f\left(\sqrt{\frac{x}{n}} e^{i\theta}\right),$$

$$f(s) = (1 - e^s)^{-\alpha-1} \exp\left\{-x\left(\frac{e^s}{1 - e^s} + \frac{1}{s}\right)\right\}.$$

The oscillatory integral is now exactly in the form to which the method of stationary phase applies. The stationary points, where $\phi'(\theta) = 0$, occur at $\theta = \theta_1 = \frac{\pi}{2}$ and $\theta = \theta_2 = \frac{3\pi}{2}$. The contribution from θ_1 is

$$\psi(\theta_1)\sqrt{\frac{2\pi}{\lambda\phi''(\theta_1)}} e^{i[\lambda\phi(\theta_1) + \frac{1}{4}\pi]} = \frac{e^{\frac{1}{2}x} n^{\frac{1}{2}\alpha - \frac{1}{4}}}{2\sqrt{\pi} x^{\frac{1}{2}\alpha + \frac{1}{4}}} e^{(\frac{1}{2}\alpha\pi + \frac{1}{4} - 2\sqrt{nx})i}.$$

The contribution from θ_2 is simply the complex conjugate of the previous expression, thus proving (5.4.12).

18. (a) Use (5.5.3) to derive the representation

$$P_n^{(\alpha,\beta)}(x)$$

$$= \binom{n+\alpha}{n}\left(\frac{x+1}{2}\right)^n \sum_{k=0}^{n} \frac{(n-k+1)_k}{(\alpha+1)_k} \binom{n+\beta}{k}\left(\frac{x-1}{x+1}\right)^k$$

$$= \binom{n+\alpha}{n}\left(\frac{x+1}{2}\right)^n F\left(-n, -n-\beta, \alpha+1; \frac{x-1}{x+1}\right).$$

(b) Show that the formula in (a) can be written as

$$P_n^{(\alpha,\beta)}(z)$$

$$= \frac{1}{2\pi i} \int_C \left(1 + \frac{x+1}{2} z\right)^{n+\alpha} \left(1 + \frac{x-1}{2} z\right)^{n+\beta} z^{-n-1} \, dz,$$

where we assume $x \neq \pm 1$, and where C is a closed curve that encircles the origin in the positive sense and excludes the points $-2(x \pm 1)^{-1}$.

(c) Prove that the formula in (a) can also be written in the form

$$P_n^{(\alpha,\beta)}(x) = \frac{1}{2\pi i} \int_{C'} \left(\frac{1}{2}\frac{t^2-1}{t-x}\right)^n \left(\frac{1-t}{1-x}\right)^\alpha \left(\frac{1+t}{1+x}\right)^\beta \frac{dt}{t-x},$$

where $x \neq \pm 1$ and C' is a contour encircling the point $t = x$ in the positive sense but not the points $t = \pm 1$. The functions $(1 - t/1 - x)^\alpha$ and $[(1 + t)/(1 + x)]^\beta$ are assumed to reduce to 1 for $t = x$.

(d) Let $t - x = w$, and establish the representation

$$2^{-\frac{1}{2}(\alpha+\beta+1)}(1+x)^\beta (1-x)^\alpha P_n^{(\alpha,\beta)}(x) = \frac{1}{2\pi i} \int_{C'} g(w) e^{Nf(w,x)} dw,$$

where

$$N = n + \frac{1}{2}(\alpha + \beta + 1),$$

$$f(w,x) = \log \frac{(w+x+1)(w+x-1)}{2w},$$

$$g(w) = \frac{(1-w-x)^\alpha}{(w+x-1)^{(\alpha+\beta+1)/2}} \frac{(w+x+1)^{(-\alpha+\beta-1)/2}}{w^{(1-\alpha-\beta)/2}}.$$

The integration path is the steepest descent curve passing through the relevant saddle points. For $x \in (0,1)$, we write $x = \cos\theta$, with $\theta \in (0, \frac{1}{2}\pi)$. The two saddle points of $f(w,x)$ are located at $w_\pm = \pm\sqrt{1-x^2} = \pm i\sin\theta$. Show that

$$g(w_+) = 2^{(\alpha+\beta-3)/2} e^{-\alpha\pi i/2 - \pi i/2} \left(\sin \frac{1}{2}\theta\right)^{\alpha-1} \left(\cos \frac{1}{2}\theta\right)^{\beta-1} e^{-i\theta/2},$$

$$e^{Nf(w_+,x)} = e^{N\theta i}, \qquad f''(w_+,x) = \frac{-ie^{-\theta i}}{2\sin\frac{1}{2}\theta \cos\frac{1}{2}\theta}.$$

From these, deduce the asymptotic formula (5.5.12).

19. Use Exercise 5 of Chapter 10 to show that

$$F(a+v,-v,c;x) = e^{v\pi i} \frac{\Gamma(c)\Gamma(1+v)}{\Gamma(c+v)} \frac{1}{2\pi i} \int_C g(t) e^{vh(t)} dt,$$

where

$$h(t) = \log \frac{1+t}{t(1+t-xt)}, \qquad g(t) = \frac{(1+t)^{a-c}}{t(1+t-xt)^a},$$

and C comes from $+\infty$ to 0 along the upper edge of the cut $[0,\infty)$ and returns along the lower edge. Use the steepest descent method to derive the asymptotic formula

$$F(a+v,-v,c;x) = \frac{\Gamma(c)\Gamma(1+v)}{\Gamma(c+v)\sqrt{v\pi}} \mathrm{Re}\left[e^{v\pi i} i^{\frac{1}{2}+a-c} x^{\frac{1}{4}-\frac{1}{2}c}(1-x)^{\frac{1}{2}c-\frac{1}{2}a-\frac{1}{4}} \right.$$

$$\times (\sqrt{x}+i\sqrt{1-x})^{-a}(\sqrt{x}-i\sqrt{1-x})^{2v}\Big][1+O(v^{-1})].$$

Deduce (10.1.12) from the last equation.

20. Use the change of variables

$$t = \sqrt{-ax + \frac{1}{2}cxs}$$

to show that the integral in Exercise 3 of Chapter 8 is equal to

$$\left(-ax + \frac{1}{2}cx\right)^{-\frac{1}{2}c+\frac{1}{2}} \int_{C'} \exp\left\{-a\log\left(1 - \frac{\sqrt{x}}{(-a+\frac{1}{2}c)^{\frac{1}{2}}s}\right)\right.$$

$$\left. + \left(-ax + t\frac{1}{2}cx\right)^{\frac{1}{2}} s\right\} s^{-c}\, ds,$$

where C' is the image of C. Show that the quantity inside the braces is

$$\left(-ax + \frac{1}{2}cx\right)^{\frac{1}{2}}\left(s - \frac{1}{s}\right) - \frac{x}{2s^2} + O\left(\left[-a + \frac{1}{2}c\right]^{-\frac{1}{2}}\right)$$

and

$$M(a,c;x) = \Gamma(c)\left(-ax + \frac{1}{2}cx\right)^{\frac{1}{2}-\frac{1}{2}c}$$

$$\times \frac{1}{2\pi i}\int_{C'} e^{Nf(s)}g(s)\,ds\left[1 + O\left(\left[-a + \frac{1}{2}c\right]^{-\frac{1}{2}}\right)\right],$$

where $N = (-ax + \frac{1}{2}cx)^{\frac{1}{2}}$, $f(s) = s - 1/s$, and $g(s) = e^{-x/2s^2}s^{-c}$. The saddle points are at $\pm i$. Show that the contribution from the saddle point at $s = i$ is

$$\frac{\Gamma(c)}{2\sqrt{\pi}}e^{\frac{1}{2}x}\left(-ax + \frac{1}{2}cx\right)^{\frac{1}{4}-\frac{1}{2}c}\exp\left\{i\left[(2cx - 4ax)^{\frac{1}{2}} - \frac{1}{2}c\pi + \frac{1}{4}\pi\right]\right\}.$$

Deduce from this the asymptotic formula (6.1.10).

21. (a) Use Exercise 6 of Chapter 10 and (6.5.6) to prove that for $\mathrm{Re}\,(b+x) > 0$, the Meixner polynomial

$$m_n(x;b,c) = \frac{(b)_n}{c^n}\frac{\Gamma(x+1)\Gamma(b)}{\Gamma(b+x)}\frac{1}{2\pi i}$$

$$\times \int_C \frac{t^{b+x-1}}{(t-1)^{x+1}}[1 - (1-c)t]^n\, dt,$$

where C is a counter-clockwise loop through $t = 0$ that encloses $t = 1$. Therefore

$$m_n(na;b,c) = \frac{(b)_n}{c^n}\frac{\Gamma(x+1)\Gamma(b)}{\Gamma(b+x)}\frac{1}{2\pi i}\int_C \frac{t^{b-1}}{t-1}e^{-nf(t)}\,dt,$$

where

$$x = na, \qquad f(t) = -\alpha \log t + \alpha \log(t-1) - \log[1 - (1-c)t].$$

(b) Show that for $0 < c < 1$, fixed x (i.e. $\alpha = O(n^{-1})$), and large n, the phase function $f(t)$ has two saddle points

$$t_+ \sim 1 - \frac{c\alpha}{1-c}, \qquad t_- \sim \frac{\alpha}{1-c}$$

in the interval $(0,1)$. Note that for large n (i.e. small α), the two saddle points are well separated. Deform the contour in (a) to run from $t = 0$ to $t = t_+$ through t_- and with $\arg(t-1) = -\pi$, and from t_+ in the lower half-plane $\operatorname{Im} t < 0$ to the $t_c = 1/1 - c$, where the integrand vanishes; from t_c the contour runs in the half-plane $\operatorname{Im} t > 0$ to t_+, and returns to the origin with $\arg(t-1) = \pi$; see the figure below.

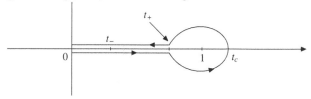

Denote the sum of the integrals along $[0, t_+]$ by I_1 and the remaining portion of the contour by I_2. Show that

$$m_n(n\alpha; b, c) = \frac{(b)_n}{c^n} \frac{\Gamma(x+1)\Gamma(b)}{\Gamma(b+x)} (I_1 + I_2),$$

where

$$I_1 = -\frac{\sin \pi x}{\pi} \int_0^{t_+} \frac{t^{b+x-1}}{(1-t)^{x+1}} [1 - (1-c)t]^n \, dt,$$

$$I_2 = \frac{1}{2\pi i} \int_{t_+}^{(1+)} \frac{t^{b+x-1}}{(t-1)^{x+1}} [1 - (1-c)t]^n \, dt.$$

(c) With $x = n\alpha$, show that

$$I_1 = -\frac{\sin \pi x}{\pi} \int_0^{t_+} \frac{t^{b-1}}{1-t} e^{-nf(t)} \, dt,$$

where

$$f(t) = -\alpha \log t + \alpha \log(1-t) - \log(1 - (1-c)t).$$

This function has the same saddle points as the phase function given in (a). Note that on the interval $(0, t_+)$, this function has only one saddle point which occurs at $t = t_-$. Furthermore, as $\alpha \to 0$, this point coalesces with the end point $t = 0$. In fact, this point disappears when

$\alpha = 0$, since the function $\log(1 - (1 - c)t)$ has no saddle point in the interval $(0, t_+)$. Thus, the steepest descent method does not work. Show that the integral

$$\int_0^\infty s^{n\alpha} e^{-ns} ds = \int_0^\infty e^{-n(s - \alpha \log s)} ds$$

has the same asymptotic phenomenon as I_1, namely, the phase function $\phi(s) = s - \alpha \log s$ also has a saddle point (i.e. $s = \alpha$) which coincides with the endpoint $s = 0$ as $\alpha \to 0$. This suggests the change of variable $t \to s$ defined implicitly by

$$f(t) = -\alpha \log s + s + A,$$

where A is a constant independent of t or s. Show that to make the mapping $t \to s$ one-to-one and analytic in the interval $0 \le t < t_+$, one should choose A so that $t = t_-$ corresponds to $s = \alpha$, i.e.

$$A = f(t_-) + \alpha \log \alpha - \alpha.$$

This gives

$$I_1 = -\frac{\sin \pi x}{\pi} e^{-nA} \int_0^{s_+} h(s) s^{b+x-1} e^{-ns} ds,$$

where

$$h(s) = \left(\frac{t}{s}\right)^{b-1} \frac{1}{1-t} \frac{dt}{ds},$$

and s_+ is implicitly defined via the equation $g(t) = -\alpha \log s + s + A$ by substituting $t = t_+$ and $s = s_+$.

(d) Show that the function $t = t(s)$ defined in (c) satisfies

$$\frac{dt}{ds}\bigg|_{s=\alpha} = \frac{1}{\sqrt{\alpha f''(t_-)}}.$$

Furthermore, the integral I_1 in (c) has the leading order approximation

$$I_1 \sim -\frac{\sin \pi x}{\pi} e^{-nA} h(\alpha) \frac{\Gamma(b+x)}{n^{b+x}}$$

as $n \to \infty$, uniformly with respect to $\alpha \in [0, \alpha_0]$, $\alpha_0 > 0$. Deduce from this the asymptotic formula (6.5.14):

$$m_n(x, b, c) \sim -\frac{\Gamma(b+n) \Gamma(x+1)}{c^n (1 - c)^{x+b} n^{b+x}} \frac{\sin \pi x}{\pi}.$$

The derivation of this formula, as presented in this exercise, is due to N. M. Temme (private communication, 2009).

22. By following the same argument as in Exercise 21, prove the asymptotic formula for the Charlier polynomial $C_n(x; a)$ given in (6.3.12).

13.7 Remarks

Some asymptotic results for special functions are treated in detail by Olver [311]. The book by Erdélyi [115] is a concise introduction to the general methods. For more detail, see Bleistein and Handelsman [48], Copson [88], van der Corput [408], and Wong [443]. Asymptotic expansions for solutions of general ordinary differential equations are treated by Wasow [422].

The method used in Sections 13.1–13.3 and in Exercise 12 goes back at least to Carlini [64] in 1817. In the mathematical literature it is sometimes called the Liouville or Liouville–Green approximation [261, 173] (1837). In the physics literature it is called the the WKB or WKBJ approximation, referring to papers by Jeffreys [203] (1924), and by Wentzel [432], Kramers [231], and Brillouin [53] (1926), who also developed connection formulas. The method was adapted by Steklov [378] and Uspensky [406] to obtain asymptotics of classical orthogonal polynomials.

The method of steepest descents is an adaptation of the method of Laplace that we used to obtain the asymptotics of the Legendre functions. It was developed by Cauchy and Riemann, and adapted by Debye [97] to study the asymptotics of Bessel functions.

The method of stationary phase was developed by Stokes [387] and Kelvin [214] for the asymptotic evaluation of integrals that occur in the study of fluid mechanics.

14

Elliptic functions

Integrating certain functions of x that involve expressions of the form $\sqrt{P(x)}$, P a quadratic polynomial, leads to trigonometric functions and their inverses. For example, the sine function can be defined implicitly by inverting a definite integral:

$$\theta = \int_0^{\sin\theta} \frac{ds}{\sqrt{1-s^2}}.$$

Integrating functions involving expressions of the form $\sqrt{P(x)}$, where P is a polynomial of degree 3 or more, leads to new types of transcendental functions. When the degree is 3 or 4, the functions are called elliptic functions. For example, the Jacobi elliptic function $\text{sn}(u) = \text{sn}(u, k)$ is defined implicitly by

$$u = \int_0^{\text{sn}u} \frac{ds}{\sqrt{(1-s^2)(1-k^2 s^2)}}. \tag{14.0.1}$$

As functions of a complex variable, the trigonometric functions are periodic with a real period. Elliptic functions are doubly periodic, having two periods whose ratio is not real.

This chapter begins with the question of integrating a function $R(z, \sqrt{P(z)})$, where R is a rational function of two variables and P is a polynomial of degree 3 or 4 with no repeated roots. The general case can be reduced to

$$P(z) = (1-z^2)(1-k^2 z^2).$$

We sketch Legendre's reduction of the integration of $R(z, \sqrt{P(z)})$ to three cases, called elliptic integrals of the first, second, and third kind. The elliptic integral of the first kind is (14.0.1), leading to the Jacobi elliptic function $\text{sn}u$ and then to the associated Jacobi elliptic functions $\text{cn}u$, $\text{dn}u$.

Jacobi eventually developed a second approach to elliptic functions as quotients of certain entire functions that just miss being doubly periodic: the theta functions. After covering the basics of theta functions and their relation to the Jacobi elliptic functions, we turn to the Weierstrass approach, which is based on a single function $\wp(u)$ that is doubly periodic with double poles.

14.1 Integration

Unlike differentiation, integration of relatively simple functions is not a mechanical process.

Any rational function (quotient of polynomials) can be integrated, in principle, by factoring the denominator and using the partial fraction decomposition. The result is a sum of a rational function and a logarithmic term. Any function of the form

$$f(z) = \frac{p(z, \sqrt{P(z)})}{q(z, \sqrt{P(z)})}, \tag{14.1.1}$$

where p and q are polynomials in two variables and P is a quadratic polynomial, can be integrated by reducing it to a rational function. In fact, after a linear change of variable (possibly complex) we may assume that $P(z) = 1 - z^2$. Setting $z = 2u/(1 + u^2)$ converts the integrand to a rational function of u.

This process breaks down at the next degree of algebraic complexity: integrating a function of the form (14.1.1) when P is a polynomial of degree 3 or 4 with no multiple roots. The functions obtained by integrating such rational functions of z and $\sqrt{P(z)}$ are known as elliptic functions. The terminology stems from the fact that calculating the arc length of an ellipse, say as a function of the angle in polar coordinates, leads to such an integral. The same is true for another classical problem – calculating the arc length of a lemniscate (the locus of points, the product of whose distances from two fixed points at distance $2d$ is equal to d^2; Jacob Bernoulli [39]).

In this section, we prove Legendre's result [252] that any such integration can be reduced to one of three basic forms:

$$\int \frac{dz}{\sqrt{P(z)}}, \quad \int \frac{z^2 \, dz}{\sqrt{P(z)}}, \quad \int \frac{dz}{(1 + az^2)\sqrt{P(z)}}. \tag{14.1.2}$$

The omitted steps in the proof are included as exercises.

The case of a polynomial of degree 3 can be reduced to that of a polynomial of degree 4 by a linear fractional transformation and vice versa; see the exercises for this and for subsequent statements for which no argument is supplied. Suppose that P has degree 4, with no multiple roots. Up to a linear fractional transformation, we may assume for convenience that the roots are ± 1 and $\pm 1/k$. Thus we may assume that

$$P(z) = (1 - z^2)(1 - k^2 z^2).$$

Suppose that $r(z, w)$ is a rational function in two variables. It can be written as the sum of two rational functions, one that is an even function of w and one that is an odd function of w:

$$r(z, w) = \frac{1}{2}[r(z, w) + r(z, -w)] + \frac{1}{2}[r(z, w) - r(z, -w)]$$

$$= r_1(z, w^2) + r_2(z, w^2)w.$$

Therefore, in integrating $r(z, w)$ with $w^2 = P(z)$, we may reduce to the case when the integrand has the form $r(z)\sqrt{P(z)}$, where r is a rational function of z.

At the next step we decompose r into even and odd parts, so that we are considering

$$\int r_1(z^2)\sqrt{P(z)}\, dz + \int r_2(z^2)\sqrt{P(z)}\, z\, dz.$$

Since P is a function of z^2, the substitution $s = z^2$ converts the integral on the right to the integral of a rational function of s and $\sqrt{Q(s)}$, with Q quadratic. As for the integral on the left, multiplying the numerator and denominator by $\sqrt{P(z)}$ converts it to the form

$$\int \frac{R(z^2)\, dz}{\sqrt{P(z)}},$$

where R is a rational function. We use the partial fraction decomposition of R to express the integral as a linear combination of integrals with integrands

$$J_n(z) = \frac{z^{2n}}{\sqrt{P(z)}}, \quad n = 0, \pm 1, \pm 2, \ldots;$$

$$K_m(z) = K_m(a, z) = \frac{1}{(1 + az^2)^m \sqrt{P(z)}}, \quad a \neq 0, \ m = 1, 2, 3, \ldots$$

(This is a temporary notation, not to be confused with the notation for cylinder functions.)

At the next step, we show that integration of J_{n+2}, $n \geq 0$, can be reduced to that of J_{n+1} and J_n. This leads recursively to the first two cases of (14.1.2). The idea is to relate these terms via a derivative:

$$\left[z^{2n+1}\sqrt{P(z)}\right]' = (2n+1)z^{2n}\sqrt{P(z)} + \frac{1}{2}z^{2n+1}\frac{P'(z)}{\sqrt{P(z)}}$$

$$= \frac{(2n+1)z^{2n}P(z) + \frac{1}{2}z^{2n+1}P'(z)}{\sqrt{P(z)}}$$

$$= \frac{(2n+3)k^2 z^{2n+4} - (2n+2)(1+k^2)z^{2n+2} + (2n+1)z^{2n}}{\sqrt{P(z)}}$$

$$= (2n+3)k^2 J_{n+2} - (2n+2)(1+k^2)J_{n+1} + (2n+1)J_n.$$

Therefore, up to a multiple of $z^{2n+1}\sqrt{P(z)}$, the integral of J_{n+2} is a linear combination of the integrals of J_{n+1} and J_n. The same calculation can be used to move indices upward if $n < 0$, leading again to J_0 and J_1.

A similar idea is used to reduce integration of K_{m+1} to integration of K_k, $k \leq m$, together with J_0 if $m = 2$, and also J_1 if $m = 1$:

$$\left[\frac{z\sqrt{P(z)}}{(1+az^2)^m}\right]' = \frac{\sqrt{P(z)}}{(1+az^2)^m} + \frac{zP'(z)}{2(1+az^2)^m\sqrt{P(z)}} - \frac{2maz^2\sqrt{P(z)}}{(1+az^2)^{m+1}}$$

$$= \frac{P(z)(1+az^2) + \frac{1}{2}zP'(z)(1+az^2) - 2amz^2P(z)}{(1+az^2)^{m+1}\sqrt{P(z)}}. \qquad (14.1.3)$$

The numerator in the last expression is a polynomial in z^2 of degree 3. Writing it as a sum of powers of $1 + az^2$, we may rewrite the last expression as a combination of K_{m+1}, K_m, K_{m-1}, and K_{m-2} if $m \geq 3$. For $m = 2$, we have K_3, K_2, K_1, and J_0, and for $m = 1$, we have K_2, K_1, J_0, and J_1. This completes the proof of Legendre's result.

This result leads to the following classification: *the elliptic integral of the first kind*

$$F(z) = F(k,z) = \int_0^z \frac{d\zeta}{\sqrt{(1-\zeta^2)(1-k^2\zeta^2)}}, \qquad (14.1.4)$$

the elliptic integral of the second kind

$$E(z) = E(k,z) = \int_0^z \sqrt{\frac{1-k^2\zeta^2}{1-\zeta^2}}\, d\zeta, \qquad (14.1.5)$$

and the elliptic integral of the third kind

$$\Pi(a,z) = \Pi(a,k,z) = \int_0^z \frac{d\zeta}{(1+a\zeta^2)\sqrt{(1-\zeta^2)(1-k^2\zeta^2)}}. \qquad (14.1.6)$$

Note that the integrand in (14.1.5) is just $J_0(\zeta) - k^2 J_1(\zeta)$.

14.2 Elliptic integrals

We begin with the properties of the elliptic integral of the first kind

$$F(z) = F(k,z) = \int_0^z \frac{d\zeta}{\sqrt{(1-\zeta^2)(1-k^2\zeta^2)}}. \qquad (14.2.1)$$

The parameter k is called the *modulus*. For convenience we assume that $0 < k < 1$; the various formulas to follow extend by analytic continuation to all values $k \neq \pm 1$.

Since the integrand is multiple-valued, we shall consider the integral to be taken over paths in the Riemann surface of the function $\sqrt{(1-z^2)(1-k^2z^2)}$.

This surface can be visualized as two copies of the complex plane with slits on the intervals $[-1/k, -1]$ and $[1, 1/k]$, the two copies being joined across the slits. In fact, we adjoin the point at infinity to each copy of the plane. The result is two copies of the Riemann sphere joined across the slits. The resulting surface is a torus.

The function $\sqrt{(1-z^2)(1-k^2z^2)}$ is single-valued and holomorphic on each slit sphere; we take the value at $z = 0$ to be 1 on the "upper" sphere and -1 on the "lower" sphere. The function F is multiple-valued, the value depending on the path taken in the Riemann surface. The integral converges as the upper limit goes to infinity, so we also allow curves that pass through infinity in one or both spheres.

Up to homotopy (continuous deformation on the Riemann surface), two paths between the same two points can differ by a certain number of circuits from -1 to 1 through the upper (resp. lower) sphere and back to -1 through the lower (resp. upper) sphere, or by a certain number of circuits around one or both of the slits. The value of the integral does not change under homotopy. Taking symmetries into account, this means that determinations of F can differ by integer multiples of $4K$, where

$$K = \int_0^1 \frac{dz}{\sqrt{(1-z^2)(1-k^2z^2)}}, \tag{14.2.2}$$

or by integer multiples of $2iK'$, where

$$K' = \int_1^{1/k} \frac{dz}{\sqrt{(z^2-1)(1-k^2z^2)}}. \tag{14.2.3}$$

Here the integrals are taken along line segments in the upper sphere, so that K and K' are positive. In fact $4K$ corresponds to integrating from -1 to 1 in the upper sphere and returning to -1 in the lower sphere, while $2iK'$ corresponds to integrating from 1 to $1/k$ along the upper edge of the slit in the upper sphere and back along the lower edge of the slit.

The change of variables

$$z = \frac{1}{k}\sqrt{1-k'^2\zeta^2}$$

in (14.2.3) shows that the constants K' and K are related by

$$K' = \int_0^1 \frac{d\zeta}{\sqrt{(1-\zeta^2)(1-k'^2\zeta^2)}} = K(k'),$$

where $k' = \sqrt{1-k^2}$ is called the *complementary modulus*.

Because of these considerations, we consider values of F to be determined only up to the addition of elements of the *period lattice*

$$\Lambda = \{4mK + 2inK', \quad m,n = 0, \pm 1, \pm 2, \dots\}.$$

It follows from the definitions that

$$F(k,1) = K, \quad F(k,1/k) = K + iK'.$$

Integrating along line segments from 0 in the upper sphere shows that

$$F(k,-z) = -F(k,z).$$

Integrating from 0 in the upper sphere to 0 in the lower sphere and then to z in the lower sphere shows that

$$F(k,z-) = 2K - F(k,z+), \tag{14.2.4}$$

where $z+$ and $z-$ refer to representatives of z in the upper and lower spheres, respectively.

Integrating from 0 to $1/k$ in the upper sphere and then to 0 in the lower sphere and from there to z in the lower sphere shows that

$$F(k,z-) = 2K + 2iK' - F(k,z+). \tag{14.2.5}$$

Integrating along the positive imaginary axis in the upper sphere gives

$$F(k,\infty) = i \int_0^\infty \frac{ds}{\sqrt{(1+s^2)(1+k^2s^2)}}. \tag{14.2.6}$$

The change of variables

$$s = \frac{\zeta}{\sqrt{1-\zeta^2}}$$

in the integral in (14.2.6) shows that

$$F(k,\infty) = iK(k') = iK'. \tag{14.2.7}$$

Three classical transformations of F can be accomplished by changes of variables. Let $k_1 = (1 - k')/(1 + k')$. The change of variables

$$\zeta = \varphi(t) \equiv (1 + k')t \sqrt{\frac{1 - t^2}{1 - k^2 t^2}}$$

in the integrand for $F(k_1, \cdot)$ leads to the identity

$$F(k_1, \varphi(z)) = \int_0^{\varphi(z)} \frac{d\zeta}{\sqrt{(1 - \zeta^2)(1 - k_1^2 \zeta^2)}}$$

$$= (1 + k') \int_0^z \frac{dt}{\sqrt{(1 - t^2)(1 - k^2 t^2)}}.$$

This is *Landen's transformation* [240]:

$$F(k_1, z_1) = (1 + k') F(k, z); \tag{14.2.8}$$

$$k_1 = \frac{1 - k'}{1 + k'}, \quad z_1 = (1 + k')z \sqrt{\frac{1 - z^2}{1 - k^2 z^2}}.$$

Now take $k_1 = 2\sqrt{k}/(1 + k)$. Then the change of variables

$$\zeta = \varphi(t) = \frac{(1 + k)t}{1 + kt^2}$$

in the integrand for $F(k_1, \cdot)$ leads to the identity

$$F(k_1, z_1) = (1 + k) F(k, z); \qquad k_1 = \frac{2\sqrt{k}}{1 + k}, \quad z_1 = \frac{(1 + k)z}{1 + kz^2}. \tag{14.2.9}$$

This is known as *Gauss's transformation* [152], or the *descending Landen transformation*.

These two transformations lend themselves to computation. To use (14.2.9), let $k_2 = 2\sqrt{k_1}/(1 + k_1)$; then

$$k_2' = (1 - k_1)/(1 + k_1) = \frac{(k_1')^2}{(1 + k_1)^2}.$$

Continuing, let $k_{n+1} = 2\sqrt{k_n}/(1 + k_n)$. So long as $0 < k_1 < 1$, it follows that the sequence $\{k_n'\}$ decreases rapidly to zero, so $k_n \to 1$. If, for example, $|z_1| < 1$, the corresponding sequence z_n will converge rapidly to a limit Z, giving

$$F(k_1, z_1) = \prod_{n=1}^{\infty} (1 + k_n)^{-1} F(1, Z),$$

where

$$F(1, z) \equiv \lim_{k \to 1} F(k, z) = \tanh^{-1} z. \tag{14.2.10}$$

Similar considerations apply to (14.2.8) and lead to an evaluation involving

$$F(0, z) \equiv \lim_{k \to 0} F(k, z) = \sin^{-1} z. \tag{14.2.11}$$

The change of variables

$$\zeta = \frac{is}{\sqrt{1 - s^2}}$$

in the integrand for $F(k, \cdot)$ leads to the identity

$$F(k, ix) = i \int_0^{x/\sqrt{1+x^2}} \frac{ds}{\sqrt{(1-s^2)(1-k'^2 s^2)}}.$$

This is *Jacobi's imaginary transformation*:

$$F(k, ix) = iF\left(k', \frac{x}{\sqrt{1+x^2}}\right),$$

or equivalently,

$$iF(k, y) = F\left(k', \frac{iy}{\sqrt{1-y^2}}\right). \qquad (14.2.12)$$

Our final result concerning the function F is Euler's *addition formula* [122]:

$$F(k, x) + F(k, y) = F\left(k, \frac{x\sqrt{P(y)} + y\sqrt{P(x)}}{1 - k^2 x^2 y^2}\right), \qquad (14.2.13)$$

where as before $P(x) = (1 - x^2)(1 - k^2 x^2)$. To prove (14.2.13), it is enough to assume that $0 < x, y < 1$ and that we have chosen a parameterized curve $\{(x(t), y(t))\}$, with $y(0) = 0$ and $x(T) = x$, $y(T) = y$, such that

$$F(k, x(t)) + F(k, y(t)) = F(k, C_1), \quad \text{constant}. \qquad (14.2.14)$$

Differentiating with respect to t gives

$$\frac{x'}{\sqrt{P(x)}} + \frac{y'}{\sqrt{P(y)}} = 0.$$

(We are abusing notation and writing x and y for $x(t)$ and $y(t)$.) We follow an argument of Darboux. After reparameterizing, we may assume

$$x'(t) = \sqrt{P(x(t))}, \quad y'(t) = -\sqrt{P(y(t))}. \qquad (14.2.15)$$

Often it is helpful to consider the Wronskian $W = xy' - yx'$. Differentiating (14.2.15) gives x'' and y'' as functions of x and y and leads to

$$W' = xy'' - yx'' = 2k^2 xy(y^2 - x^2).$$

Using (14.2.15) again,

$$W(xy)' = (xy')^2 - (yx')^2 = (x^2 - y^2)(1 - k^2 x^2 y^2).$$

Thus

$$\frac{W'}{W} = -\frac{2k^2 (xy)(xy)'}{1 - k^2 x^2 y^2} = \left[\log(1 - k^2 x^2 y^2)\right]',$$

so

$$\frac{x(t)\sqrt{P(y(t))} + y(t)\sqrt{P(x(t))}}{1 - k^2 x(t)^2 y(t)^2} = C_2, \quad \text{constant}. \qquad (14.2.16)$$

Taking $t = 0$ in (14.2.14) and (14.2.16) shows that

$$C_1 = x(0) = C_2.$$

This proves (14.2.13). The case $x = y$ is known as *Fagnano's duplication formula* [128]. Fagnano's formula inspired Euler to find the full addition formula; see Ayoub [25] and D'Antonio [92].

The definite integrals over the unit interval

$$K(k) = \int_0^1 \frac{dt}{\sqrt{(1-t^2)(1-k^2t^2)}}, \qquad E(k) = \int_0^1 \frac{\sqrt{1-k^2t^2}\,dt}{\sqrt{1-t^2}}$$

are known as the *complete elliptic integrals of the first and second kind*, respectively. They can be expressed as the hypergeometric functions

$$K(k) = \frac{\pi}{2} F\left(\frac{1}{2},\frac{1}{2},1;k^2\right), \qquad E(k) = \frac{\pi}{2} F\left(-\frac{1}{2},\frac{1}{2},1;k^2\right); \qquad (14.2.17)$$

see Exercise 7.

14.3 Jacobi elliptic functions

The integral defining F is analogous to the simpler integral

$$\int_0^x \frac{dt}{\sqrt{1-t^2}}.$$

This is also multiple-valued: values differ by integer multiples of 2π. One obtains a single-valued entire function by taking the inverse

$$u = \int_0^{\sin u} \frac{dt}{\sqrt{1-t^2}}.$$

Jacobi defined the function $\mathrm{sn}\, u = \mathrm{sn}(u,k)$ by

$$u = \int_0^{\mathrm{sn}\, u} \frac{d\zeta}{\sqrt{(1-\zeta^2)(1-k^2\zeta^2)}}. \qquad (14.3.1)$$

It follows from the discussion in Section 14.2 that sn is *doubly periodic*, with a real period $4K$ and a period $2iK'$:

$$\mathrm{sn}(u+4K) = \mathrm{sn}(u+2iK') = \mathrm{sn}\, u.$$

Moreover, sn is odd: $\mathrm{sn}(-u) = \mathrm{sn}\, u$. It follows that sn is odd around $z = 2K$ and around $z = iK'$. The identity (14.2.4) implies that sn is even around $z = K$. It follows from this, in turn, that sn is even around $z = K + iK'$. Note that a function f is even (resp. odd) around a point $z = a$ if and only if $f(z + 2a) = f(-z)$ (resp. $f(z+2a) = -f(-z)$).

In summary,

$$\mathrm{sn}u = \mathrm{sn}(u+4K) = \mathrm{sn}(u+2iK')$$
$$= \mathrm{sn}(2K-u) = \mathrm{sn}(2K+2iK'-u)$$
$$= -\mathrm{sn}(-u) = -\mathrm{sn}(4K-u) = -\mathrm{sn}(2iK'-u). \qquad (14.3.2)$$

Because of the periodicity, it is enough to compute values of $\mathrm{sn}u$ for u in the *period rectangle*

$$\Pi = \{u \mid 0 \le \mathrm{Re}\,u < 4K, \, 0 \le \mathrm{Im}\,u < 2K'\}. \qquad (14.3.3)$$

The various calculations of values of F give

$$\mathrm{sn}(0) = \mathrm{sn}(2K) = 0;$$
$$\mathrm{sn}(K) = -\mathrm{sn}(3K) = 1;$$
$$\mathrm{sn}(iK') = -\mathrm{sn}(2K+iK') = \infty; \qquad (14.3.4)$$
$$\mathrm{sn}(K+iK') = -\mathrm{sn}(3K+iK') = k^{-1}.$$

It follows from (14.2.6) and (14.2.7) that as $t \to +\infty$,

$$iK' - F(it) \sim i\int_t^\infty \frac{ds}{ks^2} = \frac{i}{kt},$$

or, setting $\varepsilon = -1/kt$,

$$\mathrm{sn}(iK'+i\varepsilon) \sim \frac{1}{ik\varepsilon}.$$

Therefore sn has a simple pole at $u = iK'$ with residue $1/k$. Consequently, it also has a simple pole at $u = 2K + iK'$ with residue $-1/k$.

Differentiating (14.3.1) gives

$$\mathrm{sn}'u = \sqrt{1-\mathrm{sn}^2u}\sqrt{1-k^2\mathrm{sn}^2u}. \qquad (14.3.5)$$

This leads naturally to the introduction of the two related functions

$$\mathrm{cn}u = \sqrt{1-\mathrm{sn}^2u},$$
$$\qquad (14.3.6)$$
$$\mathrm{dn}u = \sqrt{1-k^2\mathrm{sn}^2u}.$$

The three functions sn, cn, and dn are the *Jacobi elliptic functions*.

The only zeros of $1-\mathrm{sn}^2$ (resp. $1-k^2\mathrm{sn}^2$) in the period rectangle (14.3.3) are at K and $3K$ (resp. $K+iK'$ and $3K+iK'$). These are double zeros, so we may choose branches of the square roots that are holomorphic near these points. We choose the branches with value 1 at $u = 0$. The resulting functions are, like sn itself, meromorphic in the complex plane.

Since sn is even or odd around each of the points 0, K, $2K$, iK', and $K+iK'$, it follows that the functions cn and dn are each even or odd around each of these points. Since neither function vanishes at $u = 0$ or $u = 2K$, they are even around 0 and $2K$. Similarly, cn is even around $K + iK'$ and dn is even around K. Since cn has a simple zero at $u = K$ and a simple pole at $u = iK'$, it is odd around these points. Similarly, dn is odd around iK' and $K+iK'$. It follows that cn has periods $4K$ and $2K+2iK'$, while dn has periods $2K$ and $4iK'$. Therefore cn is even around $u = 2K$ and dn is even around $u = 2iK'$. (See Exercises 10 and 11.)

Combining these observations with the computations (14.3.4), we obtain the following:

$$\mathrm{cn} = \mathrm{cn}(u + 4K) = \mathrm{cn}(u + 2K + 2iK')$$

$$= \mathrm{cn}(-u) = -\mathrm{cn}(2K - u) = -\mathrm{cn}(2iK' - u);$$

$$\mathrm{dn} = \mathrm{dn}(u + 2K) = \mathrm{dn}(u + 4iK') = \mathrm{dn}(2K - u) \qquad (14.3.7)$$

$$= \mathrm{dn}(-u) = -\mathrm{dn}(2iK' - u) = -\mathrm{dn}(2K + 2iK' - u).$$

We have established most of the values in the table

	0	K	$2K$	$3K$	iK'	$K+iK'$	$2K+iK'$	$3K+iK'$
sn	0	1	0	-1	∞	k^{-1}	∞	$-k^{-1}$
cn	1	0	-1	0	∞	$-ik'k^{-1}$	∞	$ik'k^{-1}$
dn	1	k'	1	k'	∞	0	∞	0

It follows from (14.3.5) and (14.3.6) that

$$\mathrm{sn}'(u) = \mathrm{cn}\,u\,\mathrm{dn}\,u;$$

$$\mathrm{cn}'(u) = -\mathrm{sn}\,u\,\mathrm{dn}\,u; \qquad (14.3.8)$$

$$\mathrm{dn}'(u) = -k^2\mathrm{sn}\,u\,\mathrm{cn}\,u.$$

Since

$$\mathrm{sn}\,u = \sqrt{1 - \mathrm{cn}^2 u} = \frac{1}{k}\sqrt{1 - \mathrm{dn}^2 u}$$

and

$$\mathrm{dn}\,u = \sqrt{1 - k^2 + k^2\mathrm{cn}^2 u}; \qquad \mathrm{cn}\,u = \frac{1}{k}\sqrt{k^2 - 1 + \mathrm{dn}^2 u},$$

it follows that cn and dn can be defined implicitly by the integrals

$$u = \int_{\mathrm{cn}u}^{1} \frac{d\zeta}{\sqrt{(1-\zeta^2)(1-k^2+k^2\zeta^2)}};$$

$$(14.3.9)$$

$$u = \int_{\mathrm{dn}u}^{1} \frac{d\zeta}{\sqrt{(1-\zeta^2)(\zeta^2-k'^2)}}.$$

From the point of view of the Jacobi elliptic functions, the Landen transformation, the Gauss transformation, Jacobi's imaginary transformation, and the addition formula take the following forms. The identity (14.2.8) is equivalent to

$$\mathrm{sn}\left([1+k']u,k_1\right) = (1+k')\frac{\mathrm{sn}(u,k)\,\mathrm{cn}(u,k)}{\mathrm{dn}(u,k)}; \quad k_1 = \frac{1-k'}{1+k'}. \quad (14.3.10)$$

It follows that

$$\mathrm{cn}\left([1+k']u,k_1\right) = (1+k')\frac{\mathrm{dn}^2(u,k)-k'}{k^2\,\mathrm{dn}(u,k)};$$

$$(14.3.11)$$

$$\mathrm{dn}\left([1+k']u,k_1\right) = (1-k')\frac{\mathrm{dn}^2(u,k)+k'}{k^2\,\mathrm{dn}(u,k)}.$$

The identity (14.2.9) is equivalent to

$$\mathrm{sn}\left([1+k]u,k_1\right) = \frac{(1+k)\mathrm{sn}(u,k)}{1+k\,\mathrm{sn}^2(u,k)}; \quad k_1 = \frac{2\sqrt{k}}{1+k}. \quad (14.3.12)$$

It follows that

$$\mathrm{cn}\left([1+k]u,k_1\right) = \frac{\mathrm{cn}(u,k)\,\mathrm{dn}(u,k)}{1+k\,\mathrm{sn}^2(u,k)};$$

$$(14.3.13)$$

$$\mathrm{dn}\left([1+k]u,k_1\right) = \frac{1-k\,\mathrm{sn}^2(u,k)}{1+k\,\mathrm{sn}^2(u,k)}.$$

The identity (14.2.12) is equivalent to

$$\mathrm{sn}(iu,k) = i\frac{\mathrm{sn}(u,k')}{\mathrm{cn}(u,k')}. \quad (14.3.14)$$

It follows that

$$\mathrm{cn}(iu,k) = \frac{1}{\mathrm{cn}(u,k')};$$

$$(14.3.15)$$

$$\mathrm{dn}(iu,k) = \frac{\mathrm{dn}(u,k')}{\mathrm{cn}(u.k')}.$$

The addition formula (14.2.13) is equivalent to

$$\operatorname{sn}(u+v) = \frac{\operatorname{sn} u \operatorname{cn} v \operatorname{dn} v + \operatorname{sn} v \operatorname{cn} u \operatorname{dn} u}{1 - k^2 \operatorname{sn}^2 u \operatorname{sn}^2 v}. \tag{14.3.16}$$

It follows that

$$\operatorname{cn}(u+v) = \frac{\operatorname{cn} u \operatorname{cn} v - \operatorname{sn} u \operatorname{sn} v \operatorname{dn} u \operatorname{dn} v}{1 - k^2 \operatorname{sn}^2 u \operatorname{sn}^2 v};$$

$$\tag{14.3.17}$$

$$\operatorname{dn}(u+v) = \frac{\operatorname{dn} u \operatorname{dn} v - k^2 \operatorname{sn} u \operatorname{sn} v \operatorname{cn} u \operatorname{cn} v}{1 - k^2 \operatorname{sn}^2 u \operatorname{sn}^2 v}.$$

These formulas imply the product formulas

$$\operatorname{sn}(u+v)\operatorname{sn}(u-v) = \frac{\operatorname{sn}^2 u - \operatorname{sn}^2 v}{1 - k^2 \operatorname{sn}^2 u \operatorname{sn}^2 v};$$

$$\tag{14.3.18}$$

$$\operatorname{cn}(u+v)\operatorname{cn}(u-v) = \frac{1 - \operatorname{sn}^2 u - \operatorname{sn}^2 v + k^2 \operatorname{sn}^2 u \operatorname{sn}^2 v}{1 - k^2 \operatorname{sn}^2 u \operatorname{sn}^2 v}.$$

A commonly used notation for reciprocals and quotients of the Jacobi elliptic functions is due to Glaisher [165]:

$$\operatorname{ns} = \frac{1}{\operatorname{sn}}, \quad \operatorname{nc} = \frac{1}{\operatorname{cn}}, \quad \operatorname{nd} = \frac{1}{\operatorname{dn}};$$

$$\operatorname{sc} = \frac{\operatorname{sn}}{\operatorname{cn}}, \quad \operatorname{sd} = \frac{\operatorname{sn}}{\operatorname{dn}}, \quad \operatorname{cd} = \frac{\operatorname{cn}}{\operatorname{dn}};$$

$$\operatorname{cs} = \frac{\operatorname{cn}}{\operatorname{sn}}, \quad \operatorname{ds} = \frac{\operatorname{dn}}{\operatorname{sn}}, \quad \operatorname{dc} = \frac{\operatorname{dn}}{\operatorname{cn}}.$$

To complete this section, we note that the change of variables $\zeta = \operatorname{sn} s$ converts the elliptic integrals of the first, second, and third kind, (14.1.4), (14.1.5), and (14.1.6), to

$$F(z) = \int_0^{\operatorname{sn}^{-1} z} ds = \operatorname{sn}^{-1} z;$$

$$E(z) = \int_0^{\operatorname{sn}^{-1} z} (1 - k^2 \operatorname{sn}^2 s)\, ds = \int_0^{\operatorname{sn}^{-1} z} \operatorname{dn}^2(s)\, ds;$$

$$\Pi(a,z) = \int_0^{\operatorname{sn}^{-1} z} \frac{ds}{1 + a \operatorname{sn}^2 s}.$$

14.4 Theta functions

The Jacobi elliptic functions are examples of the general notion of an *elliptic function* – a function f that is meromorphic in the complex plane and doubly periodic:

$$f(z) = f(z + 2\omega_1) = f(z + 2\omega_2),$$

with periods $2\omega_j \neq 0$, such that ω_2/ω_1 is not real. Such a function is determined by its values in any *period parallelogram*

$$\Pi_a = \{z \mid z = a + 2s\omega_1 + 2t\omega_2, \ 0 \leq s, t < 1\}.$$

If f is entire, then it is bounded on a period parallelogram, therefore bounded on the plane, and therefore constant, by Liouville's theorem. Otherwise it has at least two poles, counting multiplicity, in each Π_a. To see this, note first that by changing a slightly we may assume that there is no pole on the boundary. Periodicity implies that the integral over the boundary C_a of Π_a vanishes:

$$\int_{C_a} f(z) \, dz = 0.$$

Therefore the sum of the residues is zero, so there are at least two simple poles or one multiple pole.

A nonconstant elliptic function takes each complex value the same number of times in each Π_a. To see this, suppose that f does not take the value c on the boundary and has no poles on the boundary. Again, periodicity implies that

$$\frac{1}{2\pi i} \int_{C_a} \frac{f'(z)}{f(z) - c} \, dz = 0.$$

However, the integral is equal to the number of times (counting multiplicity) that f takes the value c in Π_a, minus the number of poles (counting multiplicity) of f in Π_a. By continuity, this number is independent of c, and by varying a we may ensure that any given value c is not taken on the boundary. The Jacobi elliptic functions illustrate this. For example, in the period rectangle (14.3.3) sn takes the value 0 twice, takes the values 1 and -1 once each but with multiplicity 2, and has two simple poles ($c = \infty$).

One consequence of the Weierstrass factorization theorem is that any function meromorphic in the plane is a quotient of entire functions. Since a doubly periodic entire function is constant, a nonconstant elliptic function cannot be the quotient of doubly periodic entire functions. However, it can be expressed as a quotient of entire functions that are "nearly" periodic. The basic such function is a *theta function*.

Up to a linear transformation of the independent variable, we may consider the periods of a doubly periodic function to be 1 and τ, with $\text{Im } \tau > 0$. Thus the basic period parallelogram is

$$\Pi = \{z \mid z = s + t\tau,\ 0 \le s, t < 1\}, \tag{14.4.1}$$

with oriented boundary Γ. Following Jacobi, we look for an entire function Θ that has period 1 and comes close to having period τ:

$$\Theta(z+1) = \Theta(z); \quad \Theta(z+\tau) = a(z)\,\Theta(z),$$

where a is entire, nonzero, and has period 1. This amounts to requiring that a be a constant times an integer power of $e^{2i\pi z}$. If Θ has no zeros on Γ, the number of zeros in Π is

$$\frac{1}{2\pi i}\int_\Gamma \frac{\Theta'(\zeta)}{\Theta(\zeta)}\,d\zeta = \frac{1}{2\pi i}\left\{\int_0^1 + \int_1^{1+\tau} + \int_{1+\tau}^\tau + \int_\tau^0\right\}\frac{\Theta'(\zeta)}{\Theta(\zeta)}\,d\zeta$$

$$= \frac{1}{2\pi i}\left\{\int_0^1 - \int_\tau^{1+\tau} + \int_1^{1+\tau} - \int_0^\tau\right\}\frac{\Theta'(\zeta)}{\Theta(\zeta)}\,d\zeta$$

$$= -\frac{1}{2\pi i}\int_0^1 \frac{a'(s)}{a(s)}\,ds.$$

Thus the simplest choice is $a(z) = c\,e^{-2\pi i z}$, which implies a single zero in each period parallelogram. With this choice we would have

$$\Theta(z+1) = \Theta(z); \quad \Theta(z+\tau) = c\,e^{-2i\pi z}\,\Theta(z). \tag{14.4.2}$$

To construct such a function, we note that for Θ to have period 1, it must have the form

$$\Theta(z) = \sum_{n=-\infty}^{\infty} a_n\,p(z)^{2n}, \quad p(z) = e^{i\pi z}.$$

Now $p(z+\tau) = q\,p(z)$, where $q = q(\tau) = e^{i\pi\tau}$, so the second equation in (14.4.2) implies

$$a_n\,q^{2n} = c\,a_{n+1}. \tag{14.4.3}$$

Taking $c = -1$ and $a_0 = 1$, we find that $\Theta(z)$ should be given by

$$\Theta(z) = \sum_{n=-\infty}^{\infty} (-1)^n\,p^{2n}\,q^{n(n-1)}, \quad p = e^{i\pi z},\ q = e^{i\pi\tau}. \tag{14.4.4}$$

The assumption $\operatorname{Im}\tau > 0$ implies that this series converges very rapidly, uniformly on bounded sets. Therefore Θ is an entire function and

$$\Theta(z+1) = \Theta(z); \quad \Theta(z+\tau) = -e^{-2\pi i z}\Theta(z). \tag{14.4.5}$$

It follows from (14.4.4) that

$$\Theta(0) = \sum_{n=-\infty}^{\infty} (-1)^n\,q^{n(n-1)} \tag{14.4.6}$$

and replacing n by $-m$ for $n \leq 0$ shows that $\Theta(0) = 0$. By construction, Θ has no other zeros in Π. The properties (14.4.5) and $\Theta(0) = 0$ characterize Θ up to a constant: if Ψ were another such entire function, then Ψ/Θ would be a doubly periodic entire function, thus constant.

Suppose that an elliptic function f with periods 1 and τ has zeros $\{a_1, a_2, \ldots, a_k\}$ and poles $\{b_1, b_2, \ldots, b_k\}$ in Π, repeated according to multiplicity. If any lie on the boundary of Π, we may translate slightly so that they lie in the interior. The residue theorem gives

$$\frac{1}{2\pi i} \int_\Gamma \frac{z f'(z)}{f(z)} \, dz = (a_1 + a_2 + \cdots + a_k) - (b_1 + b_2 + \cdots + b_k).$$

Because of periodicity,

$$\frac{1}{i} \int_\Gamma \frac{z f'(z)}{f(z)} \, dz = \left[\frac{1}{i} \int_1^{1+\tau} \frac{f'(z) \, dz}{f(z)} \right] + \tau \left[\frac{1}{i} \int_{1+\tau}^{\tau} \frac{f'(z) \, dz}{f(z)} \right].$$

Each integral in brackets is the change in the argument of f along a segment. Periodicity implies that the change is an integer multiple of 2π. Therefore

$$(a_1 + a_2 + \cdots + a_k) - (b_1 + b_2 + \cdots + b_k) \in \Lambda, \tag{14.4.7}$$

where Λ is the period lattice

$$\Lambda = \{m + n\tau, \quad m, n = 0, \pm 1, \pm 2, \ldots \}.$$

Conversely, suppose that the disjoint sets of points $\{a_j\}$ and $\{b_j\}$ in Π satisfy condition (14.4.7). Then there is an elliptic function with precisely these zeros and poles in Π (Abel [1]) and it can be represented as an exponential times a quotient of translates of Θ (Jacobi [198]). In fact, let

$$(a_1 + a_2 + \cdots + a_k) - (b_1 + b_2 + \cdots + b_k) = m + n\tau.$$

(Since the a_j and b_j belong to Π, this condition implies that $k > 1$.) Then

$$f(z) = e^{-2n\pi i z} \frac{\Theta(z - a_1)\Theta(z - a_2) \cdots \Theta(z - a_k)}{\Theta(z - b_1)\Theta(z - b_2) \cdots \Theta(z - b_k)} \tag{14.4.8}$$

is the desired function. It is unique up to a constant factor.

In addition to quotients like (14.4.8), it is convenient to represent elliptic functions by using the functions

$$Z(z) = \frac{\Theta'(z)}{\Theta(z)}, \quad Z'(z) = \frac{\Theta''(z)}{\Theta(z)} - Z(z)^2.$$

It follows from (14.4.5) that

$$Z(z + \tau) = Z(z) - 2\pi i,$$

so Z' and linear combinations of the translates

$$c_1 Z(z - b_1) + c_2 Z(z - b_2) + \cdots + c_n Z(z - b_n), \quad c_1 + c_2 + \cdots + c_n = 0$$

have period τ and thus are elliptic functions. Since these are derivatives, they can be integrated immediately (in terms of functions of Θ and its translates and derivatives). Note that Z has a simple pole at each lattice point, while for $m \geq 1$ the derivative $Z^{(m)}$ is an elliptic function with a pole of order $m + 1$ at each lattice point.

This leads to an integration procedure for any elliptic function f. We may suppose that f has periods 1 and τ. If f has a pole of order $k \geq 2$ at $z = b$, then there is a constant c such that $f(z) - c Z^{(k-1)}(z - b)$ has a pole of order $< k$ at $z = b$. Thus the integration problem can be reduced to the case of functions f that have only simple poles. Suppose that the poles in Π are $b_1, \ldots, b_n$, with residues $\beta_1, \ldots, \beta_n$. We know that $\sum \beta_j = 0$, so there is a linear combination of translates of Z that has the same poles and residues in Π as f. It follows that the difference between f and this linear combination of translates of Z is constant.

The function Θ has exactly one zero in the period parallelogram Π, at $z = 0$. This fact and the properties (14.4.5) imply that the zeros of Θ are precisely the points of the lattice Λ. It follows that

$$\Theta\left(z + \frac{1}{2}\tau\right) = 0$$

if and only if $z = m + (n - \frac{1}{2})\tau$ for some integers m and n, or equivalently, $p^2 q^{2n-1} = 1$ for some integer n. The product

$$\prod_{n=1}^{\infty} (1 - p^2 q^{2n-1})(1 - p^{-2} q^{2n-1}), \quad p = p(z),$$

converges for all z and has the same zeros as $\Theta\left(z + \frac{1}{2}\tau\right)$, so

$$\Theta\left(z + \frac{1}{2}\tau\right) = c(z,\tau) \prod_{n=1}^{\infty} (1 - p^2 q^{2n-1})(1 - p^{-2} q^{2n-1}), \qquad (14.4.9)$$

where $c(z, \tau)$ is an entire function of z. It can be shown that $c(z, \tau) = G(\tau)$ is independent of z and can be evaluated as a product involving powers of $q = e^{i\pi\tau}$ (see Exercises 28–34). The result is one version of Jacobi's *triple product formula*,

$$\Theta\left(z + \frac{1}{2}\tau\right) = \prod_{n=1}^{\infty} (1 - q^{2n})(1 - p^2 q^{2n-1})(1 - p^{-2} q^{2n-1}). \qquad (14.4.10)$$

This implies another version,

$$\Theta(z) = \prod_{n=1}^{\infty} (1 - q^{2n})(1 - p^2 q^{2n-2})(1 - p^{-2} q^{2n}). \tag{14.4.11}$$

14.5 Jacobi theta functions and integration

According to the results of the preceding section, the Jacobi elliptic functions can be expressed as quotients of translates of Θ, after a linear change of variables. As we shall see, it is convenient for this purpose to introduce the *Jacobi theta functions*. These are normalized versions of Θ and of translations of Θ by half-periods:

$$\theta_1(z) = i \frac{q^{\frac{1}{4}}}{p} \Theta(z) = i \sum_{n=-\infty}^{\infty} (-1)^n p^{2n-1} q^{(n-\frac{1}{2})^2};$$

$$\theta_2(z) = \frac{q^{\frac{1}{4}}}{p} \Theta\left(z + \frac{1}{2}\right) = \sum_{n=-\infty}^{\infty} p^{2n-1} q^{(n-\frac{1}{2})^2};$$

$$\tag{14.5.1}$$

$$\theta_3(z) = \Theta\left(z + \frac{1}{2} + \frac{1}{2}\tau\right) = \sum_{n=-\infty}^{\infty} p^{2n} q^{n^2};$$

$$\theta_4(z) = \Theta\left(z + \frac{1}{2}\tau\right) = \sum_{n=-\infty}^{\infty} (-1)^n p^{2n} q^{n^2}.$$

(There are various other notations and normalizations; see Whittaker and Watson [435].) Note that because of the factor p^{-1}, θ_1 and θ_2 are periodic with period 2, not 1. Also, θ_1 is an odd function of z, while θ_2, θ_3, and θ_4 are even functions of z.

The triple product formula (14.4.10) implies corresponding formulas for the θ_j:

$$\theta_1(z) = 2q^{\frac{1}{4}} \sin(\pi z) \prod_{n=1}^{\infty} (1 - q^{2n})(1 - p^2 q^{2n})(1 - p^{-2} q^{2n});$$

$$\theta_2(z) = 2q^{\frac{1}{4}} \cos(\pi z) \prod_{n=1}^{\infty} (1 - q^{2n})(1 + p^2 q^{2n})(1 + p^{-2} q^{2n});$$

$$\theta_3(z) = \prod_{n=1}^{\infty} (1 - q^{2n})(1 + p^2 q^{2n-1})(1 + p^{-2} q^{2n-1});$$

$$\theta_4(z) = \prod_{n=1}^{\infty} (1 - q^{2n})(1 - p^2 q^{2n-1})(1 - p^{-2} q^{2n-1}).$$

The identities (14.4.4) and (14.5.1) lead to the following table of values:

	0	$\tfrac{1}{2}$	$\tfrac{1}{2}\tau$	$\tfrac{1}{2}+\tfrac{1}{2}\tau$
θ_1	0	$\sum q^{(n-\frac{1}{2})^2}$	$iq^{-\frac{1}{4}}\sum(-1)^n q^{n^2}$	$q^{-\frac{1}{4}}\sum q^{n^2}$
θ_2	$\sum q^{(n-\frac{1}{2})^2}$	0	$q^{-1/4}\sum q^{n^2}$	$-iq^{-\frac{1}{4}}\sum(-1)^n q^{n^2}$
θ_3	$\sum q^{n^2}$	$\sum(-1)^n q^{n^2}$	$q^{-\frac{1}{4}}\sum q^{(n+\frac{1}{2})^2}$	0
θ_4	$\sum(-1)^n q^{n^2}$	$\sum q^{n^2}$	0	$q^{-\frac{1}{4}}\sum q^{(n-\frac{1}{2})^2}$

Consider now the Jacobi elliptic functions with modulus k, and let $\tau = iK'/K$. The function

$$\frac{\theta_1(u/2K)}{\theta_4(u/2K)}$$

is meromorphic as a function of u with simple zeros at $u = 0$ and $u = 2K$ and simple poles at $u = iK'$ and $u = 2K + iK'$, and has periods $4K$ and $2iK'$. It follows that it is a multiple of $\operatorname{sn}u$, and conversely,

$$\operatorname{sn}u = C\frac{\theta_1(u/2K)}{\theta_4(u/2K)}.$$

Each side of this equation may be evaluated at $u = K$, and this determines the constant C. Similar arguments apply to cn and dn. The results are

$$\operatorname{sn} = \frac{1}{\sqrt{k}}\frac{\theta_1(u/2K)}{\theta_4(u/2K)};$$

$$\operatorname{cn}u = \sqrt{\frac{k'}{k}}\frac{\theta_2(u/2K)}{\theta_4(u/2K)};$$
(14.5.2)

$$\operatorname{dn}u = \sqrt{k'}\frac{\theta_3(u/2K)}{\theta_4(u/2K)}.$$

Jacobi obtained a large number of formulas relating products of translations of theta functions, having a form like

$$\Theta(z+w)\Theta(z-w) = c_1\Theta(z+a_1)\Theta(z+a_2)\Theta(w+a_3)\Theta(w+a_4)$$
$$+ c_2\Theta(w+a_1)\Theta(w+a_2)\Theta(z+a_3)\Theta(z+a_4),$$

where the a_j belong to the set $\{0, \tfrac{1}{2}, \tfrac{1}{2}\tau, \tfrac{1}{2}+\tfrac{1}{2}\tau\}$. The constants are chosen so that the quotient of the right side by the left side is a doubly periodic function of z, for any given w, and the zeros of the denominator are cancelled by the zeros of the numerator. Extensive lists are given in Whittaker and Watson [435] and in Rainville [339]. (There are various other notations and normalizations; see Whittaker and Watson [435].)

A deeper result is Jacobi's remarkable identity

$$\theta_1' = \pi\,\theta_2\theta_3\theta_4. \tag{14.5.3}$$

(The factor π is due to the normalization we have chosen here.) See [435, 16].

The normalizations of the Jacobi theta functions have the consequence that each one has the form

$$\theta(z,\tau) = \sum_{n=-\infty}^{\infty} a_n\,e^{2(n+c)i\pi z}\,e^{(n+c)^2 i\pi\tau}$$

for some value of c, and therefore is a solution of the partial differential equation

$$\theta_{zz}(z,\tau) = 4\pi\,i\theta_\tau(z,\tau).$$

If we take as variables $x = z$ and $t = -i\tau/4\pi$, this is the heat equation

$$\psi_t = \psi_{xx}. \tag{14.5.4}$$

The theta functions are periodic in x. Periodic solutions of the heat equation can be obtained in two different ways, a fact that provides an approach to obtaining Jacobi's imaginary transformation (14.2.12) or (14.3.14), (14.3.15) in terms of theta functions.

The fundamental solution for the heat equation on the line, i.e. the solution ψ of (14.5.4) that has the property that

$$\lim_{t\to 0+} \int_{-\infty}^{\infty} \psi(x-y,t)f(y)\,dy = f(x) \tag{14.5.5}$$

for every bounded, continuous function f, is

$$\psi(x,t) = \frac{e^{-x^2/4t}}{\sqrt{4\pi t}}; \tag{14.5.6}$$

see Exercise 41. One way to obtain the fundamental solution for the *periodic* problem is to periodize ψ, which gives

$$\sum_{n=-\infty}^{\infty} \frac{e^{-(x+n)^2/4t}}{\sqrt{4\pi t}} = \frac{e^{-x^2/4t}}{\sqrt{4\pi t}} \sum_{n=-\infty}^{\infty} e^{-nx/2t}\,e^{-n^2/4t}. \tag{14.5.7}$$

A second way to find the periodic fundamental solution is to expand in Fourier series (or to separate variables), which leads to

$$\sum_{n=-\infty}^{\infty} e^{2ni\pi x}\,e^{-4n^2\pi^2 t}; \tag{14.5.8}$$

see Exercise 42.

Let

$$z = x, \quad \tau = 4i\pi t, \quad z_1 = \frac{ix}{4\pi t} = -\frac{z}{\tau},$$

$$\tau_1 = \frac{i}{4\pi t} = -\frac{1}{\tau}, \quad q = e^{i\pi \tau}, \quad q_1 = e^{i\pi \tau_1}.$$

Then the equality of the two expressions (14.5.7) and (14.5.8) for the periodic fundamental solution takes the form

$$\sum_{n=-\infty}^{\infty} p(z)^{2n} q^{n^2} = \frac{e^{-i\pi z^2/\tau}}{\sqrt{-i\tau}} \sum_{n=-\infty}^{\infty} p(z_1)^{2n} q_1^{n^2},$$

or

$$\theta_3(z|\tau) = \frac{e^{-i\pi z^2/\tau}}{\sqrt{-i\tau}} \theta_3\left(-\frac{z}{\tau}\Big| -\frac{1}{\tau}\right), \qquad (14.5.9)$$

where the notation makes explicit the dependence on the parameter τ.

Finally, let us return to the integration problem of Section 14.1. We have seen that the integrals to be evaluated are

$$F(z) = \mathrm{sn}^{-1} z;$$

$$E(z) = \int_0^{\mathrm{sn}^{-1} z} \mathrm{dn}^2(\zeta) d\zeta;$$

$$\Pi(a, z) = \int_0^{\mathrm{sn}^{-1} z} \frac{d\zeta}{1 + a\,\mathrm{sn}^2 \zeta} = \mathrm{sn}^{-1} z - a \int_0^{\mathrm{sn}^{-1} z} \frac{\mathrm{sn}^2 \zeta\, d\zeta}{1 + a\,\mathrm{sn}^2 \zeta}.$$

It follows from previous results that the residue of dn at $u = iK'$ is $-i$. Also, dn is odd around iK', so dn^2 is even around iK' and it follows that

$$\mathrm{dn}^2(iK' + u) = -\frac{1}{u^2} + O(1).$$

The adapted theta function $\theta(u)$ is defined by

$$\theta(u) = \theta_4(u/2K) = \Theta\left((u + iK')/2K\right).$$

Jacobi's Z-function is

$$Z(u) = \frac{\theta'(u)}{\theta(u)}.$$

The residue at iK' is 1, so

$$Z'(iK' + u) = -\frac{1}{u^2} + O(1).$$

Since Z' is elliptic and its only poles are at the lattice points, it follows that $\mathrm{dn}^2 - Z'$ is a constant, which is customarily written E/K. Thus

$$E(z) = \frac{\theta'\left(\mathrm{sn}^{-1}z\right)}{\theta\left(\mathrm{sn}^{-1}z\right)} + \frac{E}{K}\,\mathrm{sn}^{-1}z.$$

Since $\mathrm{sn}^{-1}1 = K$ and $\theta'(K) = 0$, it follows that the constant E is

$$E = E(1) = \int_0^1 \sqrt{\frac{1 - k^2\zeta^2}{1 - \zeta^2}}\, d\zeta,$$

the complete elliptic integral of the second kind.

For the elliptic integral of the third kind, we use the addition formula

$$Z(u) + Z(v) - Z(u+v) = k^2 \mathrm{sn}\,u\,\mathrm{sn}\,v\,\mathrm{sn}(u+v). \qquad (14.5.10)$$

This is proved in the usual way, by establishing that the two sides are elliptic functions of u with the same poles and residues, and that they agree at $u = 0$. A consequence of this and (14.3.16) is that

$$Z(u-v) - Z(u+v) + 2Z(v)$$
$$= k^2 \mathrm{sn}\,u\,\mathrm{sn}\,v\,[\mathrm{sn}(u+v) + \mathrm{sn}(u-v)]$$
$$= 2k^2 \frac{\mathrm{sn}\,v\,\mathrm{cn}\,v\,\mathrm{dn}\,v\,\mathrm{sn}^2 u}{1 - k^2\mathrm{sn}^2 v\,\mathrm{sn}^2 u}. \qquad (14.5.11)$$

Finding $\Pi(a, z)$ reduces to finding

$$\int \frac{\mathrm{sn}^2 u}{1 + a\,\mathrm{sn}^2 u}\, du.$$

If $a = 0$, -1, ∞, or $-k^2$, the integral can be reduced to integrals of the first and second kind. Otherwise, we choose b such that $a = -k^2\mathrm{sn}^2 b$. Up to multiplication by a constant, we are left with

$$\int_0^{\mathrm{sn}^{-1}z} 2k^2 \frac{\mathrm{sn}\,b\,\mathrm{cn}\,b\,\mathrm{dn}\,b\,\mathrm{sn}^2 u}{1 - k^2\mathrm{sn}^2 b\,\mathrm{sn}^2 u}\, du$$
$$= \int_0^{\mathrm{sn}^{-1}z} [Z(u-b) - Z(u+b) + 2Z(b)]\, du$$
$$= \log\left[\frac{\theta(w-b)}{\theta(w+b)}\right] + 2Z(b)\,w, \qquad w = \mathrm{sn}^{-1}z.$$

14.6 Weierstrass elliptic functions

We return to the general notion of an elliptic function: a meromorphic function f with periods $2\omega_1$, $2\omega_2$, where we assume that $\text{Im}(\omega_2/\omega_1) > 0$. As noted earlier, unless f is constant it has at least two poles, counting multiplicity, in the parallelogram

$$\Pi = \Pi(\omega_1, \omega_2) = \{u \mid u = 2s\omega_1 + 2t\omega_2, \ -1 \le s, t < 1\}.$$

Thus in some sense the simplest such function would have a double pole at each point of the period lattice

$$\Lambda = \Lambda(\omega_1, \omega_2) = \{2n_1\omega_1 + 2n_2\omega_2 \mid n_1, n_2 = 0, \pm 1, \pm 2, \ldots\},$$

and no other poles. A consequence is that for any complex c, $f(u) = c$ would have exactly two solutions u, counting multiplicity, in Π.

We show below that there is such a function, the *Weierstrass $\wp$-function* $\wp(u) = \wp(u, \Lambda)$, which is even and has the property

$$\wp(u) = \wp(u, \omega_1, \omega_2) = \frac{1}{u^2} + O(u^2) \quad \text{as } u \to 0. \tag{14.6.1}$$

This condition determines $\wp$ uniquely, since the difference between two such functions would be a bounded entire function that vanishes at the origin. The function $\wp$ satisfies a differential equation analogous to (14.3.5) satisfied by Jacobi's function sn. The property (14.6.1) implies that $(\wp')^2 - 4\wp^3$ is an even elliptic function with at most a double pole at $u = 0$, so there are constants g_2 and g_3 such that

$$(\wp')^2 = 4\wp^3 - g_2\wp - g_3. \tag{14.6.2}$$

Let $\omega_3 = -\omega_1 - \omega_2$. Since $\wp$ is even with period $2\omega_j$, $j = 1, 2, 3$, it is even around ω_j, and therefore $\wp'(\omega_j) = 0$, and it follows from this and (14.6.2) that $e_j = \wp(\omega_j)$ is a root of the cubic

$$Q(t) = 4t^3 - g_2 t - g_3. \tag{14.6.3}$$

The function $\wp(u) - e_j$ has a double root at ω_j, $j = 1, 2$, and at $u = -\omega_3 = \omega_1 + \omega_2$, $j = 3$. Since each of these points is in Π, it follows that the e_j are distinct. Therefore they are simple roots of Q. It follows that

$$e_1 + e_2 + e_3 = 0;$$

$$4(e_2 e_3 + e_3 e_1 + e_1 e_2) = -g_2;$$

$$4e_1 e_2 e_3 = g_3.$$

Any elliptic function f with periods $2\omega_1$, $2\omega_2$ can be expressed as a rational function of $\wp$ and the derivative $\wp'$. To see this, suppose first that f is even. If the origin is a pole of f, we may subtract a linear combination of powers of $\wp$ so that the resulting function g is regular at the origin. The zeros and poles of g in Π can be taken to be $\pm a_1, \ldots, \pm a_n$ and $\pm b_1, \ldots, \pm b_n$, respectively, repeated according to multiplicity. The product

$$\prod_{j=1}^{n} \frac{\wp(u) - \wp(a_j)}{\wp(u) - \wp(b_j)}$$

has the same zeros and poles as g, so g is a constant multiple. Thus f is a rational function of $\wp$. If f is odd, then $f = g\wp'$, where $g = f/\wp'$ is even.

A second representation of elliptic functions can be obtained by using the *Weierstrass zeta function* ζ, which is characterized by

$$\zeta'(u) = -\wp(u), \quad \zeta(-u) = -\zeta(u). \tag{14.6.4}$$

Since $\wp$ is periodic with period $2\omega_j$, the integral

$$-\int_u^{u+2\omega_j} \wp(s)\,ds = 2\eta_j$$

is a constant. It follows that

$$\zeta(u + 2\omega_j) = \zeta(u) + 2\eta_j \tag{14.6.5}$$

Setting $u = -\omega_j$ shows that $\eta_j = \zeta(\omega_j)$.

Suppose now that f is an elliptic function with periods $2\omega_j$ and distinct poles a_k in Π. Let c_k be the residue at a_k; then $\sum c_k = 0$. The function

$$g(u) = f(u) - \sum c_k \zeta(u - a_k)$$

has no simple poles, and

$$g(u + 2\omega_j) = g(u) - 2\eta_j \sum c_k = g(u).$$

Therefore g has only multiple poles and is, up to an additive constant, a linear combination of derivatives of translates of ζ. Thus

$$f(u) = C + \sum c_k \zeta(u - a_k) + \sum_{v>0} c_{vk} \zeta^{(v)}(u - a_k).$$

This reduces the problem of integrating f to the problem of integrating ζ. It is convenient at this point to introduce the *Weierstrass sigma function* $\sigma(u)$, which is characterized by the conditions

$$\frac{\sigma'}{\sigma} = \zeta; \quad \lim_{u \to 0} \frac{\sigma(u)}{u} = 1.$$

Then an integral of ζ is $\log \sigma$.

It follows from the construction below that

$$\sigma(u + 2\omega_j) = e^{2\eta_j(u+\omega_j)}\sigma(u). \qquad (14.6.6)$$

Thus σ is analogous to the theta function.

The function σ allows a third representation of an elliptic function f, this time as a quotient of entire functions. Suppose that the zeros and poles of f in Π are $a_1, \ldots, a_n$ and $b_1, \ldots, b_n$. As noted at the beginning of Section 14.4, $w = \sum(a_j - b_j)$ belongs to Λ. Therefore we may replace a_1 by $a_1 - w$ and assume that $\sum(a_j - b_j) = 0$. The function

$$g(u) = \prod_{j=1}^{n} \frac{\sigma(u - a_j)}{\sigma(u - b_j)}$$

has the same zeros and poles as f and is doubly periodic by (14.6.6). Therefore f is a constant multiple of g.

The function $\wp$ has an addition formula. Given u and v in Π such that $\wp(u) \neq \wp(v)$, determine constants such that

$$\wp'(u) - B = A\,\wp(u), \quad \wp'(v) - B = A\wp(v).$$

Then

$$A = \frac{\wp'(u) - \wp'(v)}{\wp(u) - \wp(v)}.$$

The function $\wp' - B - A\wp$ has a unique pole, of order 3 in Π, so it has three zeros in Π. The sum of the zeros is an element of Λ (see the argument leading to (14.4.7)). By construction, u and v are zeros, so $-(u + v)$ is also a zero. Therefore $\wp(u)$, $\wp(v)$, and $\wp(u+v) = \wp(-u-v)$ are three roots of the cubic $Q(t) - (At + B)^2$. For most values of u and v, these roots are distinct, so their sum is $A^2/4$:

$$\wp(u + v) = \frac{1}{4}\left[\frac{\wp'(u) - \wp'(v)}{\wp(u) - \wp(v)}\right]^2 - \wp(u) - \wp(v). \qquad (14.6.7)$$

Up to this point, we have been *assuming* that there is a function $\wp$ with the property (14.6.1). To construct $\wp$, we begin with the (formal) derivative

$$\wp'(u) = \sum_{p \in \Lambda} \frac{2}{(p - u)^3} = \sum_{m,n=-\infty}^{\infty} \frac{2}{(2m\omega_1 + 2n\omega_2 - u)^3}. \qquad (14.6.8)$$

The series converges uniformly near any point not in Λ and defines a function that is odd, meromorphic, has a triple pole at each point of Λ, and has periods $2\omega_1$ and $2\omega_2$. Near $u = 0$,

$$\wp'(u) = -\frac{2}{u^3} + O(1).$$

Therefore we may define a function $\wp$ by

$$
\wp(u) = \frac{1}{u^2} + \int_0^u \left[\wp'(s) + \frac{2}{s^3} \right] ds
$$

$$
= \frac{1}{u^2} + \sum_{p \in \Lambda, p \neq 0} \left[\frac{1}{(u-p)^2} - \frac{1}{p^2} \right]. \tag{14.6.9}
$$

This is an even meromorphic function that satisfies (14.6.1) and has a double pole at each point of Λ. The function $\wp(u + 2\omega_j)$ has the same derivative, so

$$
\wp(u + 2\omega_j) - \wp(u) = c_j, \quad \text{constant.}
$$

Since $\wp$ is even, setting $u = -\omega_j$ shows that the constant is zero. Thus $\wp$ has periods $2\omega_j$ and is the desired elliptic function.

Similarly, we may define

$$
\zeta(u) = \frac{1}{u} + \int_0^u \left[\frac{1}{s^2} - \wp(s) \right] ds = \frac{1}{u} + \sum_{p \in \Lambda, p \neq 0} \left[\frac{1}{u-p} + \frac{1}{p} + \frac{u}{p^2} \right]
$$

and

$$
\log \sigma(u) = \log u + \int_0^u \left[\zeta(s) - \frac{1}{s} \right] ds
$$

$$
= \log u + \sum_{p \in \Lambda, p \neq 0} \left[\log \left(1 - \frac{u}{p} \right) + \frac{u}{p} + \frac{u^2}{2p^2} \right],
$$

so

$$
\sigma(u) = u \prod_{p \in \Lambda, p \neq 0} \left(1 - \frac{u}{p} \right) \exp \left(\frac{u}{p} + \frac{u^2}{2p^2} \right).
$$

14.7 Exercises

1. Show that letting $z = \sin\theta$ converts the integral of (14.1.1) to the integral of a rational function of $\sin\theta$ and $\cos\theta$. Show that setting $u = \tan(\frac{1}{2}\theta)$ converts this integral to the integral of a rational function of u.

2. Consider an integral $\int r(z, \sqrt{Q(z)}) dz$, where Q is a polynomial of degree 3 (resp. 4) with distinct roots.
   (a) Show that there is a linear fractional transformation (Möbius transformation), i.e. a transformation $z = \varphi(w) = (aw + b)/(cw + d)$ that converts the integral to one of the same type, $\int r_1(w, \sqrt{Q_1(w)}) dw$, with Q_1 of degree 4 (resp. 3). Hint: one can take $Q_1(w) = (cw+d)^4 Q(\varphi(w))$, and $\varphi(\infty)$ may or may not be a root of Q.

(b) Show that if Q has degree 4, there is a linear fractional transformation that converts the integral to one of the same type but with the polynomial $(1 - \zeta^2)(1 - k^2\zeta^2)$. Hint: map two roots to ± 1; this leaves one free parameter.

3. Compute the constants in the representation (14.1.3) of K_2 in terms of integrals of K_1, J_0, J_1.

4. Verify that the indicated change of variables leads to (14.2.8).

5. Verify that the indicated change of variables leads to (14.2.9).

6. Verify (14.2.10) and (14.2.11).

7. Verify (14.2.17): let $t = \sin\theta$ and integrate the series expansion of the resulting integrands term by term, using Exercise 2 of Chapter 1.

8. Show that as $\varepsilon \to 0$, $F(k, 1+\varepsilon) \sim K - i\sqrt{2\varepsilon}/k'$. Deduce that $1 - \mathrm{sn}^2$ has double zeros at K and $3K$.

9. Show that as $\varepsilon \to 0$, $F(k, 1/k+\varepsilon) \sim K + iK' - \sqrt{2k\varepsilon}/k'$. Deduce that $1 - k^2\mathrm{sn}^2$ has double zeros at $K + iK'$ and $3K + iK'$.

10. Suppose that a function $f(z)$ is even (resp. odd) around $z = a$ and even (resp. odd) around $z = b$. Show that it has period $2(b - a)$.

11. Suppose that a function $f(z)$ is odd around $z = a$ and even around $z = b$. Show that it has period $4(b - a)$.

12. Use (14.3.10) to obtain (14.3.11).

13. Prove that $\mathrm{sn}(u, k_1)$ in (14.3.10) has periods $2K(1 + k')$, $2iK'(1 + k')$.

14. Use (14.3.12) to obtain (14.3.13).

15. What are the periods of $\mathrm{sn}(u, k_1)$ in (14.3.12)?

16. What are the limiting values as $k \to 0$ and as $k \to 1$ of the functions sn, cn, and dn? (See (14.2.11) and (14.2.10).) What do the formulas (14.3.10) and (14.3.11) reduce to in these limits? What about (14.3.12) and (14.3.13)?

17. Use (14.3.14) to obtain (14.3.15).

18. Use (14.3.16) to obtain (14.3.17).

19. Show that the Gauss transformation (14.3.12), (14.3.13) is the composition of the Jacobi transformation, the Landen transformation, and the Jacobi transformation.

20. Show that the Landen transformation (14.3.10), (14.3.11) is the composition of the Jacobi transformation, the Gauss transformation, and the Jacobi transformation.

21. Prove (14.3.18).

22. Use (14.3.16) and (14.3.17) to verify

$$\mathrm{sn}(u + K) = \frac{\mathrm{cn}\, u}{\mathrm{dn}\, u};$$

$$\mathrm{cn}(u + K) = -\frac{k'\,\mathrm{sn}\, u}{\mathrm{dn}\, u};$$

$$\mathrm{dn}(u + K) = \frac{k'}{\mathrm{dn}\, u}$$

and find corresponding formulas for translation by iK' and by $K + iK'$.

23. In each of the following integrals, a substitution like $t = \text{sn}^2 u$ converts the integrand to a rational function of t and $\sqrt{Q(t)}$, with Q quadratic, so that the integral can be found in terms of elementary functions of t. Verify the results

$$\int \text{sn}\, u\, du = -\frac{1}{k} \cosh^{-1}\left(\frac{\text{dn}\, u}{k'}\right) + C$$

$$= \frac{1}{k} \log(\text{dn}\, u - k\, \text{cn}\, u) + C;$$

$$\int \text{cn}\, u\, du = \frac{1}{k} \cos^{-1}(\text{dn}\, u) + C;$$

$$\int \text{dn}\, u\, du = \frac{1}{k} \sin^{-1}(\text{sn}\, u) + C;$$

$$\int \frac{du}{\text{sn}\, u} = \log\left(\frac{\text{dn}\, u - \text{cn}\, u}{\text{sn}\, u}\right) + C;$$

$$\int \frac{du}{\text{cn}\, u} = \frac{1}{k'} \log\left(\frac{k'\, \text{sn}\, u + \text{dn}\, u}{\text{cn}\, u}\right) + C;$$

$$\int \frac{du}{\text{dn}\, u} = \frac{1}{k'} \cos^{-1}\left(\frac{\text{cn}\, u}{\text{dn}\, u}\right) + C;$$

$$\int \frac{\text{cn}\, u}{\text{sn}\, u}\, du = \log\left(\frac{1 - \text{dn}\, u}{\text{sn}\, u}\right) + C;$$

$$\int \frac{\text{dn}\, u}{\text{sn}\, u}\, du = \log\left(\frac{1 - \text{cn}\, u}{\text{sn}\, u}\right) + C;$$

$$\int \frac{\text{sn}\, u}{\text{cn}\, u}\, du = \frac{1}{k'} \log\left(\frac{\text{dn}\, u + k'}{\text{cn}\, u}\right) + C;$$

$$\int \frac{\text{dn}\, u}{\text{cn}\, u}\, du = \log\left(\frac{1 + \text{sn}\, u}{\text{cn}\, u}\right) + C;$$

$$\int \frac{\text{sn}\, u}{\text{dn}\, u}\, du = -\frac{1}{kk'} \sin^{-1}\left(\frac{k\, \text{cn}\, u}{\text{dn}\, u}\right) + C;$$

$$\int \frac{\text{cn}\, u}{\text{dn}\, u}\, du = \frac{1}{k} \log\left(\frac{1 + k\, \text{sn}\, u}{\text{dn}\, u}\right) + C.$$

24. In each of the following integrals, a substitution like $v = (\text{sn}^{-1}t)^2$ converts the integrand to a rational function of v and $\sqrt{Q(v)}$, with Q quadratic. Verify

$$\int \text{sn}^{-1}t\,dt = t\,\text{sn}^{-1}t + \frac{1}{k}\cosh^{-1}\left(\frac{\sqrt{1-k^2t^2}}{k'}\right) + C$$

$$= t\,\text{sn}^{-1}t + \frac{1}{k}\log\left(\sqrt{1-k^2t^2} + k\sqrt{1-t^2}\right) + C;$$

$$\int \text{cn}^{-1}t\,dt = t\,\text{cn}^{-1}t - \frac{1}{k}\cos^{-1}\left(\sqrt{k'^2 + k^2t^2}\right) + C;$$

$$\int \text{dn}^{-1}t\,dt = t\,\text{dn}^{-1}t - \sin^{-1}\left(\frac{\sqrt{1-t^2}}{k}\right) + C.$$

25. Deduce (14.4.3) from (14.4.2).
26. Deduce (14.4.4) from (14.4.3).
27. Prove that (14.4.8) has period τ.
28. Show that the product in (14.4.9) has period 1 as a function of z. Show that the effect of changing z to $z + \tau$ in the product is to multiply the product by $-p^{-2}q^{-1}$. Deduce that the function $c(z, \tau) = G(\tau)$ depends only on τ.
29. Show that the function G in Exercise 28 has limit 1 as $\text{Im}\,\tau \to +\infty$.
30. Show that

$$\Theta\left(\frac{1}{4} + \frac{1}{2}\tau\right) = G(\tau)\prod_{n=1}^{\infty}(1 + q^{4n-2})$$

$$= \sum_{n=-\infty}^{\infty}(-1)^n i^n q^{n^2} = \sum_{m=-\infty}^{\infty}(-1)^m q^{(2m)^2}.$$

31. Use Exercise 30 to show that

$$\frac{G(\tau)}{G(4\tau)} = \prod_{n=1}^{\infty}\frac{(1 - q^{8n-4})^2}{1 + q^{4n-2}} = \prod_{n=1}^{\infty}(1 - q^{8n-4})(1 - q^{4n-2}).$$

32. Show that for $|w| < 1$,

$$\prod_{n=1}^{\infty}(1 - w^{2n-1}) = \frac{\prod_{n=1}^{\infty}(1 - w^n)}{\prod_{n=1}^{\infty}(1 - w^{2n})}.$$

33. Use Exercises 31 and 32 to show that

$$\frac{G(\tau)}{G(4\tau)} = \frac{\prod_{n=1}^{\infty}(1 - q^{2n})}{\prod_{n=1}^{\infty}(1 - q^{8n})}.$$

Iterate this identity to get

$$\frac{G(\tau)}{G(4^m \tau)} = \frac{\prod_{n=1}^{\infty}(1-q^{2n})}{\prod_{n=1}^{\infty}(1-q^{4^m \, 2n})}.$$

34. Deduce from Exercises 28 and 33 that $G(\tau) = \prod_{n=1}^{\infty}(1-q^{2n})$.
35. Find the Fourier sine expansion of θ_1 (expressing it as a combination of the functions $\sin(m\pi z)$) and the Fourier cosine expansions of $\theta_2, \theta_3, \theta_4$.
36. Use (14.4.10) and (14.5.1) to verify the product expansions of the θ_j.
37. Verify the table of values of the Jacobi theta functions.
38. Use the tables of values of the Jacobi elliptic functions and the Jacobi theta functions to obtain (14.5.2).
39. Prove that the quotient

$$\frac{\Theta(2z|2\tau)}{\Theta(z|\tau)\,\Theta\left(z+\frac{1}{2}|\tau\right)}$$

is constant. Evaluate at $z = \frac{1}{2}\tau$ to show that the constant is $\prod(1+q^{2n})/(1-q^{2n})$.
40. Show that the periods τ and τ_1 associated with the Landen transformation are related by $\tau_1 = 2\tau$. It follows that the Landen transformation relates theta functions with parameter τ to theta functions with parameter 2τ.
41. Suppose that $\psi(x,t)$ satisfies the heat equation (14.5.4) and also has the property (14.5.5), say for every continuous function f that vanishes outside a bounded interval. Assume that these properties determine ψ uniquely.
    (a) Show that

    $$u(x,t) = \int_{-\infty}^{\infty} \psi(x-y,t)f(y)\,dy, \quad t > 0,$$

    is a solution of the heat equation with $u(x,t) \to f(x)$ as $t \to 0$. Assume that these properties determine u uniquely.
    (b) Show that for $\lambda > 0$, $u(\lambda x, \lambda^2 t)$ is a solution of the heat equation with limit $f(\lambda x)$ as $t \to 0$, and deduce that

    $$\int_{-\infty}^{\infty} \psi(x-y,t)f(\lambda y)\,dy = u(\lambda x, \lambda^2 t) = \int_{-\infty}^{\infty} \psi(\lambda x - \lambda y, \lambda^2 t)f(\lambda y)\,d(\lambda y),$$

    so $\psi(\lambda x, \lambda^2 t) = \lambda^{-1}\psi(x,t)$.
    (c) Deduce from (b) that $\psi(x,t)$ has the form $t^{-\frac{1}{2}}F(x^2/t)$ and use the heat equation to show that $F(s)$ satisfies the equation

    $$\left\{ s\frac{d}{ds} + \frac{1}{2} \right\}(4F' + F) = 0,$$

    with solution $F(s) = Ae^{-s/4}$.
    (d) Deduce from the preceding steps that ψ should be given by (14.5.6). (The constant A can be determined by taking $f \equiv 1$ in (14.5.5).)

42. Suppose that $\psi_p(x,t)$ is the fundamental solution for the periodic heat equation, i.e. that ψ satisfies the heat equation, and that for any continuous periodic function f, the function

$$u(x,t) = \int_0^1 \psi_p(x-y,t)f(y)\,dy$$

is a solution of the heat equation with the property that $u(x,t) \to f(x)$ as $t \to 0$. It is not difficult to show that $u(\cdot,t)$ is continuous and periodic for each $t > 0$, and hence has a Fourier expansion

$$u(x,t) = \sum_{n=-\infty}^{\infty} a_n(t)\,e^{2n\pi ix}, \quad a_n(t) = \int_0^1 u(x,t)\,e^{-2n\pi ix}\,dx;$$

see Appendix B.

(a) Assuming that the Fourier expansion can be differentiated term by term, find the coefficients a_n. (Use the condition at $t = 0$.)

(b) Use the result from part (a) to write $u(x,t)$ as an integral and thus show that $\psi_p(x,t)$ is given by (14.5.8).

43. Verify (14.5.10) and (14.5.11).

44. Integrate the Weierstrass zeta function ζ over the boundary of Π to show that $2\eta_1\omega_2 - 2\eta_2\omega_1 = \pi i$.

45. Express sn in terms of the Weierstrass $\wp$ function, the Weierstrass zeta function, and the Weierstrass sigma function.

46. Show that

$$\frac{1}{\operatorname{sn}(u,k)^2} = \wp(u,K,iK') + \frac{1+k^2}{3}.$$

47. Use Exercise 46 to show that any elliptic function with periods $2K$ and $2iK'$ is a rational function of sn, cn, and dn.

48. Determine the coefficients of u^2 and u^4 in the McLaurin expansion (Taylor expansion at $u = 0$) of $\wp(u) - 1/u^2$. Use this to show that the coefficients g_2, g_3 in (14.6.2) are

$$g_2 = 60 \sum_{p\in\Lambda, p\neq 0} \frac{1}{p^4}; \quad g_3 = 140 \sum_{p\in\Lambda, p\neq 0} \frac{1}{p^6}.$$

14.8 Remarks

The history and theory of elliptic functions is treated in the survey by Mittag-Leffler (translated by Hille) [289], and in the books by Akhiezer [8], Appell and Lacour [15], Armitage and Eberlein [16], Chandrasekharan [66], Lang [241], Lawden [248], Neville [300], Prasolov and Solovyev [334], Tannery and Molk [395], Temme [397], Tricomi [403], and Walker [416].

(The extensive list of results in Abramowitz and Stegun [4] is marred by an idiosyncratic notation.)

There are very extensive lists of formulas in Tannery and Molk. Akhiezer discusses the transformation theory of Abel and Jacobi. Appell and Lacour give a number of applications to physics and to geometry. Chandrasekharan has several applications to number theory and Lawden has applications to geometry, while Armitage and Eberlein have applications to geometry, mechanics, and statistics. The book by Lang covers a number of modern developments of importance in algebraic number theory. By now, the applications to number theory include the proof of Fermat's last theorem by Wiles and Taylor; see [436, 396]. Applications to physics and engineering are treated by Oberhettinger and Magnus [309].

The theory of elliptic functions was developed in the eighteenth and early nineteenth centuries through the work of Euler, Legendre [253], Gauss [153], Abel [1], Jacobi [197, 198], and Liouville [262], among others. Abel and Jacobi revolutionized the subject in 1827 by studying the inverse functions and developing the theory in the complex plane. (Gauss's discoveries in this direction were made earlier but were only published later, posthumously.) For an assessment of the early history, see Mittag-Leffler [289] and also Dieudonné [103], Klein [218], and Stillwell [385]. Liouville introduced the systematic use of complex variable methods, including his theorem on bounded entire functions.

The version developed by Weierstrass [430] later in the nineteenth century is simpler than the Jacobi approach via theta functions [200]. Mittag-Leffler [289], Neville [300], and Tricomi [403] used the Weierstrass functions rather than the theta functions to develop the theory of Abel and Jacobi. On the other hand, the work of Abel and Jacobi generalizes to curves of arbitrary genus, i.e. to polynomial equations of arbitrary degree. A classic treatise on the subject is Baker [29]; see also Kempf [216] and Polishchuk [332]. Theta functions have become important in the study of certain completely integrable dynamical systems, e.g. in parameterizing special solutions of the periodic Korteweg–de Vries equation and other periodic problems; see Krichever's introduction to [29] and the survey article by Dubrovin [106].

15

Painlevé transcendents

Except for the beta and gamma functions and some close relatives, most of the special functions treated in this book are solutions of linear differential or difference equations. The elliptic functions are solutions of the nonlinear equations $(u')^2 = P(u)$, where P is a polynomial of degree 3 or 4. In the late nineteenth century, Poincaré and others considered the question of finding new special functions as solutions of more general nonlinear equations. One feature of the equations that we have considered so far is that the singularities of their solutions are fixed – they do not depend on initial conditions. For example, any solution of the hypergeometric equation is regular in the neighborhood of any point z other than (possibly) $z = 0$, 1, or ∞. Therefore any solution can be taken to be single-valued in any simply connected domain that does not contain any of these three points – fixed domains that depend only on the equation itself, not on the constants of integration. Even for equations of first order, this property can fail. The equation

$$u'(x) = u(x)^2, \qquad u(0) = \frac{1}{a} \tag{15.0.1}$$

is a benign example: the general solution $u(x) = 1/(a - x)$ has a pole at a, but is single-valued in the plane. By contrast, the equation

$$u'(x) = \frac{1}{2u(x)} \tag{15.0.2}$$

has the general solution $u(x) = \sqrt{x - a}$, which has a branch point at the point $x = a$ and is multiple-valued in any neighborhood of the branch point. The branch point itself varies with the initial condition $u(0) = \sqrt{-a}$.

It is natural to try to characterize the nonlinear equations that have the *Painlevé property*: the general solution is single-valued in any simply connected region that does not contain certain points that are independent of the initial conditions. Equation (15.0.1) is an example; (15.0.2) is not. In the case of first-order equations, this question was investigated exhaustively by

Poincaré, Briot and Bouquet, and others; for results concerning equations of the form

$$(u')^m = P(u),$$

where P is a polynomial, see Exercises 4 and 5.

Painlevé and others examined equations of the form

$$u'' = R(x, u, u') = \frac{P(x, u, u')}{Q(x, u, u')}, \tag{15.0.3}$$

where $P(x, u, v)$ and $Q(x, u, v)$ are polynomials in u and v, with coefficients that are analytic functions of x. The goal was to determine which of them have the Painlevé property. These general solutions can then be considered as functions of intrinsic interest.

The work begun by Painlevé [314] and completed by Gambier [146] and Fuchs [145] identified 50 equations that represent, up to certain transformations, every equation of the form (15.0.3) that has the Painlevé property. Of these 50 equations, 44 have the property that all solutions can be written in terms of known functions, e.g. the Weierstrass $\wp$ function. The remaining six equations have as solutions new transcendental functions, known as the *Painlevé transcendents*. These six *Painlevé equations* are:

PI $\quad u'' = 6u^2 + x;$

PII $\quad u'' = 2u^3 + xu + a;$

PIII $\quad u'' = \dfrac{(u')^2}{u} - \dfrac{u'}{x} + \dfrac{au^2 + b}{x} + cu^3 + \dfrac{d}{u};$

PIV $\quad u'' = \dfrac{(u')^2}{2u} + \dfrac{3u^3}{2} + 4xu^2 + 2(x^2 - a)u + \dfrac{b}{u};$

PV $\quad u'' = \left[\dfrac{1}{2u} + \dfrac{1}{u-1} \right] (u')^2 - \dfrac{u'}{x} + \dfrac{(u-1)^2(au^2 + b)}{x^2 u}$

$\qquad\qquad + \dfrac{cu}{x} + \dfrac{d\,u(u+1)}{u-1};$

PVI $\quad u'' = \left[\dfrac{1}{u} + \dfrac{1}{u-1} + \dfrac{1}{u-x} \right] \dfrac{(u')^2}{2} - \left[\dfrac{1}{x} + \dfrac{1}{x-1} + \dfrac{1}{u-x} \right] u'$

$\qquad\qquad + \dfrac{u(u-1)(u-x)}{x^2(x-1)^2} \left[a + \dfrac{bx}{u^2} + \dfrac{c(x-1)}{(u-1)^2} + \dfrac{dx(x-1)}{(u-x)^2} \right],$

where a, b, c, d are arbitrary constants. Each of the equations PI–PV can be derived from one or more of the succeeding equations by a certain limiting process, called "coalescence":

PII $\to$ PI: Replace u by $\varepsilon u + \varepsilon^{-5}$, x by $\varepsilon^2 x - 6\varepsilon^{-10}$, and a by $4\varepsilon^{-15}$, and let $\varepsilon \to 0$.

PIII → PII: Replace u by $1+2\varepsilon u$, x by $1+\varepsilon^2 x$, a by $-\frac{1}{2}\varepsilon^{-6}$, b by $\frac{1}{2}\varepsilon^{-6}+2a\varepsilon^{-3}$, c and $-d$ by $\frac{1}{4}\varepsilon^{-6}$, and let $\varepsilon \to 0$.

PIV → PII: Replace u by $2^{2/3}\varepsilon^{-1}u+\varepsilon^{-3}$, x by $2^{-2/3}\varepsilon x-\varepsilon^{-3}$, a by $-2a-\frac{1}{2}\varepsilon^{-6}$, and b by $-\frac{1}{2}\varepsilon^{-12}$, and let $\varepsilon \to 0$.

PV → PIII: Replace u by $1+x\varepsilon u$, x by x^2, a by $\frac{1}{4}a\varepsilon^{-1}+\frac{1}{8}c\varepsilon^{-2}$, b by $-\frac{1}{8}c\varepsilon^{-2}$, c by $\frac{1}{4}\varepsilon b$, and d by $\frac{1}{8}\varepsilon^2 d$, and let $\varepsilon \to 0$.

PV → PIV: Replace u by $2^{-1/2}\varepsilon u$, x by $1+2^{1/2}\varepsilon x$, a by $\frac{1}{2}\varepsilon^{-4}$, b by $\frac{1}{4}b$, c by $-\varepsilon^{-4}$, and d by $a\varepsilon^{-2}-\frac{1}{2}\varepsilon^{-4}$, and let $\varepsilon \to 0$.

PVI → PV: Keep u,a,b as u,a,b, replace x by $1+\varepsilon x$, c by $c\varepsilon^{-1}-d\varepsilon^{-2}$, and d by $d\varepsilon^{-2}$, and let $\varepsilon \to 0$.

See Exercise 1.

In recent decades, these equations and their solutions have been found to be closely related to certain integrable systems, and this has led to new methods for investigating their properties.

A full presentation of either the classical theory or the modern developments is beyond the scope of a single chapter. We concentrate instead on one important and representative example, equation PII. We illustrate Painlevé's methods in deriving this equation and then trace some of the modern developments, as specialized to PII. These include the realization of PII as the *compatibility condition* for an over-determined system of linear equations, the associated Bäcklund transformations and isomonodromy interpretation, and the use of the Riemann–Hilbert method to relate asymptotics at $-\infty$ and $+\infty$. Some analogous results for PI are sketched in the exercises.

15.1 The Painlevé method

As noted in the introduction, we shall consider the equations

$$u'' = R(x,u,u') = \frac{P(x,u,u')}{Q(x,u,u')}, \tag{15.1.1}$$

where $P(x,u,v)$ and $Q(x,u,v)$ are polynomials in u and v, with coefficients that are analytic functions of x. Equation (15.1.1) can be written as a system:

$$u' = v, \qquad v' = R(x,u,v). \tag{15.1.2}$$

If one adds the initial conditions $u(x_0) = u_0$, $u'(x_0) = v_0$, (15.1.2) is equivalent to the pair of integral equations

$$u(x) = u_0 + \int_{x_0}^{x} v(y)\,dy, \qquad v(x) = v_0 + \int_{x_0}^{x} R(y,u(y),v(y))\,dy. \tag{15.1.3}$$

As in Chapter 3, if $R(x_0, u_0, v_0)$ is finite, (15.1.3) can be solved by successive approximations in a neighborhood of x_0: $u_0(x) \equiv u_0$, $v_0(x) \equiv v_0$, and

$$u_{n+1}(x) = u_0 + \int_{x_0}^{x} v_n(y) \, dy,$$

$$v_{n+1}(x) = v_0 + \int_{x_0}^{x} R(y, u_n(y), v_n(y)) \, dy.$$

An analysis similar to that in Chapter 3 shows that this sequence of pairs of functions converges to a solution of (15.1.3) with the given initial conditions, and that this solution is unique. Moreover, if $R = R(x, u, v; \alpha)$ depends analytically on an additional parameter α, then the solution is an analytic function of α.

A key observation in determining which equations (15.1.1) have the Painlevé property is the following. Suppose that functions u and U are related by a transformation

$$U(x) = \frac{a(x) u(\varphi(x)) + b(x)}{c(x) u(\varphi(x)) + d(x)}, \tag{15.1.4}$$

where φ, a, b, c, d are analytic, normalized with $ad - bc = 1$. Then u satisfies an equation of the form (15.1.1) if and only if U satisfies an equation of the same form, with a suitable choice of R; see Exercise 2. But u and U have the same singularities, other than poles, so if one of the corresponding equations has the Painlevé property, then so does the other. The goal is to classify all the possibilities, up to transformations (15.1.4). Every such equation with the Painlevé property is related to one of the 50 mentioned in the introduction by such a transformation. Conversely, each of these 50 equations represents a large class of equations, parameterized by five analytic functions a, b, c, d, φ, subject to the single relation $ad - bc = 1$.

Painlevé's method was to start with an equation (15.1.1) that is *assumed* to have the Painlevé property near x_0, and introduce a scaling by a parameter α in such a way that the scaled equation has the property for $0 < |\alpha| < 1$ and simplifies at $\alpha = 0$. As we shall see, the simplified equation must also have the property. This imposes strong constraints on the original equation.

Lemma 15.1.1 *Suppose that $R(x, u, v; \alpha)$ depends analytically on α for $|\alpha| < 1$ and that all solutions of*

$$u''(x) = R(x, u, u'; \alpha) \tag{15.1.5}$$

are single-valued in x in some region Ω for $0 < |\alpha| < 1$. Then solutions of (15.1.5) for $\alpha = 0$ are also single-valued for $x \in \Omega$.

Proof Suppose that $u(x, \alpha)$ is a solution of (15.1.5), and that $s \to x(s), 0 \leq s \leq 1$, is a continuous curve in Ω that begins and ends at a point x_0. Then $u(x(s), \alpha)$

is analytic with respect to α. By assumption, $u(x(1), \alpha) \equiv u(x(0), \alpha)$ for $\alpha \neq 0$, so the same is true at $\alpha = 0$. □

Remark. A similar result holds for systems like (15.1.2), or more general systems of the form

$$u'(x) = F(x, u, v; \alpha), \qquad v'(x) = G(x, u, v; \alpha).$$

The next lemma supplies a key step in the use of the preceding lemma.

Lemma 15.1.2 *Suppose that as a function of v, $R(x_0, u_0, v)$ has a zero at $v = v_0$. Then there is a region $\Omega \subset \mathbf{C}^2$, a positive integer m, and an analytic function $r(x, u)$ such that*

$$R(x, u, v) = [v - r]^m S(x, u, v), \qquad (x, u) \in \Omega, \qquad (15.1.6)$$

and $S(x, u, v) \neq 0$.

Remark. It may be that the region Ω does not include the point (x_0, u_0). This lemma replaces a recurring, incorrect reduction used in [193].

Proof By varying x_0 and u_0, we may suppose that the zero has minimal order m: for nearby x, u, any zero near $v = v_0$ has order $\geq m$. The series expansion in v shows that $R(x_0, u_0, v) \neq 0$ on a small circle C around v_0. Denote differentiation with respect to v by a subscript. The integral

$$\frac{1}{2\pi i} \int_C \frac{R_v(x, u, v)}{R(x, u, v)} \, dv$$

gives the number of solutions of $R(x, u, v) = 0$ (counting multiplicity) inside the circle C. Since the number is m at x_0, u_0, it will be m nearby. By the minimality assumption, each such zero will have multiplicity m and occur at a unique point $v = r(x, u)$. Computing the residue in the following integral shows that

$$\frac{1}{2\pi i} \int_C \frac{v R_v(x, u, v)}{m R(x, u, v)} \, dv = r(x, u).$$

This identity shows that r is analytic in x and u. □

As a first application of the Painlevé method, we show that the function R must be a polynomial in v.

Proposition 15.1.3 *Suppose that (15.1.1) has the Painlevé property. Then R is a polynomial in v, of degree at most 2.*

Proof Suppose instead that R has a pole as a function of v. Applying Lemma 15.1.2 to the reciprocal $1/R$, we find that locally,

$$R(x, u, v) = [v - r]^{-m} S(x, u, v), \qquad S(x, u, v) \neq 0.$$

The original equation is equivalent to a system

$$u' = w + r(x, u),$$
$$w' = w^{-m} \widetilde{S}(x, u, r + w). \tag{15.1.7}$$

We introduce a parameter α by writing

$$x = x_0 + \alpha^{m+1} X, \quad u = u_0 + \alpha^{m+1} U, \quad w = \alpha W.$$

In the variables X, U, W, the system (15.1.7) has the form

$$U'(X) = r_0 + O(\alpha), \qquad W'(X) = W^{-m} S_0 + O(\alpha),$$

where $r_0 = r(x_0, u_0) = v_0$ and $S_0 = S(x_0, u_0, v_0)$. At $\alpha = 0$, the system has the solutions

$$U(X) = r_0 X, \qquad W(X) = \left[(m+1) S_0 X + W(0)^{m+1} \right]^{1/(m+1)}.$$

Thus W has a movable branch point unless $m + 1 = 1$. Therefore the original equation does not have the Painlevé property unless R is a polynomial in v.

To bound the degree of this polynomial, we look at the equivalent system

$$u' = \frac{1}{w}, \qquad w' = -w^2 R\left(x, u, \frac{1}{w}\right). \tag{15.1.8}$$

An adaptation of the previous argument shows that $w^2 R(x, u, 1/w)$ cannot have a pole if the Painlevé property is to hold. Therefore the degree of $R(x, u, v)$ in v is at most 2. $\qquad\square$

We have established that if (15.1.1) has the Painlevé property, it must be of the form

$$u'' = L(x, u) [u']^2 + M(x, u) u'(x) + N(x, u). \tag{15.1.9}$$

Setting $U(X) = u(x_0 + \alpha X)$, this equation has the form

$$U'' = L(x_0, U) [U']^2 + O(\alpha),$$

so a necessary condition for the Painlevé property is that the $\alpha = 0$ limit have the Painlevé property. The next step in the Painlevé analysis is to examine equations of the form

$$u'' = l(u) [u']^2. \tag{15.1.10}$$

Painlevé showed by similar arguments that the rational function l must have at most simple poles, and, if l is not zero, the problem can be reduced to a first-order equation

$$[u']^m = P(u); \qquad l(u) = \frac{P'(u)}{m P(u)}, \tag{15.1.11}$$

where m is a positive integer and P is a polynomial of degree $2m$. For details, see Section 14.21 of [193], or Exercises 6, 7, and 8.

In turn, if (15.1.11) has the Painlevé property, Briot and Bouquet [54] showed that it must reduce to one of the cases

I $\quad P(u) = (u - a_1)^{m+1}(u - a_2)^{m-1}$;

III $\quad P(u) = (u - a_1)(u - a_2)(u - a_3)(u - a_4)$;

IV $\quad P(u) = (u - a_1)^2(u - a_2)^2(u - a_3)^2$; $\hspace{2cm}$ (15.1.12)

V $\quad P(u) = (u - a_1)^3(u - a_2)^3(u - a_3)^2$;

VI $\quad P(u) = (u - a_1)^5(u - a_2)^4(u - a_3)^3$;

see Section 13.8 of [193], or Exercises 4 and 5. (There is a case II which can be considered as case III with $a_1 = a_2$; otherwise the a_j are assumed to be distinct.)

15.2 Derivation of PII

We focus now on the case (15.1.9) when $L \equiv M \equiv 0$:

$$u''(x) = R(x, u(x)), \hspace{2cm} (15.2.1)$$

where $R(x, u)$ is analytic in x and rational in u, and is not constant.

Proposition 15.2.1 *If* (15.2.1) *has the* Painlevé *property, then* $R(x, u)$ *is a polynomial in* u *of degree* ≤ 3.

Proof Suppose, to the contrary, that $R(x, \cdot)$ has a pole. As in the proof of Proposition 15.1.3, we may reduce to the case $R(x, u) = u^{-m}S(x, u)$, where S is a polynomial in u and $m > 0$. Setting $u = \alpha^2 U$, $x = x_0 + \alpha^{m+1}X$, where $u(x_0) = 0$, we find that the Painlevé property for (15.2.1) implies the Painlevé property for

$$U''(X) = S_0 U(X)^{-m}, \hspace{1cm} S_0 = S(x_0, 0). \hspace{1cm} (15.2.2)$$

In turn, the Painlevé property implies that any solution of (15.2.2) has a Laurent expansion around $X = 0$:

$$U(X) = a_0 X^k + a_1 X^{k+1} + a_2 X^{k+2} + \cdots, \quad a_0 \neq 0, \hspace{1cm} (15.2.3)$$

where k is an integer, $k \neq 0$. Either $k = 1$ or (15.2.2) implies that

$$k(k-1)a_0 X^{k-2} = S_0(a_0 X^k)^{-m},$$

so $k = 2/(m+1)$ is positive. Then the left-hand side of (15.2.2) is regular at $X = 0$, while the right-hand side has a pole.

We have shown that $R(x,u)$ is necessarily a polynomial in u. Suppose that at $x = x_0$ it has degree $m > 1$. Let $u = a^{-2}U$, $x = x_0 + a^{m-1}X$. Then (15.2.1) has the Painlevé property if and only if

$$U''(X) = R_0 U^m(X), \qquad R_0 = \lim_{u \to \infty} \frac{R(x_0, u)}{u^m} \tag{15.2.4}$$

has the Painlevé property. As before, we look for a Laurent expansion (15.2.3). Equation (15.2.4) requires

$$k(k-1)a_0 X^{k-2} = R_0(a_0 X^k)^m,$$

so $k = -2/(m-1)$. For k to be an integer we need $m = 2$ or $m = 3$. $\qquad\square$

We have shown that the equation to be considered here is either linear or has the form

$$u''(x) = a_3(x)u(x)^3 + a_2(x)u(x)^2 + a_1(x)u(x) + a_0(x).$$

Let us assume that the degree is 3: $a_3(x) \neq 0$. Replacing u by $u - b$ with $3a_3 b = a_2$ allows us to assume that $a_2(x) = 0$, leading to

$$u''(x) = A(x)u(x)^3 + B(x)u(x) + C(x). \tag{15.2.5}$$

This equation can, in turn, be reduced to the form

$$v''(x) = 2v(x)^3 + B(x)v(x) + C(x). \tag{15.2.6}$$

In fact, let $u(x) = a(x)v(y)$, $y = \varphi(x)$. Then (15.2.6) is equivalent to

$$v'' = -\frac{2a'\varphi' + a\varphi''}{a(\varphi')^2}v' + \frac{a^3 A}{a(\varphi')^2}v^3 + \frac{aB - a''}{a(\varphi')^2}v + \frac{C}{a(\varphi')^2}.$$

This will have the desired form (15.2.6) if

$$\varphi' \frac{2a'\varphi' + a\varphi''}{a(\varphi')^2} = 2\frac{a'}{a} + \frac{\varphi''}{\varphi} = \left[\log(a^2\varphi')\right]' = 0$$

and $a^2 A = 2(\varphi')^2$. Thus we may take $a^2\varphi' = 1$, $a = (2/A)^{1/6}$.

Remark. A slight modification in the argument shows that the more general equation

$$u''(x) = M(x)u'(x) + A(x)u(x)^3 + B(x)u(x) + C(x) \tag{15.2.7}$$

can be reduced to the form (15.2.6); see Exercise 3.

The principal result of this section is the following.

Theorem 15.2.2 *If* (15.2.6) *has the Painlevé property, then C is constant and B is either constant or a linear function of x.*

Proof Let $v = \alpha^{-1}u$, $x = x_0 + \alpha X$, so that (15.2.6) is

$$u''(X, \alpha) = 2u^3(X, \alpha) + \alpha^2 B(x)u(X, \alpha) + \alpha^3 C(x). \tag{15.2.8}$$

Let us develop the various terms in powers of α:

$$u(X, \alpha) = u_0(X) + \alpha^2 u_2(X) + \alpha^3 u_3(X) + \alpha^4 u_4(X) + \cdots;$$

$$B(x) = b_0 + \alpha X b_1 + \alpha^2 X^2 b_2 + \cdots, \qquad b_n = \frac{1}{n!}B^{(n)}(x_0);$$

$$C(x) = c_0 + \alpha X c_1 + \alpha^2 X^2 c_2 + \cdots, \qquad c_n = \frac{1}{n!}C^{(n)}(x_0).$$

Expanding (15.2.8) and gathering like powers of α results in the sequence of equations

$$u_0'' = 2u_0^3;$$
$$u_2'' = 6u_0^2 u_2 + b_0 u_0;$$
$$u_3'' = 6u_0^2 u_3 + u_0 b_1 X + c_0;$$
$$u_4'' = 6u_0^2 u_4 + 6u_0 u_2^2 + u_0 b_2 X^2 + u_2 b_0 + c_1 X; \ldots$$

We take $u_0(X) = \pm 1/X$ as the solution of the first of these equations. The operator $(d/dX)^2 - 6/X^2$ takes X^n to $[n(n-1) - 6]X^{n-2}$, so we may take $u_2(X) = \mp b_0 X/6$ and u_3 to be $-(c_0 \pm b_1)X^2/4$. The next equation becomes

$$u_4'' - \frac{6}{X^2}u_4 = (c_1 \pm b_2)X.$$

This has the solution $u_4 = (c_1 \pm b_2)(X^3 \log X)/5$. If the solution $u(X, \alpha)$ has the Painlevé property for small $|\alpha|$, then each of the terms u_j must also be single-valued, so a necessary condition is $c_1 \pm b_2 = 0$. Thus we need

$$B''(x_0) = 2b_2 = 0; \quad C'(x_0) = c_1 = 0.$$

Since the starting point x_0 was arbitrary, this completes the proof. $\qquad\square$

This result leads to various simplifications. If B and C are both constants, the general solution of (15.2.6) is an elliptic function; see Exercise 9. Otherwise, by a linear change of coordinate $x = ay + \beta$, $v = \gamma u$, we reduce to PII:

$$u''(x) = 2u(x)^3 + xu(x) - c, \qquad c \text{ constant.} \tag{15.2.9}$$

(Replacing a, in the normalization PII given in the introductory section, by $-c$ will be convenient later on.) We shall refer to (15.2.9) in this form as PII(c).

15.3 Solutions of PII

As indicated at the start of Section 15.1, equation PII(c) has a regular solution with initial conditions $u(x_0) = u_0$, $u'(x_0) = v_0$ in the neighborhood of x_0. Let us look for a solution with a pole at x_0, having the Laurent expansion

$$u(x) = \frac{a_{-m}}{(x-x_0)^m} + \frac{a_{1-m}}{(x-x_0)^{m-1}} + \frac{a_{2-m}}{(x-x_0)^{m-2}} + \cdots,$$

with $m > 0$. Looking at the highest powers of $(x-x_0)^{-1}$ on each side of (15.2.9), we see that necessarily $m = 1$ and $a_{-1} = \pm 1$. (This accounts for the choice of 2 as the coefficient of u^3. The same consideration accounts for the choice of 6 as the coefficient of u^2 in PI.) Further comparison of the coefficients on the two sides of (15.2.9) shows that one must have

$$u(x) = \frac{\varepsilon}{x-x_0} - \frac{x_0\varepsilon}{6}(x-x_0) - \frac{\varepsilon-c}{4}(x-x_0)^2 + h(x-x_0)^3 + a_4(x-x_0)^4 + \cdots,$$
(15.3.1)

where $\varepsilon = \pm 1$, h is an arbitrary constant and the remaining coefficients a_4, $a_5, \ldots,$ are determined by

$$[(m+2)(m+1) - 6]\, a_{m+2}$$
$$= x_0 a_m + a_{m-1} + 6\varepsilon \sum_{j+k=m+1} a_j a_k + 2 \sum_{j+k+l=m} a_j a_k a_l, \quad (15.3.2)$$

where the sums are over $j,k > 0$ and $j,k,l > 0$. This can be used to prove convergence of the series near $x = x_0$, and thus the existence of a one-parameter family of solutions with a simple pole and residue $\varepsilon = \pm 1$ at $x = x_0$. In fact, suppose that $|a_k| \le M^k$ for $k = 1,2,3,4$, with $M \ge |x_0| + 3$. Suppose $m \ge 2$. There are m summands in the first sum and $(m-1)(m-2)/2$ summands in the second sum in (15.3.2). If $|a_j| \le M^j$ for $0 < j \le m$, then (15.3.2) implies that for $m > 1$,

$$|a_{m+2}| \le |x_0|M^m + M^{m-1} + \frac{6m}{(m-1)(m+4)}M^{m+1} + \frac{(m-2)}{(m+4)}M^m. \quad (15.3.3)$$

Both fractional coefficients are ≤ 2 when $m > 2$, so $|a_m| \le M^m$ for all m, and the series converges when $|x-x_0| < 1/M$.

We consider now the possibility of rational solutions $u = P/Q$, where P and Q are polynomials. The following result is due to Airault [5].

Theorem 15.3.1 *Equation* PII(c) *has a rational solution if and only if c is an integer. If c is an integer, the rational solution is unique.*

Existence and uniqueness will be proved in the next sections. Here we show that c must be an integer and obtain some information about the form of a rational solution.

If u is rational, it has the form

$$u(x) = p(x) + O(x^{-1}) \quad \text{as } x \to \infty,$$

where p is a polynomial. Comparison of the two sides of (15.2.9) as $x \to \infty$ shows that the polynomial p is zero. We know that the poles of u are simple, so u has the form

$$u(x) = \frac{\varepsilon_1}{x - x_1} + \frac{\varepsilon_2}{x - x_2} + \cdots + \frac{\varepsilon_n}{x - x_n}, \qquad \varepsilon_j^2 = 1, \qquad (15.3.4)$$

where the x_j are distinct. As $x \to \infty$, it follows from (15.2.9) and from (15.3.4) that

$$0 = \lim_{x \to \infty} x u(x) - c = \varepsilon_1 + \varepsilon_2 + \cdots + \varepsilon_n - c. \qquad (15.3.5)$$

Therefore c must be an integer.

The expansion (15.3.1) shows that u^2 has no simple poles. Therefore

$$u(x)^2 = \frac{1}{(x - x_1)^2} + \frac{1}{(x - x_2)^2} + \cdots + \frac{1}{(x - x_n)^2} \qquad (15.3.6)$$

and it follows that

$$c^2 = \lim_{x \to \infty} x^2 u(x)^2 = n. \qquad (15.3.7)$$

Therefore PII(0) has only the trivial polynomial $u \equiv 0$ as a rational solution.

Comparing (15.3.5) and (15.3.7) shows that if k of the ε_j are positive, then $c = 2k - n$ and $n = c^2$, so $2k = c^2 + c$. We have established the following.

Proposition 15.3.2 *If* PII(c) *has a rational solution* u, *then* c *is an integer, and* u *has the form* (15.3.4) *where* $n = c^2$. *Moreover,* $(c^2 + c)/2$ *of the* ε_j *equal* $+1$ *and* $(c^2 - c)/2$ *of the* ε_j *equal* -1.

As is shown in the next section, the rational solutions for integer c can be calculated sequentially, starting from the zero solution for $c = 0$. The fact that the degree is c^2 shows that there is a rapid increase in complexity: already for $c = \pm 3$, u is a ratio of polynomials of degree 8 and 9.

To complete the Painlevé program, after identifying the 50 canonical reduced forms mentioned in the introduction, one needs to show explicitly that each of the 50 equations has the Painlevé property, that 44 have general solutions that are expressible via previously known functions, and that none of the remaining ones, PI–PVI, has such a general solution. In view of the limiting operations that lead from PVI through each of the other equations to PI, it is enough to show that PVI has the Painlevé property and that the general solution of PI is a genuinely new transcendental function. See Section 14.43 of Ince [193] for the argument that the general solution of PI is new. We show later that PII has the Painlevé property.

Assuming the Painlevé property, it follows that every solution u of PII is meromorphic in the plane. Therefore it is the quotient of entire functions $u = v/w$ (see the last paragraph of Appendix A). Painlevé showed how to find such a representation, using the form of u near a pole. If u has a pole at x_j, then for x near $x = x_j$,

$$u(x)^2 = \frac{1}{(x - x_j)^2} + O(1). \tag{15.3.8}$$

Choose a point x_0 that is not a pole of u and let

$$w(x) = \exp\left\{ -\int_{x_0}^{x} \int_{x_0}^{y} u^2(z)\,dz\,dy \right\}. \tag{15.3.9}$$

The paths of integration are taken so as to avoid the poles of u. In view of (15.3.8), for x near x_j,

$$-\int_{x_0}^{x} \int_{x_0}^{y} u^2(z)\,dz\,dy = \log(x - x_j) + O(1).$$

The branch of the logarithm varies according to the path, but exponentiation removes the ambiguity. It follows from this construction that w has a simple zero at each pole of u, and no other zeros. Therefore $v = uw$ is an entire function, and we have obtained a representation of u as a quotient of the entire functions v/w.

Differentiating (15.3.9) twice and multiplying by w gives

$$w w'' = (w')^2 - v^2. \tag{15.3.10}$$

Putting v/w into PII(c) and using (15.3.10) leads to

$$w^2 v'' + v(w')^2 = 2v' w w' + v^3 + xv w^2 - cw^3. \tag{15.3.11}$$

Conversely, if functions v, w satisfy the system (15.3.10), (15.3.11), then $u = v/w$ is a solution of (15.2.9).

Finally, let us consider the asymptotics of a general solution of PII as $x \to \infty$. We look for a renormalization of the form $u(x) = a(x)U(y)$, $y = \varphi(x)$ in such a way that the term xu in PII is replaced by U alone. This leads to the choice $a(x) = \varphi'(x) = x^{1/2}$, so $y = 2x^{3/2}/3$ and

$$U''(y) = 2U(y)^3 + U(y) - \frac{U'(y)}{y} - \frac{2c}{3y} + \frac{U(y)}{9y^2}.$$

For large y, this can be viewed as a perturbation of the equation $U'' = 2U^3 + U$. Multiplying by U' and integrating converts this latter equation to

$$(U')^2 = U^4 + U^2 + a$$

for some constant a. The solutions of this equation are elliptic functions of the form $\mathrm{sn}(\alpha y + \beta)$ with zeros and poles on translates of a lattice: $y = m\omega_1 + n\omega_2$. This gives a good qualitative description of u for large x, as shown by Boutroux [52] in 1913–1914; see Hille [185] for a sketch of the argument in the case of PI. More recently, refinements of Boutroux's results have been obtained by exploiting connections to Riemann–Hilbert problems. The details are beyond the scope of this chapter. (The two-part memoir by Boutroux [52] comes to 280 pages, and the paper of Deift and Zhou [101] to 100 pages. Fokas, Its, Kapaev, and Novokshenov [137] devote 140 pages to PII(0) and 60 more pages to general PII(c).) A brief treatment of some results for PII(0) is given in Section 15.8.

15.4 Compatibility conditions and Bäcklund transformations

The Painlevé equations can be obtained as *compatibility conditions* for certain systems of equations [136, 205]. This fact can be used effectively in studying the Painlevé transcendents. For PII, consider the system of equations in 2×2 matrix form

$$\Psi_x(x, \lambda) = [\lambda J + Q(x)]\, \Psi(x, \lambda); \qquad (15.4.1)$$

$$\lambda\, \Psi_\lambda(x, \lambda) = A(x, \lambda)\Psi(x, \lambda), \qquad (15.4.2)$$

where the subscripts denote partial derivatives and

$$J = \begin{bmatrix} 1 & 0 \\ 0 & -1 \end{bmatrix}, \qquad Q(x) = \begin{bmatrix} 0 & u(x) \\ u(x) & 0 \end{bmatrix};$$

$$A(x, \lambda) = \sum_{j=0}^{3} \lambda^j A_j(x), \qquad A_3 = -4J.$$

It follows that

$$\{\lambda\, \Psi_\lambda\}_x = \{A(x, \lambda)\Psi\}_x = A_x(x, \lambda)\Psi + A(x, \lambda)(\lambda J + Q)\Psi$$

$$= \lambda\, \{\Psi_x\}_\lambda = \lambda\, \{(\lambda J + Q)\Psi\}_\lambda = \lambda J\Psi + (\lambda J + Q)A(x, \lambda)\Psi.$$

The matrix solution Ψ is assumed to be a fundamental solution, i.e. invertible, so the compatibility condition is

$$A_x(x, \lambda) - [Q(x), A(x, \lambda)] = [\lambda J, A(x, \lambda)] + \lambda J, \qquad (15.4.3)$$

where $[A, B]$ denotes the commutator $AB - BA$. Since $A_3 = -4J$ commutes with J, collecting terms of like degree leads to four nontrivial equations that can be written as

$$-[Q,A_3] = [J,A_2]; \tag{15.4.4}$$

$$A_2' - [Q,A_2] = [J,A_1]; \tag{15.4.5}$$

$$A_1' - [Q,A_1] = [J,A_0] + J; \tag{15.4.6}$$

$$A_0' - [Q,A_0] = 0, \tag{15.4.7}$$

where the primes denote derivatives with respect to x. Equation (15.4.4) determines the off-diagonal part of A_2 and (15.4.5) then shows that the diagonal part of A_2 is constant, and we take it to be zero. Thus

$$A_2 = \begin{bmatrix} 0 & -4u \\ -4u & 0 \end{bmatrix} = -4Q, \tag{15.4.8}$$

and similarly,

$$A_1 = \begin{bmatrix} x+2u^2 & -2u' \\ 2u' & -x-2u^2 \end{bmatrix}, \tag{15.4.9}$$

$$A_0 = (-u'' + 2u^3 + xu)\begin{bmatrix} 0 & 1 \\ 1 & 0 \end{bmatrix}. \tag{15.4.10}$$

Therefore (15.4.7) is equivalent to PII(c):

$$u''(x) = 2u(x)^3 + xu(x) - c, \tag{15.4.11}$$

so

$$A_0 = c\begin{bmatrix} 0 & 1 \\ 1 & 0 \end{bmatrix}.$$

The system (15.4.1), (15.4.2) can be exploited to investigate properties of PII. We start with some preliminary observations.

Proposition 15.4.1 (a) *Suppose that* Ψ *and* Ψ_1 *are solutions of* (15.4.1) *and that* Ψ *is invertible. Then the determinants are constant with respect to x and there is a matrix C, constant with respect to x such that* $\Psi_1 = \Psi C$.
(b) *Suppose that* Ψ *and* Ψ_1 *are solutions of* (15.4.2) *and* Ψ *is invertible. Then the determinants are constant with respect to λ and there is a matrix C, constant with respect to λ such that* $\Psi_1 = \Psi C$.
(c) *If* $\Psi(x,\lambda)$ *is a solution of one or both of equations* (15.4.1) *and* (15.4.2), *then so is the function*

$$\begin{bmatrix} 0 & 1 \\ 1 & 0 \end{bmatrix} \Psi(x,-\lambda) \begin{bmatrix} 0 & 1 \\ 1 & 0 \end{bmatrix}. \tag{15.4.12}$$

Proof The statements about the determinants follow from the fact that $\lambda J + Q$ and $A(\lambda)$ have trace zero; see Exercise 11. The statements about the relationship between Ψ_1 and Ψ follow from differentiating $\Psi^{-1}\Psi_1$; see Exercise 11. The statement about (15.4.12) follows from the (easily checked) identities

$$\begin{bmatrix} 0 & 1 \\ 1 & 0 \end{bmatrix} [\lambda J + Q(x)] \begin{bmatrix} 0 & 1 \\ 1 & 0 \end{bmatrix} = -\lambda J + Q(x);$$

$$\begin{bmatrix} 0 & 1 \\ 1 & 0 \end{bmatrix} A(x,\lambda) \begin{bmatrix} 0 & 1 \\ 1 & 0 \end{bmatrix} = A(x,-\lambda). \qquad \square$$

For some calculations, it is convenient to select a basis for the space of 2×2 complex matrices. We let $\mathbf{1}$ denote the identity matrix and set

$$J_1 = J = \begin{bmatrix} 1 & 0 \\ 0 & -1 \end{bmatrix}, \qquad J_2 = \begin{bmatrix} 0 & 1 \\ 1 & 0 \end{bmatrix}, \qquad J_3 = \begin{bmatrix} 0 & 1 \\ -1 & 0 \end{bmatrix}. \quad (15.4.13)$$

Then

$$\begin{aligned} J_1^2 = J_2^2 &= \mathbf{1} = -J_3^2; \\ J_1 J_2 &= -J_2 J_1 = J_3; \\ J_1 J_3 &= -J_3 J_1 = J_2; \\ J_2 J_3 &= -J_3 J_2 = -J_1. \end{aligned} \qquad (15.4.14)$$

It follows that a matrix B commutes with J_2 if and only if it has the form

$$B = a\mathbf{1} + bJ_2 = \begin{bmatrix} a & b \\ b & a \end{bmatrix}, \qquad (15.4.15)$$

while B anticommutes with J_2, i.e. $BJ_2 = -J_2 B$, if and only if B has the form

$$B = aJ_1 + bJ_3 = \begin{bmatrix} a & b \\ -b & -a \end{bmatrix}. \qquad (15.4.16)$$

Moreover, if B has this latter form then

$$B^2 = (aJ_1 + bJ_3)^2 = (a^2 - b^2)\mathbf{1}. \qquad (15.4.17)$$

In view of part (c) of Proposition 15.4.1, we usually normalize in such a way that Ψ satisfies the *symmetry condition*

$$\Psi(x,-\lambda) = J_2 \Psi(x,\lambda) J_2. \qquad (15.4.18)$$

As a first application of these ideas, we discuss the question of transforming a solution of PII(c) into a solution with a different constant. One transformation is trivial: if u satisfies PII(c), then $-u$ satisfies PII($-c$). We can construct a more subtle transformation as follows. Suppose that the system (15.4.1), (15.4.2) is satisfied with $Q = uJ_2$, and suppose that u satisfies PII(c). We may assume that Ψ satisfies the symmetry condition. We look for a new system with

$$\widetilde{\Psi}(x,\lambda) = \left[\mathbf{1} - \lambda^{-1}B(x)\right]\Psi(x,\lambda). \tag{15.4.19}$$

(The general form of this transformation is dictated by a choice of asymptotics as $\lambda \to \infty$.) The symmetry condition implies that $B = -J_2BJ_2$. We will also want $B^2 = 0$, so B has the form $B = b(J_1 \pm J_3)$. Let us choose

$$B = b(J_1 + J_3) = b\begin{bmatrix} 1 & 1 \\ -1 & -1 \end{bmatrix}.$$

Then $\mathbf{1} - \lambda^{-1}B$ has inverse $\mathbf{1} + \lambda^{-1}B$, so

$$\begin{aligned}
\lambda\widetilde{\Psi}_\lambda &= \lambda^{-1}B\widetilde{\Psi} + \left[\mathbf{1} - \lambda^{-1}B\right]A(\lambda)\Psi \\
&= \left\{\lambda^{-1}B + \left[\mathbf{1} - \lambda^{-1}B\right]A(\lambda)\left[\mathbf{1} + \lambda^{-1}B\right]\right\}\widetilde{\Psi}.
\end{aligned}$$

We need the terms of order λ^{-2} and λ^{-1} to vanish:

$$BA_0B = 0; \qquad B + [A_0, B] - BA_1B = 0.$$

The first equation follows from the form of B. Since $A_0 = cJ_2$ and $[J_2, B] = -2B$, the second equation is

$$(1 - 2c)B = BA_1B. \tag{15.4.20}$$

We assume $c \neq 1/2$. Now A_1 has the form $\alpha J_1 + \beta J_3$, so a calculation shows that $BA_1B = 2(\alpha - \beta)bB$. Therefore (15.4.20) yields

$$b = \frac{1 - 2c}{2(\alpha - \beta)} = \frac{1 - 2c}{2(x + 2u^2 + 2u')}.$$

There appears to be a problem if $u' = -u^2 - x/2$, but it is easily seen that this identity implies that u satisfies PII(1/2) – the case that we have excluded.

This shows that $\widetilde{\Psi}$ satisfies an equation of the form $\lambda\Psi_\lambda = \widetilde{A}(x,\lambda)\Psi$, where $\widetilde{A}$ is a polynomial in λ with leading term $\lambda^3 A_3$. The constant term is

$$\widetilde{A}_0 = A_0 + [A_1, B] - BA_2B.$$

The third term on the right-hand side is zero, and a calculation shows that $[A_1, B] = (1 - 2c)J_2$. Thus the constant term is $\widetilde{A}_0 = (1 - c)J_2$.

Next, we want to show that $\widetilde{\Psi}$ satisfies an equation of the form (15.4.1). But

$$\begin{aligned}
\widetilde{\Psi}_x & \widetilde{\Psi}^{-1} \\
&= -\lambda^{-1}B'(1 + \lambda^{-1}B) + (1 - \lambda^{-1}B)(\lambda J + Q)(1 + \lambda^{-1}B) \\
&= -\lambda^{-2}B'B + \lambda^{-1}\left(-B' + [Q, B] - BJB\right) + Q + [J, B] + \lambda J.
\end{aligned}$$

The coefficient of λ^{-2} is $(b'/b)B^2 = 0$. Since $Q = uJ_2$ and $BJB = 2bB$, the coefficient of λ^{-1} is B multiplied by

$$
-\frac{b'}{b} - 2u - 2b = b\left[\left(\frac{1}{b}\right)' - \frac{2u}{b} - 2\right]
$$

$$
= \frac{2b}{1 - 2c}\left[(x + 2u^2 + 2u')' - 2u(x + 2u^2 + 2u') - (1 - 2c)\right]
$$

$$
= \frac{2b}{1 - 2c}\left[1 + 4uu' + 2u'' - 2ux - 4u^3 - 4uu' - 1 + 2c\right]
$$

$$
= \frac{4b}{1 - 2c}\left[u'' - 2u^3 - ux + c\right] = 0. \tag{15.4.21}
$$

Thus $\widetilde{\Psi}$ satisfies (15.4.1) with

$$
\widetilde{Q} = Q + [J, B] = Q + 2bJ_2 = (u + 2b)J_2.
$$

It will be shown in the next section that if u is any solution to PII(c), then there is a symmetric $\Psi(x, \lambda)$ that satisfies (15.4.1) and (15.4.2). We have just shown that if the constant c is not $\frac{1}{2}$, then the function (15.4.19) satisfies a system of the same form, with the associated constant term $\widetilde{A}_0 = (1 - c)J_2$ and the associated function

$$
\widetilde{u} = u + \frac{1 - 2c}{x + 2u^2 + 2u'}. \tag{15.4.22}
$$

Therefore $\widetilde{u}$ is a solution of PII($1 - c$). This transformation and the transformation $u \to -u$ are examples of *Bäcklund transformations* for PII. By composing these two transformations, we can obtain others. Taking the negative of $\widetilde{u}$ gives a solution of PII($c - 1$). First taking the negative and then applying the transformation (15.4.22) gives a solution of PII($c + 1$). Thus we have the following result.

Theorem 15.4.2 *Suppose that* $u = u(x, c)$ *is a solution of* PII(c). *Then* $u(x, -c) = -u(x, c)$ *is a solution of* PII($-c$). *Moreover,* $u(x, c - 1)$ *and* $u(x, c + 1)$ *defined by*

$$
u(x, c - 1) = -u - \frac{1 - 2c}{x + 2u^2 + 2u'}, \qquad c \neq \frac{1}{2}; \tag{15.4.23}
$$

$$
u(x, c + 1) = -u + \frac{1 + 2c}{x + 2u^2 - 2u'}, \qquad c \neq -\frac{1}{2}, \tag{15.4.24}
$$

are solutions of PII($c - 1$) *and* PII($c + 1$), *respectively.*

Remark. In the argument so far, we have assumed the existence of solutions Ψ of (15.4.1) and (15.4.2). However one can, in principle, verify directly that if

u is a solution of PII(c), then (15.4.23) is a solution of PII($c - 1$) and (15.4.24) is a solution of PII($c + 1$).

Starting from the trivial solution $u_0 = 0$ to PII(0), this procedure produces the solutions

$$u_1(x) = \frac{1}{x}, \qquad u_2(x) = -\frac{1}{x} + \frac{3x^2}{x^3 + 4}$$

to PII(1) and PII(2), respectively. Continuing, and taking negatives, we have proved the existence part of Theorem 15.3.1.

The computation can be simplified somewhat. We obtain a relation between three consecutive terms u_{n-1}, u_n, and u_{n+1} by eliminating u_n':

$$2x + 4u_n^2 = \frac{1 + 2n}{u_n + u_{n+1}} - \frac{1 - 2n}{u_n + u_{n-1}}. \qquad (15.4.25)$$

Applying the transformation (15.4.22) twice converts the constant c first to $1 - c$ and then to $1 - (1 - c) = c$. This is an example of a *self-Bäcklund transformation*. However, it turns out to be the identity transformation.

Proposition 15.4.3 *The transformation* (15.4.22) *is its own inverse.*

See Exercise 13.

Corollary 15.4.4 *The tranformations* (15.4.23) *and* (15.4.24) *are invertible.*

Since PII(0) has only the trivial rational solution $u \equiv 0$, the following is a consequence.

Corollary 15.4.5 *For each integer n, the rational solution of* PII(n) *is unique.*

Let us look briefly at some special cases of PII(c). We noted above that $u' = -u^2 - x/2$ implies that u is a solution to PII(1/2). The equation $u' + u^2 + x/2 = 0$ is an example of a *Riccati equation*, and the standard approach is to take $u = \varphi'/\varphi$. This leads to

$$\varphi'' + \frac{x}{2}\varphi = 0.$$

A further transformation, $\varphi(x) = \psi(i2^{-1/2}x)$, leads to Airy's equation (9.9.3). This fact and the previous constructions imply the following.

Proposition 15.4.6 *Any* PII($n + \frac{1}{2}$), n *an integer, has a solution that is a rational function of Airy functions and their derivatives.*

As noted above, the remaining Painlevé equations can also be realized as compatibility conditions for systems like (15.4.1), (15.4.2), and studied by similar methods. For example, the 2×2 matrix system

$$\Psi_x(x, \lambda) = [\lambda J_1 + v(x)J_2] \Psi(x, \lambda);$$

$$\lambda \Psi_\lambda(x, \lambda) = \sum_{j=0}^{5} \lambda^j A_j(x) \Psi(x, \lambda); \qquad (15.4.26)$$

$$v'(x) = 2u(x) - v(x)^2,$$

where $A_5 = -4J_1$, has the compatibility condition PI; see Exercise 12.

The system (15.4.1), (15.4.2) is the simplest of these systems, and the results for PII are the most complete.

15.5 Construction of Ψ

We assume throughout this section that the compatibility condition (15.4.3) is satisfied. Rewrite it as a commutation condition for differential operators:

$$\left[\frac{\partial}{\partial x} - \lambda J - Q(x)\right] \left[\lambda \frac{\partial}{\partial \lambda} - A(x, \lambda)\right]$$

$$= \left[\lambda \frac{\partial}{\partial \lambda} - A(x, \lambda)\right] \left[\frac{\partial}{\partial x} - \lambda J - Q(x)\right]. \qquad (15.5.1)$$

Near $\lambda = 0$, (15.4.2) can be considered as a small perturbation of the equation $\lambda \Psi_\lambda = A_0 \Psi = cJ_2 \Psi$, which has the solution

$$\lambda^{A_0} = \exp(c \log \lambda J_2) = \frac{\lambda^c + \lambda^{-c}}{2} 1 + \frac{\lambda^c - \lambda^{-c}}{2} J_2.$$

Proposition 15.5.1 *If $A_0 = cJ_2$ and $c + \frac{1}{2}$ is not an integer, then the system* (15.4.1), (15.4.2) *has a solution Ψ of the form*

$$\Psi(x, \lambda) = V(x, \lambda) \lambda^{A_0}, \qquad (15.5.2)$$

where $V(x, \cdot)$ is an entire function of λ.

Proof Equation (15.4.2) for Ψ of the form (15.5.2) is equivalent to

$$\lambda V_\lambda(x, \lambda) = A(x, \lambda)V(x, \lambda) - V(x, \lambda)A_0. \qquad (15.5.3)$$

We look for a series solution $V(x) = \sum_{k=0}^{\infty} \lambda^k v_k(x)$. Equation (15.5.3) implies the sequence of equations

$$k v_k - [A_0, v_k] = A_3 v_{k-3} + A_2 v_{k-2} + A_1 v_{k-1},$$

where we have taken $v_j = 0$ if $j < 0$. Let us also impose the symmetry condition, so that

$$v_k(x) = \begin{cases} a_k(x)J_2 + d_k(x)\mathbf{1}, & k = 2,4,6,\ldots; \\ b_k(x)J_1 + c_k(x)J_3, & k = 1,3,5,\ldots \end{cases}$$

Thus $[A_0, v_k] = 0$ for k even and $[A_0, v_k] = 2cJ_2 v_k$ for k odd. Inverting $k\mathbf{1} - 2cJ_2$, we find, for even and odd k, respectively,

$$v_k = \frac{1}{k}[A_3 v_{k-3} + A_2 v_{k-2} + A_1 v_{k-1}];$$

$$v_k = \frac{1}{k^2 - 4c^2}(k\mathbf{1} + 2cJ_2)[A_3 v_{k-3} + A_2 v_{k-2} + A_1 v_{k-1}].$$

Since we have assumed that $2c$ is not an odd integer, there is a constant $C \geq 1$ such that $C|k^2 - 4c^2| \geq k(k + |2c|)$ for all odd integers k. Let $C_A = \max\{1, C\|A_j\|, j = 1,2,3\}$. It follows inductively that

$$\|v_k(x)\| \leq \frac{C_A^k}{\Gamma(\frac{1}{3}k + 1)}\|v_0(x)\|,$$

where the norm is any matrix norm that satisfies $\|AB\| \leq \|A\|\|B\|$. Therefore the formal series $\sum_{k=0}^{\infty} \lambda^k v_k$ converges to an entire function. We now specify $v_0(x)$ by choosing it to be an invertible solution of the matrix equation

$$v_0'(x) = Q(x)v_0(x). \tag{15.5.4}$$

Condition (15.5.1) implies that the function

$$U(x, \lambda) \equiv \Psi_x(x, \lambda) - [\lambda J + Q(x)]\Psi(x, \lambda)$$

is a solution of $\lambda U_\lambda = A(\lambda)U$ and is therefore a multiple of Ψ. Multiplying by λ^{-cJ_2}, we have the analogous result for V:

$$V_x(x, \lambda) - [\lambda J + Q(x)]V(x, \lambda) = V(x, \lambda)C(x)$$

for some matrix function C. But (15.5.4) implies that at $\lambda = 0$ the left-hand side of this equation is identically zero. Thus V satisfies (15.4.1), and so does Ψ. □

For large λ, (15.5.3) can be viewed as a perturbation of the equation

$$\Psi_\lambda = \lambda^2 A_3 \Psi = -4\lambda^2 J\Psi,$$

which has the solution $\exp(\frac{1}{3}\lambda^3 A_3) = \exp(-\frac{4}{3}\lambda^3 J)$, so we look for Ψ as a multiple of this function. To take into account (15.4.1) as well, we set

$$\Phi(x, \lambda) = -\frac{4}{3}\lambda^3 J + \lambda x \tag{15.5.5}$$

and consider solutions of the form

$$\Psi(x,\lambda) = m(x,\lambda)\exp[\Phi(x,\lambda)J]. \tag{15.5.6}$$

Note that Ψ satisfies the symmetry condition (15.4.18) if and only if the associated m satisfies it.

The equations for m that correspond to (15.4.1) and (15.4.2) are

$$m_x(x,\lambda) = [\lambda J, m(x,\lambda)] + Q(x)m(x,\lambda); \tag{15.5.7}$$

$$\lambda m_\lambda(x,\lambda) = A(x,\lambda)m(x,\lambda) + (4\lambda^3 - \lambda x)m(x,\lambda)J. \tag{15.5.8}$$

The following argument is a more direct version of the construction in the paper [35] of Beals and Sattinger. We begin by looking for an asymptotically approximate solution of (15.5.8): a matrix-valued function f of the form

$$f(x,\lambda) = \mathbf{1} + \lambda^{-1}f_1(x) + \lambda^{-2}f_2(x) + \lambda^{-3}f_3(x),$$

such that

$$\lambda f_\lambda(x,\lambda) - A(x,\lambda)f(x,\lambda) - (4\lambda^3 - x\lambda)f(x,\lambda)J = R(x,\lambda) = O(\lambda^{-1}). \tag{15.5.9}$$

This is equivalent to the conditions

$$4[J,f_k] = A_2f_{k-1} + A_1f_{k-2} - xf_{k-2}J + A_0f_{k-3}, \quad k = 0,1,2,3; \tag{15.5.10}$$

see Exercise 14. The solution is not unique, but it can be checked directly that one solution is

$$f(x,\lambda) = \mathbf{1} - \frac{1}{2\lambda}u(x)J_3 - \frac{1}{4\lambda^2}u'(x)J_2 + \frac{1}{8\lambda^3}[c - xu(x) - u^3(x)]J_3. \tag{15.5.11}$$

For a derivation from (15.5.10), see Exercise 15.

The behavior of an actual solution of (15.5.8) as $\lambda \to \infty$ depends on the way in which λ approaches ∞. We begin by distinguishing the rays where $\mathrm{Re}(\lambda^3) = 0$, starting with the fourth quadrant:

$$\Sigma_\nu = \{\lambda : \arg\lambda = -\pi/6 + \nu\pi/3\}, \qquad \nu = 0,1,2,3,4,5. \tag{15.5.12}$$

It will be convenient to continue the numbering cyclically: $\Sigma_{\nu+6} = \Sigma_\nu$. (These rays are called *Stokes lines* by mathematicians, and *anti-Stokes* lines by physicists; their importance was emphasized by Stokes [387].)

Proposition 15.5.2 *For each x in the domain of u, and each sector Ω_ν with opening $2\pi/3$ bounded by rays $\Sigma_{\nu-1}$ and $\Sigma_{\nu+1}$, there is a unique solution $m_\nu(x,\lambda)$ of (15.5.8) with the property that for each closed subsector, $m(x,\lambda)$ is defined and holomorphic for sufficiently large λ and satisfies*

$$\lim_{\lambda \to \infty} m_\nu(x,\lambda) = \mathbf{1}. \tag{15.5.13}$$

Moreover, $m_\nu(x, \lambda)$ also satisfies (15.5.7).

Proof We look for a solution m of (15.5.8) in the form $m = fg$, where f is the approximate solution (15.5.11). Equation (15.5.8) is equivalent to the equation

$$\lambda g_\lambda(x, \lambda) - (-4\lambda^3 + x\lambda)[J, g] = \lambda s(x, \lambda) g(x, \lambda), \qquad (15.5.14)$$

where

$$\lambda s(x, \lambda) = f(x, \lambda)^{-1} R(x, \lambda),$$

and $R(x, \lambda)$ denotes the remainder term in (15.5.9). Note that $s(x, \lambda) = O(\lambda^{-2})$.

Equation (15.5.14) with condition $g(x, \infty) = \mathbf{1}$ is equivalent to the integral equation

$$g(x, \lambda) = \mathbf{1} + \int_\infty^\lambda e^{[\Phi(x,\lambda) - \Phi(x,\zeta)]J} s(x, \zeta) g(x, \zeta) e^{[\Phi(x,\zeta) - \Phi(x,\lambda)]J} \, d\zeta. \qquad (15.5.15)$$

We need to specify the path of integration in such a way that a solution exists. For any fixed λ in the sector, the curve

$$C_\lambda = \{\zeta : \operatorname{Re} \Phi(x, \lambda) = \operatorname{Re} \Phi(x, \zeta)\}$$

has a branch that is asymptotic to the ray Σ_ν. We take the integration in (15.5.15) along this branch, so that the exponents that occur are purely imaginary. In a fixed closed subsector, given $r_0 > 0$, there is $\delta > 0$ such that if λ belongs to the subsector and $|\lambda| \geq r_0/\delta$, then $|\zeta| \geq r_0$ along this branch of the curve. In particular, r_0 may be chosen so that for the curves in this subsector that start with $|\lambda| \geq r_0/\delta$,

$$\int^\lambda |s(x, \zeta)| \, |d\zeta| \leq \frac{c}{|\lambda|}$$

for some constant c. It follows from this that for λ in the closed sector and $|\lambda|$ greater than the larger of r_0/δ and c, the integral equation can be solved by successive approximations: $g = \sum_{k=0}^\infty g_k$ with $g_0 = \mathbf{1}$ and $g_{k+1} = Wg_k$, where W denotes the integral operator

$$Wh(\lambda) \equiv \int_\infty^\lambda e^{[\Phi(x,\lambda) - \Phi(x,\zeta)]J} s(x, \zeta) h(\zeta) e^{[\Phi(x,\zeta) - \Phi(x,\lambda)]J} \, d\zeta. \qquad (15.5.16)$$

This construction implies that in the given closed subsector $g(x, \lambda) = \mathbf{1} + O(\lambda^{-1})$, and g has an asymptotic expansion in powers of λ^{-1}, so the same is true of m:

$$m(x, \lambda) \sim \mathbf{1} + \lambda^{-1} m_1(x) + \lambda^{-2} m_2(x) + \cdots \qquad (15.5.17)$$

Uniqueness follows from the fact that two solutions m and $\widetilde{m}$ are related by

$$\widetilde{m} = m e^{\Phi J} C e^{-\Phi J}$$

with matrix $C = C(x)$ independent of λ. Using the asymptotic condition along a curve where $\operatorname{Re}\Phi = 0$, we can conclude that $C = 1$.

By construction, $\Psi_\nu = m_\nu e^{\Phi J}$ satisfies (15.4.2). The compatibility condition implies that $(\Psi_\nu)_x - (\lambda J + Q)\Psi_\nu$ also satisfies (15.4.2), and therefore

$$m_x - \lambda[J,m] - Qm = me^{\Phi J}C_1 e^{-\Phi J},$$

where C_1 is independent of λ. Again the asymptotic conditions imply that C_1 is diagonal. However, it follows from the expansion (15.5.17) that

$$m_x - \lambda[J,m] - Qm = -[J,m_1] - Q + O(\lambda^{-1}),$$

so $C_1 = -[J,m_1] - Q$, which has diagonal part zero. Therefore $C_1 = 0$ and m satisfies (15.5.7). □

Remark. The assumption that $c + \frac{1}{2}$ is not an integer can be omitted, but at the cost of complicating the statements and proofs of Proposition 15.5.1, and of results in the next section.

15.6 Monodromy and isomonodromy

In this section, we consider the *monodromy* of an equation that has the form of (15.4.2):

$$\lambda \Psi_\lambda(\lambda) = A(\lambda)\Psi(\lambda), \tag{15.6.1}$$

where

$$A(\lambda) = -4\lambda^3 J_1 + a\lambda^2 J_2 + \lambda(\alpha J_1 + \beta J_3) + cJ_2. \tag{15.6.2}$$

As in the previous section, we can construct (unique) solutions Ψ_ν such that

$$\lim_{\lambda \to \infty} \Psi_\nu(\lambda)\exp(-\Phi(\lambda)J) = 1 \tag{15.6.3}$$

along rays in the sector Ω_ν, where $\Phi(\lambda) = \Phi(0,\lambda) = -4\lambda^3/3$. (The subsequent discussion does not change if we take $\Phi(\lambda) = \Phi(x_0,\lambda)$ instead.) The functions Ψ_ν extend holomorphically to the punctured plane $\mathbf{C} \setminus \{0\}$. According to Proposition 15.4.1, the functions Ψ_ν and $\Psi_{\nu+1}$ are necessarily related by a constant matrix S_ν:

$$\Psi_{\nu+1}(\lambda) = \Psi_\nu(\lambda)S_\nu. \tag{15.6.4}$$

If $A_0 = cJ_2$ and $c + \frac{1}{2}$ is not an integer, we know that there is a (unique) solution Ψ that has the form

$$\Psi(\lambda) = V(\lambda)\lambda^{A_0}, \qquad V \text{ entire}, \quad V(0) = 1.$$

Then Ψ is related to Ψ_ν by

$$\Psi(\lambda) = \Psi_\nu(\lambda)B_\nu. \tag{15.6.5}$$

The collection $\{S_\nu, B_\nu\}$ is the *monodromy data* for the problem (15.6.1). It characterizes the problem.

Proposition 15.6.1 *Suppose $A_0 = cJ_2$ where $c + \frac{1}{2}$ is not an integer. Then the rest of the matrix-valued polynomial $A(\lambda)$ in (15.6.1) is uniquely determined by the monodromy data $\{S_\nu, B_\nu\}$.*

Proof First, let us take m to be the piecewise holomorphic function defined to coincide with

$$m_\nu(\lambda) = \Psi_\nu e^{-\Phi(0,\lambda)J}$$

on the intersection

$$\Omega_{\nu+1} \cap \Omega_\nu \cap \{\lambda \,|\, |\lambda| > 1\},$$

and with the entire function $\Psi_0 \lambda^{-A_0}$ on the unit disc $\{|\lambda| < 1\}$.

Suppose that (15.6.1) with $\widetilde{A}(\lambda)$ has the same monodromy data. Construct the corresponding functions $\widetilde{\Psi}$, $\widetilde{m}_\nu$, and $\widetilde{m}$. Consider the function $\varphi = \widetilde{m} m^{-1}$. This function has limits from each side on the intersections of rays Σ_ν with $\{|\lambda| > 1\}$. The fact that the S_ν are the same implies that φ is continuous across these rays and therefore holomorphic in the complement of the closed unit disc. Similarly, φ is continuous up to the unit circle except possibly at its intersection with the rays Σ_ν. Therefore φ is holomorphic except at these six points. But φ is bounded, so the singularities are removable and φ is constant. Taking the limit along a ray shows that $\varphi = 1$. Therefore $\widetilde{m} = m$ and, outside the unit disc,

$$\widetilde{A}(\lambda) = [\lambda \widetilde{m}_\lambda(\lambda) + \widetilde{m}(\lambda)\lambda \, \Phi_\lambda(0, \lambda)J] \widetilde{m}(\lambda)^{-1}$$
$$= [\lambda m_\lambda(\lambda) + m(\lambda)\lambda \, \Phi_\lambda(0, \lambda)J] m(\lambda)^{-1} = A(\lambda). \qquad \square$$

There is considerable redundancy in the monodromy data. The differential equation (15.6.1) implies that $\det \Psi$ is constant, and it follows from the normalizations at 0 and ∞ that $\det \Psi = \det \Psi_\nu = 1$. Therefore

$$\det S_\nu = 1 = \det B_\nu. \tag{15.6.6}$$

Moreover,

$$S_0 S_1 S_2 S_3 S_4 S_5 = B_0 e^{-2\pi i c J_2} B_0^{-1}; \tag{15.6.7}$$

see Exercise 19. The symmetry property implies that

$$S_{\nu+3} = J_2 S_\nu J_2. \tag{15.6.8}$$

Moreover,

$$e^{\Phi J} S_\nu e^{-\Phi J} = e^{\Phi J} \Psi_\nu^{-1} \Psi_{\nu+1} e^{-\Phi J} = m_\nu^{-1} m_{\nu+1}. \tag{15.6.9}$$

Also, it follows from (15.6.4) and (15.6.5) that

$$B_{\nu+1} = S_\nu^{-1} B_\nu. \tag{15.6.10}$$

Taking the limit along a ray in $\Omega_\nu \cap \Omega_{\nu+1}$, we find that the diagonal part of S_ν is $\mathbf{1}$. Moreover, along such a ray, $\text{Re}\,\Phi \to -\infty$ if ν is even and $\text{Re}\,\Phi \to +\infty$ if ν is odd. The right-hand side of (15.6.9) is bounded, so necessarily S_ν has the form

$$S_\nu = \begin{bmatrix} 1 & a_\nu \\ 0 & 1 \end{bmatrix}, \quad \nu \text{ even}; \qquad S_\nu = \begin{bmatrix} 1 & 0 \\ a_\nu & 1 \end{bmatrix}, \quad \nu \text{ odd}. \qquad (15.6.11)$$

It follows from (15.6.8) and (15.6.7) that

$$a_{j+3} = a_j; \qquad a_1 + a_2 + a_3 + a_1 a_2 a_3 = -2i\sin(c\pi); \qquad (15.6.12)$$

see Exercise 20.

The *isomonodromy* problem associated with (15.6.1) is to determine smooth maps $x \to A(x, \lambda)$, such that $A(x, \lambda)$ has the same monodromy data as $A(\lambda)$. For associated normalized Ψ, we would have a map

$$\Psi(x, \lambda) = U(x, \lambda)[\Psi(\lambda)].$$

Assume that $U(\cdot, \lambda)$ is parameterized so as to have the group property $U(x + y, \lambda) = U(x, \lambda)U(y, \lambda)$, so it can be differentiated, giving

$$\Psi_x(x, \lambda) = \mathbf{G}(x, \lambda)[\Psi(x, \lambda)],$$

where $\mathbf{G}$ is some linear transformation of the space of 2×2 matrices. The assumption that the isomonodromy data is constant implies that the normalized solutions satisfy

$$\begin{aligned}
0 &= \Psi_\nu \left\{ \Psi_\nu^{-1} \Psi_{\nu+1} \right\}_x \Psi_{\nu+1}^{-1} \\
&= \Psi_\nu \left\{ -\Psi_\nu^{-1} \mathbf{G}[\Psi_\nu] \Psi_\nu^{-1} \Psi_{\nu+1} + \Psi_\nu^{-1} \mathbf{G}[\Psi_{\nu+1}] \right\} \Psi_{\nu+1}^{-1} \\
&= -\mathbf{G}[\Psi_\nu] \Psi_\nu^{-1} + \mathbf{G}[\Psi_{\nu+1}] \Psi_{\nu+1}^{-1}.
\end{aligned}$$

This suggests that $\mathbf{G}[B]B^{-1}$ is independent of B, which implies that $\mathbf{G}$ is left multiplication by a matrix-valued function g:

$$\Psi_x(x, \lambda) = g(x, \lambda)\Psi(x, \lambda).$$

This leads to the compatibility condition

$$A_x(x, \lambda) = [g(x, \lambda), A(x, \lambda)] + \lambda g_\lambda(x, \lambda),$$

which suggests that $g(x, \cdot)$ should be a polynomial in λ. Its highest-order term must commute with $A_3 = -4J_1$. The symmetry condition implies

$$g(x, -\lambda) = J_2 g(x, \lambda) J_2,$$

so the highest-order term must have odd degree. Thus the simplest possibility for an isomonodromy flow is given by $g(x, \lambda) = \lambda J + u(x) J_2$, and we return to a system of the form (15.4.1), (15.4.2).

Starting with the general form (15.6.2), there is a unique choice of x_0, $u(x_0)$, and $u'(x_0)$ that equates (15.6.2) to (15.4.8)–(15.4.10):

$$x_0 = a - \frac{a^2}{8}, \quad u(x_0) = -\frac{a}{4}, \quad u'(x_0) = -\frac{\beta}{2}.$$

The constant c is given, and PII(c) is the *isomonodromy equation* for (15.6.1).

The reader is invited to use the compatibility equation to verify that the higher-degree cases of $g(x, \lambda)$ trivialize: $g(x, \lambda) = 0$, so $A(x, \lambda)$ is constant.

15.7 The inverse problem and the Painlevé property

The results in the previous two sections suggest a way to solve PII(c) when $c + \frac{1}{2}$ is not an integer. We give here a brief sketch of the method and how it may be used to prove the Painlevé property.

Given initial values $u(x_0) = u_0$ and $u'(x_0) = v_0$, we construct the corresponding polynomial $A(x_0, \lambda) = \sum_{j=0}^{3} \lambda^j A_j(x_0)$:

$$A_3 = -4J_1, \quad A_2 = -4u_0 J_2, \quad A_1 = (x_0 + 2u_0^2)J_1 - 2v_0 J_3, \quad A_0 = cJ_2, \tag{15.7.1}$$

and compute the associated monodromy data $\{S_v, B_v\}$.

According to Proposition 15.6.1 and its proof, we can determine $A(x, \lambda)$, and in particular $A_2(x) = -4u(x)J_2$, by finding functions $m_v(x, \lambda)$, $m(x, \lambda) = V(x, \lambda)$, holomorphic in the correct regions, and satisfying certain jump conditions on the boundary.

To pose the problem precisely, suppose that we have chosen a constant $R > 0$. Let Γ denote the circle $\{|\lambda| = R\}$, oriented in the direction of increasing argument. Let L_v denote the ray $\{\arg \lambda = v\pi/3, |\lambda| \geq R\}$, oriented from the finite endpoint toward ∞. Let L be the union of Γ and the L_v. We want to find a matrix-valued function m, tending to $\mathbf{1}$ at ∞, holomorphic on the complement of L, and with limits m_+ from the left and m_- from the right along the circle and the oriented rays, such that

$$m_+(x, \lambda) = m_-(x, \lambda) e^{\Phi(x,\lambda)J} S_v e^{-\Phi(x,\lambda)J}, \quad \lambda \in L_v; \tag{15.7.2}$$

$$m_+(x, \lambda) = m_-(x, \lambda) e^{\Phi(x,\lambda)J} B_v \lambda^{-A_0}, \quad \lambda \in \Gamma_v, \tag{15.7.3}$$

where Γ_v is the open arc between the intersection of Γ with L_{v-1} and with L_v. As before, $\Phi(x, \lambda) = -\frac{4}{3}\lambda^3 + x\lambda$.

Assume for the moment that we have a solution m. Define f on L to be the additive jump $f = m_+ - m_-$. On L_v,

$$f(x,\lambda) = m_+(x,\lambda) - m_-(x,\lambda) = m_-(x,\lambda)e^{\Phi(x,\lambda)J}[S_v - 1]e^{-\Phi(x,\lambda)J},$$

with a similar relation on Γ. Taking the Cauchy transform Cf,

$$[Cf](x,\lambda) = \frac{1}{2\pi i}\int_L \frac{f(x,\zeta)\,d\zeta}{\zeta - \lambda},$$

gives a function that is holomorphic on the complement of L. This function has the same additive jump as m; see Appendix A. Therefore $m - Cf$ is an entire function. According to (15.6.11) and the choice of contours, the matrix $A_v = S_v - 1$ has a single nonzero entry, and $e^{\Phi J}A_v e^{-\Phi J}$ decreases like $\exp(-8|\lambda|^3/3)$ along the ray. It follows that $Cf \to 0$ at ∞, and we conclude that $m = 1 + Cf$.

The problem of finding m has been reduced to the problem of finding f on L. The multiplicative jump conditions (15.7.2) and (15.7.3) can be written in terms of f as

$$f(x,\lambda) - [Cf]_-(x,\lambda)e^{\Phi(x,\lambda)J}A_v e^{-\Phi(x,\lambda)J}$$
$$= e^{\Phi(x,\lambda)J}A_v e^{-\Phi(x,\lambda)J}, \quad \lambda \in L_v; \qquad (15.7.4)$$

$$f(x,\lambda) - [Cf]_-(x,\lambda)\left[e^{\Phi(x,\lambda)J}B_v\lambda^{-A_0} - 1\right]$$
$$= e^{\Phi(x,\lambda)J}B_v\lambda^{-A_0} - 1, \quad \lambda \in \Gamma_v. \qquad (15.7.5)$$

We seek to solve these equations for $|x - x_0| < 1$. For R sufficiently large, the function $|\Phi(x,\lambda)|$ will be large on L_v, uniformly with respect to x for $|x - x_0| \le 1$, so the factor $e^{\Phi J}A_v e^{-\Phi J}$ will be small on L_v. It is a classical result that the mapping $g \to [Cg]_-$ is a projection with norm 1 in the Hilbert space $L^2(L_v, |d\zeta|)$. Therefore the part of the operator acting on f on the left side of (15.7.4) will be a small perturbation of the identity operator in each subspace $L^2(L_v, |d\zeta|)$.

In view of (15.6.10), for large R, the discontinuities of the multiplicative jump across Γ in (15.7.3), which occur at the intersections of Γ with the L_v, will be small. As we show later, this implies that there is a rational matrix-valued function $u(x,\lambda)$ such that

$$e^{\Phi(x,\lambda)J}\widetilde{B}_v(x,\lambda)\lambda^{-A_0} = e^{\Phi(x,\lambda)J}B_v(x,\lambda)\lambda^{-A_0}u(x,\lambda)^{-1} \approx 1$$

for $|x - x_0| < 1$. If we replace B_v by $\widetilde{B}_v$ and f by $\widetilde{f}$ in (15.7.5), the operator on the left is also a small perturbation of the identity. Both this operator and the term on the right-hand side depend analytically on x, so the same is true of the solution $\widetilde{f}$.

With f also replaced by $\widetilde{f}$ on L_v, the function $\widetilde{m} = 1 + C\widetilde{f}$ satisfies (15.7.2) and satisfies (15.7.3) with jump $\widetilde{B}(x,\lambda)$. It can be shown that $\widetilde{m}$ is continuous from each side on the boundary L. The function m^* defined by

$$m^*(x,\lambda) = \begin{cases} \widetilde{m}(x,\lambda), & |\lambda| > R; \\ \widetilde{m}(x,\lambda)u(x,\lambda), & |\lambda| < R, \end{cases}$$

satisfies the multiplicative jump conditions (15.7.2), (15.7.3) and has limit **1** at infinity. However, m^* has poles at the poles $\{\lambda_k\}$ of u that are inside the circle Γ; these may be assumed to be simple:

$$m^*(x,\lambda) = \mathbf{1} + \sum_{j=1}^{N} \frac{1}{\lambda - \lambda_j} a_j(x,\lambda).$$

The final step is to remove these poles.

The determinant $h = \det m^*$ is a rational function that is invertible on the circle Γ and has N double poles inside Γ (we assume here that the a_j are invertible). Therefore h has $2N$ zeros $\mu_k(x)$, counting multiplicity, inside Γ. Suppose first that the zeros μ_k are simple, which implies that the $m^*(x,\mu_k)$ are singular but not zero. To complete the construction of m, we look for a rational function

$$r(x,\lambda) = \mathbf{1} + \sum_{k=1}^{2N} \frac{1}{\lambda - \mu_k(x)} b_k(x)$$

such that $m(x,\lambda) = r(x,\lambda)m^*(x,\lambda)$ has no poles and thus is the desired solution of the system (15.7.2), (15.7.3).

The requirement that μ_k not be a pole means that $b_k(x)m^*(x,\mu_k) = 0$, which gives two linear constraints on the entries of the matrix b_k. The requirement that λ_j not be a pole is equivalent to

$$r(x,\lambda_j) = \mathbf{1} + \sum_{k=1}^{2N} \frac{1}{\lambda_j - \mu_k(x)} b_k(x) = 0.$$

This gives $4N$ more linear conditions, for a total of $8N$ conditions on the $8N$ entries of the b_k, with coefficients that are analytic functions of x. The system has a (unique) solution at $x = x_0$, and therefore has a solution at all points with $|x - x_0| < 1$, except for any points where the determinant vanishes. At these points, the solution has poles as a function of x. A similar but more complicated argument applies if some of the zeros μ_k are multiple.

Finally, we outline the argument concerning rational approximation. We want a rational function u that approximates a given piecewise continuous function g on the circle Γ. It is enough to consider the scalar case. We need a quantitative and uniform version of the approximation result for the Cauchy transform in Appendix A.

Let Cg be the Cauchy integral of g. A calculation using the change of variables $\zeta = \lambda e^{i\theta}$ shows that for $\lambda \in \Gamma$ and $0 < \varepsilon < 1$,

$$[Cg](\lambda - \varepsilon\lambda) - [Cg](\lambda + \varepsilon\lambda) = \int_{-\pi}^{\pi} G_\varepsilon(\theta) g(\lambda e^{i\theta}) d\theta, \qquad (15.7.6)$$

where G_ε is an *approximate identity*:

$$\int_{-\pi}^{\pi} |G_\varepsilon(\theta)| \, d\theta \leq \text{constant};$$

$$\int_{-\pi}^{\pi} G_\varepsilon(\theta) \, d\theta = 1;$$

$$\lim_{\varepsilon \to 0} \int_{\delta < |\theta| < \pi} |G_\varepsilon(\theta)| \, d\theta = 0 \quad \text{for all } \delta > 0.$$

Suppose the jumps of g are smaller than δ. For small enough ε, the integral (15.7.6) will be uniformly within the distance 2δ of $g(\lambda)$ on Γ. For sufficiently large N, the Riemann sum

$$\frac{1}{2\pi i} \sum_{j=1}^{N} \left[\frac{1}{\zeta_j - \lambda(1 - \varepsilon)} - \frac{1}{\zeta_j - \lambda(1 + \varepsilon)} \right] g(\sigma_j) [\zeta_{j+1} - \zeta_j], \qquad (15.7.7)$$

where $\zeta_j = \lambda \exp(2\pi j i/N)$ and $\sigma_j = \zeta_j \exp(\pi j i/N)$, will be uniformly within 3δ of $g(\lambda)$ on Γ.

Remark. If we take N to be even, the rational approximation u obtained in this way will satisfy the symmetry condition

$$u(x, \lambda) = J_2 u(x, -\lambda) J_2.$$

Tracing through the argument, we obtain the corresponding symmetry for m.

It remains to be shown that the function defined for λ not in L by

$$\Psi(x, \lambda) = \begin{cases} m(x, \lambda) e^{\Phi(x,\lambda)J}, & |\lambda| > R, \\ m(x, \lambda) \lambda^{cJ_2}, & |\lambda| < R, \end{cases}$$

satisfies equations of the form (15.4.1), (15.4.2), showing that we have constructed a solution of PII(c). It will be convenient for the moment to relabel $m = \widehat{m}$ outside the circle Γ and let

$$\widehat{m}(x, \lambda) = m(x, \lambda) \lambda^{cJ_2} e^{-\Phi J}, \qquad |\lambda| < R.$$

Differentiating the identity (15.7.2) with respect to x and with respect to λ shows that both the functions

$$\widehat{m}_x(x, \lambda) - \lambda [J, \widehat{m}(x, \lambda)],$$

$$\lambda \widehat{m}_\lambda(x, \lambda) + \lambda \Phi_\lambda(x, \lambda) \widehat{m}(x, \lambda) J, \qquad (15.7.8)$$

satisfy the same jump conditions as $\widehat{m}$ on the rays L_v. Some more calculation shows that the multiplicative jumps of (15.7.8) on the arcs Γ_v are the same as those of $\widehat{m}$. Therefore there are matrix-valued functions $A(x, \lambda)$, $q(x, \lambda)$ such that

$$\widehat{m}_x(x, \lambda) - \lambda\,[J, \widehat{m}(x, \lambda)] = q(x, \lambda)\,\widehat{m}(x, \lambda);$$
$$\lambda\,\widehat{m}_\lambda(x, \lambda) + \lambda\Phi_\lambda(x, \lambda)\widehat{m}(x, \lambda)J = A(x, \lambda)\,\widehat{m}(x, \lambda).$$

The functions $A(x, \lambda)$ and $q(x, \lambda)$ are continuous across L and therefore are holomorphic, except possibly at $\lambda = 0$, where $\widehat{m}$, unlike m, is singular (unless $A_0 = 0$). Some more calculation shows that inside Γ,

$$q \equiv (\widehat{m}_x - \lambda[J, \widehat{m}])\,\widehat{m}^{-1} = (m_x - \lambda Jm)m^{-1}; \tag{15.7.9}$$

$$A \equiv (\lambda \widehat{m}_\lambda + \lambda\Phi_\lambda \widehat{m}J)\widehat{m}^{-1} = (\lambda m_\lambda + cmJ_2)m^{-1}. \tag{15.7.10}$$

Therefore q and A are regular at the origin, and hence are entire functions of λ. The asymptotics show that $A(x, \lambda)$ is a polynomial with respect to λ with leading term $-4\lambda^3 J$, while $q(x, \lambda) = Q(x)$ is constant with respect to λ. The symmetry condition for m implies that $A(x, -\lambda) = J_2 A(x, \lambda)J_2$ and $Q(x) = J_2 Q(x)J_2$. Moreover, $\det m$ is constant, so $A(x, \lambda)$ and $Q(x) = u(x)J_2$ have trace 0. Therefore the system of equations that is satisfied by Ψ has the form (15.4.1), (15.4.2) and the constant term $A_0 \equiv cJ_2$. It follows that $u(x)$ is a solution of PII(c). The polynomial $A(x_0, \lambda)$ was chosen in such a way that u has the initial conditions $u(x_0) = u_0$, $u'(x_0) = v_0$.

As noted, the construction works for $|x - x_0| < 1$, with possible poles at some discrete values of x. Continuing, we find that the solution $m(x, \lambda)$ is meromorphic with respect to x, and (15.7.9) implies that $u(x)$ is also meromorphic. Thus PII(c) has the Painlevé property so long as $c + \frac{1}{2}$ is not an integer. Lemma 15.1.1 allows us to conclude that PII(c) also has the Painlevé property when $c + \frac{1}{2}$ is an integer.

15.8 Asymptotics of PII(0)

If u is a solution of PII(0), i.e. $u'' = 2u^3 + xu$, and $u(x) \to 0$ as $x \to \pm\infty$, then asymptotically the equation looks like Airy's equation $v'' = xv$. One might expect the solution to be asymptotic to a solution of Airy's equation:

$$u(x) \sim \begin{cases} k\,\mathrm{Ai}(x) & \text{as } x \to +\infty; \\[2mm] \dfrac{d}{|x|^{1/4}}\,\sin\left\{\tfrac{2}{3}|x|^{3/2} + o(x)\right\} & \text{as } x \to -\infty. \end{cases}$$

The following result is due to Hastings and McLeod [179].

Theorem 15.8.1 (a) *If u is a solution of* PII(0) *such that* $u(x) \to 0$ *as* $x \to +\infty$, *then u is asymptotic as* $x \to +\infty$ *to* $k \operatorname{Ai}(x)$ *for some constant k. For any k there is a unique solution of* PII(0) *that is asymptotic to* $k \operatorname{Ai}(x)$ *as* $x \to +\infty$.

(b) *If* $|k| < 1$, *the solution asymptotic to* $k \operatorname{Ai}(x)$ *exists for all x and is asymptotic as* $x \to -\infty$ *to*

$$\frac{d}{|x|^{1/4}} \sin\left\{ \frac{2}{3}|x|^{3/2} - \frac{3}{4}d^2 \log|x| + \gamma \right\} \tag{15.8.1}$$

for some constants d and γ.

See [179] for the proof.

Theorem 15.8.2 *For* $|k| < 1$, *the constants d and* γ *of* (15.8.1) *are related by*

$$d^2 = -\frac{\log(1 - k^2)}{\pi}; \tag{15.8.2}$$

$$\gamma = \frac{3}{4}\pi - \frac{3}{2}d^2 \log 2 - \arg \Gamma\left(-\frac{i}{2}d^2\right). \tag{15.8.3}$$

This was proved by Deift and Zhou [101]. We sketch here a simpler argument due to Bassom, Clarkson, Law, and McLeod [32]; calulations may be checked by the reader. The argument uses the asymptotics of the solutions of (15.4.2) that were constructed in Section 15.5. Specifically, we use the solutions Ψ_1 and Ψ_2, connected by

$$\Psi_2(x, \lambda) = \Psi_1(x, \lambda) \begin{bmatrix} 1 & 0 \\ a_1 & 1 \end{bmatrix}. \tag{15.8.4}$$

The goal is to relate the asymptotics of u as $x \to -\infty$ and the asymptotics as $x \to +\infty$ to the coefficient a_1, and thus to each other. We sketch the argument for $x \to +\infty$, which is based on finding a good asympotic solution to the second-order equation that is satisfied by $\psi \equiv [\Psi_1]_{21}$. We are working in the case PII(0), so

$$\Psi_\lambda(x, \lambda) = B(x, \lambda)\Psi(x, \lambda), \tag{15.8.5}$$

where

$$B(x, \lambda) = \frac{1}{\lambda} A(x, \lambda) = \begin{bmatrix} -4\lambda^2 + x + 2u^2 & -4\lambda u - 2u_x \\ -4\lambda u + 2u_x & 4\lambda^2 - x - 2u^2 \end{bmatrix}.$$

It follows that

$$\Psi_{\lambda\lambda}(x, \lambda) = B_\lambda(x, \lambda)\Psi(x, \lambda) + B(x, \lambda)^2 \Psi(x, \lambda). \tag{15.8.6}$$

Equation (15.8.6) allows us to write the second derivative of the entry $\psi = [\Psi_1]_{21}$ as a linear combination of ψ itself and $\psi_1 = [\Psi_1]_{11}$. Equation (15.8.5) allows us to write ψ_1 as a linear combination of ψ and ψ_λ. The resulting second-order equation is

$$\psi_{\lambda\lambda} = \frac{2u}{2\lambda u - u_x}\psi_\lambda + \Big[(4\lambda^2 - x)^2 + 4(xu^2 + u^4 - u_x^2)$$

$$+ 8\lambda - \frac{2u(4\lambda^2 - x - 2u^2)}{2\lambda u - u_x}\Big]\psi. \tag{15.8.7}$$

We use a gauge transformation $g\,\psi = \varphi$ to remove the first-order term. The requirement on g is that

$$\frac{g_\lambda}{g} = -\frac{u}{2\lambda u - u_x} = -\frac{1}{2}\Big[\log\Big(\lambda - \frac{u_x}{2u}\Big)\Big]_\lambda,$$

so we take

$$\psi(x,\lambda) = \Big(\lambda - \frac{u_x}{2u}\Big)^{1/2}\varphi(x,\lambda). \tag{15.8.8}$$

Consider φ as a function of the rescaled variables

$$\xi = x^{3/2}, \quad \eta = x^{-1/2}\lambda, \qquad x > 0. \tag{15.8.9}$$

The equation for φ is

$$\varphi_{\eta\eta} = \Big[\xi^2(4\eta^2 - 1)^2 + \xi\Big(8\eta - \frac{2u(4\eta^2 - 1)}{2u\eta - x^{-1/2}u_x}\Big) + o(\xi)\Big]\varphi.$$

Assume that $u(x) \sim k\,\mathrm{Ai}(x)$ as $x \to +\infty$. Then

$$u(x) \sim \frac{k}{2\sqrt{\pi}\,x^{1/4}}e^{-\frac{2}{3}x^{3/2}}, \qquad \frac{u'(x)}{u(x)} \sim -\sqrt{x}. \tag{15.8.10}$$

Therefore the equation for φ can be rewritten as

$$\varphi_{\eta\eta} = \xi^2\Big[(4\eta^2 - 1)^2 + \frac{4\eta + 2}{\xi} + o\Big(\frac{1}{\xi}\Big)\Big]\varphi$$

$$= \xi^2 F(\xi,\eta)\varphi. \tag{15.8.11}$$

We consider (15.8.11) in the half-plane $\mathrm{Re}\,\eta > 0$. For large ξ, $F(\eta,\xi)$ has two zeros $\eta_\pm$ in this plane:

$$\eta_\pm = \frac{1}{2} \pm \frac{i}{2\sqrt{\xi}} + o\Big(\frac{1}{\sqrt{\xi}}\Big). \tag{15.8.12}$$

Following Olver [310], we look for a transformation

$$w(\zeta,\xi) = \rho(\zeta,\xi)\varphi\,(\eta(\zeta,\xi))$$

such that (15.8.11) is equivalent to the simpler equation

$$w_{\zeta\zeta} = \xi^2\big[\zeta^2 + \theta^2 + o(\xi^{-1})\big]w. \tag{15.8.13}$$

The equation for w will not have a first-order term if

$$\rho(\zeta,\xi) = [\eta_\zeta(\zeta,\eta)]^{-1/2},$$

and then the equation is

$$w_{\zeta\zeta} = \left[\xi^2 \eta_\zeta^2 F + \frac{\rho_{\zeta\zeta}}{\rho}\right] w. \tag{15.8.14}$$

Equations (15.8.13) and (15.8.14) lead to

$$\frac{d\eta}{d\zeta} = \frac{(\zeta^2 + \theta^2)^{1/2}}{F(\xi,\eta)^{1/2}} \sim \frac{\zeta}{4\eta^2}. \tag{15.8.15}$$

As we shall see, $\theta^2 \sim 1/\xi$, so we consider the modified equation

$$w_{\zeta\zeta} = \xi^2 \left(\zeta^2 + \frac{1}{\xi}\right) w.$$

This can be scaled to Weber's equation with index -1,

$$W''(y) = \left(\frac{y^2}{4} + \frac{1}{2}\right) W(y), \tag{15.8.16}$$

by setting $w(\zeta) = W(\sqrt{2\xi}\,\zeta)$. It follows from the results in [32] that the solutions of (15.8.14) are asymptotic to functions $W(\sqrt{2\xi}\,\zeta)$, with W a solution of (15.8.16).

Tracing backward, the conclusion so far is that for large x and η,

$$\psi(x,\lambda) \sim \left(\lambda + \frac{\sqrt{x}}{2}\right)^{1/2} \frac{\sqrt{\zeta}}{2\eta} W\left(\sqrt{2\xi}\,\zeta\right), \tag{15.8.17}$$

where W is a solution of (15.8.16). Now $\Psi_1 = m_1 e^{\Phi J}$, and the construction of m_1 in Section 15.5 shows that along rays in the open sector bounded by Σ_0 and Σ_1, i.e. with $|\arg\lambda| < \frac{1}{6}\pi$,

$$m_1(x,\lambda) = 1 + \lambda^{-1} f_1 + O(\lambda^{-2}), \qquad Q = -[J, f_1].$$

Therefore, asymptotically along these rays,

$$\psi \sim \frac{u}{2\lambda} e^{\Phi}, \qquad \Phi(x,\lambda) = -\frac{4}{3}\lambda^3 + x\lambda. \tag{15.8.18}$$

It follows from (15.8.18) and (15.8.15) that W is bounded along rays in the sector $|\arg\zeta| < \frac{1}{4}\pi$. The only such solutions of (15.8.16) are multiples of D_{-1}. Thus we have

$$[\Psi_1]_{21} = \psi \sim h(x,\lambda) D_{-1}(\sqrt{2\xi}\,\zeta). \tag{15.8.19}$$

On the sector $\frac{\pi}{6} < \arg \lambda < \frac{\pi}{2}$,

$$[\Psi_1]_{21} = [\Psi_2]_{21} - a_1[\Psi_2]_{22}$$

$$\sim \frac{u}{2\lambda} e^{\Phi} - a_1 e^{-\Phi} = \frac{u}{2\lambda} e^{\Phi} \left[1 - \frac{2\lambda a_1}{u} e^{-2\Phi} \right]$$

$$\sim \frac{u}{2\lambda} e^{\Phi} \left[1 - \frac{a_1}{k} 4\lambda\sqrt{\pi} x^{1/4} e^{\frac{2}{3}x^{3/2}} e^{-2\Phi} \right]. \tag{15.8.20}$$

This also holds along the branch of the curve $\operatorname{Re} \Phi = 0$ that is asymptotic to the ray $\arg \lambda = \frac{1}{2}\pi$. Let us compare this to (15.8.19). On the ray $\arg y = \arg \zeta = \frac{3}{4}\pi$, which corresponds to $\arg \lambda = \frac{1}{2}\pi$,

$$D_{-1}(y) \sim y^{-1} e^{-y^2/4} + \sqrt{2\pi}\, e^{y^2/4}.$$

Thus along the branch of $\operatorname{Re} \Phi = 0$ that is asymptotic to $\arg \lambda = \frac{1}{2}\pi$, (15.8.19) gives

$$[\Psi_1]_{21} \sim h(x,\lambda) \frac{e^{-\xi\zeta^2/2}}{\sqrt{2\xi}\,\zeta} \left[1 + 2\sqrt{\pi\xi}\,\zeta\, e^{\xi\zeta^2} \right]. \tag{15.8.21}$$

The last step in this part of the argument is to determine $\xi\zeta^2$ asymptotically. In view of (15.8.15), we take

$$\int_{\eta_+}^{\eta} \sqrt{F(\xi,s)}\,ds = \int_{\zeta_+}^{\zeta} \sqrt{t^2 + \theta^2}\,dt. \tag{15.8.22}$$

The zeros $\eta = \eta_\pm$ of F must correspond to the zeros $\zeta_\pm = \pm i\theta$ of $\zeta^2 + \theta^2$, so

$$-i \int_{\eta_-}^{\eta_+} [F(\xi,\eta)]^{1/2}\,d\eta = \int_{-\theta}^{\theta} (\theta^2 - s^2)^{1/2}\,ds = \frac{\pi}{2}\theta^2.$$

It follows from (15.8.12) and the form of F that the left-hand side of the previous equation is $\approx \pi/2\xi$, so $\theta \approx \xi^{-1/2}$.

Integrate (15.8.22), using the first two terms of the binomial expansion of the square roots, to get, formally,

$$\frac{4}{3}\eta^3 - \eta + \frac{1}{3} + \frac{1}{2\xi}\log(2\eta\sqrt{\xi}) = \frac{\zeta^2}{2} + \frac{1}{2\xi}\log(\zeta\sqrt{\xi}) + o(\xi^{-1}). \tag{15.8.23}$$

Multiplying by 2ξ gives

$$\xi\zeta^2 \sim -2\Phi(x,\lambda) + \log\left(\frac{2\eta}{\zeta}\right) + \frac{2}{3}\xi. \tag{15.8.24}$$

Comparing (15.8.20) and (15.8.21) in light of (15.8.9) and (15.8.24) gives

$$k = -a_1. \tag{15.8.25}$$

An analogous computation for $x \to -\infty$ connects a_1 to these asymptotics, completing the proof of Theorem 15.8.2.

15.9 Exercises

1. Verify the statements in the introduction about reduction by coalescence:
   (a) PII $\to$ PI,
   (b) PIII $\to$ PII,
   (c) PIV $\to$ PII,
   (d) PV $\to$ PIII,
   (e) PV $\to$ PIV,
   (f) PVI $\to$ PV.
2. Suppose U is related to u by (15.1.4):
$$U(x) = \frac{a(x)u(\varphi(x)) + b(x)}{c(x)u(\varphi(x)) + d(x)},$$
   where φ, a, b, c, d are analytic and normalized with $ad - bc = 1$. Show that u satisfies an equation of the form (15.1.1) if and only if U satisfies an equation of the same form, with a suitable choice of R.
3. Adapt the proof, reducing (15.2.5) to (15.2.6) to show that (15.2.7) can also be reduced to (15.2.6).
4. Consider the equation $(u')^m = P(u)$, where P is a polynomial of degree $2m$ with distinct roots a_j:
$$P(u) = C \prod_{j=1}^{k} [u - a_j]^{d_j}, \qquad d_j > 0, \ d_1 + d_2 + \cdots + d_k = 2m.$$
   Assume that m and the d_j have no common factor $l > 0$, since otherwise the equation can be reduced to one with m replaced by m/l. Therefore some d_j must be less than m. Suppose that the equation has the Painlevé property. The goal of this exercise and the next is to show that the equation has the Painlevé property if and only if it is one of those listed in (15.1.12).
   (a) Suppose u is a solution with $u(0) = a_j$:
$$u(x) = a_j + c_1 x^r + c_2 x^{r+1} + \cdots, \qquad c_1 \neq 0,$$
   where r is an integer. Show that the equation implies that $d_j/m = 1 - 1/r$ and conclude that $d_j < m$ implies $d_j \geq m/2$.
   (b) Conclude from (a) that P has at most four distinct roots.
   (c) Suppose $d_1 > m$. Use (a) to show that there is only one other root. Use (a) and the assumption about common factors to show that $d_1 = m+1$, $d_2 = m-1$.

5. Consider the equation in Exercise 4 under the assumptions that m and the d_j have no common factor and that each d_j is less than or equal to m.
   (a) Suppose $m = 2$. Use (a) to conclude that the d_j are either $\{2,1,1\}$ or $\{1,1,1,1\}$.
   (b) Suppose $m = 3$. Use (a) to conclude that the d_j are $\{2,2,2\}$.
   (c) Suppose $m = 4$. Show that the d_j are $\{3,3,2\}$.
   (e) Show that $m \neq 5$.
   (f) Suppose $m = 6$. Show that the d_j are $\{5,4,3\}$.
   (g) Show that m cannot exceed 6. (Hint: how many distinct roots are possible? How large can the smallest of the d_j be?)

6. Consider (15.1.10): $u'' = l(u)[u']^2$, where $l(u)$ is rational. Show that any pole of $l(u)$ is simple. Hint: suppose $l(u)$ has a pole of order $r > 1$, which we may assume to be at the origin. Write the equation as a system for u and $v = u'$, and then scale with $u = \alpha U$, $v = \alpha^r V$ so that

$$U' = \alpha^{r-1}V, \qquad V' = \frac{kV^2}{U^r} + O(\alpha),$$

with $k \neq 0$. Show that this has the solution

$$U = U_0 - \frac{\alpha^{r-1}}{k}U_0^r \log\left(\frac{x+C}{x_0+C}\right) + O(\alpha^r);$$

$$V = -\frac{U_0^r}{k(x+C)} + O(\alpha), \qquad C = -x_0 - \frac{U_0^r}{kV_0}.$$

Thus there is a movable branch point.

7. Consider (15.1.10) and assume that $l(u)$ has a simple pole, which we may assume to be at the origin. Scaling this as in Exercise 6, with $r = 1$, leads to

$$U' = V, \qquad V' = \frac{kV^2}{U} + O(\alpha).$$

Show that when $\alpha = 0$, the solution has a movable branch point unless $k = 1$ or $k - 1 = 1/n$ where n is an integer.

8. Writing (15.1.10) in the form $u''/u' = l(u)u'$ shows that it has a first integral

$$u'(x) = C \exp \int_{u_0}^{u} l(w)\,dw. \qquad (15.9.1)$$

According to Exercises 6 and 7, at any pole of l, l has the form

$$l(u) = \frac{1+\frac{1}{n}}{u-a} + O(1).$$

(a) Show that this contributes a factor $(u-a)^{1+\frac{1}{n}}$ to the term on the right in (15.9.1).

(b) Let m be the least common denominator of the exponents of all such factors. Write

$$[u']^m = \phi(u). \tag{15.9.2}$$

Show that ϕ has no singularities in $\mathbf{C}$ except poles.
(c) Setting $U = 1/u$, show that ϕ has at most a pole at infinity, and hence is a rational function.
(d) Differentiate (15.9.2) and show that it becomes (15.1.10).

9. Suppose that B and C are constants. Show that the general solution of

$$v''(x) = 2v(x)^3 + Bv(x) + C$$

is an elliptic function. Hint: multiply both sides by $v'(x)$ and integrate once.

10. Adapt the method at the beginning of Section 15.3 to show that PI has a one-parameter family of solutions with a double pole at $x = 0$.

11. Suppose that $U'(x) = A(x)U(x)$, where U and A are 2×2 matrix-valued functions and U is invertible.
(a) Show that $\det U$ is constant if and only if A has trace zero.
(b) Show that $(U(x)^{-1})' = -U(x)^{-1}A(x)$.
(c) If U_1 is a second solution, show that $U_1(x) = U(x)C$ for some constant matrix C.

12. Adapt the procedure used for PII to show that the compatibility for the system (15.4.26) is PI.

13. Prove Proposition 15.4.3. Hint: the composition of the transformation with itself takes u to $u + 2b + 2\tilde{b}$. Using the previous notation for A_1, show that

$$2b + 2\tilde{b} = \frac{1 - 2c}{(\alpha - \beta)(\tilde{\alpha} - \tilde{\beta})} \left[(\tilde{\alpha} - \tilde{\beta}) - (\alpha - \beta) \right],$$

and that

$$\tilde{\alpha} - \tilde{\beta} = \alpha - \beta + 4(b' + 2ub + 2b^2).$$

Use the identity (15.4.21) to show that $b + \tilde{b} = 0$.

14. Verify that the conditions (15.5.10) are equivalent to (15.5.9).

15. This exercise sketches a derivation of the approximate solution (15.5.11). Fix x. To ease notation, write $A_1 = \alpha J_1 + \beta J_3$, so $\alpha = x + 2u^2$, $\beta = -2u'$, and recall that $A_2 = -4uJ_2$, $A_0 = cJ_2$. In view of symmetry, look for the f_j in the form

$$f_2 = aJ_2, \qquad f_j = b_j J_1 + c_j J_3, \quad j = 1, 3.$$

Here the J_j are the matrices in (15.4.13). The conditions (15.5.10) are trivially satisfied at $k = 0$.
(a) Show that at steps $k = 1$ and $k = 2$, respectively, the conditions are equivalent to

$$c_1 = -u/2, \qquad 8a - 4ub_1 = \beta.$$

(b) With these conditions satisfied, show that the condition at step $k = 3$ is

$$8c_3 = -xu - u^3 + c - b_1\beta.$$

(c) Take $b_1 = b_3 = 0$ and show that the result is (15.5.11).

16. Given a smooth function $u(x)$, let $Q(x) = u(x)J_2$.

(a) Show that the formal series

$$m(x, \lambda) \sim 1 + \lambda^{-1}m_1(x) + \lambda^{-2}m_2(x) + \cdots$$

is a formal solution of (15.5.7) if and only if

$$[J, m_{k+1}] = m_k' - Qm_k, \quad k = 0, 1, 2, \ldots$$

(b) Show that there is a solution whose coefficients have the form

$$m_j = a_j 1 + b_j J_2, \quad j \text{ even};$$
$$m_j = c_j J_1 + d_j J_3, \quad j \text{ odd}.$$

(c) Show that the coefficients can be chosen to depend smoothly on u, so that, for a given x_0, $u \equiv 0$ implies that $m_j(x_0) = 0$ for $j > 0$.

17. This exercise and the next sketch a more abstract construction of a solution of (15.5.9), following [35]. Suppose that m and $\widehat{m}$ are formal solutions as in Exercise 16.

(a) Prove that $g = m^{-1}\widehat{m}$ is a formal solution of $g_x = \lambda[J, g]$ and deduce from the recursion relations that the coefficients g_j of g are diagonal.

(b) Use part (a) to show that

$$F = -4mJm^{-1} \sim -4J + \lambda^{-1}F_1 + \lambda^{-2}F_2 + \cdots$$

is independent of the choice of formal solution m.

(c) Show that F is a formal solution of

$$F' = [\lambda J + Q, F].$$

(d) Deduce from (c) that

$$[J, F_{j+1}] = F_j' - [Q, F_j], \quad j = 0, 1, 2, \ldots$$

(e) Use (b) and Exercise 16 to show that F depends smoothly on u and that $u \equiv 0$ implies $F_j(x) \equiv 0$ for $j > 0$.

18. Suppose that F is as in Exercise 17. Set

$$A_3(x) = F_0(x) = -4J, \qquad A_2(x) = F_1(x);$$
$$A_1(x) = F_2(x) + xJ, \qquad A_0(x) = F_3(x) + xQ(x).$$

(a) Show that the conditions in part (d) of Exercise 17 are equivalent to equations (15.4.4)–(15.4.6), and deduce that the A_j here are identical to the A_j given by (15.4.8)–(15.4.10).

(b) Suppose that u satisfies PII(c) for some c. Let m be an asymptotic solution of $m' = \lambda[J,m] + Qm$ as in Exercise 16 and let f be the truncated series

$$f(x,\lambda) = 1 + \lambda^{-1} m_1 + \lambda^{-2} m_2 + \lambda^{-3} m_3.$$

Show that

$$f' = \lambda[J,f] + Qf + O(\lambda^{-3}).$$

(c) Let $F = -4mJm^{-1}$. Then $fFf^{-1} = mFm^{-1} + O(\lambda^{-4})$. Deduce that

$$(\lambda^3 F + x\lambda J + xQ)f = -4\lambda^3 fJ + x\lambda fJ + O(\lambda^{-1}).$$

(d) Use parts (a) and (c) to show that f is a solution to (15.5.9).
19. Prove (15.6.7).
20. Verify (15.6.12).
21. Verify various calculations in the section on asymptotics.

15.10 Remarks

The classification of second-order equations of the form (15.1.1) was begun by Painlevé in the 1890s in lectures in Stockholm, a series of notes in the *Comptes Rendus* and two memoirs [315, 316, 317], with additions by Gambier [146] and Fuchs [145]. Equations PI, PII, and PIII were treated completely by Painlevé. The study of PIV and PV was completed by Gambier, and the study of PVI was completed by Fuchs and Gambier. Boutroux [52] made an extensive study of the asymptotics of the solutions; some indications for PI are given in [185]. The treatise by Ince [193] outlines the methods and lists all 50 canonical equations. The observation that Painlevé equations describe isomonodromy goes back to Fuchs [145] and Garnier [148]. For much more information, see Chapter 32 of [312], [313], the survey article by Clarkson [82], and the books by Steeb and Euler [377], Gromak, Laine, and Shimomura [174], Iwasaki, Kimura, Shimomura, and Yoshida [196], and Conte and Musette [85]. For more on the isomonodromy approach and its applications, see Deift and Zhou [101], and the book by Fokas, Its, Kapaev, and Novokshenov [137]. For a modern approach to the Painlevé property for isomonodromy equations, see Miwa [290].

The realization of Painlevé equations as compatibility conditions grew out of the inverse scattering approach to nonlinear PDEs, that started with the work of Gardner, Greene, Kruskal, and Miura [147], who showed that the Korteweg–de Vries equation described an isospectral flow of a one-dimensional Schrödinger equation. This connection was elucidated by Lax [247] as a compatibility condition. Ablowitz and Segur [3] showed that PI, PII, and PIII arise in looking for "self-similar" solutions of some integrable equations. Flaschka and Newell [136] found systems for which PII, and a special case of PIII, are the compatibility conditions. Jimbo and Miwa [205] found such systems for all the Painlevé equations and developed the theory of Bäcklund transformations in this connection. (Bäcklund transformations were first introduced into differential geometry by Bäcklund [26] in 1883, but the idea of changing the coefficients of an equation like (15.4.2) by changing a solution goes back at least to Moutard in 1875 [294].)

Painlevé equations come up in a number of contexts in mathematics and physics. For differential geometry, see the book by Bobenko and Eitner [49]. McCoy, Tracy, and Wu [278] showed that PIII arises in connection with the two-dimensional Ising model; for further developments in statistical physics, see the survey by Tracy and Widom [400].

For other applications, including string theory, quantum field theory, and random matrix theory, see Moore [291], Kanzieper [213], and the conference proceedings [55].

Appendix A

Complex analysis

This section contains a brief review of terminology and results from complex analysis that are used in the text.

If $z = x + iy$ is a complex number, with x and y real, then

$$z = x + iy = r\cos\theta + ir\sin\theta = r\,e^{i\theta}, \quad \bar{z} = x - iy,$$

where $r = \sqrt{x^2 + y^2}$ is the *modulus* $|z|$ of z and θ is the *argument* $\arg z$ of z. The logarithm

$$\log z = \log r + i\theta$$

is multiple-valued, that is, defined only up to integer multiples of $2\pi i$. The power

$$z^a = \exp(a\log z)$$

is also multiple-valued, unless a is an integer.

Typically, one makes these functions single-valued by restricting the domain, usually by choosing a range for the argument. The resulting domain is the complex plane minus a ray from the origin. Examples are

$$\mathbf{C} \setminus (-\infty, 0] = \{z: -\pi < \arg z < \pi\};$$
$$\mathbf{C} \setminus [0, +\infty) = \{z: 0 < \arg z < 2\pi\}.$$

This is referred to as choosing a *branch* of the logarithm or the power. The *principal branch* is the one with $\arg x = 0$ for $x > 0$.

A *region* is an open, nonempty subset of the plane which is *connected*: any two points in the set can be joined by a continuous curve that lies in the set. A region Ω is said to be *simply connected* if any closed curve lying in Ω can be continuously deformed, within Ω, to a point. The plane $\mathbf{C}$ and the disc $\{z : |z| < 1\}$ are simply connected. The annulus $\{z : 1 < |z| < 2\}$ and the punctured plane $\mathbf{C} \setminus \{0\}$ are not simply connected.

A.1 Holomorphic and meromorphic functions

A function $f(z)$ is said to be *analytic* or *holomorphic* in a region Ω if the derivative

$$f'(z) = \lim_{h \to 0} \frac{f(z+h) - f(z)}{h}$$

exists for each point z in Ω. An equivalent condition is that for each point z_0 in Ω, the function can be represented at nearby points by its Taylor series: if $|z - z_0| < \varepsilon$, then

$$f(z) = \sum_{n=0}^{\infty} a_n(z - z_0)^n, \quad a_n = \frac{f^{(n)}(z_0)}{n!}.$$

Conversely, a function that is defined in a disc by a convergent power series can often be extended to a larger region as a holomorphic function. For example,

$$f(z) = \sum_{n=0}^{\infty} z^n = \frac{1}{1 - z}, \quad |z| < 1,$$

extends to the complement of the point $z = 1$. This is an example of *analytic continuation*.

It can be deduced from the local power series representation, using connectedness, that if two functions f and g are holomorphic in a region Ω and coincide in some open subset of Ω (or on a sequence of points that converges to a point of Ω), then they coincide throughout Ω. This is one version of the principle of *uniqueness of analytic continuation*. This principle is used several times above, often in the following form. Suppose functions $u_j(a,x), j = 1,2,3$, are holomorphic with respect to a parameter a in a region Ω and satisfy a linear relation

$$u_3(a,x) = A_1(a)u_1(a,x) + A_2(a)u_2(a,x),$$

with holomorphic or meromorphic coefficients. Then, to determine the coefficients A_j throughout Ω, it is enough to determine A_1 on a subregion Ω_1 and A_2 on a subregion Ω_2. (In the cases encountered here, the form of a coefficient throughout Ω is clear once one knows the form on any subregion.)

A function f that is holomorphic in a punctured disc $\{0 < |z - z_0| < \varepsilon\}$ is said to have a *pole of order n* at z_0, n a positive integer, if

$$f(z) = \frac{g(z)}{(z - z_0)^n}, \quad 0 < |z - z_0| < \varepsilon,$$

where $g(z)$ is holomorphic in the disc $|z - z_0| < \varepsilon$ and $g(z_0) \neq 0$. An equivalent condition is that $(z - z_0)^n f(z)$ has a nonzero limit at $z = z_0$. The function f is said to have a *removable singularity* at $z = z_0$ if it has a limit at $z = z_0$. In that case, taking $f(z_0)$ to be the limit, the resulting extended function is holomorphic in the disc. A function f that is holomorphic in a region Ω except at isolated points, each of which is a pole or removable singularity, is said to be *meromorphic* in Ω. In particular, if f and g are holomorphic in Ω and g is not identically zero, then the quotient f/g is meromorphic in Ω.

A.2 Cauchy's theorem, the Cauchy integral theorem, and Liouville's theorem

A basic result of complex analysis is the *Cauchy integral theorem*. Suppose that C is a closed curve that bounds a region Ω, and suppose that f is holomorphic on Ω and continuous up to the boundary C. Then the integral vanishes:

$$\int_C f(z)\,dz = 0.$$

A typical use of the Cauchy integral theorem occurs in Appendix B: the integral

$$\int_{-\infty}^{\infty} e^{-(x+iy)^2/2}\,dx$$

is independent of y. To see this, take values $a < b$ for y, and consider the integral of this integrand over the rectangle C_R, two of whose sides are $\{x + ia \,:\, |x| \le R\}$ and $\{x + ib \,:\, |x| \le R\}$, oriented counter-clockwise. By Cauchy's theorem, the integral is zero. As $R \to \infty$, the integral over the vertical sides approaches zero, while the integral over the other sides approaches

$$\int_{-\infty}^{\infty} e^{-(x+ia)^2/2}\,dx - \int_{-\infty}^{\infty} e^{-(x+ib)^2/2}\,dx.$$

Therefore the integral above is independent of y.

One can use the Cauchy integral theorem to derive the *Cauchy integral formula*. Suppose that C is oriented so that Ω lies to the left, for example, if C is a circle oriented counter-clockwise and Ω is the enclosed disc. Then for any $z \in \Omega$,

$$f(z) = \frac{1}{2\pi i} \int_C \frac{f(\zeta)}{\zeta - z}\,d\zeta.$$

The idea of the proof is very simple: use Cauchy's theorem to replace the original curve by a very small circle centered at the point z, so that f is essentially constant on the circle and the change of variables $\zeta = z + \varepsilon e^{i\theta}$ shows that

$$\frac{1}{2\pi i} \int_{|\zeta - z| = \varepsilon} \frac{d\zeta}{\zeta - z} = \frac{1}{2\pi} \int_0^{2\pi} d\theta = 1.$$

A consequence is *Liouville's theorem*: a bounded entire function f is constant. (An *entire function* is one that is holomorphic in the entire plane $\mathbf{C}$.) To see this, observe that the Cauchy integral formula for f can be differentiated under the integral sign. The derivative is

$$f'(z) = \frac{1}{2\pi i} \int_C \frac{f(\zeta)}{(\zeta - z)^2}\,d\zeta.$$

We may take C to be a circle of radius $R > |z|$, centered at the origin. Taking $R \to \infty$, the integrand is $O(1/R^2)$ and the length of the curve is $2\pi R$, so $|f'(z)|$ is at most $O(1/R)$. Therefore $f'(z) = 0$ for every $z \in \mathbf{C}$, so f is constant.

A.3 The residue theorem and counting zeros

The Cauchy integral formula is one instance of the *residue theorem*. Suppose that f has a pole at z_0. Then near z_0 it has the Laurent expansion

$$f(z) = \frac{a_{-n}}{(z - z_0)^n} + \frac{a_{1-n}}{(z - z_0)^{n-1}} + \cdots + \frac{a_{-1}}{z - z_0} + a_0 + a_1(z - z_0) + \cdots$$

The *residue* of f at z_0, denoted by $\mathrm{res}\,(f, z_0)$, is the coefficient a_{-1} of the $1/(z - z_0)$ term in the Laurent expansion

$$\mathrm{res}\,(f, z_0) = a_{-1} = \frac{1}{2\pi i} \int_C f(\zeta)\, d\zeta,$$

where C is a sufficiently small circle centered at z_0.

Suppose as before that C is an oriented curve that bounds a region Ω lying to its left. Suppose that f is meromorphic in Ω and continuous up to the boundary C, and that the poles of f in Ω are $z_1, \ldots, z_m$. Then the residue theorem says that the integral of f over C is $2\pi i$ times the sum of the residues:

$$\int_C f(z)\, dz = 2\pi i\, [\mathrm{res}\,(f, z_1) + \cdots + \mathrm{res}\,(f, z_m)].$$

Suppose that f has no zeros on the boundary curve C. Then the quotient $g = f'/f$ is continuous on the boundary and meromorphic inside. It is easy to see that if z_0 is a zero of f with multiplicity m, then the residue of f'/f at z_0 is m. If z_0 is a pole of order n, then the residue of f'/f at z_0 is $-n$. Therefore the number of zeros (counting multiplicity) minus the number of poles (counting multiplicity) enclosed by C is

$$\frac{1}{2\pi i} \int_C \frac{f'(z)}{f(z)}\, dz.$$

In particular, if f is holomorphic in the enclosed region, then this integral counts the number of zeros. The proof is an elaboration of the proof of the Cauchy integral theorem: reduce to a sum of integrals around small circles around the singular points.

The first use of the residue theorem in the main text is in the proof of Theorem 2.2.3, where we evaluated

$$\int_C \frac{t^{z-1}\, dt}{1+t}, \quad 0 < \mathrm{Re}\, z < 1. \tag{A.3.1}$$

Here C was the curve from $+\infty$ to 0 with $\arg t = 2\pi$, returning to $+\infty$ with $\arg t = 0$.

This is taken as a limiting case of the curve C_R, where the part of C with $t \geq R > 1$ is replaced by the circle $\{|t| = R\}$, oriented counter-clockwise, and the part with $0 \leq t \leq 1/R$ is replaced by the circle $\{|t| = 1/R\}$, oriented clockwise. (This is a typical example of how the Cauchy integral theorem, the Cauchy integral formula, and the residue theorem can be extended beyond our original formulation, which assumed a bounded curve and continuity of the integrand at each point of the boundary.) The residue calculus applies to each curve C_R, and the contribution of the integration over the circles goes to zero as $R \to \infty$, so the value of (A.3.1) is $2\pi i$ times the (unique) residue at $t = -1$. With our choice of branch, the residue is $\exp[i(z-1)\pi] = -\exp(iz\pi)$. On the other hand, over the first part of C, the value of t^z differs from the value on the second part by a factor $\exp(2\pi iz)$. This gives the result

$$(1 - e^{2\pi iz}) \int_0^\infty \frac{t^{z-1}\, dt}{1+t} = -2\pi i\, e^{i\pi z}.$$

In Section 5.7, we use the following consequence of the residue theorem: if P and Q are polynomials and C encloses all zeros $\{z_1, z_2, \ldots, z_n\}$ of Q (counting multiplicity), then

$$\frac{1}{2\pi i} \int_C P(z) \frac{Q'(z)}{Q(z)}\, dz = \sum_{j=1}^{n} P(z_j).$$

Another use is in the proof of Lemma 15.1.2.

A.4 Linear fractional transformations

A *linear fractional transformation*, or *Möbius transformation*, is a function of the form

$$f(z) = \frac{az+b}{cz+d}, \quad ad - bc \neq 0.$$

It may be assumed that $ad - bc = 1$. The inverse and the composition of Möbius transformations are Möbius transformations, so the set of Möbius transformations is a group. Given any two ordered triples of distinct points in the Riemann sphere $\mathbf{S} = \mathbf{C} \cup \{\infty\}$, there is a unique Möbius transformation that takes one triple to the other. For example, the transformation

$$f(z) = \frac{az - az_0}{z - z_2}, \quad a = \frac{z_1 - z_2}{z_1 - z_0}$$

takes the triple (z_0, z_1, z_2) to $(0, 1, \infty)$, and its inverse is

$$g(w) = \frac{z_2(z_1 - z_0)w + (z_2 - z_1)z_0}{(z_1 - z_0)w + z_2 - z_1}.$$

The group of Möbius transformations that permute the points $\{0, 1, \infty\}$ is generated by the two transformations

$$z \;\to\; 1 - z = \frac{-z+1}{0z+1}, \quad z \;\to\; \frac{1}{z} = \frac{0z+1}{z+0}$$

and consists of these two transformations and

$$z \to z, \quad z \to \frac{z}{z-1}, \quad z \to \frac{1}{1-z}, \quad z \to 1 - \frac{1}{z}.$$

A.5 Weierstrass factorization theorem

The *Weierstrass factorization theorem*, mentioned in Chapter 14, implies that for every sequence of points $\{z_n\}$ in the complex plane with limit ∞, there is an entire function that has zeros (repeated according to multiplicity) at these points and no others. In particular, if f is meromorphic in $\mathbf{C}$ and h is chosen so that its zeros match the poles of f, then $g = fh$ is entire. (More precisely, g has only removable singularities, so when they are removed it is an entire function.) Thus $f = g/h$ is the quotient of two entire functions.

A.6 Cauchy and Stieltjes transformations and the Sokhotski–Plemelj formula

The *Cauchy transform* of a continuous function f defined on an oriented curve Γ is the function Cf that is defined on the complement of Γ by

$$Cf(z) = \frac{1}{2\pi i} \int_\Gamma \frac{f(s)\, ds}{s-z}.$$

(Note that the Cauchy integral formula amounts to the following statement: if Γ encloses a region Ω and f is holomorphic on Ω and continuous on the closure, then f on Ω is the Cauchy transform of the restriction of f to the boundary.)

For a smooth curve, such as a line or circle, the function f can be recovered from Cf as follows: at a given point s in Γ, for convenience, choose coordinates so that the tangent to Γ is real and points $s + i\varepsilon$ for small $\varepsilon > 0$ lie to the left of the oriented curve. Then

$$f(s) = \lim_{\varepsilon \to 0+} [Cf(s + i\varepsilon) - Cf(s - i\varepsilon)].$$

In other words, f is the jump of Cf from left to right across the curve. The general case can be reduced to the case $\Gamma = \mathbf{R}$ and $f(s) = 0$ for $|s| \geq c$. Then for real x,

$$
\begin{aligned}
&Cf(x + i\varepsilon) - Cf(x - i\varepsilon) \\
&= \frac{1}{2\pi i} \int_{-\infty}^{\infty} \left[\frac{1}{s - (x + i\varepsilon)} - \frac{1}{s - (x - i\varepsilon)} \right] f(s)\, ds \\
&= \frac{1}{\pi} \int_{-\infty}^{\infty} \frac{\varepsilon}{(s - x)^2 + \varepsilon^2} f(s)\, ds = \frac{1}{\pi} \int_{-\infty}^{\infty} \frac{f(x + \varepsilon t)\, dt}{1 + t^2} \\
&\to \left[\frac{1}{\pi} \int_{-\infty}^{\infty} \frac{dt}{1 + t^2} \right] f(x) = f(x).
\end{aligned}
$$

For historical reasons, the same transform with a different normalization is known as the *Stieltjes transform*. The curve in question is normally the real axis, so the Stieltjes transform has the form

$$Sf(z) = -2\pi i\, Cf(z) = \int_{-\infty}^{\infty} \frac{f(x)\, dx}{z - x}.$$

Remarks. 1. For either transform, a general measure $d\mu(x)$ may replace the density $f(x)\, dx$; see Section 4.4.

2. The same idea occurs in the inversion formulas for the Fourier and Mellin transforms and in the proof of the theorem of Weierstrass on approximation by trigonometric polynomials; see Appendix B. The idea is explained clearly in Weierstrass's paper [431]. In a slightly more general formulation: to approximate a continuous function f on the line, say, find an approximate identity – a family of continuous functions $\{G_\varepsilon(x,y)\}_{\varepsilon>0}$ with the properties

$$\int_{-\infty}^{\infty} |G_\varepsilon(x,y)|\, dy < C < \infty;$$

$$\int_{-\infty}^{\infty} G_\varepsilon(x,y)\, dy = 1;$$

$$\lim_{\varepsilon \to 0} \int_{|y-x|>\delta} G_\varepsilon(x,y)\,dx = 0.$$

Then

$$\lim_{\varepsilon \to 0} f_\varepsilon(x) \equiv \lim_{\varepsilon \to 0} \int_{-\infty}^{\infty} G_\varepsilon(x,y) f(y)\,dy = f(x).$$

If G has certain smoothness properties, these properties will carry over to f_ε.

The *Sokhotski–Plemelj formula*, proved by Sokhotski in 1873 and rediscovered by Plemelj in 1908, deals with the one-sided limits of the Cauchy or Stieltjes transform. Again, the case of the real line is typical. Suppose that $f : \mathbf{R} \to \mathbf{C}$ is smooth (to some extent) at x and integrable. Then

$$\lim_{\varepsilon \to 0+} \int_{-\infty}^{\infty} \frac{f(s)\,ds}{(x \pm i\varepsilon) - s} = \mp i\pi f(x) + \mathrm{P.\,V.} \int_{-\infty}^{\infty} \frac{f(x-t)\,dt}{t},$$

where P.V. denotes the *Cauchy principal value*

$$\mathrm{P.\,V.} \int_{-\infty}^{\infty} f(x-s)\frac{ds}{s} = \lim_{\varepsilon \to 0+} \int_{|s|>\varepsilon} \frac{f(x-s)\,ds}{s}. \tag{A.6.1}$$

This follows from the identity

$$\frac{1}{(x \pm i\varepsilon) - s} = \frac{(x-s) \mp i\varepsilon}{(x-s)^2 + \varepsilon^2}.$$

In fact, the identity leads to two integrals, one like that encountered above for the Cauchy transform, and one that transforms to

$$\int_{-\infty}^{\infty} \frac{f(x-t)\,t\,dt}{t^2 + \varepsilon^2}.$$

The transform Hf defined by (A.6.1) is known as the *Hilbert transform* [184]. The smoothness condition mentioned above can be very weak:

$$\int_{\varepsilon < |s| < 1} \frac{f(x-s)\,ds}{s} = \int_{\varepsilon < |s| < 1} \frac{f(x-s) - f(x)}{s}\,ds,$$

so if $|f(x-s) - f(x)| \le |s|^a$ for some $a > 0$, then the limit in (A.6.1) exists. In particular, differentiability at x is sufficient.

Appendix B

Fourier analysis

This section contains a brief account of the facts from classical Fourier analysis and their consequences that are used at various points above. These include the Fourier transform and its inverse, and the Mellin transform and its inverse.

B.1 Fourier and inverse Fourier transforms

Suppose that $f(x)$ is a (real- or) complex-valued function that is absolutely integrable:

$$\int_{-\infty}^{\infty} |f(x)| \, dx < \infty. \tag{B.1.1}$$

The Fourier transform $\widehat{f}$ is defined by

$$\widehat{f}(\xi) = \frac{1}{\sqrt{2\pi}} \int_{-\infty}^{\infty} e^{-ix\xi} f(x) \, dx, \quad \xi \in \mathbf{R}.$$

The condition (B.1.1) implies that $\widehat{f}$ is bounded and continuous. It can also be shown that $\widehat{f}(\xi) \to 0$ as $|\xi| \to \infty$, so $\widehat{f}$ is uniformly continuous.

A particularly useful example is $f(x) = \exp(-\frac{1}{2}x^2)$, which is its own Fourier transform:

$$\frac{1}{\sqrt{2\pi}} \int_{-\infty}^{\infty} e^{-ix\xi} e^{-\frac{1}{2}x^2} \, dx = \frac{e^{-\frac{1}{2}\xi^2}}{\sqrt{2\pi}} \int_{-\infty}^{\infty} e^{-\frac{1}{2}(x+i\xi)^2} \, dx.$$

To see this, take $z = x + i\xi$ in the integral on the right, so that it is an integral over the line $\{\operatorname{Im} z = \xi\}$. By the Cauchy integral theorem, the path of integration can be changed to the real line, so

$$\frac{1}{\sqrt{2\pi}} \int_{-\infty}^{\infty} e^{-ix\xi} e^{-\frac{1}{2}x^2} \, dx = \frac{e^{-\frac{1}{2}\xi^2}}{\sqrt{2\pi}} \int_{-\infty}^{\infty} e^{-\frac{1}{2}x^2} \, dx = e^{-\frac{1}{2}\xi^2}.$$

(See Appendix A.) Extensive tables of Fourier transforms are in [308].

If $g(\xi)$ is absolutely integrable, then the inverse Fourier transform is defined by

$$\check{g}(x) = \frac{1}{\sqrt{2\pi}} \int_{-\infty}^{\infty} e^{ix\xi} g(\xi) \, d\xi, \quad x \in \mathbf{R}.$$

The argument just given shows that $\exp\left(-\frac{1}{2}\xi^2\right)$ has the inverse Fourier transform $\exp\left(-\frac{1}{2}x^2\right)$.

The terminology "inverse Fourier transform" is justified as follows. Suppose that f is a bounded, uniformly continuous, and absolutely integrable function, and suppose that its Fourier transform $\widehat{f}$ itself is absolutely integrable. We show now that $(\widehat{f})^\vee = f$:

$$f(x) = \frac{1}{2\pi} \int_{-\infty}^{\infty} e^{ix\xi} \left[\int_{-\infty}^{\infty} e^{-iy\xi} f(y)\,dy \right] d\xi.$$

We introduce a convergence factor $\exp\left(-\frac{1}{2}(\varepsilon\xi)^2\right)$, $\varepsilon > 0$:

$$\int_{-\infty}^{\infty} e^{ix\xi} \left[\int_{-\infty}^{\infty} e^{-iy\xi} f(y)\,dy \right] d\xi = \lim_{\varepsilon\to 0} \int_{-\infty}^{\infty} \int_{-\infty}^{\infty} e^{i(x-y)\xi - \frac{1}{2}(\varepsilon\xi)^2} f(y)\,dy\,d\xi.$$

The convergence factor allows us to change the order of integration. By what we have just shown about the function $\exp(-\frac{1}{2}\varsigma^2)$,

$$\frac{1}{2\pi} \int_{-\infty}^{\infty} e^{i(x-y)\xi - \frac{1}{2}(\varepsilon\xi)^2}\,d\xi = \frac{1}{2\pi} \int_{-\infty}^{\infty} e^{i[(x-y)/\varepsilon]\varsigma}\, e^{-\frac{1}{2}\varsigma^2} \frac{d\varsigma}{\varepsilon}$$

$$= \frac{1}{\varepsilon\sqrt{2\pi}} \exp\left[-\frac{(x-y)^2}{2\varepsilon^2} \right] \equiv G_\varepsilon(x-y).$$

The functions $\{G_\varepsilon\}$ are easily seen to have the properties of an approximate identity (see Appendix A):

$$G_\varepsilon(x) > 0;$$

$$\int_{-\infty}^{\infty} G_\varepsilon(x)\,dx = 1; \qquad\qquad\qquad (B.1.2)$$

$$\lim_{\varepsilon\to 0} \int_{|x|>\delta} G_\varepsilon(x)\,dx = 0, \quad \delta > 0.$$

According to the calculations above,

$$\frac{1}{2\pi} \int_{-\infty}^{\infty} e^{ix\xi} \left[\int_{-\infty}^{\infty} e^{-iy\xi} f(y)\,dy \right] d\xi - f(x)$$

$$= \lim_{\varepsilon\to 0} \int_{-\infty}^{\infty} G_\varepsilon(x-y)\,[f(y) - f(x)]\,dy. \qquad\qquad (B.1.3)$$

The assumptions on f and the conditions (B.1.2) imply that the limit of (B.1.3) is zero.

The Fourier inversion result used in Section 9.7 is the two-dimensional version of this result. The preceding proof can be adapted easily, or the result can be proved in two steps by taking the transform in one variable at a time.

B.2 Proof of Theorem 4.1.5

Suppose that w is a positive weight function on the interval (a,b) and

$$\int_a^b e^{2c|x|}\, w(x)\,dx \; < \; \infty \qquad\qquad\qquad (B.2.1)$$

for some $c > 0$. Note that this implies that all moments are finite, so the orthonormal polynomials $\{P_n\}$ necessarily exist. Given $f \in L_w^2$, let

$$f_n(x) = \sum_{k=0}^{n} (f, P_k) P_k(x).$$

Then for $m < n$,

$$\|f_n - f_m\|^2 = \sum_{k=m+1}^{n} |(f, P_n)|^2.$$

By (4.1.10),

$$\sum_{k=0}^{\infty} |(f, P_n)|^2 \le \|f\|^2 < \infty,$$

so the sequence $\{f_n\}$ is a Cauchy sequence in L_w^2. Therefore it has a limit g, and we need to show that $g = f$. For any m, $(f_n, P_m) = (f, P_m)$ for $n \ge m$, and it follows that $(g, P_m) = (f, P_m)$, for all m. Thus $h = f - g$ is orthogonal to every P_m and, therefore, orthogonal to every polynomial. We want to show that $h \equiv 0$, or equivalently, that $hw \equiv 0$.

Extend h and w to the entire real line if necessary, by taking them to vanish outside (a, b). Note that $|hw|$ is absolutely integrable, by the Cauchy–Schwarz inequality, since it is the product of square integrable functions $|h|\sqrt{w}$ and $\sqrt{w}$. By (B.2.1), the Fourier transform

$$H(\xi) = \widehat{hw}(\xi) = \frac{1}{\sqrt{2\pi}} \int_{-\infty}^{\infty} e^{-ix\xi} h(x) w(x) dx$$

has an extension to the strip $\{|\operatorname{Im}\xi| < c\}$. Moreover, H is holomorphic in the strip, with derivatives

$$\frac{d^n H}{d\xi^n}(0) = (-i)^n \int_{-\infty}^{\infty} x^n h(x) w(x) dx.$$

Since h is orthogonal to polynomials, all derivatives of H vanish at $\xi = 0$. Since H is holomorphic in the strip, this implies that $H \equiv 0$, and therefore the inverse Fourier transform hw is 0. $\qquad\square$

B.3 Riemann–Lebesgue lemma

In Section 5.8, we used the Riemann–Lebesgue lemma in the following form: if f is a bounded function on a finite interval (a, b), then

$$\lim_{\lambda \to +\infty} \int_a^b \cos(\lambda x) f(x) dx = 0.$$

Suppose first that f is differentiable and that f and f' are continous on the closed interval $[a, b]$. Integration by parts gives

$$\int_a^b \cos(\lambda x) f(x) dx = \frac{\sin(\lambda x)}{\lambda} f(x)\Big|_a^b - \frac{1}{\lambda} \int_a^b \sin(\lambda x) f'(x) dx$$

$$= O(\lambda^{-1}).$$

For general f, given $\varepsilon > 0$, choose a continuously differentiable function g such that

$$\int_a^b |f(x) - g(x)|\,dx \; < \; \varepsilon$$

and apply the previous argument to g to conclude that

$$\left| \int_a^b \cos(\lambda x) f(x)\,dx \right| \; < \; 2\varepsilon$$

for large λ.

B.4 Fourier series and the Weierstrass approximation theorem

In Section 14.4, we tacitly used the finite interval analogue of the Fourier inversion result, namely that the functions

$$e_n(x) = e^{2n\pi ix} = e_1(x)^n, \quad n = 0, \pm 1, \pm 2, \ldots,$$

which are orthogonormal in the space $L^2(I)$, $I = (0,1)$ with weight 1, are complete in this space. To prove this, we use the following construction of an approximate identity, due to Weierstrass [431]. Define

$$G_n(x) = c_n 4^n \cos^{2n}(\pi x) = c_n 2^n [\cos 2\pi x + 1]^n = c_n(e_1 + e_{-1} + 2)^n,$$

where c_n is chosen so that $\int_0^1 G_n(x)\,dx = 1$. The G_n have period 1, are nonnegative, and satisfy the analogue of (B.1.2):

$$\lim_{n \to \infty} \int_\delta^{1-\delta} G_n(x)\,dx = 0, \quad 0 < \delta < \frac{1}{2}.$$

If follows that for any continuous function g with period 1, the sequence

$$g_n(x) = \int_0^1 G_n(x - y) g(y)\,dy$$

converges uniformly to g. Now G_n is in the span of $\{e_k\}_{|k| \le n}$, so g_n is a linear combination of these functions, i.e. a *trigonometric polynomial*. This proves the first version of the *Weierstrass approximation theorem*.

Theorem B.4.1 *Any continuous periodic function with period 2π can be approximated uniformly by trigonometric polynomials.*

The second version is the following.

Theorem B.4.2 *Any continuous function on a closed bounded interval can be approximated uniformly by polynomials.*

This second version of the theorem can be derived from the first. Rescale the interval to $[0, \pi]$ and extend the function to be an even periodic function on the line. It can be approximated uniformly by trigonometric polynomials. Any trigonometric polynomial is an entire function, so it is, on $[0, \pi]$, the uniform limit of its Taylor polynomials.

Let us return to the proof of completeness. Any function h in $L^2(I)$ that is orthogonal to each e_k is also orthogonal to each g_n. Taking limits, h is orthogonal to each continous periodic g. The function h itself can be approximated in the L^2 norm by continuous periodic functions, so $||h||^2 = (h,h) = 0$. As in the proof of Theorem 4.1.5, this implies that the $\{e_n\}$ are dense in $L^2(I)$. Note that in this case, since the functions $\{e_n\}$ are complex-valued, we use the complex inner product

$$(f,g) = \int_0^1 f(x)\overline{g(x)}\,dx.$$

The Fourier series expansion of $f \in L^2(I)$ takes the form

$$f = \sum_{n=-\infty}^{\infty} a_n e_n, \quad a_n = (f,e_n) = \int_0^1 f(x)e^{-2n\pi ix}\,dx.$$

The partial sums

$$f_n(x) = \sum_{m=-n}^{n} a_m e^{2m\pi ix}$$

converge to f in the $L^2(I)$ norm. They can also be shown to converge to f at any point at which f is differentiable, by an argument similar to the arguments used in Section 5.8.

B.5 The Mellin transform and its inverse

In Section 12.6, we used the Mellin transform and its inverse. These results are equivalent to Fourier inversion, after a change of variable, but we give here a direct proof. Again,

$$\mathcal{M}g(z) = \int_0^\infty x^z g(x)\,\frac{dx}{x}$$

whenever the integral is absolutely convergent. Suppose that this is true for a given real value $z = c$, and that $\mathcal{M}g(c + it)$ is also abolutely integrable with respect to $t \in \mathbf{R}$:

$$\int_{-\infty}^\infty |\mathcal{M}g(c+it)|\,dt < \infty.$$

To verify the inverse formula

$$g(x) = \frac{1}{2\pi i}\int_{c-i\infty}^{c+i\infty} x^{-s}\mathcal{M}g(s)\,ds,$$

we write the right-hand side as a double integral and then introduce a convergence factor that allows a change in the order of integration: set

$$\begin{aligned}
f(x) &= \int_{c-i\infty}^{c+i\infty} x^{-s}\mathcal{M}g(s)\,ds \\
&= \frac{1}{2\pi}\int_{-\infty}^\infty x^{-c-it}\left\{\int_0^\infty y^{c+it}g(y)\,\frac{dy}{y}\right\}dt \\
&= \frac{1}{2\pi}\lim_{\varepsilon\to 0+}\int_{-\infty}^\infty e^{-\varepsilon|t|}x^{-c-it}\left\{\int_0^\infty y^{c+it}g(y)\,\frac{dy}{y}\right\}dt \\
&= \frac{1}{2\pi}\int_0^\infty \left(\frac{y}{x}\right)^c\left\{\lim_{\varepsilon\to 0+}\int_{-\infty}^\infty e^{-\varepsilon|t|+it\log(y/x)}\,dt\right\}g(y)\,\frac{dy}{y}.
\end{aligned}$$

The inner integral is

$$\int_0^\infty \left[e^{-[\varepsilon + i \log(y/x)]t} + e^{-[\varepsilon - i \log(y/x)]t} \right] dt = \frac{2\varepsilon}{\varepsilon^2 + \log^2(y/x)}.$$

Inserting this value above and letting $y = ux$, we obtain

$$f(x) = \frac{1}{\pi} \lim_{\varepsilon \to 0+} \int_0^\infty \frac{\varepsilon\, u^c\, g(ux)}{\varepsilon^2 + \log^2 u} \frac{du}{u}.$$

Let $u = \exp(\varepsilon\tau)$, so that

$$f(x) = \frac{1}{\pi} \lim_{\varepsilon \to 0+} \int_{-\infty}^\infty \frac{e^{\varepsilon c\tau}\, g(e^{\varepsilon\tau}x)\, d\tau}{1 + \tau^2}$$

$$= \left\{ \frac{1}{\pi} \int_{-\infty}^\infty \frac{d\tau}{1 + \tau^2} \right\} g(x) = g(x).$$

To show that the left inverse

$$\mathcal{M}^{-1}\varphi(x) = \int_{c-i\infty}^{c+i\infty} x^{-s}\varphi(s)\, ds$$

is also a right inverse, we suppose that the integrand here is absolutely integrable on the vertical line, and that $\mathcal{M}^{-1}\varphi$ is integrable with respect to $x^{c-1}dx$ on the half-line. We write $\mathcal{M}(\mathcal{M}^{-1}\varphi)$ as a double integral

$$\int_0^\infty x^{c+it} \left\{ \int_{c-i\infty}^{c+i\infty} x^{-z}\varphi(z)\, dz \right\} \frac{dx}{x}$$

$$= \frac{1}{2\pi} \int_0^\infty \int_{-\infty}^\infty x^{i(t-s)}\varphi(c+is)\, ds \frac{dx}{x},$$

introduce a convergence factor $\exp(-\varepsilon|\log x|)$, and reverse the order of integration. Set $x = e^u$ so the inner integral is

$$\int_{-\infty}^\infty e^{-\varepsilon|u| + i(t-s)u}\, du = \int_0^\infty \left[e^{-[\varepsilon - i(t-s)]u} + e^{-[\varepsilon + i(t-s)]u} \right] du$$

$$= \frac{2\varepsilon}{\varepsilon^2 + (t-s)^2}.$$

Thus

$$\mathcal{M}\left[\mathcal{M}^{-1}\varphi \right](c+it) = \frac{1}{\pi} \lim_{\varepsilon \to 0} \int_{-\infty}^\infty \left[\frac{\varepsilon}{\varepsilon^2 + (t-s)^2} \right] \varphi(c+is)\, ds$$

$$= \lim_{\varepsilon \to 0} \frac{1}{\pi} \int_{-\infty}^\infty \frac{\varphi(c+it - i\varepsilon\tau)\, d\tau}{1 + \tau^2}$$

$$= \varphi(c+it).$$

The integrability conditions (12.3.26) imply that the imaginary axis is in the domain of $\mathcal{M}g_j$ and of $\mathcal{M}(g_1 * g_2)$. It is easily checked that

$$\mathcal{M}[g_1 * g_2] = \mathcal{M}g_1 \cdot \mathcal{M}g_2.$$

References

[1] Abel, N. H., Recherches sur les fonctions elliptiques, *J. Reine Angew. Math.* **2** (1827), 101–181; **3** (1828), 160–190; *Oeuvres 1*, pp. 263–388.

[2] Abel, N. H., Sur une espèce de fonctions entières nées du developpement de la fonction $(1 - v)^{-1} e^{-\frac{xv}{1-v}}$ suivant les puissances de v, *Oeuvres 2* (1881), p. 284.

[3] Ablowitz, M. J. and Segur, H., Exact linearization of a Painlevé transcendent, *Phys. Rev. Lett.* **38** (1977), 1103–1106.

[4] Abramowitz, M. and Stegun, I., *Handbook of Mathematical Functions*, Dover, Mineola, NY 1965.

[5] Airault, H. Rational solutions of Painlevé equations, *Stud. Appl. Math.* **61** (1979), 31–53; (correction) **64** (1981), 183.

[6] Airy, G. B., On the intensity of light in the neighborhood of a caustic, *Trans. Camb. Philos. Soc.* **VI** (1838), 379–402.

[7] Akhiezer, N. I., *The Classical Moment Problem and Some Related Questions in Analysis*, Hafner, New York, NY 1965.

[8] Akhiezer, N. I., *Elements of the Theory of Elliptic Functions*, Translations of Mathematical Monographs, American Mathematical Society, Providence, RI 1990.

[9] Al-Salam, W. A., Characterization theorems for orthogonal polynomials, *Orthogonal Polynomials*, NATO Adv. Sci. Inst. Ser. C: Math. Phys. Sci., 294, Kluwer, Dordrecht 1990, pp. 1–24.

[10] Andrews, G. E., Askey, R., and Roy, R., *Special Functions*, Cambridge University Press, Cambridge 1999.

[11] Andrews, L. C., *Special Functions of Mathematics for Engineers*, 2nd ed., McGraw-Hill, New York, NY 1992.

[12] Aomoto, K., Jacobi polynomials associated with Selberg integrals, *SIAM J. Math. Anal.* **18** (1987), 545–549.

[13] Appell, P. Sur les fonctions hypergéométriques de deux variables. *J. Math. Pures Appl.* **8** (1882), 173–216.

[14] Appell, P. and Kampé de Fériet, J., *Fonctions Hypergéométriques de Plusieurs Variables: Polynômes d'Hermite*, Gauthier-Villars, Paris 1926.

[15] Appell, P. and Lacour, E., *Principes de la Théorie des Fonctions Elliptiques et Applications*, Gauthier-Villars, Paris 1922.

[16] Armitage, J. V. and Eberlein, W. F., *Elliptic Functions*, Cambridge University Press, Cambridge 2006.

[17] Artin, E., *The Gamma Function*, Holt, Rinehart and Winston, New York, NY 1964.

[18] Askey, R., *Orthogonal Polynomials and Special Functions*, Society for Industrial and Applied Mathematics, Philadelphia, PA 1975.

[19] Askey, R., Continuous Hahn polynomials, *J. Phys. A* **18** (1985), L1017–1019.

[20] Askey, R., Handbooks of special functions, *A Century of Mathematics in America*, vol. 3, P. Duren, ed., American Mathematical Society, Providence, RI 1989, pp. 369–391.

[21] Askey, R. and Wilson, J. A., A set of orthogonal polynomials that generalize the Racah coefficients or 6j symbols, *SIAM J. Math. Anal.* **10** (1979), 1008–1016.

[22] Askey, R. and Wilson J. A., *Some Basic Hypergeometric Orthogonal Polynomials that Generalize Jacobi Polynomials*, Memoirs of the American Mathematical Society, 54, American Mathematical Society, Providence, RI 1985.

[23] Atakishiyev, N. M., Rahman, M., and Suslov, S. K., On classical orthogonal polynomials, *Constr. Approx.* **11** (1995), 181–226.

[24] Atakishiyev, N. M. and Suslov, S. K., The Hahn and Meixner polynomials of an imaginary argument and some of their applications, *J. Phys. A* **18** (1985), 1583–1596.

[25] Ayoub, R., The lemniscate and Fagnano's contribution to elliptic integrals, *Arch. Hist. Exact Sci.* **29** (1983/84), 131–149.

[26] Bäcklund, A. V., Om ytor med konstant negativ krokning, *Lunds Univ. Arsskr.* **19** (1883), 1–41.

[27] Baik, J., Kriecherbauer, T., McLaughlin, K., and Miller, P. D., *Discrete Orthogonal Polynomials: Asymptotics and Applications*, Princeton University Press, Princeton, NJ 2007.

[28] Bailey, W. N., *Generalized Hypergeometric Series*, Cambridge University Press, Cambridge 1935.

[29] Baker, H. F., *Abelian Functions: Abel's Theory and the Allied Theory of Theta Functions*, Cambridge University Press, New York, NY 1995.

[30] Barnes, E. W., A new development of the theory of the hypergeometric functions, *Proc. London Math. Soc.* **50** (1908), 141–177.

[31] Bassett, A. B., On the potentials of the surfaces formed by the revolution of limaçons and cardioids, *Proc. Camb. Philos. Soc.* **6** (1889), 2–19.

[32] Bassom, A. P., Clarkson, P. A., Law, C. K., and McLeod, J. B., Application of uniform asymptotics to the second Painlevé transcendent, *Arch. Rational Mech. Anal.* **143** (1998), 241–271.

[33] Beals, R., Gaveau, B., and Greiner, P. C., Uniform hypoelliptic Green's functions, *J. Math. Pures Appl.* **77** (1998), 209–248.

[34] Beals, R. and Kannai, Y., Exact solutions and branching of singularities for some equations in two variables, *J. Diff. Equations* **246** (2009), 3448–3470.

[35] Beals, R., and Sattinger, D. H., Integrable systems and isomonodromy deformations, *Physica D* **65** (1993), 17–47.

[36] Beals, R., and Szmigielski, J., Meijer *G*-functions: a gentle introduction, *Notices Amer. Math. Soc.* **60** (2013), 866–872.

[37] Bernoulli, D., Solutio problematis Riccatiani propositi in Act. Lips. Suppl. Tom VIII p. 73, *Acta Erud. Publ. Lipsiae* (1725), 473–475.

[38] Bernoulli, D., Demonstrationes theorematum suorum de oscillationibus corporum filo flexili connexorum et catenae verticaliter suspensae, *Comm. Acad. Sci. Petr.* **7** (1734–35, publ. 1740), 162–173.

[39] Bernoulli, Jac., Curvatura laminae elasticae, *Acta Erud.* (1694), 276–280; *Opera I*, pp. 576–600.

[40] Bernoulli, Jac., *Ars Conjectandi*, Basel 1713.

[41] Bernoulli, Joh., Methodus generalis construendi omnes aequationes differentialis primi gradus, *Acta Erud. Publ. Lipsiae* (1694), 435–437.

[42] Bernstein, S., Sur les polynomes orthogonaux relatifs à un segment fini, *J. de Math.* **9** (1930), 127–177; **10** (1931), 219–286.

[43] Bessel, F. W., Untersuchung des Theils der planetarischen Störungen aus der Bewegung der Sonne entsteht, *Berliner Abh.* (1824), 1–52.

[44] Beukers, F., Hypergeometric functions, how special are they? *Notices Amer. Math. Soc.* **61** (2014), 48–56

[45] Binet, J. P. M., Mémoire sur les intégrales définies eulériennes et sur leur application à la théorie des suites ainsi qu'à l'évaluation des fonctions des grandes nombres, *J. École Roy. Polyt.* **16** (1838–9), 123–143.

[46] Birkhoff, G. D., Formal theory of irregular linear difference equations, *Acta Math.* **54** (1930), 205–246.

[47] Birkhoff, G. D. and Trjitzinsky, W. J., Analytic theory of singular difference equations, *Acta Math.* **60** (1933), 1–89.

[48] Bleistein, N. and Handelsman, R. A., *Asymptotic Expansions of Integrals*, Dover, Mineola, NY 1986.

[49] Bobenko, A. I. and Eitner, U., *Painlevé Equations in the Differential Geometry of Surfaces*, Springer, Berlin 2000.

[50] Bochner, S., Über Sturm-Liouvillesche Polynomsysteme, *Math. Z.* **29** (1929), 730–736.

[51] Bohr, H. and Mollerup, J., *Laerebog i Matematisk Analyse*, vol. 3, J. Gjellerup, Copenhagen 1922.

[52] Boutroux, P., Recherches sur les transcendantes de M. Painlevé et l'étude asymptotique des équations différentielles du seconde ordre, *Ann. Sci. École Norm. Sup.* **30** (1913), 255–375; **31** (1914), 99–159.

[53] Brillouin, L., Remarques sur la mécanique ondulatoire, *J. Phys. Radium* **7** (1926), 353–368.

[54] Briot C. and J.-C. Bouquet, Mémoire sur l'intégration des équations différentielles au moyen des fonctions elliptiques, *J. École Imp. Poly.* **36** (1856), 199–254.

[55] Bruno, A. D. and Batkhin, A. B., *Painlevé Equations and Related Topics*, de Gruyter, Berlin 2012.

[56] Brychkov, Yu. A., *Handbook of Special Functions: Derivatives, Integrals, Series and Other Formulas*, CRC Press, Boca Raton, FL 2008.

[57] Buchholz, H., *The Confluent Hypergeometric Function with Special Emphasis on its Applications*, Springer, New York, NY 1969.

[58] Burchnall, J. L., Differential equations associated with hypergeometric functions. *Quart. J. Math. Oxford* **13** (1942), 90–106.

[59] Burchnall, J. L. and Chaundy, T. W., Expansions of Appell's double hypergeometric functions, II, *Quart. J. Math. Oxford* **12** (1941), 112–128.

[60] Burkhardt, H., Entwicklungen nach oscillierenden Funktionen und Integration der Differentialgleichungen der mathematischen Physik, *Deutsche Math. Ver.* **10** (1901–1908), 1–1804.

[61] Campbell, R., *Les Intégrales Euleriennes et leurs Applications*, Dunod, Paris 1966.

[62] Cao, L.-H., Li, Y.-T., and Lin, Y., Asymptotic approximations of the continuous Hahn polynomials and their zeros, *J. Approx. Theory*, to appear.

[63] Carleman, T., Sur le problème des moments, *C. R. Acad. Sci. Paris* **174** (1922), 1680.

[64] Carlini, F., Ricerche sulla convergenza della serie che serva alla soluzione del problema di Keplero, *Appendice all' Effemeridi Astronomiche di Milano per l'Anno 1818*, Milan 1817, pp. 3–48.

[65] Carlson, B. C., *Special Functions of Applied Mathematics*, Academic Press, New York, NY 1977.

[66] Chandrasekharan, K., *Elliptic Functions*, Springer, Berlin 1985.

[67] Charlier, C. V. L., Über die Darstellung willkürlichen Funktionen, *Ark. Mat. Astr. Fysic* **2** (1905–6), 1–35.

[68] Chaundy, T. W. An extension of hypergeometric functions, I, *Quart. J. Math. Oxford* **14** (1943), 55–78.

[69] Chebyshev, P. L., Théorie des mécanismes connus sous le nom de parallélogrammes, *Publ. Soc. Math. Warsovie* **7** (1854), 539–568; *Oeuvres 1*, pp. 109–143.

[70] Chebyshev, P. L., Sur les fractions continues, *Utzh. Zap. Imp. Akad. Nauk* **3** (1855), 636–664; *J. Math. Pures Appl.* **3** (1858), 289–323; *Oeuvres 1*, pp. 201–230.

[71] Chebyshev, P. L., Sur une nouvelle série, *Bull. Phys. Math. Acad. Imp. Sci. St. Pét.* **17** (1858), 257–261; *Oeuvres 1*, pp. 381–384.

[72] Chebyshev, P. L., Sur le développment des fonctions à une seule variable, *Bull. Phys. Math. Acad. Imp. Sci. St. Pét.* **1** (1859), 193–200; *Oeuvres 1*, pp. 499–508.

[73] Chebyshev, P. L., Sur l'interpolation des valeurs équidistantes, *Zap. Akad. Nauk* **4** (1864); *Oeuvres 1*, pp. 219–242.

[74] Chebyshev, P. L., Sur les valeurs limites des intégrales, *J. Math. Pures Appl.* **19** (1874), 193–200.

[75] Chester, C., Friedman, B., and Ursell, F., An extension of the method of steepest descents, *Proc. Camb. Philos. Soc.* **53** (1957), 599–611.

[76] Chihara, T. S., *An Introduction to Orthogonal Polynomials*, Gordon and Breach, New York, NY 1978.

[77] Christoffel, E. B., Ueber die Gaussische Quadratur und eine Verallgemeinerung derselben, *J. Reine Angew. Math.* **55** (1858), 61–82.

[78] Christoffel, E. B., Ueber die lineare Abhängigkeit von Funktionen einer einzigen Veränderlichen, *J. Reine Angew. Math.* **55** (1858), 281–299.

[79] Christoffel, E. B., Sur une classe particulière de fonctions entières et de fractions continues, *Ann. Mat. Pura Appl.* **8** (1877), 1–10.

[80] Chu, S.-C., *Ssu Yuan Yü Chien* (Precious Mirror of the Four Elements), 1303.

[81] Clancey, K. F. and Gohberg, I, *Factorization of Matrix Functions and Singular Integral Operators*, Birkhäuser, Basel, 1981.

[82] Clarkson, P. A., Painlevé equations – nonlinear special functions, *Orthogonal Polynomials and Special Functions: Computation and Applications*, F. Marcellán and W. Van Assche, eds., Springer, Berlin 2006, pp. 331–411.

[83] Clausen, T., Ueber die Fälle, wenn die Reihe von der Form $y = 1 + \frac{\alpha}{1} \cdot \frac{\beta}{\gamma x} + \frac{\alpha \cdot \alpha + 1}{1 \cdot 2} \cdot \frac{\beta \cdot \beta + 1}{\gamma \cdot \gamma + 1} x^2$+etc. ein Quadrat von der Form $z = 1 + \frac{\alpha'}{1} \cdot \frac{\beta'}{\gamma'} \cdot \frac{\delta'}{\varepsilon' x} + \frac{\alpha \cdot \alpha' + 1}{1 \cdot 2} \cdot \frac{\beta' \cdot \beta' + 1}{\gamma' \cdot \gamma' + 1} \cdot \frac{\delta' \cdot \delta' + 1}{\varepsilon' \cdot \varepsilon' + 1} x^2$+etc. hat, *J. Reine Angew. Mat.* **3** (1828), 93–96.

[84] Coddington, E. and Levinson, N., *Theory of Ordinary Differential Equations*, McGraw-Hill, New York, NY 1955.

[85] Conte, R., and Musette, M., *The Painlevé Handbook*, Springer, Dordrecht 2008.

[86] Copson. E. T., *An Introduction to the Theory of Functions of a Complex Variable*, Oxford University Press, Oxford 1955.

[87] Copson, E. T., On the Riemann–Green function, *Arch. Rat. Mech. Anal.* **1** (1958), 324–348.

[88] Copson, E. T., *Asymptotic Expansions*, Cambridge University Press, Cambridge 2004.

[89] Courant, R. and Hilbert, D., *Methods of Mathematical Physics*, vol. 1, Wiley, New York, NY 1961.

[90] Cuyt, A., Petersen, V. B., Verdonk, B., Waadeland, H., and Jones, W. B., *Handbook of Continued Fractions for Special Functions*, Springer, New York, NY 2008.

[91] Dai, D., Ismail, M. E. H., and Wang, X.-S., Plancherel–Rotach asymptotic expansion for some polynomials from indeterminate moment problems, *Constr. Approx.* **40** (2014), 61–104.

[92] D'Antonio, L., Euler and elliptic integrals, *Euler at 300: An Appreciation*, Mathematical Association of America, Washington, DC 2007, pp. 119–129.

[93] Darboux, G., Mémoire sur l'approximation des fonctions de très-grandes nombres et sur une classe étendue de développements en série, *J. Math. Pures Appl.* **4** (1878), 5–57.

[94] Darboux, G., *Théorie Générale des Surfaces*, vol. 2, book 4, Gauthier-Villars, Paris 1889.

[95] Davis, P. J., Leonhard Euler's integral: a historical profile of the gamma function. In memoriam: Milton Abramowitz, *Amer. Math. Monthly* **66** (1959), 849–869.

[96] de Bruijn, N. G., *Asymptotic Methods in Analysis*, North-Holland, Amsterdam 1961.

[97] Debye, P., Näherungsformeln für die Zylinderfunktionen für große Werte des Arguments und unbeschränkt veränderliche Werte des Index, *Math. Ann.* **67** (1909), 535–558.

[98] Deift, P., *Orthogonal Polynomials and Random Matrices: a Riemann–Hilbert Approach*, American Mathematical Society, Providence, RI 1999.

[99] Deift, P., Kriecherbauer, T., McLaughlin, K., Venakides, S., and Zhou, X., Strong asymptotics of orthogonal polynomials with respect to exponential weights, *Comm. Pure Appl. Math.* **70** (1999), 1491–1552.

[100] Deift, P., and Zhou, X., A steepest descent method for oscillatory Riemann–Hilbert problems: Asymptotics for the MKdV equation, *Ann. Math.* **137** (1993), 295–368.

[101] Deift, P., and Zhou, X., Asymptotics for the Painlevé II equation, *Comm. Pure Appl. Math.* **48** (1995), 277–377.

[102] Delache, S., and Leray, J., Calcul de la solution élémentaire de l'opérateur d'Euler–Poisson–Darboux et de l'opérateur de Tricomi–Clairaut hyperbolique d'ordre 2, *Bull. Soc. Math. France* **99** (1971), 313–336.

[103] Dieudonné, J., *Abrégé d'Histoire des Mathématiques*, vols. 1 and 2, Hermann, Paris 1978.

[104] Dieudonné, J., *Special Functions and Linear Representations of Lie Groups*, CBMS Series, 42, American Mathematical Society, Providence, RI 1980.

[105] Dirichlet, G. L., Sur les séries dont le terme général dépend de deux angles, et qui servent à exprimer des fonctions arbitraires entre les limites données, *J. für Math.* **17** (1837), 35–56.

[106] Dubrovin, B., Theta functions and nonlinear equations, *Russian Math. Surv.* **36** (1981), 11–80.

[107] Dunkl, C., A Krawtchouk polynomial addition theorem and wreath products of symmetric groups, *Indiana Univ. Math. J.*, **25** (1976), 335–358.

[108] Dunster, T. M., Uniform asymptotic expansions for Charlier polynomials, *J. Approx. Theory* **112** (2001), 93–133.

[109] Dutka, J., The early history of the hypergeometric function, *Arch. Hist. Exact Sci.* **31** (1984/85), 15–34.

[110] Dutka, J., The early history of the factorial function, *Arch. Hist. Exact Sci.* **43** (1991/92), 225–249.

[111] Dutka, J., The early history of Bessel functions, *Arch. Hist. Exact Sci.* **49** (1995/96), 105–134.

[112] Dwork, B., *Generalized Hypergeometric Functions*, Oxford University Press, New York, NY 1990.

[113] Edwards, H. M., *Riemann's Zeta Function*, Dover, Mineola, NY 2001.

[114] Erdélyi, A., Über eine Integraldarstellung der $M_{k,m}$-Funktion und ihre aymptotische Darstellung für grosse Werte von $\Re(k)$, *Math. Ann.* **113** (1937), 357–361.

[115] Erdélyi, A., *Asymptotic Expansions*, Dover, Mineola, NY 1956.

[116] Erdélyi, A., Magnus, W., Oberhettinger, F., and Tricomi, F. G., *Higher Transcendental Functions*, vols. I–III, Robert E. Krieger, Melbourne, FL 1981.

[117] Erdélyi, A., Magnus, W., and Oberhettinger, F., *Tables of Integral Transforms*, vols. I and II, McGraw-Hill, New York, NY 1954.

[118] Erdős, P. and Turán, P., On interpolation. III: Interpolatory theory of polynomials, *Ann. Math.* **41** (1940), 510–553.

[119] Euler, L., De progressionibus transcendentibus seu quarum termini generales algebraice dari nequent, *Comm. Acad. Sci. Petr.* **5** (1738), 36–57; *Opera Omnia I* vol. 14, pp. 1–24.

[120] Euler, L., De productis ex infinitus factoribus ortis, *Comm. Acad. Sci. Petr.* **11** (1739), 3–31; *Opera Omnia I* vol. 14, pp. 260–290.

[121] Euler, L., De motu vibratorum tympanorum, *Comm. Acad. Sci. Petr.* **10** (1759), 243–260; *Opera Omnia II*, vol. 10, pp. 344–358.

[122] Euler, L., De integratione aequationis differentialis $m\,dx/\sqrt{1-x^4}=n\,dy/\sqrt{1-y^4}$, *Nov. Comm. Acad. Sci. Petr.* **6** (1761), 37–57; *Opera Omnia I* vol. 20, pp. 58–79.

[123] Euler, L., *Institutiones Calculi Integralis*, vol. 2, St. Petersburg 1769; *Opera Omnia I* vol. 12, pp. 221–230.

[124] Euler, L., Evolutio formulae integralis $\int x^{f-1}(lx)^{m/n}dx$ integratione a valore $x=0$ ad $x=1$ extensa, *Nov. Comm. Acad. Sci. Petr.* **16** (1771), 91–139; *Opera Omnia I*, vol. 17, pp. 316–357.

[125] Euler, L., Demonstratio theorematis insignis per conjecturam eruti circa integrationem formulae $\int \frac{d\varphi \cos\varphi}{(1+aa-2a\cos\varphi)^{n+1}}$, *Institutiones Calculi Integralis*, vol. 4, St. Petersburg 1794, pp. 242–259; *Opera Omnia I*, vol. 19, pp. 197–216.

[126] Euler, L., Methodus succincta summas serierum infintarum per formulas differentiales investigandi, *Mém. Acad. Sci. St. Pét.* **5** (1815), 45–56; *Opera Omnia I*, vol. 16, pp. 200–213.

[127] Fabry, E., Sur les intégrales des équations différentielles à coefficients rationnels, Dissertation, Paris 1885.

[128] Fagnano dei Toschi, G. C., *Produzioni Matematiche*, Gavelliana, Pesaro 1750.

[129] Favard, J., Sur les polynômes de Tchebicheff, *C. R. Acad. Sci. Paris* **200** (1935), 2052–2053.

[130] Ferreira, C., López, J. L., and Pagola, P. J., Asymptotic approximations between the Hahn-type polynomials and Hermite, Laguerre and Charlier polynomials, *Acta Appl. Math.* **103** (2008), 235–252.

[131] Ferreira, C., López, J. L., and Sinusía, E. P., Asymptotic relations between the Hahn-type polynomials and Meixner–Pollaczek, Jacobi, Meixner and Krawtchouk polynomials, *J. Comput. Appl. Math.* **217** (2008), 88–109.

[132] Fejér, L., Sur une méthode de M. Darboux, *C. R. Acad. Sci. Paris* **147** (1908), 1040–1042.

[133] Fejér, L., Asymptotikus értékek meghatározáráról (On the determination of asymptotic values), *Math. Termész. Ért.* **27** (1909), 1–33; *Ges. Abh. I*, pp. 445–503.

[134] Ferrers, N. M., *An Elementary Treatise on Spherical Harmonics and Functions Connected with Them*, MacMillan, London 1877.

[135] Fine, N. J., *Basic Hypergeometric Series and Applications*, American Mathematical Society, Providence, RI 1988.

[136] Flaschka, H., and Newell, A. C., Monodromy and spectrum-preserving deformations I, *Comm. Math. Phys.* **61** (1980), 65–116.

[137] Fokas, A. S., Its, A. R., Kapaev, A. A., and Novokshenov, V. Y., *Painlevé Transcendents: the Riemann–Hilbert Approach*, American Mathematical Society, Providence, RI 2006.

[138] Fokas, A. S., Its, A. R., and Kataev, A. V., An isomondromy approach to the problem of two-dimensional quantum gravity, *Uspekhi Mat. Nauk* **45** (1990), 135–136; *Russian Math. Surv.* **45** (1990), 155–157.

[139] Forrester, P. J. and Warnaar, S. O., The importance of the Selberg integral, *Bull. Amer. Math. Soc.* **45** (2008), 489–534.

[140] Forsyth, A. R., *A Treatise on Differential Equations*, Dover, Mineola, NY 1996.

[141] Fourier, J. B. J., *La Théorie Analytique de la Chaleur*, Firmin Didot, Paris 1822.

[142] Fox, C., The G and H-functions as symmetrical Fourier kernels, *Trans. Amer. Math. Soc.* **98** (1963), 395–429.

[143] Fox, C., Integral transforms based upon fractional integration, *Proc. Camb. Philos. Soc.* **59** (1963), 63–71.

[144] Freud, G., *Orthogonal Polynomials*, Pergamon, New York, NY 1971.

[145] Fuchs, R., Uber lineare homogene Differentialgleichungen zweiter Ordnung mit drei in Endlichen gelegenen wesentlichen Singularitäten, *Math. Ann.* **63** (1907), 301–321.

[146] Gambier, B., Sur les équations du seconde ordre et du premier degré dont l'intégrale générale est à points critiques fixes, *Acta Math.* **33** (1910), 1–55.

[147] Gardner, C., Greene, J., Kruskal, M., and Miura, R., Korteweg–deVries equation and generalization. VI. Methods for exact solution. *Comm. Pure Appl. Math.* **27** (1974), 97–133.

[148] Garnier, R., Sur des équations différentielles du troisième ordre dont l'intégrale générale est uniforme et sur une classe d'équations nouvelles d'ordre supérieur dont l'intégrale générale a ses points critiques fixes, *Ann. Sci. École Norm. Sup.* **29** (1912), 1–126.

[149] Gasper, G. and Rahman M., *Basic Hypergeometric Series*, Cambridge University Press, Cambridge 2004.

450   References

[150] Gauss, C. F., Disquisitiones generales circa seriem infinitam $1 + \frac{\alpha \cdot \beta}{1 \cdot \gamma} x +$ $\frac{\alpha(\alpha+1)\beta(\beta+1)}{1 \cdot 2 \cdot \gamma(\gamma+1)} x^2 + \cdots$, *Comm. Soc. Reg. Sci. Gött.* **2** (1813), 46pp.; *Werke 3*, pp. 123–162.

[151] Gauss, C. F., Methodus nova integralium valores per approximationem inveniendi, *Comm. Soc. Reg. Sci. Gött.* **3** (1816), 39–76; *Werke 3*, pp. 163–196.

[152] Gauss, C. F., Determinatio attractionis quam in punctum quodvis positionis datae exerceret platea si ejus massa per totam orbitam ratione temporis quo singulae partes discribuntur uniformiter esset dispertita, *Comm. Soc. Reg. Sci. Gött.* **4** (1818), 21–48; *Werke 3*, pp. 332–357.

[153] Gauss, C. F., Lemniscatische Funktionen I, *Werke 3*, 1876, pp. 404–412.

[154] Gauss, C. F., Kugelfunktionen, *Werke 5*, 1877, pp. 630–632.

[155] Gautschi, W., Some elementary inequalities relating to the gamma and incomplete gamma function, *J. Math. Phys.*, **38** (1959), 77–81.

[156] Gautschi, W., *Orthogonal Polynomials: Computation and Approximation*, Oxford University Press, New York, NY 2004.

[157] Gautschi, W., Leonhard Euler: his life, the man, and his works, *SIAM Review* **50** (2008), 3–33.

[158] Gegenbauer, L., Über die Bessel'schen Functionen, *Wiener Sitzungsber.* **70** (1874), 6–16.

[159] Gegenbauer, L., Über einige bestimmte Itegrale, *Wiener Sitzungsber.* **70** (1874), 434–443.

[160] Gegenbauer, L., Über die Funktionen $C_n^\nu(x)$, *Wiener Sitzungsber.* **75** (1877), 891–896; **97** (1888), 259–316.

[161] Gegenbauer, L., Das Additionstheorem der Funktionen $C_n^\nu(x)$, *Wiener Sitzungsber.* **102** (1893), 942–950.

[162] Gelfand, I. M., Graev, M. I., and Retakh, V. S., General hypergeometric systems of equations and series of hypergeometric type. *Uspekhi Mat. Nauk* **47** (1992), 3–82, 235; *Russian Math. Surv.* **47** (1992) 1–88.

[163] Gelfand, I. M., Kapranov, M. M., and Zelevinsky, A. V., Generalized Euler integrals and A-hypergeometric functions, *Adv. Math.* **84** (1990), 255–271.

[164] Geronimus, L. Y., *Orthogonal Polynomials*, American Mathematical Society, Providence, RI 1977.

[165] Glaisher, J. W. L., On elliptic functions, *Messenger Math.* **15** (1881–2), 81–138.

[166] Godefroy, M., *La Fonction Gamma*, Gauthier-Villars, Paris 1901.

[167] Gordon, W., Über den Stoss zweier Punktladungen nach der Wellenmechanik, *Z. für Phys.* **48** (1928), 180–191.

[168] Goursat, E., Sur l'équation différentielle linéaire qui admet pour intégrale la série hypergéométrique, *Ann. Sci. École Norm. Sup.* **10** (1881), 3–142.

[169] Goursat, E., Mémoire sur les fonctions hypergéométriques dordre supérieur, *Ann. Sci. École Norm. Sup.* **12** (1883), 261–286, 395–430.

[170] Gradshteyn, I. S. and Ryzhik, I. M., *Table of Integrals, Series, and Products*, A. Jeffrey and D. Zwillinger, eds., Academic Press, San Diego, CA 2007.

[171] Graf, J. H., Ueber die Addition und Subtraction der Argumente bei Bessel'schen Funktionen, *Math. Ann.* **43** (1893), 136–144.

[172] Green, G., *An Essay on the Application of Mathematical Analysis to the Theories of Electricity and Magnetism*, Nottingham 1828; *J. für Math.* **39** (1850), 74–89.

[173] Green, G., On the motion of waves in a variable canal of small depth and width, *Trans. Camb. Philos. Soc.* **6** (1837), 457–462.

[174] Gromak, V. I., Laine, I., and Shimomura, S., *Painlevé Differential Equations in the Complex Plane*, De Gruyter, Berlin 2002.

[175] Hahn, W., Bericht über die Nullstellungen der Laguerreschen und Hermiteschen Polynome, *Jahresber. Deutschen Math. Verein* **44** (1934), 215–236; **45** (1935), 211.

[176] Hahn, W., Über Orthogonalpolynomen die q-Differenzgleichungen genügen, *Math. Nachr.* **2** (1949), 4–34.

[177] Hankel, H., Die Cylinderfunktionen erster und zweiter Art, *Math. Ann.* **1** (1869), 467–501.

[178] Hansen, P. A., Ermittelung der absoluten Störungen in Ellipsen von beliebiger Excentricität und Neigung, I, *Schriften der Sternwarte Seeberg*, Carl Gläser, Gotha 1843.

[179] Hastings, S. P. and McLeod, J. B., A boundary value problem associated with the second Painlevé transcendent and the Korteweg–de Vries equation, *Arch. Rational Mech. Anal.* **73** (1980), 31–51.

[180] Heine, H. E., *Handbuch der Kugelfunktionen*, Reimer, Berlin 1878, 1881.

[181] Hermite, C., Sur une nouveau développement en série de fonctions, *C. R. Acad. Sci. Paris*, **58** (1864), 93–100.

[182] Higgins, J. R., *Completeness and Basis Properties of Sets of Special Functions*, Cambridge University Press, Cambridge 1977.

[183] Hilbert, D., Über die Diskriminanten der in endlichen abbrechenden hypergeometrischen Reihe, *J. Reine Angew. Math.* **103** (1885), 337–345.

[184] Hilbert, D., Über eine Anwendung der Integralgleichungen auf ein Problem der Funktionentheorie. *Verh. 3. Internat. Math. Kongr. Heidelberg*, 1904.

[185] Hille, E., *Ordinary Differential Equations in the Complex Domain*, Dover, Mineola, NY 1976.

[186] Hille, E., Shohat, J., and Walsh, J. L., *A Bibliography on Orthogonal Polynomials*, Bulletin National Research Council, 103, National Academy of Science, Washington, DC 1940.

[187] Hobson, E. W., On a type of spherical harmonics of unrestricted degree, order, and argument, *Phil. Trans.* **187** (1896), 443–531.

[188] Hobson, E. W., *The Theory of Spherical and Ellipsoidal Harmonics*, Chelsea, New York, NY 1965.

[189] Hochstadt, H., *The Functions of Mathematical Physics*, Wiley, New York, NY 1971.

[190] Hoëné-Wronski, J., *Réfutation de la Théorie des Fonctions Analytiques de Lagrange*, Blankenstein, Paris 1812.

[191] Horn, J., Ueber die Convergenz der hypergeometrischen Reihen zweier und dreier Vernderlichen, *Math. Ann.* **34** (1889), 544–600.

[192] Horn, J., Hypergeometrische Funktionen zweier Vernderlichen, *Math. Ann.* **105** (1931), 381–407.

[193] Ince, E. L., *Ordinary Differential Equations*, Dover, Mineola, NY 1956.

[194] Ismail, M. E. H., *Classical and Quantum Orthogonal Polynomials in One Variable*, Cambridge University Press, Cambridge 2005.

[195] Ivić, A., *The Riemann Zeta-Function: Theory and Applications*, Dover, Mineola, NY 2003.

[196] Iwasaki, K., Kimura, T., Shimomura, S., and Yoshida, M., *From Gauss to Painlevé: A Modern Theory of Special Functions*, Vieweg, Braunschweig 1991.

[197] Jacobi, C. G. J., Extraits de deux letters de M. Jacobi de l'Université de Königsberg à M. Schumacher, *Schumacher Astron. Nachr.* **6** (1827); *Werke 1*, pp. 31–36.

[198] Jacobi, C. G. J., *Fundamenta Nova Theoriae Functionum Ellipticarum*, Borntraeger, Königsberg 1829; *Werke 1*, pp. 49–239.

[199] Jacobi, C. G. J., Versuch einer Berechnung der grossen Ungleichheit des Saturns nach einer strengen Entwicklung, *Schumacher Astron. Nachr.* **28** (1849), 65–80, 81–94; *Werke 7*, pp. 145–174.

[200] Jacobi, C. G. J., Theorie der elliptischen Functionen aus den eigenschaften der Thetareihen abgeleitet, *Werke 1*, 1881, pp. 497–538.

[201] Jacobi, C. G. J., Untersuchungen über die Differentialgleichung der hypergeometrischen Reihe, *J. Reine Angew. Math.* **56** (1859), 149–165; *Werke 6*, pp. 184–202.

[202] Jahnke, E. and Emde, F., *Tables of Higher Functions*, McGraw-Hill, New York, NY 1960.

[203] Jeffreys, H., On certain approximate solutions of linear differential equations of the second-order, *Proc. London Math. Soc.* **23** (1924), 428–436.

[204] Jeffreys, H., Asymptotic solutions of linear differential equations, *Philos. Mag.* **33** (1942), 451–456.

[205] Jimbo, M. and Miwa, T., Monodromy preserving deformation of linear ordinary differential equations with rational coefficients. II, *Physica D* **2** (1981), 407–448.

[206] Jin, X.-S. and Wong, R., Uniform asymptotic expansions for Meixner polynomials, *Constr. Approx.* **14** (1998), 113–150.

[207] Jin, X.-S. and Wong, R., Asymptotic formulas for the Meixner polynomials, *J. Approx. Theory* **14** (1999), 281–300.

[208] Johnson, D. E. and Johnson, J. R., *Mathematical Methods in Engineering and Physics: Special Functions and Boundary-Value Problems*, Ronald Press, New York, NY 1965.

[209] Kamke, E., *Differentialgleichungen: Lösungsmethoden und Lösungen Band 1: Gewöhnliche Differentialgleichungen*, Chelsea, New York, NY 1959.

[210] Kampé de Fériet, J., *La Fonction Hypergéométrique*, Gauthier-Villars, Paris 1937.

[211] Kampé de Fériet, J. and Appell, P. E., *Fonctions Hypergéométriques et Hypersphériques*, Gauthier-Villars, Paris 1926.

[212] Kanemitsu, S. and Tsukada, H., *Vistas of Special Functions*, World Scientific, Hackensack, NJ 2007.

[213] Kanzieper, E., Replica field theories, Painlevé transcendents, and exact correlation functions, *Phys. Rev. Lett.* **89** (2002), 250201.

[214] Kelvin, W. T. (Lord), On the wave produced by a single impulse in water of any depth or in a dispersive medium, *Philos. Mag.* **23** (1887), 252–255.

[215] Kelvin, W. T. (Lord), Presidential address to the Institute of Electrical Engineers, 1889, *Math. and Phys. Papers III*, pp. 484–515.

[216] Kempf, G. R., *Complex Abelian Varieties and Theta Functions*, Springer, Berlin 1991.

[217] Khrushchev, S., *Orthogonal Polynomials and Continued Fractions from Euler's Point of View*, Cambridge University Press, Cambridge 2008.

[218] Klein, F., *Vorlesungen über die Entwicklung der Mathematik im 19. Jahrhundert, I*, Springer, Berlin 1927.

[219] Klein, F., *Vorlesungen Über die Hypergeometrische Funktion*, Springer-Verlag, Berlin 1981.

[220] Knopp, K., *Funktionentheorie II*, 5th ed., de Gruyter, Berlin 1941.

[221] Koekoek, R. Lesky, P. A., and Swarttouw, R. F., *Hypergeometric Orthogonal Polynomials and their q-Analogues*, Springer, Berlin 2010.

[222] Koelink, H. T., On Jacobi and continuous Hahn polynomials, *Proc. Amer. Math. Soc.* **124** (1996), 887–898.

[223] Koepf, W., *Hypergeometric Summation: An Algorithmic Approach to Summation and Special Function Identities*, Vieweg, Braunschweig 1998.

[224] Kolmogorov, A. N. and Yushkevich, A. P., eds., *Mathematics of the 19th Century: Function Theory According to Chebyshev, Ordinary Differential Equations, Calculus of Variations, Theory of Finite Differences*, Birkhäuser, Basel 1998.

[225] Koornwinder, T., Jacobi polynomials, III. An analytical proof of the addition formula, *SIAM J. Math. Anal.* **6** (1975), 533–540.

[226] Koornwinder, T., Yet another proof of the addition formula for Jacobi polynomials, *J. Math. Anal. Appl.* **61** (1977), 136–141.

[227] Korenev, B. G., *Bessel Functions and their Applications*, Taylor & Francis, London 2002.

[228] Koshlyakov, N. S., On Sonine's polynomials, *Messenger Math.*, **55** (1926), 152–160.

[229] Krall, A. M., *Hilbert Space, Boundary Value Problems and Orthogonal Polynomials*, Birkhäuser, Basel 2002.

[230] Krall, H. L. and Frink, O., A new class of orthogonal polynomials: the Bessel polynomials, *Trans. Amer. Math. Soc.* **65** (1948), 100–115.

[231] Kramers, H. A., Wellenmechanik und halbzahlige Quantisierung, *Z. für Phys.* **39** (1926), 828–840.

[232] Krawtchouk, M., Sur une généralisation des polynomes d'Hermite, *C. R. Acad. Sci. Paris* **189** (1929), 620–622.

[233] Kuijlaars, A., Riemann–Hilbert analysis for orthogonal polynomials, *Orthogonal Polynomials and Special Functions: Leuven 2002*, H. T. Koelinck and W. Van Assche, eds., Springer, Berlin 2003.

[234] Kuijlaars, A. and Van Assche, W., The asymptotic zero distribution of orthogonal polynomials with varying recurrence coefficients, *J. Approx. Theory* **99** (1999), 167–197.

[235] Kummer, E. E., Über die hypergeometrische Reihe, *J. Reine Angew. Math.* **15** (1836), 127–172.

[236] Laforgia, A., Further inequalities for the gamma function, *Math. Comp.* **42** (1984), 597–600.

[237] Lagrange, J. L., Des oscillations d'un fil fixé par une de ses extrémités et chargés d'un nombre quelconque de poids, in the memoir Solution de différents problèmes de calcul intégral, *Misc. Taur.* **3** (1762–1765); *Oeuvres 1*, pp. 471–668.

[238] Lagrange, J. L., Sur le problème de Kepler, *Hist. Acad. Sci. Berlin* **25** (1769), 204–233; *Oeuvres 3*, pp. 113–138.

[239] Laguerre, E. N., Sur l'intégrale $\int_x^\infty x^{-1} e^{-x} dx$, *Bull. Soc. Math. France* **7** (1879), 72–81.

[240] Landen, J., An investigation of a general theorem for finding the length of any arc of any conic hyperbola, by means of two elliptic arcs, with some other new and useful theorems deduced therefrom, *Phil. Trans.* **65** (1775), 283–289.

[241] Lang, S., *Elliptic Functions*, Springer, New York, NY 1987.

[242] Laplace, P. S., Théorie d'attraction des sphéroïdes et de la figure des planètes, *Mém. Acad. Roy. Sci. Paris* (1782, publ. 1785), 113–196.

[243] Laplace, P. S., *Mémoire sur les Intégrales Définies, et leurs Applications aux Probabilités*, Mém. de l'Acad. Sci., 11, Gauthier-Villars, Paris 1810–1811; *Oeuvres 12*, Paris 1898, pp. 357–412.

[244] Laplace, P. S., *Théorie Analytique des Probabilités*, 2nd ed., *Oeuvres 7*.

[245] Larsen, M. E., *Summa Summarum*, Canadian Mathematical Society, Ottawa; Peters, Wellesley, MA 2007.

[246] Lauricella, G., Sulle funzioni ipergeometriche a pi variabili, *Rend. Circ. Mat. Palermo* **7** (1893), 111–158.

[247] Lax, P. Integrals of nonlinear equations of evolution and solitary waves, *Comm. Pure Appl. Math.* **21** (1968), 467–490.

[248] Lawden, D. F., *Elliptic Functions and Applications*, Springer, New York, NY 1989.

[249] Lebedev, N. N., *Special Functions and Their Applications*, Dover, Mineola, NY 1972.

[250] Legendre, A. M., Recherches sur les figures des planètes, *Mém. Acad. Roy. Sci. Paris* (1784, publ. 1787), 370–389.

[251] Legendre, A. M., Recherches sur l'attraction des sphéroides homogènes, *Mém. Math. Phys. Acad. Sci.* **10** (1785), 411–434.

[252] Legendre, A. M., *Exercises de Calcul Intégrale*, Courcier, Paris 1811.

[253] Legendre, A. M., *Traité des Fonctions Elliptiques et des Intégrales Eulériennes*, Huzard-Courcier, Paris 1825–28.

[254] Leonard, D. A., Orthogonal polynomials, duality and association schemes, *SIAM J. Math. Anal.* **13** (1982), 656–663.

[255] Levin, A. L. and Lubinsky, D. S., *Orthogonal Polynomials for Exponential Weights*, Springer, New York, NY 2001.

[256] Li, X. and Wong, R., On the asymptotics of the Meixner–Pollaczek polynomials and their zeros, *Constr. Approx.* **17** (2001), 59–90.

[257] Lin, Y. and Wong, R., Global asymptotics of the discrete Chebyshev polynomials, *Asympt. Anal.* **82** (2013), 39–64.

[258] Lin, Y. and Wong, R., Global asymptotics of the Hahn polynomials, *Anal. Appl. (Singap.)* **11** (2013), 1–47.

[259] Lin, Y. and Wong, R., Asymptotics of the Meijer *G*-functions, *Proceedings of the Conference on Constructive Functions, 2014*, D. Hardin, D. Lubinsky, and B. Simanek, eds., *Contemp. Math.*, to appear.

[260] Liouville, J., Mémoire sur la théorie analytique de la chaleur, *Math. Ann.* **21** (1830–1831), 133–181.

[261] Liouville, J., Second mémoire sur de développement des fonctions ou parties de fonctions en séries dont les divers termes sont assujéties à satisfaire à une même équation de seconde ordre, contenant un paramètre variable, *J. de Math.* **2** (1837), 16–35.

[262] Liouville, J., Lectures published by C. W. Borchardt, *J. für Math. Reine Angew.* **88** (1880), 277–310.

[263] Lommel, E. C. J. von, *Studien über die Bessel'schen Funktionen*, Teubner, Leipzig 1868.

[264] Luke, Y. L., *The Special Functions and Their Approximations*, 2 vols., Academic Press, New York, NY 1969.

[265] Lützen, J., The solution of partial differential equations by separation of variables: a historical survey, *Studies in the History of Mathematics*, Mathematical Association of America, Washington, DC 1987.

[266] Macdonald, H. M., Note on Bessel functions, *Proc. London Math. Soc.* **22** (1898), 110–115.

[267] Macdonald, I. G., *Symmetric Functions and Orthogonal Polynomials*, American Mathematical Society, Providence, RI 1998.

[268] MacRobert, T. M., *Spherical Harmonics: An Elementary Treatise on Harmonic Functions with Applications*, Pergamon, Oxford 1967.

[269] Magnus, W. and Oberhettinger, F., *Formulas and Theorems for the Special Functions of Mathematical Physics*, Chelsea, New York, NY 1949.

[270] Magnus, W., Oberhettinger, F., and Soni, R. P., *Formulas and Theorems for the Special Functions of Mathematical Physics*, 3rd ed., Springer, New York, NY 1966.

[271] Mainardi, F. and Pagnini, G., Salvatore Pincherle: the pioneer of the Mellin–Barnes integrals, *J. Comput. Appl. Math.* **153** (2003), 332–342.

[272] Marcellán, F. and Álvarez-Nodarse, R., On the "Favard theorem" and its extensions, *J. Comput. Appl. Math.* **127** (2001), 231–254.

[273] Mathai, A. M. and Saxena, R. K., *Generalized Hypergeometric Functions with Applications in Statistics and Physical Sciences*, Lect. Notes Math., 348, Springer, Berlin 1973.

[274] Mathai, A. M., Saxena, R. K., and Haubold, H. J., *The H-Function*, Springer, New York, NY 2010.

[275] Matsuda, M., *Lectures on Algebraic Solutions of Hypergeometric Differential Equations*, Kinokuniya, Tokyo 1985.

[276] McBride, E. B., *Obtaining Generating Functions*, Springer, New York, NY 1971.

[277] McCoy, B., Spin systems, statistical mechanics and Painlevé functions, *Painlevé Transcendents, their Asymptotics and Physical Applications*, NATO Adv. Sci. Inst. Ser. B: Phys., 278, Plenum, New York, NY 1992, pp. 377–391.

[278] McCoy, B., Tracy, C., and Wu, T., Painlevé functions of the third kind. *Math. Phys.* **18** (1977), 1058–1092.

[279] Mehler, F. G., Notiz über die Dirichlet'schen Integralausdrucke für die Kugelfuntion $P_n(\cos\theta)$ und über einige analoge Integralform für die Cylinderfunktion $J(z)$, *Math. Ann.* **5** (1872), 141–144.

[280] Mehta, M. L., *Random Matrices*, Elsevier, Amsterdam 2004.

[281] Meijer, C. S., Über Whittakersche bezw. Besselsche Funktionen und deren Produkten, *Nieuw Arch. Wisk.* **18** (1936), 10–39.

[282] Meijer, C. S., Multiplikationstheoreme für die Funktion $G_{p,q}^{m,n}(z)$, *Nederl. Akad. Wetensch., Proc. Ser. A* **44** (1941), 1062–1070.

[283] Meijer, C. S., On the G-functions, I–VII, *Nederl. Akad. Wetensch., Proc. Ser. A* **49** (1946), 344–356, 457–469, 632–641, 765–772, 936–943, 1063–1072, 1165–1175.

[284] Meijer, C. S., Expansion theorems for the G-function, I–XI, *Nederl. Akad. Wetensch., Proc. Ser. A* **55** (1952), 369–379, 483–487; **56** (1953), 43–49, 187–193, 349–357; **57** (1954), 77–82, 83–91, 273–279; **58** (1955), 243–251, 309–314; **59** (1956), 70–82.

[285] Meixner, J., Orthogonale Polynomsysteme mit einer besonderen Gestalt der erzeugenden Funktion, *J. London Math. Soc.* **9** (1934), 6–13.

[286] Mellin, Hj., Abriss einer einheitlichen Theorie der Gamma- und der Hypergeometrischen Funktionen, *Math. Ann.* **68** (1910), 305–337.

[287] Miller, W., Jr., *Lie Theory and Special Functions*, Academic Press, New York, NY 1968.

[288] Miller, W., Jr., *Symmetry and Separation of Variables*, Addison-Wesley, Reading, MA 1977.

[289] Mittag-Leffler, G., An introduction to the theory of elliptic functions, *Ann. Math.* **24** (1923), 271–351.

[290] Miwa, T., Painlevé property of monodromy preserving deformation equations and the analyticity of τ functions, *Publ. Res. Inst. Math. Sci.* **17** (1981), 703–721.

[291] Moore, G., Geometry of the string equations, *Comm. Math. Phys.* **133**, 261–304.

[292] Motohashi, Y., *Spectral Theory of the Riemann Zeta-Function*, Cambridge University Press, Cambridge 1997.

[293] Mott, N. F., The solution of the wave equation for the scattering of a particle by a Coulombian centre of force, *Proc. R. Soc. A* **118** (1928), 542–549.

[294] Moutard, Th. F., Sur les équations différentielles linéares du second ordre, *C.R. Acad. Sci. Paris* **80** (1875), 729–733; *J. École Polyt.* **45** (1878), 1–11.

[295] Müller, C., *Analysis of Spherical Symmetries in Euclidean Spaces*, Springer, New York, NY 1998.

[296] Murphy, R., *Treatise on Electricity*, Deighton, Cambridge 1833.

[297] Murphy, R., On the inverse method of definite integrals with physical applications, *Trans. Camb. Philos. Soc.* **4** (1833), 353–408; **5** (1835), 315–393.

[298] Neumann, C. G., *Theorie der Bessel'schen Funktionen. Ein Analogon zur Theorie der Kugelfunctionen*, Teubner, Leipzig 1867.

[299] Nevai, P. G., *Orthogonal Polynomials*, Memoirs of the American Mathematical Society, 213, American Mathematical Society, Providence, RI 1979.

[300] Neville, E. H., *Jacobian Elliptic Functions*, Oxford University Press, Oxford 1951.

[301] Newman, F. W., On $\Gamma(a)$ especially when a is negative, *Camb. Dublin Math. J.* **3** (1848), 59–60.

[302] Nielsen, N., *Handbuch der Theorie der Zylinderfunktionen*, Teubner, Leipzig 1904.

[303] Nielsen, N., *Handbuch der Theorie de Gammafunktion*, Chelsea, New York, NY 1965.

[304] Nikiforov, A. F., Suslov, S. K., and Uvarov, V. B., *Classical Orthogonal Polynomials of a Discrete Variable*, Springer, Berlin 1991.

[305] Nikiforov, A. F. and Uvarov, V. B., *Special Functions of Mathematical Physics: A Unified Introduction with Applications*, Birkhäuser, Basel 1988.

[306] Nørlund, N. E., The logarithmic solutions of the hypergeometric equation, *K. Dan. Vidensk. Selsk. Mat. Fys. Skr.* **5** (1963), 1–58.

[307] Nørlund, N. E., Hypergeometric functions, *Acta Math.* **94** (1955), 289–349.

[308] Oberhettinger, F., *Tables of Fourier Transforms and Fourier Transforms of Distributions*, Springer, Berlin 1990.

[309] Oberhettinger, F. and Magnus, W., *Anwendung der Elliptische Funktionen in Physik und Technik*, Springer, Berlin 1949.

[310] Olver, F. W. J., Second-order linear differential equations with two turning points, *Philos. Trans. R. Soc. A* **20** (1975), 131–174.

[311] Olver, F. W. J., *Asymptotics and Special Functions*, Peters, Wellesley, MA 1997.

[312] Olver, F. W. J., Lozier, D. W., Clark, C. W., and Boisvert, R. F., *Digital Library of Mathematical Functions*, National Institute of Standards and Technology, Gaithersburg, MD 2007.

[313] Olver, F. W. J., Lozier, D. W., Clark, C. W., and Boisvert, R. F., *Handbook of Mathematical Functions*, Cambridge University Press, Cambridge 2010.

[314] Painlevé, P., Gewöhnliche Differentialgleichungem: Existenz der Lösungen, *Enzykl. Math. Wiss.* **2** no. 2/3 (1900), 189–229.

[315] Painlevé, P., Mémoire sur les équations différentielles dont l'intégrale générale est uniforme, *Bull. Soc. Math. France* **28** (1900), 201–261.

[316] Painlevé, P., Sur les équations différentielles du seconde ordre à points critiques fixes, *C. R. Acad. Sci. Paris* **143** (1906), 1111–1117.

[317] Painlevé, P., *Oeuvres 1, 3*, Centre Nationale de la Recherche Scientifique, Paris 1972, 1975.

[318] Pan, J. H. and Wong, R., Uniform asymptotic expansions for the discrete Chebyshev polynomials, *Stud. Appl. Math.* **128** (2011), 337–384.

[319] Patterson, S. J., *An Introduction to the Theory of the Riemann Zeta-Function*, Cambridge University Press, Cambridge 1988.

[320] Perron, O., *Die Lehre von den Kettenbruchen*, 2nd ed., Teubner, Leipzig 1929.

[321] Penson, K. A., Blasiak, P., Duchamp, G. H. E., Horzela, A., and Solomon, A. I., On certain non-unique solutions of the Stieltjes moment problem, *Disc. Math. Theor. Comp. Sci.* **12** (2010), 295–306.

[322] Petiau, G., *La Théorie des Fonctions de Bessel Exposée en vue de ses Applications à la Physique Mathématique*, Centre National de la Recherche Scientifique, Paris 1955.

[323] Petkovšek, M., Wilf, H. S., and Zeilberger, D., $A = B$, Peters, Wellesley, MA 1996.

[324] Pfaff, J. F., Nova disquisitio de integratione aequationes differentio-differentialis, *Disquisitiones Analyticae*, Helmstadt 1797.

[325] Picard, É., Sur l'application des méthodes d'approximations successives à l'étude de certaines équations différentielles, *Bull. Sci. Math. 2* **12** (1888), 148–156.

[326] Pincherle, S., Sopra una trasformazione delle equazioni differenziali lineari in equazioni lineari alle differenze, e viceversa, *R. Ist. Lomb. Sci. Lett. Rend.* **19** (1886), 559–562.

[327] Pincherle, S. Sulle funzioni ipergeometriche generalizzate, *Atti R. Accad. Lincei, Rend. Cl. Sci. Fis. Mat. Natur.* **4** (1888) 694–700, 792–799.

[328] Plancherel, M. and Rotach, W., Sur les valeurs asymptotiques des polynomes d'Hermite Hn(x) = (1)nex2/2dn(ex2/2)/dxn, *Comm. Math. Helvetici* **1** (1929), 227–254.

[329] Plemelj, J., Riemannsche Formenscharen mit gegebener Monodromiegruppe. *Monatsh. Math. Physik* **19** (1908), 211–246.

[330] Pochhammer, L., Über hypergeometrische Functionen n-ter Ordnung, *J. Reine Angew. Math.* **71** (1870), 316–352.

[331] Poisson, S. D., Mémoire sur la distribution de la chaleur dans les corps solides, *J. École Roy. Polyt.* **19** (1823), 1–162, 249–403.

[332] Polishchuk, A., *Abelian Varieties, Theta Functions and the Fourier Transform*, Cambridge University Press, Cambridge 2003.

[333] Pollaczek, F., Sur une généralisation des polynômes de Legendre, *C. R. Acad. Sci. Paris* **228** (1949), 1363–1365.

[334] Prasolov, V. and Solovyev, Yu., *Elliptic Functions and Elliptic Integrals*, American Mathematical Society, Providence, RI 1997.

[335] Prosser, R. T., On the Kummer solutions of the hypergeometric equation, *Amer. Math. Monthly* **101** (1994), 535–543.

[336] Prudnikov, A. P., Brychkov, Yu. A., and Marichev, O. I., *Integrals and Series*, vols. 1–5. Gordon and Breach, New York, NY 1986–2004.

[337] Qiu, W.-Y. and Wong, R., Asymptotic expansion of the Krawtchouk polynomials and their zeros, *Comput. Methods Funct. Theory* **4** (2004), 189–226.

[338] Qiu, W.-Y. and Wong, R., Global asymptotic expansions of the Laguerre polynomials – a Riemann–Hilbert approach, *Numer. Algorithms* **49** (2008), 331–372.

[339] Rainville, E. D., *Special Functions*, Chelsea, New York, NY 1971.

[340] Remmert, R., Wielandt's theorem about the Γ-function, *Amer. Math. Monthly* **103** (1996), 214–220.

[341] Riccati, J. F. (Count), Animadversiones in aequationes differentiales secundi gradus, *Acta Erud. Publ. Lipsiae Suppl.* **8** (1724), 66–73.

[342] Riemann, B., Zwei allgemeine Lehrsätze für lineäre Differentialgleichungen mit algebraischen Koefficienten, *Ges. Math. Werke*, pp. 357–369.

[343] Riemann, B., Beiträge zur Theorie der durch Gauss'she Reihe $F(\alpha, \beta, y, x)$ darstellbaren Funktionen, *Kön. Ges. Wiss. Gött.* **7** (1857), 1–24; *Werke*, pp. 67–90.

[344] Robin, L., *Fonctions Sphériques de Legendre et Fonctions Sphéroïdales*, vols. I–III, Gauthier-Villars, Paris 1957–1959.

[345] Rodrigues, O., Mémoire sur l'attraction des sphéroides, *Corr. École Roy. Polyt.* **3** (1816), 361–385.

[346] Routh., E., On some properties of certain solutions of a differential equation of the second-order, *Proc. London Math. Soc.* **16** (1885), 245–261.

[347] Roy, R., *Sources in the Development of Mathematics. Series and Products from the Fifteenth to the Twenty-first Century*, Cambridge University Press, Cambridge 2011.

[348] Russell, A., The effective resistance and inductance of a concentric main, and methods of computing the Ber and Bei and allied functions, *Philos. Mag.* **17** (1909), 524–552.

[349] Sachdev, P. L., *A Compendium on Nonlinear Ordinary Differential Equations*, Wiley, New York, NY 1997.

[350] Saff, E. B. and Totik, V., *Logarithmic Potentials with External Fields*, Springer, New York, NY 1997.

[351] Sansone, G., *Orthogonal Functions*, Dover, Mineola, NY 1991.

[352] Šapiro, R. L., Special functions: related to representations of the group $SU(n)$ of class I with respect to $SU(n-1)$ $n \geq 3$ (in Russian), *Izv. Vysš. Učebn. Zaved. Mat.* **4** (1968), 97–107.

[353] Sasvari, Z., An elementary proof of Binet's formula for the Gamma function, *Amer. Math. Monthly* **106** (1999), 156–158.

[354] Schläfli, L., Sull'uso delle linee lungo le quali il valore assoluto di una funzione è constante, *Annali Mat.* **2** (1875), 1–20.

[355] Schläfli, L., *Ueber die zwei Heine'schen Kugelfunktionen*, Bern 1881.

[356] Schlömilch. O., *Analytische Studien*, Engelmann, Leipzig 1848.

[357] Schlömilch, O., Ueber die Bessel'schen Funktion, *Z. Math. Phys.* **2** (1857), 137–165.

[358] Schmidt, H., Über Existenz und Darstellung impliziter Funktionen bei singulären Anfangswerten, *Math. Z.* **43** (1937/38), 533–556.

[359] Schwarz, H. A., Über diejenigen Fälle, in welchen die Gaussische hypergeometrische Reihe eine algebraische Function ihres vierten Elementes darstellt, *J. Reine Angew. Math.* **75** (1873), 292–335.

[360] Seaborn, J. B., *Hypergeometric Functions and Their Applications*, Springer, New York, NY 1991.

[361] Seaton, M. J., Coulomb functions for attractive and repulsive potentials and for positive and negative energies, *Comput. Phys. Comm.* **146** (2002), 225–249.

[362] Selberg, A., Bemerkninger om et multipelt integral, *Norsk. Mat. Tidsskr.* **24** (1944), 159–171.

[363] Sharapudinov, I. I., Asymptotic properties and weighted estimates for orthogonal Chebyshev–Hahn polynomials, *Mat. Sb.* **182** (1991), 408–420; *Math. USSR Sb.* **72** (1992), 387–401.

[364] Sherman, J., On the numerators of the Stieltjes continued fractions, *Trans. Amer. Math. Soc.* **35** (1933) 64–87.

[365] Shohat, J., The relation of classical orthogonal polynomials to the polynomials of Appell, *Amer. J. Math.* **58** (1936), 453–464.

[366] Simon, B., The classical moment problem as a self-adjoint finite difference operator, *Adv. Math.* **137** (1998), 82–203.

[367] Simon, B., *Orthogonal Polynomials on the Unit Circle, Parts 1 and 2*, American Mathematical Society, Providence, RI 2005.

[368] Slater, L. J., *Confluent Hypergeometric Functions*, Cambridge University Press, New York, NY 1960.

[369] Slater, L. J., *Generalized Hypergeometric Functions*, Cambridge University Press, Cambridge 1966.

[370] Slavyanov, S. Yu. and Lay, W., *Special Functions: A Unified Theory Based on Singularities*, Oxford University Press, Oxford 2000.

[371] Sneddon, I. N., *Special Functions of Mathematical Physics and Chemistry*, 3rd ed., Longman, London 1980.

[372] Sokhotskii, Y. W., On definite integrals and functions used in series expansions, Thesis, St. Petersburg 1873.

[373] Sommerfeld, A. J. W., Mathematische Theorie der Diffraction, *Math. Ann.* **47** (1896), 317–374.

[374] Sonine, N. J., Recherches sur les fonctions cylindriques et le développment des fonctions continues en séries, *Math. Ann.* **16** (1880), 1–80.

[375] Stahl, H. and Totik, V., *General Orthogonal Polynomials*, Cambridge University Press, Cambridge 1992.

[376] Stanton, D., Orthogonal polynomials and Chevalley groups, *Special Functions: Group Theoretic Aspects and Applications*, R. Askey, T. Koornwinder, and W. Schempp, eds., Reidel, New York, NY 1984, pp. 87–128.

[377] Steeb, W.-H. and Euler, N., *Nonlinear Evolution Equations and Painlevé Test*, World Scientific Publishing, Singapore 1988.

[378] Steklov, V. A., Sur les expressions asymptotiques de certaines fonctions, définies par les équations différentielles linéaires de seconde ordre, et leurs applications au problème du développement d'une fonction arbitraire en séries procédant suivant les-dites fonctions, *Comm. Soc. Math. Kharkhow* **10** (1907), 197–200.

[379] Sternberg, W. J. and Smith, T. L., *The Theory of Potential and Spherical Harmonics*, University of Toronto Press, Toronto 1944.

[380] Stieltjes, T. J., Quelques recherches sur la théorie des quadratures dites mécaniques, *Ann. Sci. École Norm. Sup.* **1** (1884), 409–426.

[381] Stieltjes, T. J., Sur quelques théorèmes d'algèbre, *C. R. Acad. Sci. Paris* **100** (1885), 439–440; *Oeuvres 1*, pp. 440–441.

[382] Stieltjes, T. J., Sur les polynômes de Jacobi, *C. R. Acad. Sci. Paris* **100** (1885), 620–622; *Oeuvres 1*, pp. 442–444.

[383] Stieltjes, T. J., Sur les polynômes de Legendre, *Ann. Fac. Sci. Toulouse* **4** (1890), G1–G17; *Oeuvres 2*, pp. 236–252.

[384] Stieltjes, T. J., Recherches sur les fractions continues, *Ann. Fac. Sci. Toulouse* **9** (1894), J1–122; **9** (1895), A1–10.

[385] Stillwell, J., *Mathematics and its History*, Springer, New York, NY 2002.

[386] Stirling, J., *Methodus Differentialis*, London 1730.

[387] Stokes, G. G., On the numerical calculation of a class of definite integrals and infinite series, *Trans. Camb. Philos. Soc.* **9** (1850), 166–187.

[388] Stokes, G. G., On the discontinuity of arbitrary constants which appear in divergent developments, *Trans. Camb. Philos. Soc.* **10** (1857), 105–128.

[389] Stone, M. H., *Linear Transformations in Hilbert Space and their Applications to Analysis*, American Mathematical Society, Providence, RI 1932.

[390] Sturm, J. C. F., Mémoire sur les équations différentielles linéaires du second ordre, *J. Math. Pures Appl.* **1** (1836), 106–186.

[391] Szegő, G., On an inequality of P. Turán regarding Legendre polynomials, *Bull. Amer. Math. Soc.* **54** (1949), 401–405.

[392] Szegő, G., *Orthogonal Polynomials*, 4th ed., American Mathematical Society, Providence, RI 1975.

[393] Szegő, G., An outline of the history of orthogonal polynomials, *Orthogonal Expansions and their Continuous Analogues*, Southern Illinois University Press, Carbondale, IL 1968, pp. 3–11.

[394] Talman, J. D., *Special Functions: A Group Theoretic Approach*, Benjamin, New York, NY 1968.

[395] Tannery, J. and Molk, J., *Éléments de la Théorie des Fonctions Elliptiques*, vols. 1–4, Gauthier-Villars, Paris 1893–1902; Chelsea, New York, NY 1972.

[396] Taylor, R. and Wiles, A., Ring-theoretic properties of certain Hecke algebras, *Ann. Math.* **141** (1995), 553–572.

[397] Temme, N. M., *Special Functions: An Introduction to the Classical Functions of Mathematical Physics*, Wiley, New York, NY 1996.

[398] Thomae, J., Ueber die höheren hypergeometrische Reihen, insbesondere über die Reihe $1 + \frac{a_0 a_1 a_2}{1 \cdot b_1 b_2} x + \frac{a_0(a_0+1)a_1(a_1+1)a_2(a_2+1)}{1 \cdot 2 \cdot b_1(b_1+1)b_2(b_2++1)} x^2 + \dots$, *Math. Ann.* **2** (1870), 427–441.

[399] Titchmarsh, E. C., *The Theory of the Riemann Zeta-Function*, 2nd. ed., Oxford University Press, New York, NY 1986.

[400] Tracy, C. A. and Widom, H., Painlevé functions in statistical physics, *Publ. RIMS Kyoto Univ.* **47** (2011), 361–374.

[401] Tricomi, F., *Serie Ortogonale di Funzioni*, S. I. E. Istituto Editoriale Gheroni, Torino 1948.

[402] Tricomi, F., *Funzioni Ipergeometrichi Confluenti*, Ed. Cremonese, Rome 1954.

[403] Tricomi, F., *Funzioni Ellittiche*, Zanichelli, Bologna 1951.

[404] Truesdell, C., *An Essay Toward a Unified Theory of Special Functions Based upon the Functional Equation* $(\partial/\partial z)F(z,\alpha) = F(z,\alpha+1)$, Annals of Mathematics Studies, 18, Princeton University Press, Princeton, NJ 1948.

[405] Turán, P., On the zeros of the polynomials of Legendre, *Čas. Pěst. Mat. Fys.* **75** (1950), 113–122.

[406] Uspensky, J. V., On the development of arbitrary functions in series of orthogonal polynomials, *Ann. Math.* **28** (1927), 563–619.

[407] Van Assche, W., *Asymptotics for Orthogonal Polynomials,* Lect. Notes Math., 1265, Springer, Berlin 1987.

[408] van der Corput, J. G., *Asymptotic Expansions. I–III*, Department of Mathematics, University of California, Berkeley, CA 1954–1955.

[409] van der Laan, C. G. and Temme, N. M., *Calculation of Special Functions: The Gamma Function, the Exponential Integrals and Error-like Functions*, Centrum voor Wiskunde en Informatica, Amsterdam 1984.

[410] Vandermonde, A., Mémoire sur des irrationelles de différens ordres avec une application au cercle, *Mém. Acad. Roy. Sci. Paris* (1772), 489–498.

[411] Varadarajan, V. S., Linear meromorphic differential equations: a modern point of view, *Bull. Amer. Math. Soc.* **33** (1996), 1–42.

[412] Varchenko, A., *Multidimensional Hypergeometric Functions and Representation Theory of Lie Algebras and Quantum Groups*, World Scientific Publishing, River Edge, NJ 1995.

[413] Varchenko, A., *Special Functions, KZ Type Equations, and Representation Theory*, American Mathematical Society, Providence, RI 2003.

[414] Vilenkin, N. Ja., *Special Functions and the Theory of Group Representations*, American Mathematical Society, Providence, RI 1968.

[415] Vilenkin, N. Ja. and Klimyk, A. U., *Representation of Lie Groups and Special Functions*, 3 vols., Kluwer, Dordrecht 1991–1993.

[416] Walker, P. L., *Elliptic Functions*, Wiley, Chichester 1996.

[417] Wallis, J., *Arithmetica Infinitorum*, Oxford 1656.

[418] Wang, X.-S. and Wong, R., Asymptotics of orthogonal polynomials via recurrence relations, *Anal. Appl.* **10** (2012), 215–235.

[419] Wang, X.-S. and Wong, R., Asymptotics of the Racah polynomials with varying parameters, *J. Math. Anal. Appl.*, to appear.

[420] Wang, Z. and Wong, R., Asymptotic expansions for second-order linear difference equations with a turning point, *Numer. Math.* **94** (2003), 147–194.

[421] Wang, Z. and Wong, R., Linear difference equations with transition points, *Math. Comp.* **74** (2005), 629–653.

[422] Wang, Z.-X. and Guo, D. R., *Special Functions*, World Scientific, Teaneck, NJ 1989.

[423] Wasow, W., *Asymptotic Expansions for Ordinary Differential Equations*, Dover, Mineola, NY 1987.

[424] Watson, G. N., Bessel functions and Kapteyn series, *Proc. London Math. Soc.* **16** (1917), 150–174.

[425] Watson, G. N., *A Treatise on the Theory of Bessel Functions*, Cambridge University Press, Cambridge 1995.

[426] Wawrzyńczyk, A., *Group Representations and Special Functions*, Reidel, Dordrecht 1984.

[427] Weber, H., Über die Integration der partiellen Differentialgleichung: $\frac{\partial^2 u}{\partial x^2} + \frac{\partial^2 u}{\partial y^2} + k^2 u^2 = 0$, *Math. Ann.* **1** (1869), 1–36.

[428] Weber, H., Über eine Darstellung willkürlicher Funktionen durch Bessel'sche Funktionen, *Math. Ann.* **6** (1873), 146–161.

[429] Weber, M. and Erdélyi, A., On the finite difference analogue of Rodrigues' formula, *Amer. Math. Monthly* **59** (1952), 163–168.

[430] Weierstrass, K. L., *Formeln und Lehrsätze zum Gebrauch der Elliptische Funktionen*, Kaestner, Göttingen 1883–1885.

[431] Weierstrass, K. L., Über die analytische Darstellbarkeit sogenannter willkürlicher Functionen einer reellen Veränderlichen, *Sitzungsber. Königl. Preuss. Akad. Wissensch.* **II** (1885), 633–639, 789–805.

[432] Wentzel, G., Eine Verallgemeinerung der Quantenbedingungen für die Zwecke der Wellenmechanik, *Z. für Phys.* **38** (1926), 518–529.

[433] Whitehead, C. S., On the functions $\mathrm{ber}\,x$, $\mathrm{bei}\,x$, $\mathrm{ker}\,x$, and $\mathrm{kei}\,x$, *Quarterly J.* **42** (1909), 316–342.

[434] Whittaker, E. T., An expression of certain known functions as generalized hypergeometric functions, *Bull. Amer. Math. Soc.* **10** (1903), 125–134.

[435] Whittaker E. T. and Watson, G. N., *A Course of Modern Analysis*, Cambridge University Press, Cambridge 1969.

[436] Wiles, A., Modular elliptic curves and Fermat's last theorem, *Ann. Math.* **141** (1995), 443–551.

[437] Wilf, H. S. and Zeilberger, D., An algebraic proof theory for geometric (ordinary and "q") multisum/integral identities, *Invent. Math.* **108** (1992), 575–633.

[438] Wilson, J. A., Some hypergeometric orthogonal polynomials, *SIAM J. Math. Anal.* **11** (1980), 690–701.

[439] Wilson, J. A., Asymptotics for the $_4F_3$ polynomials, *J. Approx. Theory* **66** (1991), 58–71.

[440] Wimp, J., Recursion formulae for hypergeometric functions, *Math. Comp.* **22** (1968), 363–373.

[441] Wintner, A., *Spektraltheorie der Unendlichen Matrizen, Einführung in den Analytischen Apparat der Quantenmechanik*, Hitzel, Leipzig 1929.

[442] Wolfram, S., Festschrift for Oleg Marichev, http://www.stephenwolfram.com/publications/history-future-special-functions/intro.html

[443] Wong, R., *Asymptotic Approximations of Integrals*, Academic Press, Boston, MA 1989; SIAM, Philadelphia, PA 2001.

[444] Wong, R. and Li, H., Asymptotic expansions for second-order linear difference equations, *J. Comp. Appl. Math.* **41** (1992), 65–94.

[445] Wong, R. and Zhang, L., Global asymptotics of Hermite polynomials via Riemann–Hilbert approach, *Discrete Cont. Dyn. Syst. B* **7** (2007), 661–682.

[446] Yost, F. L., Wheeler, J. A., and Breit, G., Coulomb wave functions in repulsive fields, *Phys. Rev.* **49** (1936), 174–189.

[447] Zhang, S. and Jin, J., *Computation of Special Functions*, Wiley, New York, NY 1996.

[448] Zhou, X., The Riemann–Hilbert problem and inverse scattering. *SIAM J. Math. Anal.* **20** (1989), 966–986.

[449] Zwillinger, D., *Handbook of Differential Equations*, Wiley, New York, NY 1998.

Author index

Notation index

Subject index

Printed in the United States
By Bookmasters